PERGAMON INTERNATIONAL LIBRARY
of Science, Technology, Engineering and Social Studies
The 1000-volume original paperback library in aid of education, industrial training and the enjoyment of leisure
Publisher: Robert Maxwell, M.C.

Communication for Command and Control Systems

International Series on

SYSTEMS AND CONTROL, Volume 5

Other Titles in the Series

Communication for Command and Control Systems

By

D J MORRIS

Senior Systems Analyst, Elbit Computers Ltd., Haifa, Israel

PERGAMON PRESS

OXFORD · NEW YORK · TORONTO · SYDNEY · PARIS · FRANKFURT

U.K.	Pergamon Press Ltd., Headington Hill Hall, Oxford OX3 0BW, England
U.S.A.	Pergamon Press Inc., Maxwell House, Fairview Park, Elmsford, New York 10523, U.S.A.
CANADA	Pergamon Press Canada Ltd., Suite 104, 150 Consumers Road, Willowdale, Ontario M2J 1P9, Canada
AUSTRALIA	Pergamon Press (Aust.) Pty. Ltd., P.O. Box 544, Potts Point, N.S.W. 2011, Australia
FRANCE	Pergamon Press SARL, 24 rue des Ecoles, 75240 Paris, Cedex 05, France
FEDERAL REPUBLIC OF GERMANY	Pergamon Press GmbH, Hammerweg 6, D-6242 Kronberg-Taunus, Federal Republic of Germany

First edition 1983

Library of Congress Cataloging in Publication Data

Morris, David Joseph.
Communication for command & control systems.
(International series on systems and control: v. 5) (Pergamon international library of science, technology, engineering, and social sciences)
Includes biolographical references and indexes.
1. Command and control systems. 2. Telecommunication. I. Title. II. Title: Communication for command and control systems. III. Series. IV. Series: Pergamon international library of science, technology, engineering, and social studies.
TK5102.5.M668 1983 621.38 81-23385

British Library Cataloguing in Publication Data

Morris, D. J.
Communication for command & control systems.
(International series on system and control: 5).
(Pergamon international library)
1. Command and control systems
I. Title II. Series
631.38 TK5102.5

ISBN 0-08-027597-4 (Hardcover)
ISBN 0-08-027596-6 (Flexicover)

In order to make this volume available as economically and as rapidly as possible the author's typescript has been reproduced in its original form. This method unfortunately has its typographical limitations but it is hoped that they in no way distract the reader.

Printed in Great Britain by A. Wheaton & Co. Ltd., Exeter

This book is dedicated to the memory of

RAFI MOR

who encouraged me to write this book but did not live to see it published

Contents

Preface

The main purpose of this book is to provide a sound exposition of the basic theoretical and practical features involved in the design of communication networks for command and control systems. By extending the range of the topics to include many relevant aspects of the subject not usually dealt with in a single work, the book may well serve both as a text book and as a handy source of reference. Furthermore, as the book is concerned primarily with the practical side of computer controlled communication, it should be useful in particular to anyone involved in planning, designing and implementing communication systems.

This book may be considered as an extension of the treatment of the subject recorded in a previous book of mine entitled 'Introduction to Communication Command and Control Systems', to cover additional aspects. It concentrates on the communication sides of the subject by surveying the means of transferring data between the various processing points and appraising their potential advantages and possible defects in implementation. In this respect the book should prove particularly useful for the practising engineer engaged in command and control system design in civil, military and administrative spheres.

For the system designer, a prime consideration is the selection of blocks with which to construct his system. Though normally not involved in the design of the blocks but only in their selection, he must have adequate knowledge of the various types available, particularly their operational characteristics. This work claims to give the designer a unique source of reference whereby to appraise the worthiness of each block and the problems arising in interfacing them into the system. Where necessary the book presents comparison tables which can be of practical assistance in the selection of the blocks for specific operations. It must be noted, however, that the suggested procedure stresses the logical design of the complete system and does not purport to furnish the designer with the required mathematical tools.

The design material introduced in the book derives in the main from the author's practical experience in the planning of command and control systems and subsequently organized for presentation at post graduate university courses.

Attention is drawn to the extensive lists appended to each chapter acknowledging the works consulted by the author. An original feature of the book, which should prove of particular assistance to the student and a possible reference for the practising engineer are the charts and illustrations accompanying the material.

I cannot omit from this preface my indebtedness to my father for his assistance in preparing the final draft of the book.

D.J.M.

CHAPTER 1

Command and Control Systems

1.1 SYSTEM DEFINITION

Automated Command and Control (C^2) systems are a relatively new field of activity, calling insistently for fresh ideas and sophisticated skills. While the application of these systems may be claimed to be virtually still in its initial stages, it has already made its impact felt both in industry and military operations. Furthermore, although the concept of command and control has long been known, it is only now, with the comparatively recent introduction of automated means, has it become a matter of vital universal consideration.

The term 'Command and Control' derives from the military sphere, where it is used in connection with the regulations determining the procedure and function of the officers issuing commands necessary for controlling the actions of their soldiers under battle conditions.

Command and Control, as the dual name implies, must comprise both a command and a controlled action element as shown in Fig. 1.1. The main function of the command element is decision making, based on the assessments of the reported action of the controlled element, and the controlled element is to carry out the orders issued by the command element. In other words, the characteristie feature of the command and control system is that it is based on a closed loop operation. In consequence, any interruption of the loop would immediately destroy the functioning of the system. The purpose of the automated command and control system is therefore to introduce efficient means of communication between the two elements and to provide the best means of utilizing both elements.

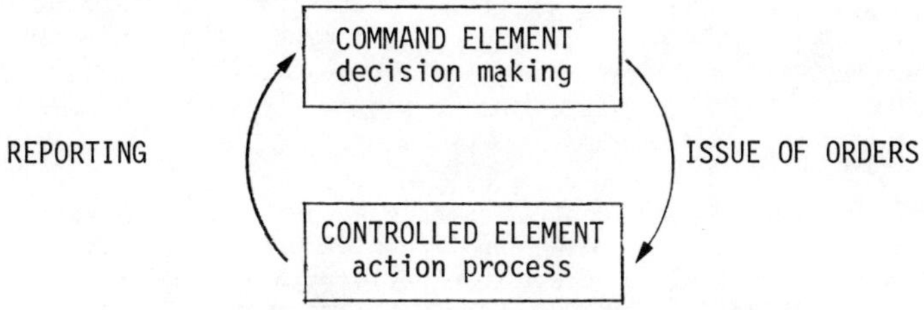

Fig. 1.1 Command and control system closed loop application

A closer study of the closed loop application diagram indicates that it could easily be applicable to any action process which necessitates decision making. In fact a command and control system is required wherever a commander, a manager or an operator is responsible for controlling the actions of people, machines or an industrial process. It is a management tool used in implementing functions of control, supervision, warning, situation assessment, decision making and decision execution.

Decision making being the major task of command and control systems can only be implemented if an accurate situation picture of the action process is first presented. The situation picture must be based on the updated information status of the actions, while the commands must be able to guide the performance of these actions in real-time. This means that the whole loop operation must be performed in real-time; in other words, the command and control system must be designed for real-time operation for otherwise the whole system would be useless.

The term real-time has many meanings but for command and control systems there is only one specific interpretation: it relates the time performance of the decision making to that actual time of the action process, to serve in guiding the process. A real-time operation is thus one which presents an answer to a continuing problem for a particular set of values while these values are still available. The term real-time cannot be measured by any absolute scale, as it must always be related to the action process it is used for. Some systems may regard its real-time function in terms of microseconds, while others will accept delays of hours and will still call their system operations in real-time. It is therefore necessary for each command and control system to define, as accurately as possible, what in absolute time, they mean by real-time.

It is not necessary for the decision element to be within eye distance of the action process; in fact they can be miles apart. This means that command and control systems require sophisticated communication networks so as to preserve the real-time function of the system. For this reason it has become popular to call these systems command, control and communication systems, or in short C^3 systems. Furthermore, since many command and control operations are performed today with computers, these systems have assumed the title of command, control, communication and computer systems, or in short C^4 systems. Such terms are definitely wrong and misleading, for the command and control systems are the process doctrines, while both the communication and computers are tools necessary for assisting in implementing these doctrines. In other words, the command and control are the systems' operational features, while communication and computers are the means furnished to achieve these operations. It would be impossible efficiently to design the tools necessary for the command and control system without the prior development of the basic principles of command and control.

The use of the abbreviated terms of C^3 or C^4 is therefore an interbreeding of two unrelated topics and should be avoided. What is suggested is that only the term command and control or the term C^2 should be used. If one wants to stress the employment of both communication and computers for the command and control systems, then the terms C^2 for C^2, or C^2 through C^2 are preferable.

1.2 SYSTEM DESCRIPTION

Automated command and control systems can be applied in many diverse spheres as they can be used to control complex processes with extreme efficiency. Examples of these systems can be located in many far apart fields, some of which are listed below:

a. Medical centres for automatice patient care

b. Supermarket control for accounting and stock ordering

c. Air-traffic control.

d. Traffic monitoring and control.

e. Centralized banking accounting.

f. Seat/room reservation.

g. Electricity supply.

h. Military manoeuvres.

Although the many types of applied C^2 systems are varied, they are designed on corresponding basic principles, having their organization and the tools for implementation similar. It should be further noted that the C^2 systems are intended for use by both large and small organizations, their purpose being to provide efficient means for planning, managing and controlling their operations, using on-line information and implementing decisions in real-time.

The principal task of all these systems is to establish and provide accurate real-time situation pictures of the problems confronting the organization and thereby provide the means for decision making. These problems are usually not known in advance and so the situation picture is the only means by which the process actions are made known to the decision makers. The situation picture can only be built-up if data is continuously acquired of the performance of the 'Controllee', as shown in Fig. 1.2. The term 'controllee' is used here as a general title for all processes, people and equipment which are controlled by the system.

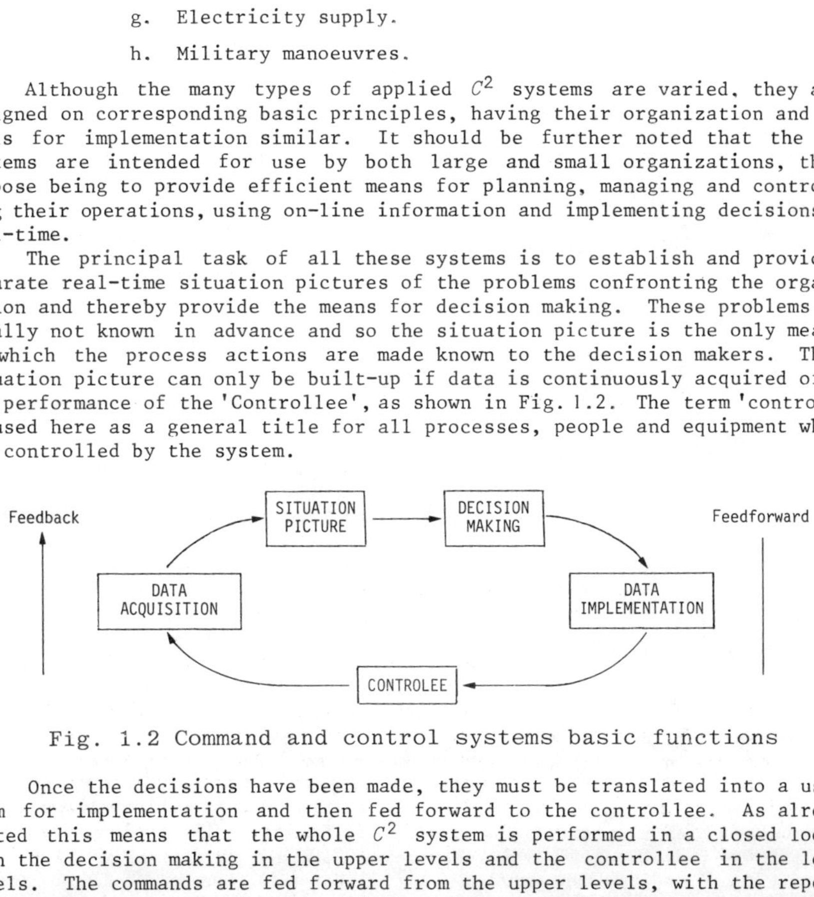

Fig. 1.2 Command and control systems basic functions

Once the decisions have been made, they must be translated into a usable form for implementation and then fed forward to the controllee. As already stated this means that the whole C^2 system is performed in a closed loop, with the decision making in the upper levels and the controllee in the lower levels. The commands are fed forward from the upper levels, with the reports fed back from the lower levels to indicate the execution of these commands.

Data acquisition, correlated with information already available in the system, provides the upper levels with a real-time situation picture of what is actually occurring in the lower levels, thus enabling decisions to be made on the spot within the course of the operation. The principle of this closed loop can only be applicable if all the connections between any of the elements in the loop are attached 'on-line'. The term 'on-line' does not only mean that two sets of equipment are directly connected but also that the operation of the equipment which receives the data is a direct reaction to that which occurs in the first set.

Command to control systems are characterized by the coordination of resources located over scattered geographical areas. They comprise data acquisition, data correlation, data processing, data updating, information displaying, decision making, command dispatching and implementation of commands. Taking all these factors as a single entity, one can visualize a system

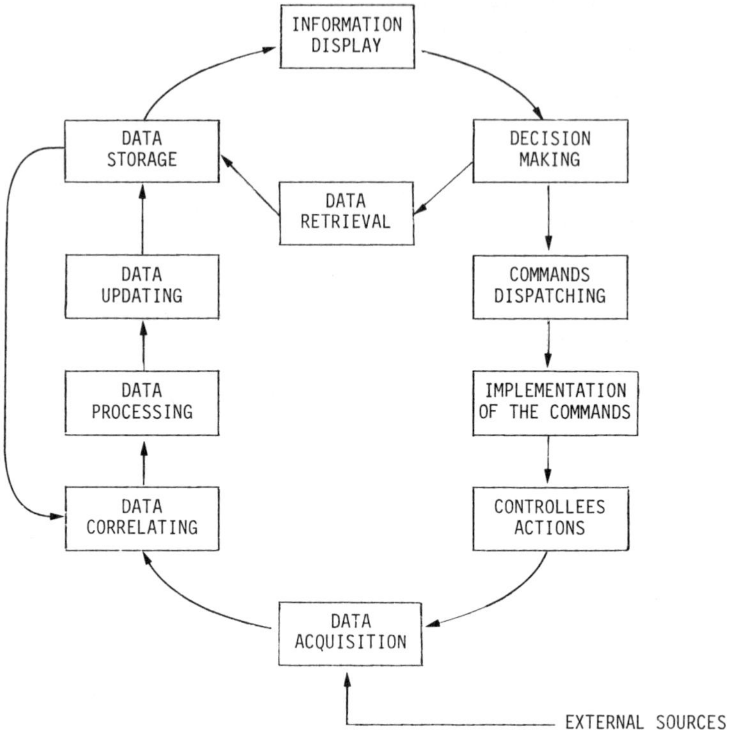

Fig. 1.3 The basic operational blocks of a C^2 system loop

as shown in Fig. 1.3. Such a schematic model characterises all command and control systems, whether they are military, industrial or commercial.

Command and control systems are much more complicated than the basic model shown above, and may comprise a number of extra blocks arranged in a hierarchy of responsibilities. Furthermore, a single system may consist of a number of loops operating independently, with some of the blocks common to a number of loops. While each loop can be regarded as a separate command and control system, they are all part of a large master command and control system. Take for example a hospital C^2 system with an independent command and control subsystem for each department. The master system may be used for the patients' accounts, medical history and other records, while the independent department subsystem may be used for the patient's daily scheduling and treatment. Take also, for example, a banking command and control system which has a national control office for major investments but allows the local branches throughout the country to have their own C^2 systems for smaller transactions. Efficiency is only achieved if each subsystem can operate independently and at the same time as part of a larger system controlled by the central office.

When designing a command and control system the whole system is defined as comprising the master system and all the subsystems. Although the development of each of the subsystems may be concluded at different times with possible gaps of years between them; all are considered as part of the master system, aiming at a common goal. Many elements of the master system may be shared by the smaller subsystems, some of which may share all or only a few of their facilities with others. These could be the same communication networks, display terminals and data base facilities. While each of the smaller command and control systems organized in the master command and control system may operate independently, they must all be regarded as an integral part of the master system, since they all belong to one organization, are designed along the same operational plan and may employ common functional units. The operational relationship between the susbsystems within the master system is that of interaction. System interaction requires coherent operation between all the subsystems achieved by means of common procedures adopted to provide direct communication in passing information and instruction. When initially designing the master C^2 system operation, the interaction operation is defined. That is, the system standards are specified for all the subsystem operations to warrant for their efficient mutual performance.

Master command and control systems may be organized to operate in centralized, hierarchical or distributed configurations, as seen in Fig. 1.4.

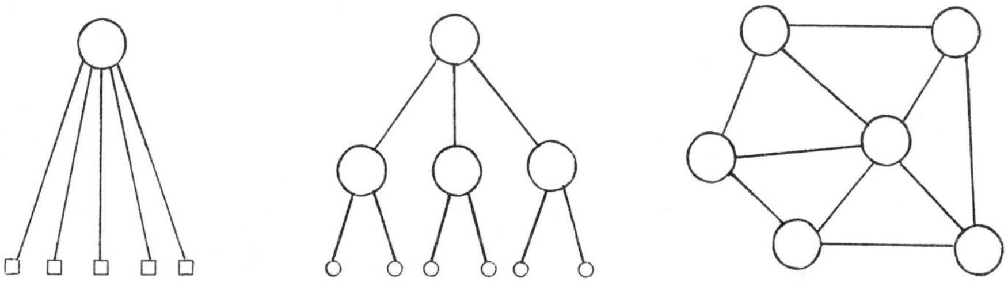

(a) CENTRALIZED (b) HIERARCHICAL (c) DISTRIBUTED

Fig. 1.4 Different operation configurations for subsystem interaction

In the centralized system all the decision making is performed in the centre location, where all the data is evaluated and commands are dispatched to their desired locations. Such an organization has the advantage of simplicity and efficiency although it is vulnerable to a single failure; it is therefore generally suited only to smaller systems. In the hierarchical system, the basic organization is similar to that of the centralized system, with all the major management decisions centralized in one location. Nevertheless, each of the subsystems could be regarded as a C^2 system operating independently with its one decision making functions. In practice each subsystem is responsible for the operation of all its lower levels and gets its policy instructions from its upper levels. In the distributed system, each subsystem operation is independent with its own decision making facility, but it must coordinate its operations with all other subsystems.

It is most difficult to classify the various C^2 systems in any of the above three forms of organization. In very large C^2 systems, the distributed system organization is preferred, although even there parts of the system may be classified as centralized while other parts may be hierarchical. Take for example the inter state electricity supply with each area employing its own command and control system and yet able, in cases of supply shortage, to get assistance from other areas. For the control of the inter area supply, master command and control systems are introduced. In practice, industrial processes generally prefer centralized systems, and military organization are based on the hierarchical system, while commercial bodies incline to the distributed system; so no rules can be adopted for the best advisable choice. The intricate problems confronting the construction of C^2 systems are further complicated by the distance between the decision centres and action locations of the controllees; there are, in addition the type of equipment used throughout the system, the volume of data flow and the diversity of the different processing required.

The most essential feature of the C^2 system, *viz.* communication, is based on the means by which the data can be transferred between the various elements of the vastly spread system. In other words, the dominating factor in the success of achieving all the system functions are the communication network and the means by which it is used. In this book, therefore, concentration is on the communication technology necessary for real-time automated command and control systems. Nevertheless, it must be stressed once more, that the communication networks are only tools to achieve the required C^2 performance. While the aims of C^2 systems may generally be well defined, it is communication technology which must provide the most suitable solutions and it devolves, therefore, on the transmission designer to suggest the most efficient operational communication system.

1.3 SYSTEM INTEROPERABILITY AND INTERACTION

The initial command and control systems were designed to solve a particular set of problems confronting an organization, but it was soon found that no C^2 system can operate completely independently without relying on external information sources. Examples of these can be found in all types of C^2 systems, whether military, commercial or industrial. We may take for example, seat reservation of the airline companies, which are in association with other companies to transfer of passengers. It is not for the passenger to book his flight itinerary with each individual company; it is the due of the first company to provide this service. Likewise, banking C^2 system serves to transfer money from one bank to the other; nor has the customer to deposit his expeditionary checks only directly to the issuing banks. These examples show the mutual relation between independent systems performing similar functions. There are even further examples where the two associated systems have no common functional relationship. There can be, for example, the C^2 system of an air transport which is associated with that of a catering system, or the system of air traffic control which must be in touch with the C^2 system of a weather bureau. Examples can also be brought from the military sphere where a ground force has to be associated with an air support through a C^2 system. In all these examples, each C^2 system is not only operating independently but may be designed in a completely different configuration with different techniques and procedures from the others.

When two unrelated C^2 systems have to operate together the common operation is termed 'interoperability'. System interoperability is defined as the ability of a C^2 system to transmit and receive information to and from other C^2 systems. For the interoperability of two systems to succeed, the two systems must be able to communicate with each other; the output of one system must be the input of the other system. Interoperability is much more than the communication between two independent computers. With it, one system can regard the other as a source of one of its data input required for its decision making. Both systems must be able to send and receive information as they need it with an on-line connection and with a real-time response. In other words, the two systems must not only be able to communicate with each other but also to operate together.

The system interaction (as described in the previous section) must not be confused with system interoperability. While in both cases (as seen in Fig. 1.5) the systems operate independently, with interaction the systems are related and have both a direct or indirect effect on each other; using similar techniques and common equipment. With interoperability, however, the systems are completely unrelated, where the only common media between them is that each system possesses information that can assist the operation of the other system.

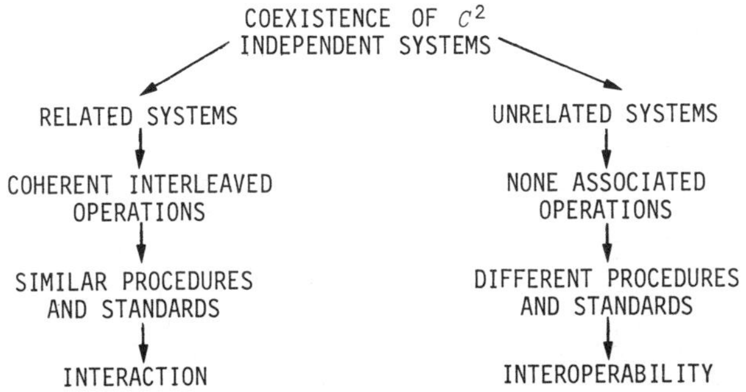

Fig. 1.5 The common operation of independent C^2 systems

The design of system interaction is much easier and simpler than that of interoperability. With master system design, the data transfer between each of the smaller systems is defined by one organization. That is, the organization determines what information is to pass between the subsystem; how it is to be transferred, who is authorized to update the files and who is authorized to see what is in the files. In a banking C^2 system, for instance, the head office has full control of the branches' decisions and is provided with means for noting all the local transactions.

With system interoperability the situation is completely different. Each system does not only operate independently but is also controlled by different organizations and may even be located in different countries. It is accepted that with interoperability there is a need for interchange of information between the systems, but this does not allow one system to update the files of another, not to mention that one system must not be allowed access to the files of the other system. The interoperability systems may even belong to competitive organizations. Take for example two airline companies which are associated in providing the passenger with a service of extended flights with both companies. This will not mean that one air company has the right to change the operations of the other company.

With system interoperability, each system must define what information can flow between them and introduce strict security measures to ensure that this right is not unduly exploited. Furthermore, special recording logs must be inserted in both systems to register all transactions for possible later analysis. The security measures in both systems need not be identical, as each can follow its own procedures.

System interoperability constitutes a major design issue, since the designer knows only what information he needs to receive from or transmit to the other systems but knows presumably very little of how the other systems operate; as each system may keep its own procedures secret from the associated systems. It follows that while the designer must provide means for free information transfer between the systems, he must at the same time ensure that the transfer is limited only to the information agreed upon by both systems. This indicates another difference between interaction and interoperability. Whereas designs for interaction are based on standards, those for interoperability are based only on mutual agreements.

1.4 THE COMMAND SUPERVISORY POST

The most important element in the command and control system is the command supervisory post where all the decision making is performed. Although the term command post is taken from the military sphere it could adequately be used for industrial and commercial systems. It is to this post that all the data is directed; it is at this post that all the decisions are made,

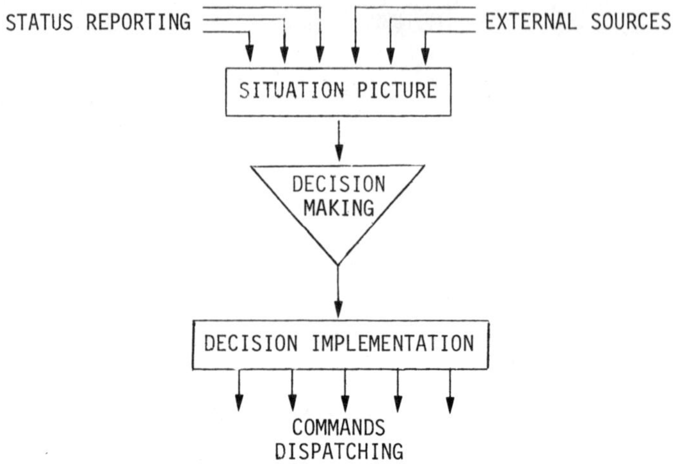

Fig. 1.6 The basic function of the command supervisory post

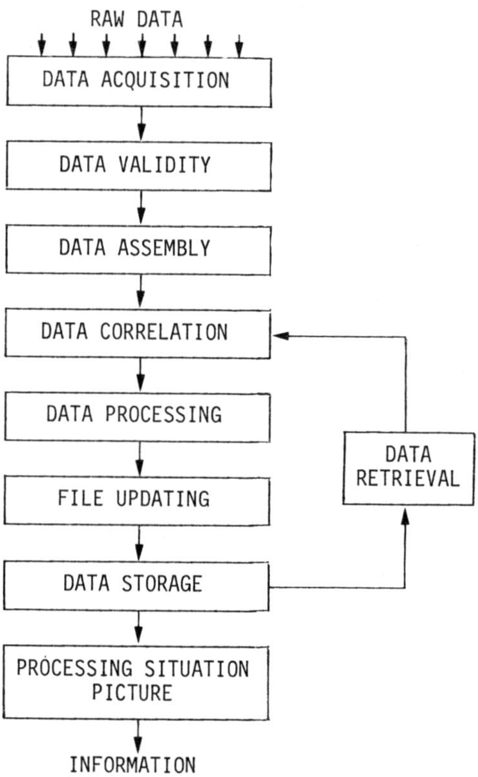

Fig. 1.7 The processing stages of converting raw data into meaningful information

and it is from this post that all the commands are dispatched. It is not important whether this post is used for management or for supervisory decision, as the function of the post is the same for both.

The command post is kept supplied with updated information and thus is enabled to adjust to the systems' situation status. The supervisor of the command post must first evaluate the situation before any decision can be made and these decisions must be based on the situation picture of the controllees' actions displayed in real-time, as shown in Fig. 1.6. Only then may the decisions by translated into commands and dispatched to the field.

Decisions are made on the basis of accurate information that is presented to the command post, and can only be made correctly if this information presents the controllees' actions in real-time. The data collected in the field is nevertheless meaningless if it is presented in its raw form. It requires first to be processed in a number of stages before it can be accepted as meaningful information, as seen in Fig. 1.7. In this respect it is important to distinguish between data and information. Data representation relates to quantities to which meaning might be assigned, while information is the meaning assigned to the data.

As the amount of data collected in the field is enormous and much of it is redundant, it is beyond the ability of any human being both to absorb this data and make the proper evaluation in the very short time available. To achieve both purposes he may employ computers or other processing equipment and proceed on the following lines. The raw data must first be checked for its validity and then assembled for the purpose of correlating it with previous acquired data. After removing most of the useless and repetitive items, the data can then be processed by associating each of the remaining data items with the relative controllees' action. The result of this process is that the corresponding files are updated. This ensures that the data files represent the updated real-time status of the controllee. The situation picture can now be processed by the combination of the different updated data items. The information acquired by this transaction represents the true meaning assigned to all the collected data items.

Although the information of the situation picture presents the true status of the controllees' actions, it is still not sufficient for decision making. While the information represents the collected and assembled dynamic actions of the controllee, this dynamic information is useless if it is not visually associated with some static situation picture. Take for example a traffic control system where the decision cannot be based on the dynamic real-time traffic situation picture unless it is directly related to a static background overlaid map.

The efficiency of the decision can only be operative if the command post presents the supervisor with various types of additional information, as seen in Fig. 1.8. That is, most the decisions are based on the situation picture comprising dynamic information displayed over a static map. Yet the supervisor must still refer to additional information before he can make his final decisions. What is meant by additional information in decision making is the previous information. For example, the treatment of a patient is based not only on his present, but also on his previous temperature. Other additional information could be statistical background presented in the form of records and tables.

In the case of emergencies the supervisor has no time to trace the cause of the new situation in books; instead he can employ the assistance of the system. That is, the system can present him with options both of the cause of the trouble and of the actions he may take. Nevertheless, it is left to the supervisor to make the final decision for action.

An important information item that must be presented in the command post is that of the system operational constraints. These constraints need not necessarily be a result of the controllees' performance but may be

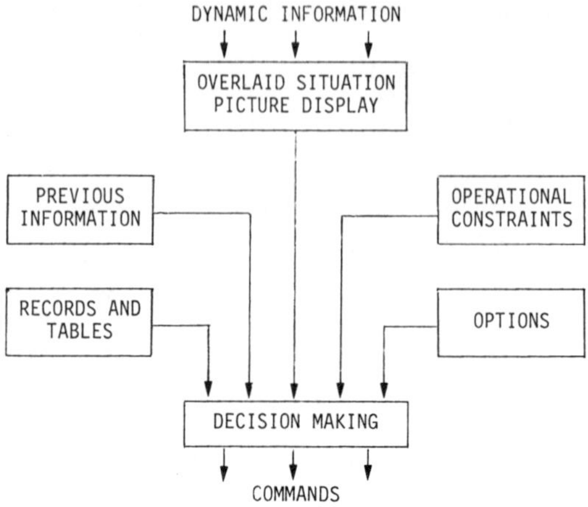

Fig. 1.8 The information presented in the command post

completely divorced from the system, although their effects have a direct influence on the controllees' action. Take for example the effects of the weather condition on an air traffic control system or the effects of the stock market on a banking C^2 system. These constraints cannot always be directly related to the situation picture, although they may have a direct effect on it. Furthermore, some constraint items may be constructed of static information while others may be dynamic information. Some of these constraints may be supplied by other systems. Decision making requires the human operator to know how and what information he can use for his decision.

1.5 THE HUMAN POSITION IN THE COMMAND AND CONTROL SYSTEM

The essential objects of command and control systems are sometimes overlooked. The system is not introduced for the purpose of restraining the actions of the controllee, which is one of the main features of C^2 systems. The ultimate aim of command and control systems is to make the human activities more efficient and more effective. It must therefore be emphasized from the outset that the centre of the command and control systems is always the MAN. Command and control systems are intended to make the human task easier. They are designed to assist the human agent in evaluating the performance of a process and controlling its action.

In an operating command and control system, the position of the human agent must always be at the top levels of the hierarchical upside-down pyramid, as seen in Fig. 1.9. Although most of the systems' functions are performed by the intelligent machines and therefore should be regarded as the pyramids' base, the whole operation will collapse if there is no human supervision and no human decision. The machine operation may be designed in such a way that there is a need for only

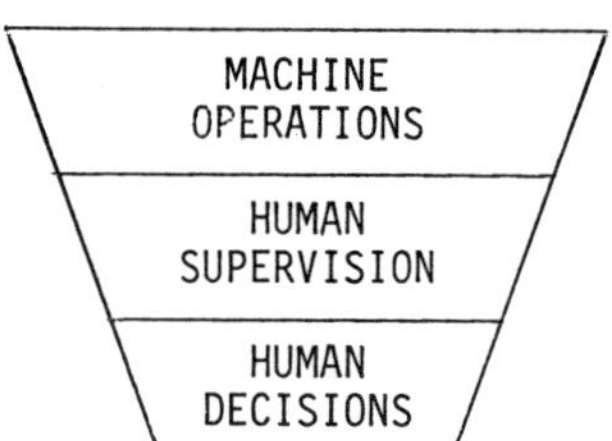

Fig. 1.9 The hierarchical levels of an operating system

minor supervision but even then unexpected problems could arise. Human agents must always be at hand to sort out ambiguities which were not considered during the design stages. However well a system is designed, there is always an operational scenario which was not thought of. Take for example the NORAD system, designed by the world's leading authorities, which had a number of false alarms in 1980. Only the human intervention prevented these alarms from probably causing a third world war.

The important position of the human factor in the hierarchical pyramid should not imply disparagement of machine operation. Machines can ideally control and interpret the operation of a process; nevertheless, they can only perform what they were designed to do. Computers, which are regarded as intelligent machines, can extend the operations of the machines and further assist the human agent in finding solutions to the operational problems. No doubt computers can perform an enormous task, therefore releasing the human agent from much of their control functions, but, despite all its attractive features, no computer can be designed to replace the human decisiveness. Humans decision can be based on intuition, while computers can only perform according to what they were programmed to do.

Computers and other intelligent equipment are used extensively throughout the C^2 system, but only as tools. That is, they are used as instruments generally employed to relieve the human agents of the onerous load of repetitive and routine operations and thereby freeing them for analytical operations.

No command and control system as yet may be fully automatic without the need for human intervention. There are always ambiguous operational situations which inevitably cause operational errors. It must be definitely assumed that there is no command and control system which does not require the human supervision of the processes actions and the human decision to override if necessary the conclusions of the machines. Without the human intervention, the system, however well designed, may not operate properly or even not at all. The human element should not only be regarded as an important part of the system, but must be considered as one of the essential elements of the systems' operational loop. It must, however, be conceded that the humans are limited in the amount of information they can observe and evaluate in a given time. There is always a danger, then, of a bottleneck occurring in the system if too many tasks are devolving in the human element.

As already stated, it is most dangerous to rely only on human reactions in the operating system. Automated command and control systems are intended to relieve and certainly not to increase the human tasks. The system operational loop must therefore be designed with both machines and humans operating in harmony, each complementing the functions of the other.

In the previous section, the function of the command post was discussed with the human factor playing a major part. It was stated that all data flows in the direction of the command post, where it presents there a situation picture providing the grounds for decision making. The function of the computer therefore is to process the data into meaningful data, while it is the function of the humans to evaluate and analyse the information. It is most important to stress that not all the information should be presented even if it is all relevant and needed for the human decisions. The loads of information the humans can carry being limited, the human agent can become exhausted after a time and his reaction will slow down. While it is essential to transfer to the computers as many tasks as possible, it should be left to the humans only the mental duties of evaluating and analysing the situations. As stated, the computers must be employed to perform all the routine and repetitive operations, and beyond, to present a summary of the results for the human supervision. The humans may adminster operations with the aid of displays which enable them to obtain more details, although only on request as required. The system must be so designed that the humans need not be bothered with routine details but may be alerted when a suspicious hazardous situation occurs. It is the function then of the computer to present this ambiguous situation as clearly as possible with all the relevant information available leaving it always to the humans to sort out the problems and make the decisions.

The question remains, what are the respective roles to be given to the human and to the computer elements. For this is is necessary to define the qualification and limitation of both for their potential mutual assistance. In the following table some of the relevant points of the characteristics of both humans and machines are compared.

FUNCTION	HUMANS	COMPUTERS
Ideal missions	Evaluating and analysing	Routine and Repetitive
Central tasks	Supervision and clearing ambiguities	Acquiring enormous amounts of data and converting them to meaningful information
Analysing situations	Sophisticated	Superficial
Reaction	Intuition	Strictly according to program
Speed of reaction	Relatively slow	Extremely fast
Deterioration with time	Prone to fatigue	No change
Function in the operational loop	Decision making	Controlling process actions & displaying situation picture

When designing a command and control system, consideration must be given to the complete operational system loop involved. The design must therefore regard not only the computers and other machines as the elements of the system but also the human factor as of correspondingly equal importance as the computer. In fact one must regard both the humans and the computers as integrated parts of the overall system where each is intended to supplement the other. Owing to this, it is extremely important to stress not only the functions of the two elements but also the interaction points between them. The main function of the machine is to prepare the information for the decision making, but it can still be relied upon to perform minor decisions of particular situations. The information, when displayed before the humans must be precise with no ambiguities. The main humans contribution is decision making, but this should not imply that it is essentially always an evaluating task, as far too many decisions can still be transferred to the computers. The human function should be concentrated on analytical problems with the decisions based on intuition resulting from the precise information displayed before them.

1.6 COMPUTERS FOR COMMAND AND CONTROL SYSTEMS

It is no exaggerated statement to assert that no automated command and control system can operate today without computers. In fact, it is nearly impossible today to convert the large amounts of data items collected in the field into useful information needed for decision making without the use of computers. Nevertheless, it must be stressed again that the computers are only used as tools in implementing the C^2 functions.

In general, computers involved in C^2 systems have two major tasks, controlling and assisting in the appraisal of the action of the process. The automated C^2 systems can therefore be regarded as combining two system functions, process control and data processing, with computers employed as the intermediate bridge between these functions. In other words, the computers are the essential links between the human world and the outer world, as seen in Fig. 1.10. The object of the computers, if so, is to make the human

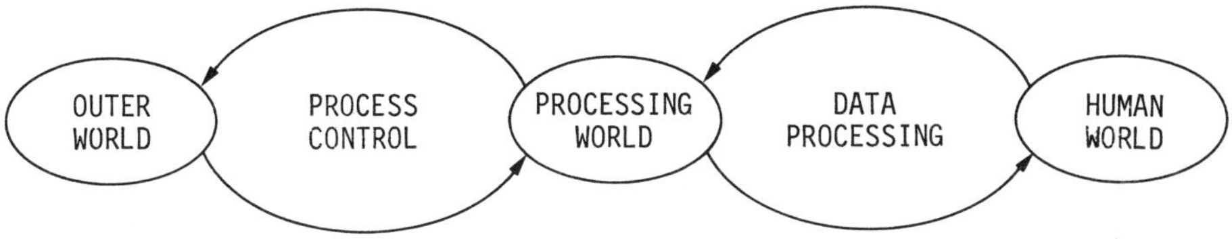

Fig. 1.10 The function of computers in C^2 systems

job easier. In both these C^2 system functions, the computers must be connected on-line and operate in a real-time environment.

Computers in C^2 systems are also involved in the data conveyance functions. That is, most of the communication tasks in the C^2 system are performed with the assistance of computers. Nevertheless, as these communication computers do not include the C^2 tasks of processing, they will be discussed separately.

As it is not at all expected of the processing computers in the C^2 systems' real-time operational loop to perform extended mathematical calculations; it is not the arithmetical ability of the computers that makes them attractive for C^2 systems but completely different qualifications. What the adopted computers must possess is the capability of handling masses of data items swiftly and efficiently. For this they must be able to handle simultaneously a large number of input transactions and respond to them all expenditiously. Furthermore, the computers must be able to store large quantities of data items and reach them in real-time for sorting them out when required for conversion into information. In other words, the computers are used here more as sorting and collating than as calculating machines.

The two operational features required from these computers are that many 'users' may communicate with them and that they have the availability of large data-bases. (Data base is defined as an entire collection of data items stored in a computer system.) This can only be implemented by sharing the computer resources in time-division and providing each input link with a direct access to all the data-base files. Each user of such a system may then be able to carry on a dialogue with the computer without being disturbed by other users.

The operation of the computers which permits direct access to the computer at the same time by a large number of users, as shown in Fig. 1.11, is known as time-sharing computing. Time sharing computers are distinctive and must be differentiated from batch processing computers. In batch processing computing, all the programs are fed separately and operated in sequence. The computer then runs a complete program without stopping, and this usually causes the other programs to wait long periods before they can gain access to the processor. This is not the case in time-sharing computing, where it is essential that a number of programs should be able to run simultaneously without being affected by one another. Each user can have a continuous run with the computer, thus viewing the computer as exclusively his and not being bothered by queues or by delays. It can now duly be claimed that the development of time sharing computers is the main factor that has determined the advancement of command and control systems.

Time-sharing computers makes demands on both hadware and software. It generally requires a large computer configuration to enable a sumber of short consecutive programs to operate simultaneously at a high speed. The computer must be fast enough to respond to any task within a defined short time interval.

Although time sharing computers are said to perform several different

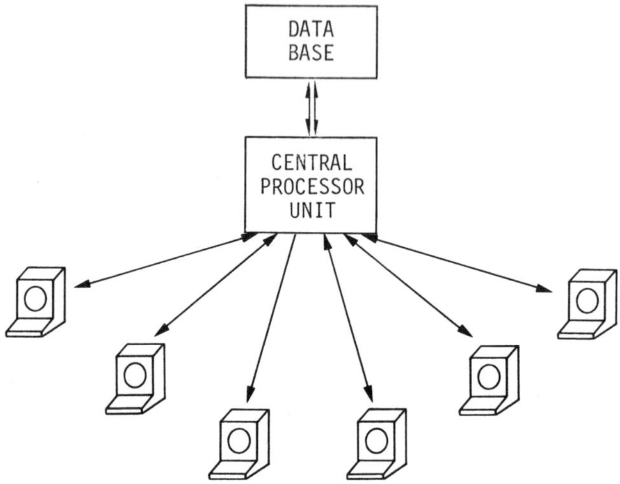

Fig. 1.11 Basic configuration of a time-sharing computing system

tasks simultaneously, this is not strictly accurate. It is obvious that a single processor can perform only one task at a time. What actually happens is that the programs are run not simultaneously but in sequence. The interval between the consecutive programs will not, however, be felt by the users, because the slow reaction of the human operator in conjunction with the high speed of the computer will give the operator the feeling that the whole system is completely devoted to solving his tasks. Nevertheless, if too many users have access to the computer simultaneously, delays may possibly occur between the introduction of a query on the terminal and the reception of the computer answer. Such a situation must clearly be avoided in a C^2 system, for such delays could be critical to the real-time reaction of the system. A situation of this sort could be avoided, however, by limiting the length and/or the number of programs being processed at any given period of time.

Each program to be processed is segmented into uncoupled pages which can then be dealt with separately, a page at a time. After running a page of one program for a short period of time, the central-processing-unit (CPU) will transfer its attention to the next consecutive program. Thus all the segments of all the programs being processed will run in sequence one page at a time, with each program segment being provided with a fixed maximum processing time. A single program will be processed a page at a time with time gaps between each page, in which other programs segments could be processed. The program segments are all held in the computer memory between the processing stages and are returned there after each processing stage. The program therefore remains in the system till the final segment has been processed. A queue is arranged of all the program segments as seen in Fig. 1.12, with the uncompleted process program returned to the queue after each run, to be handled again when its turn comes up a second time. The queue itself handles the program segments consecutively according to the rule of first-in-first-out (FIFO) principles. In this way each program may be interrupted a number of times and then be provided with a number of runs. In some systems a priority scale may be introduced which provides some of the programs with a priority advantage in the processing queue whereby they may require to be

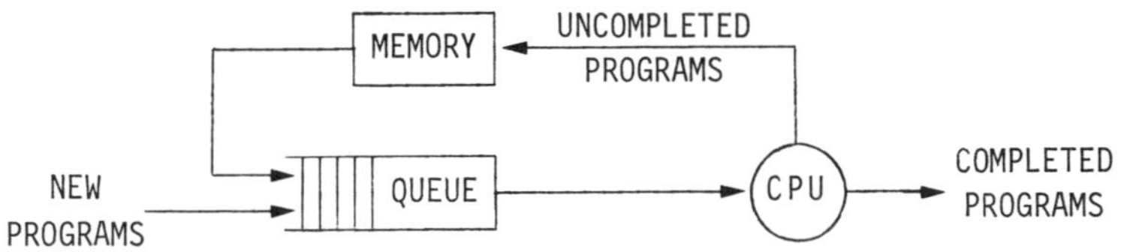

Fig. 1.12 A model of the feedback queue of processing segment programs

held for shorter periods in the system.

The whole efficiency of the time-sharing system is based on the scheduling algorithm which defines the sequence of computation operations and is thereby responsible for assigning to each program its' processing period. It is extremely important in C^2 systems that the reaction time of the system, in its worst operational conditions, will be less than the real-time factor defined for this system. It is therefore the scheduling of the system which is the most critical factor in the design of C^2 systems, as this is the factor which defines the time period a program will remain in the system before a reaction can be provided.

As a program enters the system it is assigned a priority factor in accordance with the real-time requirements of the system. The considerations that determine the priority scale are the program length, the processing complexity and the overall time it is allowed to remain in the processing facility. In consequence the processor will always handle the highest priority program segments waiting in the queue before it will deal with lower program priorities. Giving the higher priority programs an overruling advantage may sound reasonable but it is not a sound arrangement, since there is always a danger then that a lower priority program will be neglected and remain unduly long in the system. All programs must be given an opportunity of being processed; accordingly the lower priority programs are also given time factor assignments. A record of the processing and waiting time is logged within the system. If the accumulated waiting time reaches a given threshold, the priority factor of the program is automatically increased. Such an arrangement ensures that all the programs are completely processed within a given scale of time period. This may, of course, place a cost burden on the system, but this is unavoidable.

Another means of increasing the systems' efficiency is the dynamic memory allocation of the program segments. As already stated, the aim of time-sharing is to provide each user with the feeling that the computer is completely reserved for himself. This presents a basic problem of memory allocation, since the processor has no means of knowing, at the outset, the length of the programs. In batch processing and in some time-sharing systems each user is provided with a memory zone for his exclusive use. This solution, however, is not advisable for C^2 systems, since it will call for an extended main memory capacity which is not efficiently exploited. So instead of providing each program with a given memory space, each segment of the program is given a different memory location, thus resulting in one program being located in various positions of the memory. Furthermore, the program segments may be assigned different locations after each processing stage. For this purpose, the program segments are given a permanent logical address which could easily be translated into the physical address in the memory. As the program enters the system, a special address table is set up for it, as seen in Fig. 1.13. The position of the segment page within each program does not change and so it can be given a fixed page number, although its dynamic physical address may change each time the page is returned to the memory. The logical address is therefore constructed by the program number,

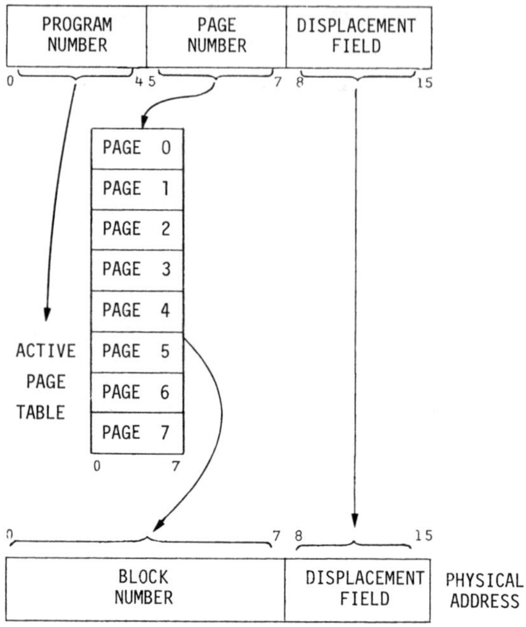

Fig. 1.13 Dynamic address of program segments

the page and the displacement field. The true physical address is then obtained by the translation of the page number for a given program number into the memory block address. This coupled with the displacement field indicated the actual location of the segment in the memory. In this way, the logical address of a page is never changed, while the active page table is modified according to the actual physical location of the segment in the memory.

In principle any large computer may operate in a time-sharing mode, although this is not always practical, as it would require certain changes in the computer. The basic requirements of a time-sharing computer are that it must always have enough memory capacity to store the incoming programs and their corresponding data-bases. The dynamic memory allocation explained above assists in increasing the efficiency control of the memory control. Nevertheless, it is impossible to ensure that there is always availability spacing in the main memory. Memory spacing within the computers' main frame is definitive; however, this could be compensated by adding external memory sources. These additional memory units are usually much slower than the main memory, but the employment of the dynamic memory addressing scheme assists in overcoming this problem. Such a technique, known as virtual memory, gives the effect of an available much larger main memory.

The requirement for computer storage capacity is great in any processing system, but in time-sharing systems this requirement is even greater since a larger number of programs of variable lengths may be processed simultaneously. This very large storage capacity is generally provided by introducing into the system various levels of memory media (as shown in Fig. 1.14) with the scheduling algorithm providing the means of fast transfer of the program segments between the various memory levels. The main memory holds the

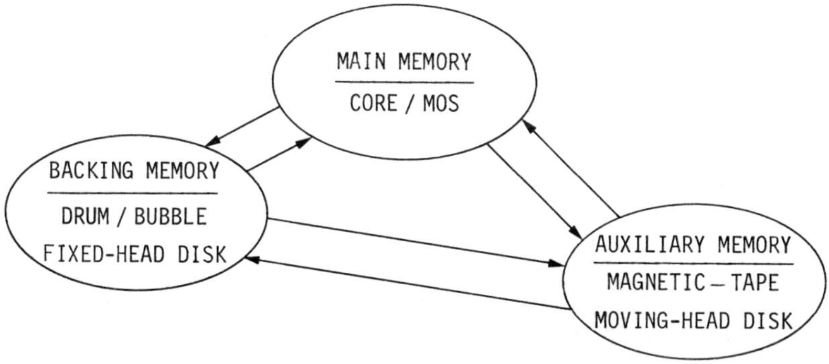

Fig. 1.14 The hierarchical three levels of memory media

scheduling instructions, the active page tables of the program and the program segment being concurrently processed, while the rest of the program bulk, together with the other programs and the data base fields, are retained in the backing and auxiliary memories.

The question could very well be asked why it is necessary to employ a number of levels of memory when one could suffice. Since speed of processing is the most essential factor to be considered in system design the ideal means of achieving it is to employ large main memories. Technically such a solution is not only possible but quite feasible. Storage capacity, however, is expensive and is increased with the requirement of faster access times. This means in effect that the cost of the system could increase with the expansion of the data base capacity. To meet this condition, various memory levels are introduced. The most essential data items are then stored in the expensive main memory media, while the rest of the items are stored in cheaper media. This is satisfactory, however, provided that the access times to all the items does not affect the real-time operation of the system. Figure 1.15 presents a comparison graph of the various memory technologies available as a function of access time versus relative cost. This sketch should be regarded only as general information and one always must bear in mind that both prices and technologies are subject to rapid changes.

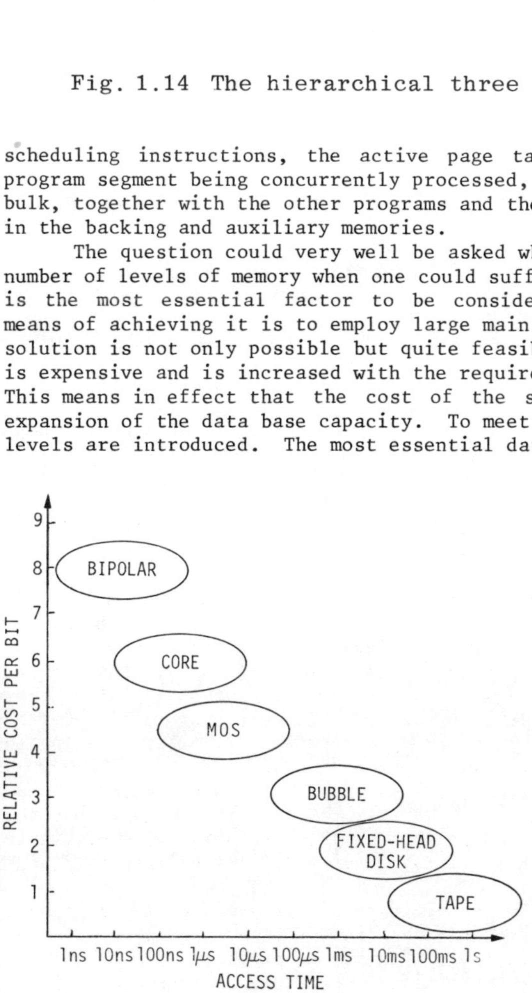

Fig. 1.15 A general graph presenting the various storage medias available

The main memory, being the fastest, keeps the operating program, the subroutine programs which are frequently referred to, and the running program segments. The backing memory holds the bulk of the programs which are currently operated by the various users. The pages of each program in the backing memory can be transferred quickly to the main memory for processing, and the full programs, after they have been completely processed, may then be transferred to the auxiliary memory for later reference. The auxiliary memory holds all the programs which are seldom referred to and the bulk of the data base. These include all the historical data files and all the data which do not require real-time handling. In brief, the main memory holds the processing data, the backing memory holds real-time data and the auxiliary holds all the historical and background data.

The efficiency of the memory organization is a direct function of the space allocation and the speed of transfer. When a program enters the system the space requirement is not known. The program is divided into pages of logical task units and each page is then transferred into the backing memory where ever space is available. The scheme is made operative by keeping in the main memory an address log book of the space allocation given to each of the page program segments. This paging arrangement allows only one active page to remain in the main memory and returns it to the backing memory as soon as the processing is completed. If the main memory requires data held in the auxiliary memory for reference, it can obtain it directly without first needing to transfer the data to the backing memory. The memory spacing can be extended by adding modular memory units to each level. The dynamic addressing scheme is that which provides the efficient indicator of the whereabouts of all the program page segments.

Despite their relatively slow access time, the backing and auxiliary memories could still operate in fast real-time systems. By predetermined transfer of the pages between the various memory levels, it is possible to ensure that no processing time is being wasted and no computer waiting time is necessary for bringing the pages forward from the backing and auxiliary memories.

1.7 COMPUTERS IN COMMAND AND CONTROL SYSTEMS

As explained in the previous section, the computing power is generally used in C^2 systems for processing the data collected in the field into meaningful information required for decision making. In fact, one can regard the computer power as responsible for the formation of large data bases which provide the real-time situation picture of the controllee's performance. Such data-bases consist of enormous background information which may be needed for the control operation. In this section the position of the computers in the complex system configuration will be discussed. It should be pointed out that although they are an essential part of C^2 systems, the organization of the data-bases and the means of obtaining the required data items are not regarded as subjects calling for elaborate treatment in this book.

The design of processing system configuration would have been simple provided all the equipment was situated in one encampment. In that case the time-sharing system could regard all the data terminals in the same way that it regards its peripheral equipment. C^2 systems, however, are spread over wide areas with the data acquisition sources located in positions which are far from the data bases. This means in effect that special communication channels must be introduced for transporting the data to-and-from the computers. It is a major object of this book to discuss the problems involved in this data communication.

The handling of the data transmission over the distant communication channels is different from that of local communication channels. With local terminals, such as peripheral equipment, the main frame of the computers is close be and therefore the number of wires between them is not limited. The data flow between the computer and its local terminals can be via a multi-wire bus channel and be controlled by the computers' interrupt scheme. That is, the computer can transmit and receive the data from the local terminals in parallel form, a character at a time. With remote data terminals the data must be transmitted in serial form, a bit at a time. This means in effect that the data transfer is much slower and calls for completely different communication control techniques to avoid impeding the computers' processing operation. These control techniques may require special processing equipment which could then be employed in special network configurations.

The full exploitation of the potential power of C^2 systems can only be achieved by merging both the processing equipment and the communication network in a single operational configuration, that is, blending them into a single sophisticated system. Nevertheless, the system's objectives may be attained if the data base computers are used also to control the data traffic in the network. This is possible in small systems, but as the system expands it is essential to separate the processing function from that of controlling the data flow in the communication network. In other words, system efficiency can only be reached by freeing the data processing computers from most of the communication tasks.

Communication tasks involve the gathering of data from isolated terminals and directing them to the processing centre. These tasks, although they may seem simple, include various postal functions, such as destination addressing recognition, traffic control, transmission authorization, speed control, message format, etc. These postal functions are generally referred to as formating, communication handshaking and protocols. They may be so complicated as to require special computing facilities to implement them. The obvious solution is to perform these communication functions of the data base computers, which would, however, have the effect of causing delays in the data sorting processing functions. The solution, must therefore be in a direction of freeing the data processing computers from most, if not all, the communication tasks. This consequently calls for the introduction of special dedicated computer processors. In this way the communication processor will serve as interface equipment between the data base processor and the communication network. In this application the data base processors are referred to as host computers while the communication computers are nodes of the network. The host computer is then isolated from the network and is capable of sorting a vast amount of data without being bothered by any of the slow communication processes. In this way, each processor in the blended system is selected on the basis of their ability to perform specific tasks.

There are various hierarchical levels (as seen in Fig. 1.16) by which the host computer can be released from its communication tasks. The decision as to what degree of separation is needed between the host computer and the network depends more on the diversity of the traffic requirements than on the amount of data processing required in the system. Each hierarchical level defines the magnitude of communication functions which are transferred to the interface node. Some systems may only use one level while others may employ a number of them.

The lowest level in the communication interface hierarchical is the line-controller, as shown in Fig. 1.17, which performs the physical selection of the communication lines. The line controller provides all the line equipment (such as modems) with all the control signals which are required to implement the functions of receiving or dispatching the data messages. It prepares the data received from the lines so that they can be fed into the CPU. Whereas it receives the data on a two-wire line, it feeds the data to the CPU on a multi-wire bus. That is, it converts the serial data bit stream received from the communication lines into parallel characters to be fed into the processor. This is implemented in the line-controller by buffer registers which receive the data a bit at a time and transmit them a character at a time, and provides speed conversion.

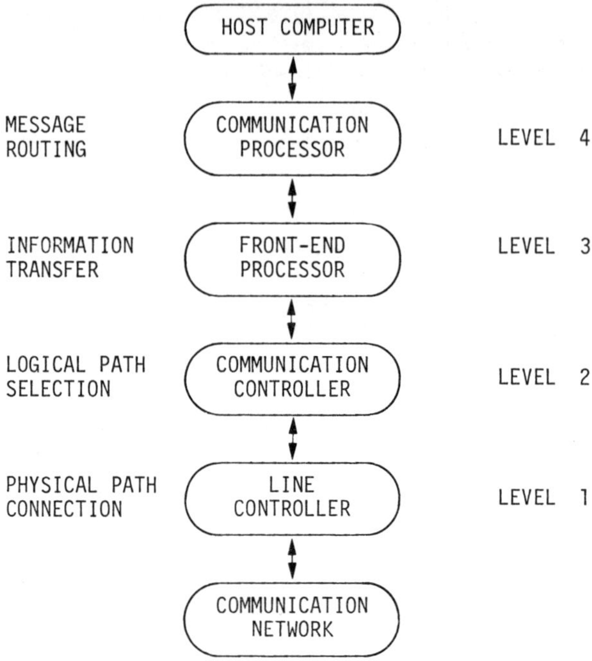

Fig. 1.16 The various hierarchical levels of host computer interface with the communication network

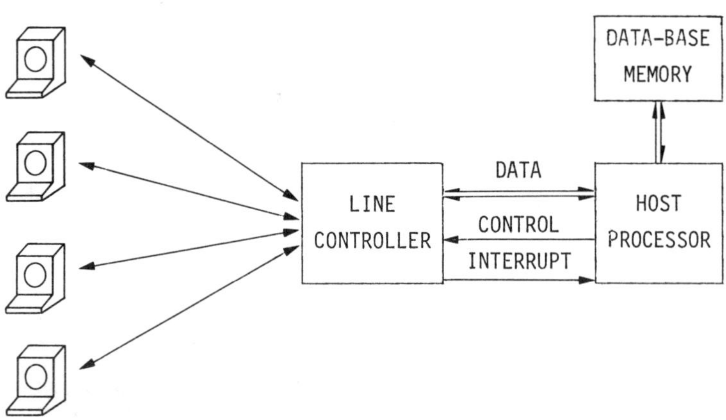

Fig. 1.17 The line controller, the lowest level of communication network interface with the host computer

The line controller is in fact part of the host computer and is usually supplied as an additional optional plug-in card. It must have enough logic to recognize the start or the synchronous bit and any other communication bit codes. After the full character has been received in the buffer, the line controller sends the CPU an interrupt flag signal. Then, when permission has been granted, it transmits to the CPU the relevant character together with its associated address but without any of the communication control bits. When the CPU sends out a message to a terminal, the line controller transmits it on the line in serial form, bit by bit, a character at a time. The line controller must first select the required line, in accordance with the CPU instructions, and then organize the data in the right timing sequence of the line, adding all the relevant communication protocol bit signals.

The line controller only handles the data one character at a time, while all the main communication tasks are still left to be performed by the central host processor. Furthermore, the central processor must first clear the buffer register in the line-controller before another character can arrive. This limits the usefulness of the line controller to only a small volume of traffic. In other words, it is slow and does not relieve the CPU of its communication tasks.

In order to cope with a larger volume of data traffic and reduce the load off the central processor, a separate processor is needed for controlling the traffic (as seen in Fig. 1.18). This processor, known both as communication controller and as input/output processor, performs additional communication functions and has the facility of initiating transmission without receiving direct instructions from the host processor. It has direct access to the main memory modules and can communicate with the host processor only through the memory.

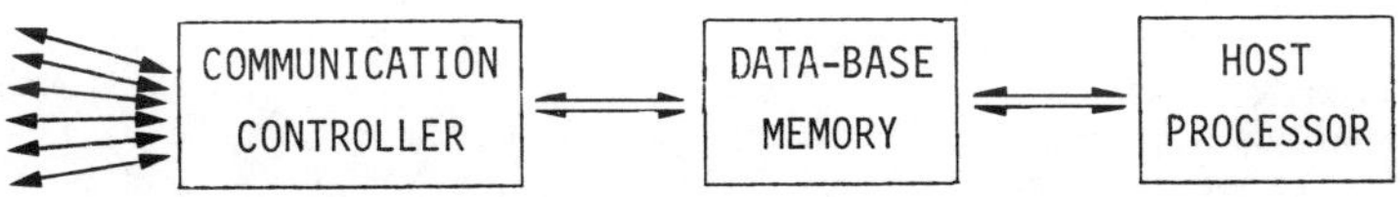

Fig. 1.18 Input/output processor controlling the communication traffic to and from the host processor

The communication controller initiates transmission procedures such as polling and selection, and therefore can communicate with intelligent terminals. In contrast to those of the line-controller, the functions of the communication controller are accomplished mostly in software. This is practical, since the communication controller has its own arithmetic unit and is provided with larger buffer registers that may hold full blocks of characters. The serial data bits, which arrive a character at a time in a random order, are built into blocks of associated pieces of information for block transmission to the next stage. The function of the block assembling, which was previously the task of the central host processor, has now been shifted to the communication controller. In this case, the assembled block can be placed directly into the main memory with no need for the host processors' intervention. Before the block is transmitted to the main memory, the communication controller performs error detection and/or correction of the received data. If an error is detected, the communication controller may either correct it or may require retransmission of the block. The communication controller also performs basic editing functions such as deletion and code conversion. In the transmission mode, the information is brought into the correct format with the required error checking codes and all the necessary communication protocol characters. The communication controller is a very primitive processor which acts only as a basic interface so as to assist the host processor, although it sill leaves all the actual message handling and further communication processoring tasks to the host processor.

With the increase of the traffic volume to and from the host computer, better efficiency can be reached by completely isolating the communication processing tasks from the central data processing tasks. This is achieved by the introduction of a stand alone computer (as shown in Fig. 1.19) which has as its prime task the communication interface processing. The host processor is now completely relieved of any communication tasks and can devote all its resources to information processing. The communication interface processor, known as a front-end processor, is now responsible for all routine communication chores, such as terminal polling, message assembly/disassembly, and error detection/correction, required to enhance the main frame performance. It provides the necessary flexibility for economic interfacing to a wide range of communication services and terminal devices.

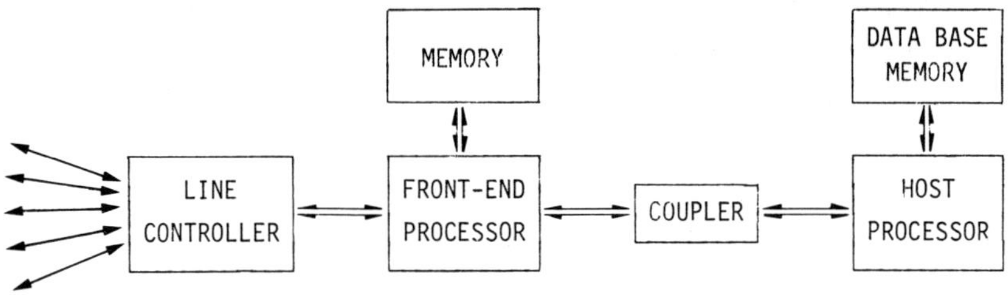

Fig. 1.19 Stand alone computer as an interface communication service to the host computer

The front-end processor (FEP) provides efficient operation of the communication network and the required throughput to the central processor by off loading from it all communication type functions, thus freeing it for purely computer processing tasks. Although the front-end processor functions are clearly communication control, its alliances are only with the host computer. In other words, the FEP controls all the traffic to and from the host computer only and is therefore intended for centralized system configurations with the host computer in the centre of the star network.

The two processors, the FEP and HP, can communicate with each other via an interface coupler which can initiate high speed parallel data transfer so as to fully exploit both processors. This double configuration is a form of parallel processing which is a much cheaper system than with a multi-processing capability where all the processors are of an identical construction. In fact, in many cases, the front-end processor may be a minicomputer while the host processor may be a very large computer installation. The front-end processor can handle all the incoming messages, performing both priority and queueing functions which were previously performed by the host processor. It can also perform basic processing, such as communication statistics, selection of alternative routings, and message heading analysis.

With the increase in the size of the C^2 system, having a number of command posts spread over wide geographical areas, a more sophisticated communicated network is required. This network must provide the facility for each data-base centre to communicate with all the others. Each computer installation may be adjacent to the other computers or many be miles apart; nevertheless, they all have the characteristic of being able to share the resources entailed in the other computer installations as if they were part of the same computer installation. Such potential resource sharing systems employ distributed system configurations. The sharing of computer resources is possibly the largesr contribution to the advancement of large and geographically spread command and control systems, for it permits the sharing of both the programming facilities and the various data-base files. This sophisticated system allows the transfer of information needed both for system interaction and interoperability.

Computer resource sharing can only be operative if each processing centre operates independently of the others but when required, can take advantage of the resources available elsewhere. This facility is provided by a complex communication network which connects all the computers together. Front-end computers may serve the system if each host computer is directly connected to all the others. Nevertheless, the system will not be cost-effective if each and every computer in the system is connected to all the others. Instead, messages from one centre may exploit the other centres as intermediate stations for reaching their final centre destination. This means additional communication functions in each centre. In order to implement this, special communication switching processors are employed at each centre (as seen in Fig. 1.20) which separate the centre processing functions from the communication functions. These communication processors now have double alliance functions: they act as interface junctions to their host processors and as nodes in the communication network. In their latter function they are known as message or packet switching centres.

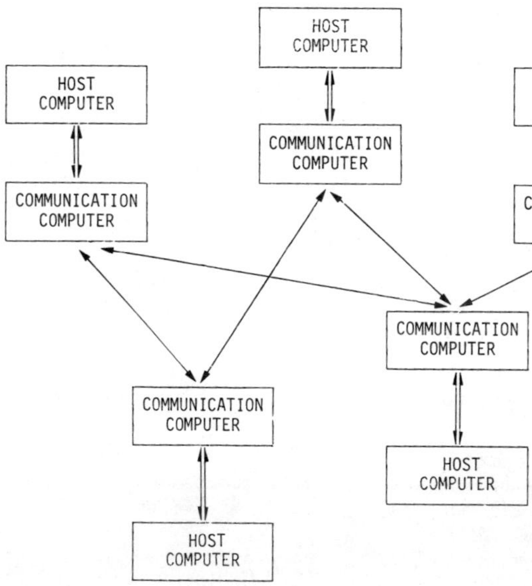

Fig. 1.20 Shared computer resources using a distribution communication network with communication computers as interfaces with the host computers

The communication processors are stand-alone computers which may serve one or two host computers. Their major task now is not servicing the host computer but the efficient control of the traffic in the network.

With large C^2 systems today it has been found necessary to add a front-end computer between the host computer and the communication computer, so as to release it from any local host duties other than the dealing with network functions. In this way the communication computers act only as the network nodes performing the transmission of the messages between the various other nodes, without them being concerned with what they are transmitting. It should be remembered that front-end processors have also the duty of assembling and disassembling the messages addressed to their host computer, a function which is now therefore removed from the communication processors. It is important to add that a front-end processor always serves a host computer, while a communication processor may operate with no association with a host computer but deal only with the message transfer in a network.

1.8 THE FUNCTION OF COMMUNICATION IN C^2 SYSTEMS

The concept of modern command and control systems can only be implemented with the availability of a sophisticated communication network. In fact communication and data transfer are the key factors to effective C^2 system operation and are therefore heavily emphasized in any C^2 system planning. An excellently designed C^2 system, with all the modern decision aids provided, would not be effective at all if the communication network did not transfer the data at the required speed. A situation picture display that is out of date or commands which arrive in the field too late, can have harmful rather than favourable effects. On the other hand, a simply designed C^2 system with only the bare decision aids will still be very useful if the whole system operation is performed in real-time. Commands issued on the basis of the minimal but true status picture have an enormous contribution to the system function.

It is not only the function of transferring raw data between two or more points that defines a communication scheme, it is rather the ability to convey large numbers of formated messages, directed to numerous varied destinations, at high speeds and within a given time limit. A message transmitted from one point must reach its final destination on time and with no changes occurring to its contents. Since the messages are transmitted in series, that is, a bit at a time, the message might easily be impaired en route. The switching centres along the route must be able to process the message destination but must not be able to change the message itself. Such communication facilities makes severe demands on the network design. Even the simplest C^2 system must employ some sort of communication network between the action points and the decision points, as seen in Fig. 1.21. In this simple network, two separate transmission channels provide the data conveyance in each direction. It is the efficiency of these channels that is the key to the successful operation of the system.

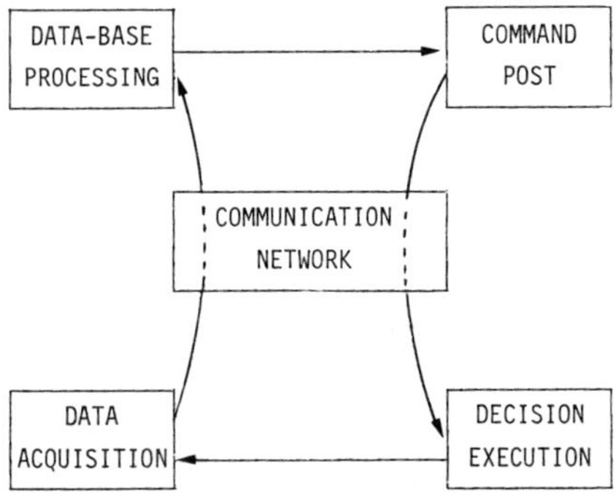

Fig. 1.21 The communication network in the centre of the C^2 system operational loop

The stage has passed where C^2 system designers may regard the communication network as a black box. Now, the whole communication network has a direct effect on the C^2 system operation and it is impossible to design a command and control operation without considering the specific effectiveness of the various communication technologies. In fact, the communication technologies have an extreme range of options to offer the system designer, as seen in Fig. 1.22. It can be seen that the design of the communication network is a far more complicated matter than to define the transmission speed, since the selection of each branch of technology has a direct influence on the others. The table presented shows only the important branches which can easily be extended, although it is offered here only to demonstrate the problem.

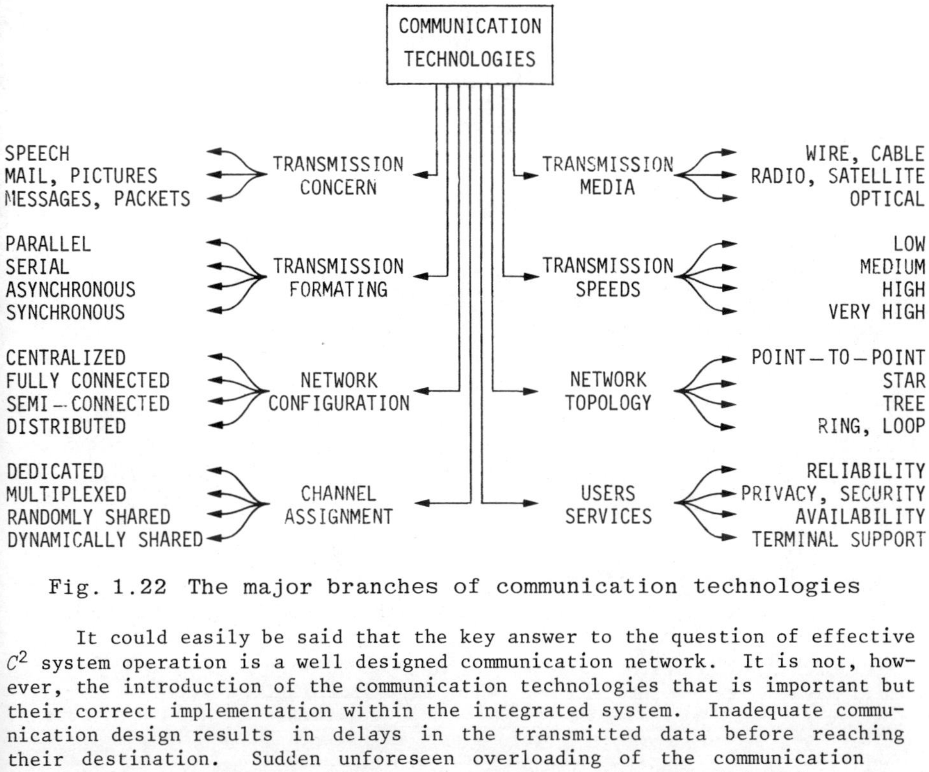

Fig. 1.22 The major branches of communication technologies

It could easily be said that the key answer to the question of effective C^2 system operation is a well designed communication network. It is not, however, the introduction of the communication technologies that is important but their correct implementation within the integrated system. Inadequate communication design results in delays in the transmitted data before reaching their destination. Sudden unforeseen overloading of the communication channels not only causes delays or errors but could easily bring the system operation to a standstill.

1.9 COMMUNICATION DESIGN CONSIDERATION IN C^2 SYSTEMS

Communication system design, however important, cannot be considered without regarding it as an integral part of the C^2 performance and bearing in mind that the C^2 system design is not as straight forward as other technical fields, as for example circuit design. Whereas circuit design is based on mathematical equations related to the physical performance of the components, in command and control systems, unfortunately, there are a number of hypothetical elements, such as the human factor, which cannot be formulated mathematically. It is admittedly possible to formulate equations related to human behaviour by taking the average reaction time of a typical operation for a specific problem, but this step would be of no avail to the system designer who is concerned with human reaction to problems of decision making and intuitive behaviour in relation to the random operations of unpredicted processes. Statistical or average calculation of a system performance will not do here, since the C^2 system must be designed to perform in a real-time environment under all conditions.

The steps adopted for the design of large C^2 systems having already been discussed in the authors' earlier book are not repeated here. What will be discussed are the concepts of C^2 system design which have a direct bearing on the associated communication network.

Although the design steps of communication network are well established, still, the design of the communication network serving large C^2 systems requires further study. In this case the whole system configuration must be regarded as a blended configuration. This means in effect that the communication network cannot be fully designed without considering the nature of the data that flows through the system and its performance under various system operational conditions.

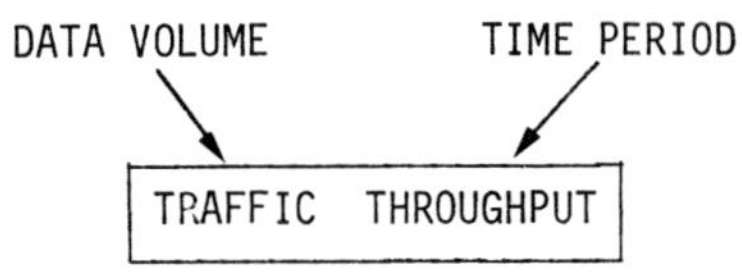

Fig. 1.23 Definition for a communication node performance

The performance of the communication switching nodes and the transmission channels is generally determined in accordance with the throughput factor. The throughput is measured by the total capability of the node (or channel) to process (or transmit) a given volume of data during a specified period of time (see Fig. 1.23). In most communication systems, the communication equipment is therefore selected according to the average traffic volume flowing through the unit. That is, the volume is first measured by the data bulk that flows through the unit over a relatively long period of time, say an hour or a day, and the result is then divided by a specified short period. This provides the throughput figure in bits/seconds.

The selected throughput figure for a given network provides an assurance that all the data can flow through the system within a defined period, it does not however, ensure a fixed transmission speed nor does it guarantee that delays do not occur in the system. Such an approach can not be applicable for C^2 systems and therefore the traffic throughput definition then needs further consideration.

The throughput factor is an important measure in determining the performance of communication nodes where processing is involved. In this case, the node may introduce delays in the transmission of the messages and this must be taken into account when designing the system. Since the traffic is usually random, the throughput is determined by the average data volume spread over a relatively long period. Even with long delays in the node, the average traffic throughput is smoothed out.

In C^2 systems the term 'average traffic' is not applicable and it is imperative that the design should be based only on the peak traffic. This means that the throughput figure too should be measured by the peak traffic and not by the average traffic. The employment of the 'average traffic' is intended for achieving the efficiency of the communication network, as it ensures an even flow of the traffic. The advantage of the communication efficiency is, however, neutralized by the introduction of delays in the data transmission whenever the traffic volume increases beyond the throughput factor. A C^2 system, operating in a real-time environment cannot afford the introduction of extra delays in the traffic flow above that which was considered in the design definition stage. C^2 system must ensure that each data item will always reach its final destination within the specified period, whatever the momentary conditions of the channels or nodes. In other words, the throughput of a C^2 system must be specified by the maximum volume of data that may be needed to be transmitted within the period defined as 'real-time'. The real-time factor is easily defined by the system operation, but the difficult question is how to measure the maximum data volume which flows during the peak period. This calls for consideration, as follows.

The volume of data passing through a communication node with processing ability is measured by the data flowing into the node and the data flowing out of it, as seen in Fig. 1.24. This design approach, however, is only feasible in a system where the type of data flowing to the node is known and, likewise, the processing activities that will be needed in the node. Therefore, before this simple node model can be used, one must specify the type of data that is to flow through the node and the relative periods when the data is needed.

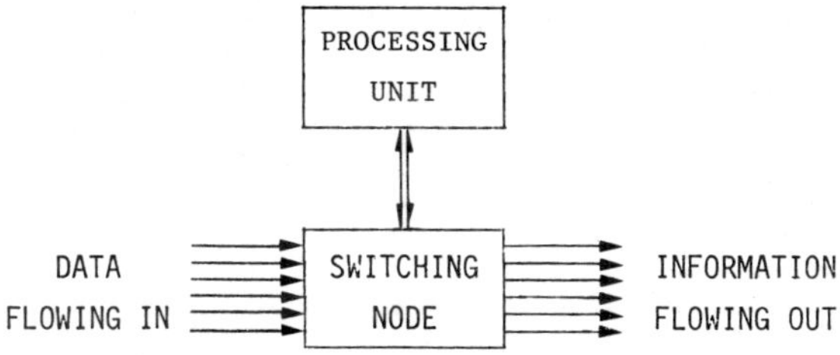

Fig. 1.24 A simple node for calculating data volume

One approach to system processing design is to copy the existing manual operations and translate them into automatic operations. Such an approach, however, can lead to a most inefficient system, since it does not try to exploit the advantages of either the computers or the operators. Nevertheless, some of the design techniques employed in the manual system could be of assistance in analysing the existing system. This design approach is based of the 5-W rules, as seen in Fig. 1.25. That is, they require answers to the following questions: who needs communication facilities ? What type of communication does he need ? When, where and why does he need it ? This approach defines the location of the nodes, the number of communication channels required and the type of network to be used. Such a system design approach could also be employed in automated command and control systems since the same locations will be used for decision making or data acquisition. Furthermore, it is the same data that flows into the systems, and whereas it was previously in speech now it is in digital form.

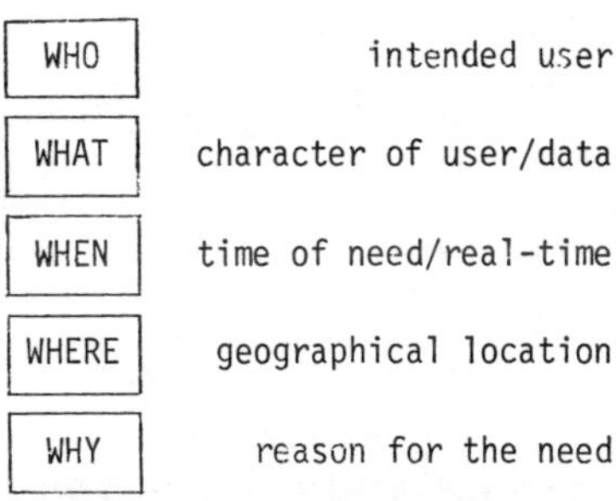

Fig. 1.25 The 5-W rules in the design of manual systems

It is a most dangerous approach to simply translate a manual system into an automatic system. It is true the system will operate, but no advantage will have been gained. For that the whole system design should be changed and be based on the operations required rather than on the functions of the people who operated the existing system. More so, the design of C^2 systems cannot be based on the translation of speech signals into digital signals even if both signals convey the same meanings. The whole design approach needs examining once more, since the approach of who needs what, when, where and why, is a misleading if not a wrong concept. With automated

C^2 systems, only the essential data needs to be flowing in the network. Most of the large volume of data flowing in a speech system are unnecessary and only required by the occupation of the people involved and not by the functions they have to perform in the system. Many of these human tasks will now be done by the automated system, and in many cases it will be found that many of their functions are unnecessary in the new system. What in fact is required is to reconsider the whole system concept and re-examine each function. The design of the decision making nodes should not be based on what data is available but on what is really necessary for making the decisions. In other words, the design of the node should start from its output and only then progress to its input, as seen in Fig. 1.26. The first design step, therefore, is to define what information is needed for the decision making, i.e., the information fed out of the node. With this information available, it is possible to calculate the data that is required to produce it. This means in effect that the basic design step is the specification of the node output, while the input is only a consequence of it. Such a design approach could also easily be applied to a data acquisition node, since there too its output is specified first. That is, one does not collect all the data that can be collected but only the data that is needed. Furthermore, such an approach also defines the volume of data to be collected and the frequency for scanning it.

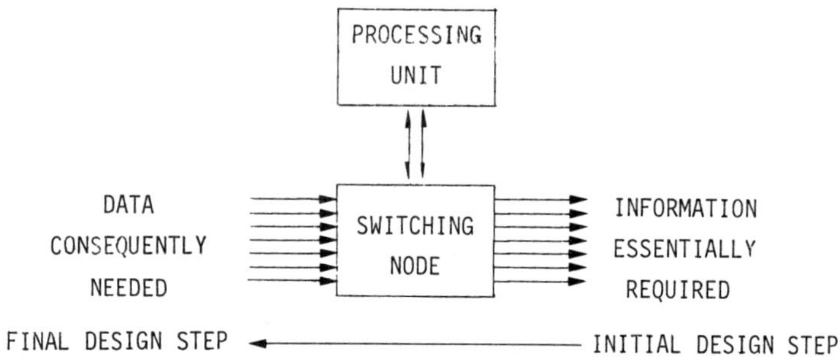

Fig. 1.26 The suggested design approach of automated C^2 systems

For the new approach the system design philosophy is modified by having the data flowing in the system specified only by what is needed by the system rather than what the people (who had operated the system) wanted to know. In other words, it can be said that the design of the system is now determined by the tasks of the people who are to perform the job and not by their hierarchical status within the organization.

While this approach serves the design of the C^2 system by specifying the type of the data flowing through the network, it does not, however, assist in the design of the communication system. In other words, the question of what flows through the nodes has been answered and likewise the question of the bulk data volume that flows in the system, but what has not yet been defined is the throughput of the communication system. That is, what is the actual momentary data volume that is to flow in the communication channels. This volume must be specified in measurable units which could then be used in specifying transmission speeds and for selecting all the communication equipment.

It was shown in Fig. 1.26 that the data collected in the field, which is fed into the processing node, is defined by the information fed out of it. One may therefore expect that the data throughput of the traffic flowing through the node is measured by the information items required to build the situation picture. This concept, though correct, could however be misleading in the design procedure, since most of the situation picture is constructed by data collected over a long period. That is, the bulk volume of the data is accumulated over a long period and is not necessarily collected in real-time. Only a small fraction of the data volume needs to be transmitted in real-time and these are the data items which cause a sudden change in the situation picture and consequently calls for the system's reaction. These data items are the result of the unpredictive and random performance of the Controllee, caused by external disturbances which affect its actions, such as natural activities, machine faults or even human errors. It is this potential random activity of the system which makes the design of the communication network so complicated. To meet possible contingencies the traffic volume in the channels of a C^2 system must be expected to include both routine and sudden random data items. The system design, nevertheless, must be able to envisage this unpredicted performance and therefore must be able to define both the routine and the random data items in measurable units which could later be embodied into the design procedure.

Although it may be obvious, it should be added that the random performance of the system operation is the major reason for introducing automated C^2 systems. If the system's performance could always be anticipated, with no ambiguous situations occurring, then there would be no need for a complex human system operation and all the control system could be transferred to a processing unit. Command and control systems, however, have emerged for the sole purpose of solving the control operations of unpredictive performances of a process or action.

The command and control system design must be directed at the possible random operational conditions of the system rather than at the normal smooth running of it. That is, theoretically it should consider the system's performance when all the unpredictive situations occur simultaneously and then design the communication network so that it can transfer in real-time all the bulk data volume created by the situation. Nevertheless, design in a system according to the worst operational conditions, though theoretically sound, is too extreme an approach. The probability that all the possible external disturbances, together with all the system faults and human errors, will occur simultaneously is so remote that it makes it impractical to design the system accordingly. What is suggested then is to envisage a practical scenario situation of the Controllees' actions under the worst possible operational conditions. Such a scenario could provide the basis for the design situation for valuing the communication system requirements under maximum load conditions. The whole C^2 system design can then be constructed around this scenario, since it is possible that such a situation may occur. If the communication network can operate under these load conditions, as envisaged by the scenario, then it will operate under much less critical conditions.

Translation of the scenario situation into useful throughput data items is performed in a number of easily designed steps, as shown in Fig. 1.27. The scenario situation is not a momentary situation but relates to a whole span of time, say a number of hours during which the critical situation can last. During this time a whole sequence of events can occur, each of them causing a maximum load traffic condition of the communication network. Nevertheless, it is possible to identify within this critical duration a period, say ten minutes, where the system operation reaches a peak. To avoid a system task failure, certain preventive decisions must be made within this period. In other words, this peak period is generally shorter than the real-time of the system. The information displayed during this period is constructed by output data volume of the nodes. That is, by formulating the output flow from the node, it is possible to calculate the data items required to produce the information displayed. Once both the nodes' input and output data traffic have been formulated, it is possible to calculate in exact values

ENVISAGE A SCENARIO SITUATION
CAUSING THE WORST LOAD CONDITIONS

↓

IDENTIFY A CRITICAL PERIOD
WITHIN THE SCENARIO

↓

IDENTIFY THE OPERATIONAL DECISIONS
NEEDED IN THIS PERIOD

↓

FORMULATE THE INFORMATION REQUIRED
FOR MAKING THE OPERATIONAL DECISIONS

↓

CALCULATE THE DATA ITEMS REQUIRED
TO PRODUCE THE INFORMATION DISPLAYED

↓

ADD ROUTINE DATA ITEMS WHICH
ARE CONTINUOUSLY COLLECTED

↓

CALCULATE THE MAXIMUM DATA ITEMS
THAT FLOW THROUGH THE SYSTEM DUR-
ING THE CRITICAL OPERATION PERIOD

Fig. 1.27 C^2 system design procedure for calculating the maximum traffic throughput

of the maximum data throughput through the node during the systems' worst operational period as envisaged by the scenario situation. One should also add to this maximum unpredicted data volume, the routine data volume that continuously flows from the field.

In conclusion it should be added that the scenario technique, as described here, is not suggested for the C^2 system design as a whole, but only for the design of the communication network within a C^2 system. The design of the C^2 system must be clarified before any initial work can be performed on the communication network. Many of these design procedures have been discussed in the author's pervious book where he presented detailed design flow charts. It should be added that one must first define the problem which the C^2 system is intended to solve and then try and analyse the difficulties involved. With the resulting information the operational specification requirements of the system can now be defined. In other words, the proposed system objectives and scope must first be classified and then followed by the system's expected qualitative and quantitative performance evaluation. Although these procedures are interesting and have an influence on the communication network, it is the latter and not the general subject of C^2 system design that is the direct concern of this book.

1.10 SYSTEM RELIABILITY CONSIDERATION

In the previous section, the design of the C^2 system was separated from the design of the communication system, although it was suggested that the command and control system architecture should be formed first and then the communication network configuration. It was further suggested that different design procedures based on different considerations should be employed for both systems. This is not suggested, however, in the case of system reliability, in fact the reverse is recommended, since the same reliability concepts and considerations should be employed for both systems. Furthermore, as a fault in either system affects the operations of the other, the reliability factor of both systems must be considered jointly. It must be taken into account that a failure in the communication channel, could result in no data flowing through the channel, and consequently cause the situation picture to be incomplete. Such a failure could nevertheless, be easily compensated for by providing alternative routes. A malfunction in the command and control system, however, may result in a sudden increase of the data load in a particular channel which could consequently saturate the whole network and affect its whole operation.

The reliability consideration of the command post in the C^2 system pursues the same standards as those needed for the network node in the communication system. In both cases the reliability requirements are for continuous operations despite any equipment failure, human error or operation malfunction that might occur in the system. For this reason, the presentation here will not be directed to any particular type of system.

Though system operations normally perform all their designed tasks, the system is still vulnerable to all kinds of external disturbances which could cause malfunction in the operational process. These disturbances are generally unforeseen and occur at random times, with no possible design techniques for avoiding the contingency. What can be designed, however, are means for preventing any damaging effects of the disturbances on the systems operations. This therefore must be considered a major purpose of system reliability design.

The vulnerability of the system operation can be classified in four types of faults: hard, intermittent, soft and design faults, as seen in Fig. 1.28. Hard faults are caused by permanent failures in the system and these are most easily detected. A typical example of such a failure is the disconnection of one of the communication channels in the network. Intermittent faults are caused by transient failures and these are very difficult to detect. A typical example of such a transient failure is an error in the data caused by channel noise while it is being transmitted between two nodes. Soft faults are generally caused by the errors of the human operators which can introduce into the communication channels wrong data which consequently cause misleading information to be transmitted and/or mistaken actions performed. Design faults are caused by programming malfunctions and are usually easily detected in the early stages of running the system. Nevertheless, in very large systems, based on stored program procedures with many subroutine loops, a number of design errors could still remain undetected even after the system is in operation.

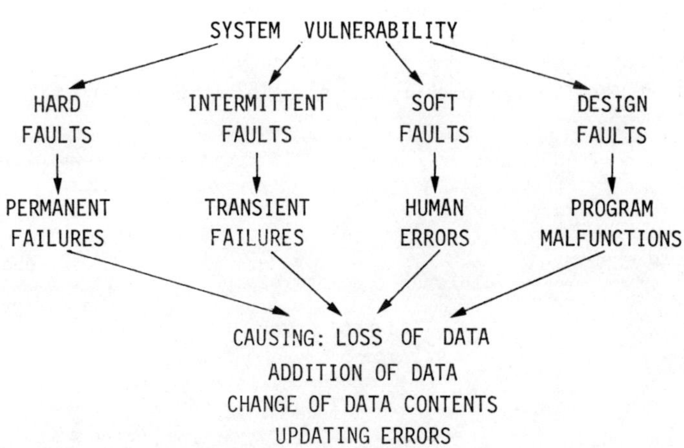

Fig. 1.28 The four categories of system faults

When designing the C^2 system, it may be impossible to foresee all the possible operational situations, and therefore the system may operate for long periods before sudden design faults may emerge. These four types of faults may result in all kinds of system malfunctions, not only a loss of a bit (or character) here or there, but the complete change in the information content. This may then consequently cause wrong decisions to be made based on what is believed to be the true situation picture.

Although four types of system faults have been identified, there are only two sources which account for them (as seen in Fig. 1.29) either by physical phenomena or by human behaviour. While it is easy to distinguish between physical faults and man-made faults their effects on the system operations are not always as easily traceable. In both cases the logical performance of the system is disrupted, and so could cause the failure of the information processing and/or the operational tasks. Physical phenomena cause changes in the component parameters but they could also affect the task implementation, or even cause software failures. Human behaviour causes operating errors but they too could cause software failures or even affect the task implementation.

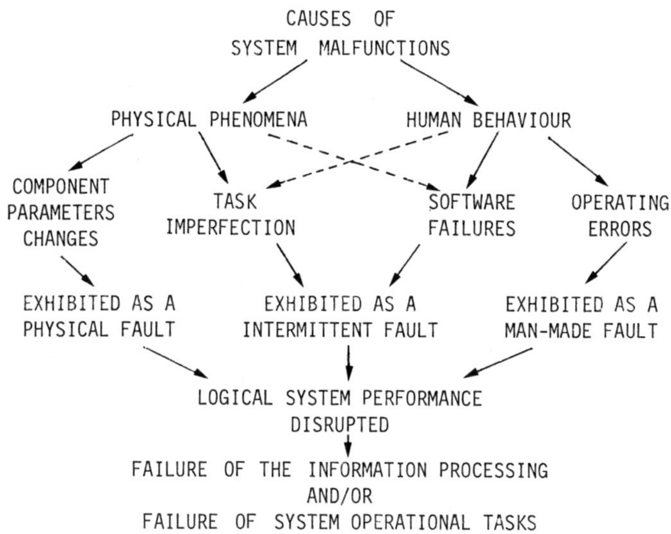

Fig. 1.29 Causes and effects of system malfunctions

Faults exhibited as intermittent failures are the most difficult to categorise, since their actual causes are difficult to pin down. It is usual to refer to intermittent fault as caused by physical phenomena although in many cases it could have been as a result of human behaviour.

It has been stated that the reliability methodology employed in communication system design is associated with that employed in C^2 system design. This is true mainly when considering the communication switching nodes; however, there are some failures which are unique applying only to communication channels, as shown in Fig. 1.30. These failures are generally produced as a result of a physical phenomenon rather than caused by human behaviour. Variation of the data contents, where a transmitted one-bit is received as a zero-bit, or vice-versa, is usually caused by external disturbances such as noise or fading. This type of communication failure is regarded as intermittent since it is generally caused by random transient

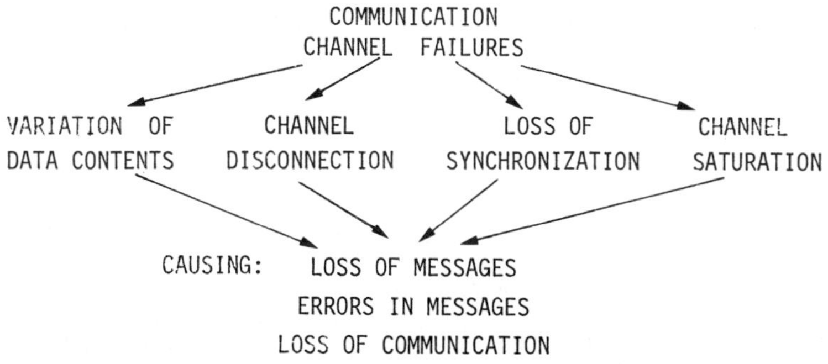

Fig. 1.30 Types of communication network failures

failures. Channel disconnection is a typical example of a 'hard' fault and can therefore be regarded as a permanent failur. Loss of synchronization, however, is difficult to classify. In a communication channel, the receiver must be fully synchronized with the transmitter before commencing high speed data transmission; else the receiver cannot relate the timing pulses with the data bit and consequently has no means of decoding the transmitted data. Loss of synchronization is usually caused by external disturbances, such as sudden change of attenuation or fading. When the disturbances are removed, synchronization is not instantaneously restored, and the receiver must once more search for synchronization. When such a failure occurs, it is exhibited as a channel disconnection fault since there will be a period of complete loss of communication and then any meaningful data which may be received during this failure has all the characteristics of noise interference. In other words, loss of synchronization failure caused by an intermittent fault appears as a hard fault. Channel saturation failures occur when too much data is transmitted through the band limited channel. One could regard this failure as a soft fault caused by human error, although this is not invariably the case. In a shared transmission channel where the channel resources are randomly assigned (as in ALOHA scheme), the channel can reach a saturation point which consequently brings all the data traffic to a complete standstill.

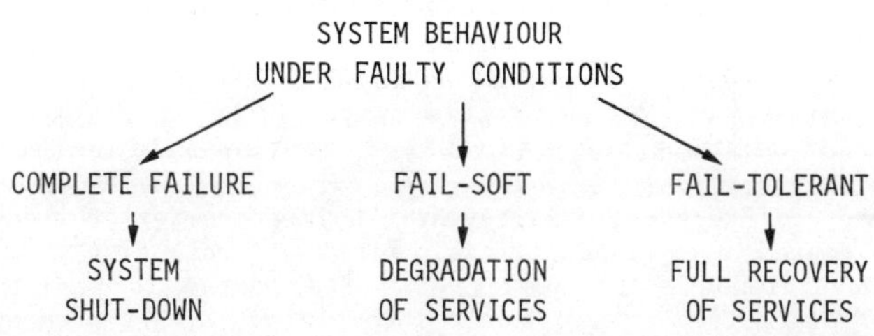

Fig. 1.31 The system behaviour after a fault is detected

As already pointed out, failures in the system operation cannot be avoided, whether in the communication or C^2 system; nevertheless, means can be introduced to reduce their effect on the system's performance. The system's behaviour under a fault condition may take one of three forms, as shown in Fig. 1.31. The system may fail outright, with complete shutdown of services, a critical situation which in C^2 systems must be avoided at all costs. The system, however, may be so designed that a failure in one of its modules will not cause any changes in the operation of the system, whereas it continues providing its service as before, despite a module failure. This type of failure is known as 'fail-tolerant' which is corrected by automatically replacing the faulty module by others, thereby allowing the full system recovery.

The term fail-safe is used in reliability circuit design, although in a different content; nevertheless, it could also apply to fail-tolerant design. It refers to the actions after the failure has been detected where a failure which causes a shutdown of a module is always designed 'safe' in avoiding damage to the remaining 'good' modules in the system.

The design for fail-tolerant reliability is expensive and some systems therefore introduce a technique known as 'fail-soft' which though not providing full recovery of services after a failure is detected prevents complete system shut down. In the case of fail-soft only parts of the tasks are performed after a failure occurs in one of the modules. That is, with a failure, either only the critical tasks are still performed, or all the tasks are performed as before although their implementation will need much longer periods than specified by the real-time requirements. Fail-soft is achieved by sharing the system's tasks among a number of parallel modules and then reassigning them between the remaining modules when one of them is faulty. Fail-soft operation is sometimes also referred to as 'graceful degradation' of services.

The principal quantitative measure generally used for weighing equipment performance and its resistance to failure is known as the reliability factor. Reliability in this context is defined as the probability that a given equipment or product will achieve its designed goal successfully under specific conditions and in a specified period of time. That is, it is the probability that the equipment will function as specified without any failure over a specified period of time. It must be stressed here that the reliability factor gives no guarantee for system operation in the time specified nor are there accurate means of measuring it. The common measures used to compare different equipment reliability performances are MTBF (mean time between failures) and MTTR (mean time to repair). The quantitative value of these factors are based on statistical measurements rather than fixed physical characteristics of the equipment. The reliability factor can then be calculated from the MTBF figures by defining it as a function of the statistical failure rate F of a set of N identical equipment units tested, over a time period t: $R(t) = e^{-Nt/F}$ and then MTBF can be expressed as

$$\text{MTBF} = \int_0^\infty R(t)\, dt \, .$$

It is needless to state that in C^2 applications, the common MTBF factor is not accepted as a measure for system performance, since its statistical measure makes it unreliable for this type of operation. MTBF is used extensively for component reliability where a large number of items can be measured and tested over long periods, and thus presenting best failure results. With equipment, the number of units tested is smaller and therefore the statistical results are incomplete. With system operation where only a few structures are implemented, no accurate MTBF figures can at all be provided. Another reason for not accepting the MTBF factor for C^2 system operation is that its major requirement is the continuity of services which must not fail even when there may be failures in some of the system modules. In other words, the reliability requirement of operating with no failure for a given time, is now extended to indefinate time. Such a requirement may be unattainable, one must nevertheless strive to achieve it, since any complete system failure could have a most critical effect on the actions of the controllee.

One of the main (but definitely not the only one) requirement of C^2

system design is that all the equipment employed in the system should have an extremely high reliability factor, i.e. $R(t) \to 1$. Although such a target may not be fully attainable, one should aim at achieving it by pursuing all the five suggestions presented in Fig. 1.32. The first essential step is to employ only the components with the most high reliability factor available and then introduce them into the circuits by allowing them the widest possible operational tolerance. This will allow the equipment to operate in extremely wide environmental conditions without a change in their functional characteristics. In C^2 systems, where critical actions are under control, one cannot afford to indulge in experiments and so only known and proven technologies are used. Even then the equipment must first be run in the laboratory for a sufficient period before it can be introduced to operate with 'live' actions. With all these four targets introduced, it is advisable to apply redundancy techniques for operation at all levels when the active units fail.

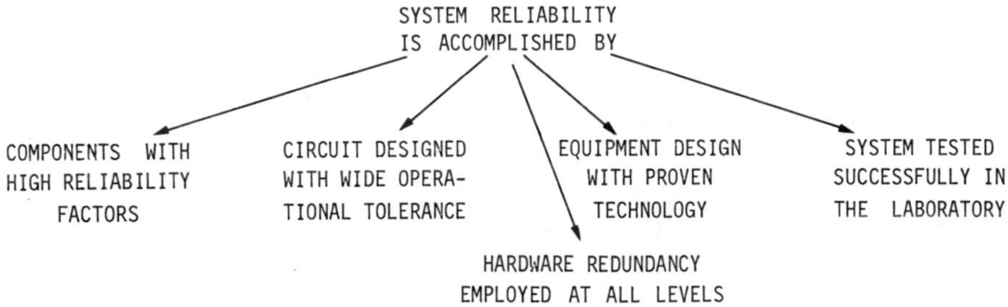

Fig. 1.32 Techniques suggested in achieving high system reliability

Redundancy being one of the essential techniques used for achieving high reliability the subject calls for elaboration. Redundancy is defined as the duplication of the equipment or modules in the system for the purpose of reliability or the continuity of operation. The duplicated units, although an integral part of the system, can be eliminated without any loss of information or without any effect on the system's functional performance. In this case each of the duplicated modules are technically identical and they can all perform the same system tasks. In this respect, one should distinguish between the term redundancy and the popular term 'backup'. Backup is defined as the existence of more than one means of performing the system's tasks; in backup, the duplicated modules need not be an integral part of the system nor be essentially identical. An example for backup could be a radio communication channel that is introduced when the wire communication channel is disconnected. Another example of backup is the resort to a human operator to replace the processing system when the equipment fails to operate. The latter example is typical of a misconception design which introduces a backup and regards it as having the same operational value as redundancy. If the designer is sure of the system reliability of the machine for its normal operation, he must also place his reliance on the redundant machines which come into operation when the main equipment fails. Though backup is nevertheless used in the military organization where there is a danger of elimination of the main system it has decidely no place in a civilian system.

The reliability factor $R(t)$ of a system is reduced when the number of modules connected in series is increased, since a failure in any of the modules soon brings it to a standstill. If each module has the reliability factor of $r(t)$ the overall reliability factor of the system will be

$R(t)_s = r(t)^n$ where n is the number of modules connected in series. Only by introducing redundancy into the system can its reliability factor be increased. If m identical modules are connected in parallel, the system $R(t)_p = 1-\{1-r(t)\}^m$. In other words as m increases $R(t)$ strives to 1, although it should be added that any introduction of redundancy, i.e., $m>1$, will double or triple the system cost. Nevertheless, redundancy is unavoidable if systen reliability is to be achieved. There are a number of redundancy techniques available to the designer, depending on the price he is ready to pay to achieve high reliability operation.

The simplest redundancy system structure is the standby redundancy where a number of modules are connected in parallel and then take over, in turn, when there is a fault in the active module. A practical example of such a standby redundancy is shown in Fig. 1.33, where two independent computers are shown as operating in symmetry. All the input data flows in parallel to both processors, thereby ensuring that the identical situation picture is built and stored in the two corresponding memory units. Nevertheless, in order to avoid conflict, only one of the processors can provide the output result. A special change-over unit continuously tests both processors and whenever a fault is located in one module it could easily change over the output control to the standby module, without any loss of data and without the system user feeling it. This system structure also has the advantage that it is possible to service one processor without disturbing the operations of the other processor, or, in other words, without affecting the systems functional operations. This type of standby redundancy is very common in communication switching nodes and other types of processing posts, and ensures smooth operation of the system despite a possible failure in one of its modules.

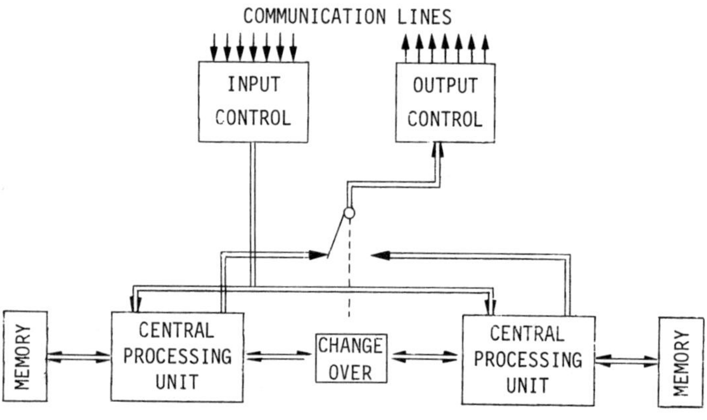

Fig. 1.33 A typical example of a standby redundancy

In systems where it is impossible to service the redundancy modules, as in the case of satellite communication nodes, a much higher reliability factor is essential. This is achieved by adding a large number of modules connected in parallel and at the same time eliminating the unreliable change over unit. Here, the output of the parallel is based on majority logic, as seen in Fig. 1.34. Accordingly the output will represent the situation as produced by the majority of the units, for at least half the modules must fail before the whole system fails. Nevertheless, the modules and their number should be so selected that a total failure should never occur. This is implemented by a number of techniques which are beyond the range of this book. As in standby redundancy, all the module which are in parallel perform the

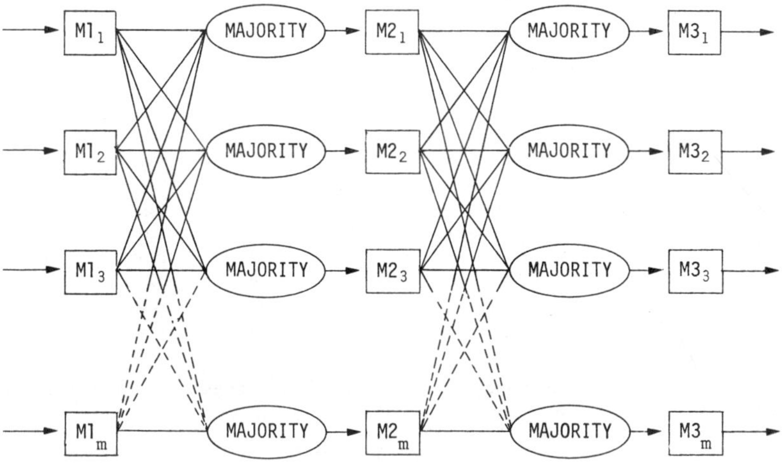

Fig. 1.34 Principle configuration of a majority redundancy system

identical function and all the output terminals produce a synonymous value. The majority logic checks all the outputs and presents its final version according to a majority vote. The majority logic has now been extended to perform more complex functions, known as threshold logic, where the circuits will produce an output if K out of m parallel modules presents the same output. Other systems include spare modules which replace the faulty ones so as not to upset the reliability based on $K>m/2$.

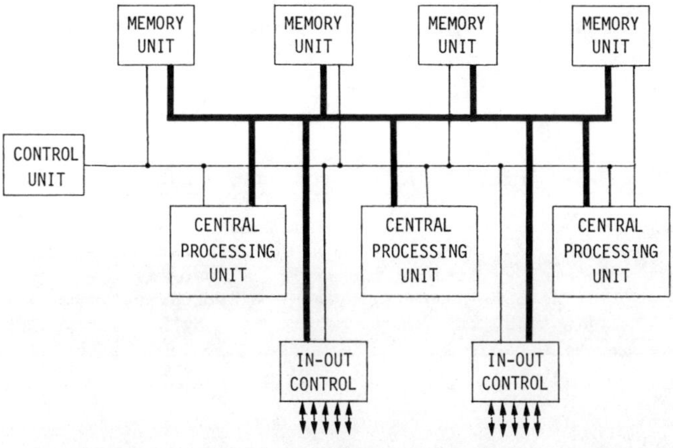

Fig. 1.35 Multi-processor configuration as an example of shared load redundancy

Many systems require high reliability but cannot afford the cost of majority or standby redundancy. If the services provided by these systems are not critical and reduction in requirements may sometimes be allowed, then a pseudo redundancy is suggested. That is, instead of employing identical parallel modules with each bearing the identical load, it is suggested that the load be shared among the parallel modules. In this technique, one used in multiprocessing (see the example given in Fig. 1.35), each module, whatever its normal task, can perform all the operational functions of the system, this enabling the system load to be shared. When a fault occurs in any one of the modules, all the remaining modules take over the tasks of the faulty modules without appreciable reduction in the services provided by the system. Nevertheless, some of the system services will be affected in some form. That is, either all the services will be provided but will require longer periods for implementation, or only the essential services will survive with no loss of time. For this reason many C^2 systems cannot employ a solution of this kind since all C^2 services must be supplied within the real-time period.

The three types of hardware redundances are summarized in Fig. 1.36. Standby and threshold redundancies provide a form of fail-tolerant operation ensuring the full recovery of the system operation after a failure is detected in one of the modules. Shared redundancy operation, however, can only be regarded as pseudo redundancy since when there is a failure in one of the modules it must lead to a degradation of services. As the system will nevertheless still continue to provide services, the degradation is regarded as graceful and can be considered as a form of fail-soft.

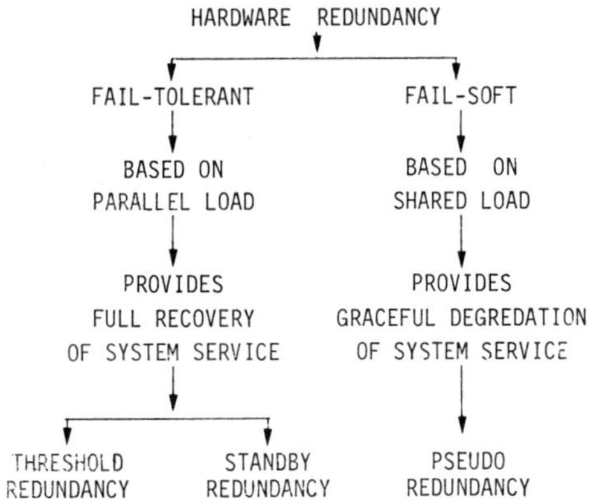

Fig. 1.36 Classification of system performance with hardware redundancy

System reliability has already been defined as the probability that the system will operate correctly without entering a failed state for a given period of time. System reliability is increased by the introduction of hardware redundancy, with its addition of extra parallel modules. Nevertheless, system survivability cannot be provided by hardware redundancy alone, as it cannot maintain the supply of full services during the failure of one of the modules and at the same time provide the means of removing the failed module. With C^2 system design one should aim at survivability by accepting the inevitability of fault occurrence and therefore provide additional facilities of maintaining the functional operations of the system even when presented with a fault. Survivability is achieved by incorporating into the system what is known as 'protective redundancy' which involves hardware, software and time redundancies. When a fault has been detected, it is the function of the hardware redundancies to replace the faulty module, and it is for the protective redundancy to nullify the effect of the fault.

With equipment operation the design target is achieved with added reliability (see Fig. 1.32); with system operation the design target can only be achieved by pursuing survivability, employing for this purpose the three suggestions presented in Fig. 1.37. Survivability calls for the exploitation

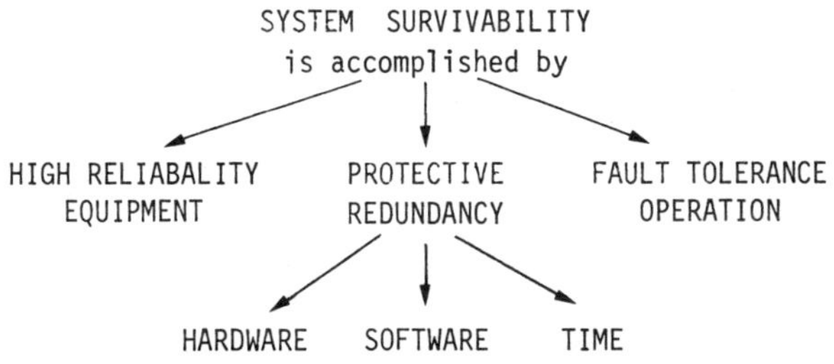

Fig. 1.37 Technique suggested in achieving high system survivability

of all the techniques needed for obtaining high reliability but above all they should also include all the techniques required for fault-tolerance operations in conjunction with protective redundancy. In fact it is the protective redundancy introduced in both circuits and in the equipment which makes possible the fault-tolerance operation. The protective redundancy is defined as the addition of hardware circuitry and modules, program software and time which are necessary not for the normal system operation but definitely needed when a fault occurs. These protective redundancy techniques provide self-checking facilities during the normal operation of the system without it affecting the services provided to the user. The hardware protective redundancy with its associated logic, provides special test points which examine the circuit and module operation and can produce a failure indication when there is a malfunction operation. The software protective redundancy provide automatic special additional test programs which examine whether the modules are functioning properly. The time protective redundancy provide the additional operational periods for testing the system operation. This additional time, although not necessary for normal system operation, must be included in the real-time calculation of the system's operation. A simple example of such a time consuming protective redundancy are the additional bits included in the data transmitted over a communication channel to enable errors to be detected.

As already stated it is not enough to introduce redundant modules which can replace the faulty ones when a failure occurs. For system survivability a search for circuit or module failure must continuously be performed so that the fault can be detected as soon as it occurs and before it can cause damage to other system operations. It is not enough to detect that there is a fault in the system, the fault must be located, diagnosed, isolated and only then replaced (as seen in Fig. 1.38). Although these steps seem obvious, a fault detected in one part of the system may be a result of a chain reaction of a fault originated in a completely different location. The fault itself must be analysed so as to diagnose whether its cause is a result of a hard, soft or an intermittent fault. The fault source must then be isolated before it can cause any further damage to the

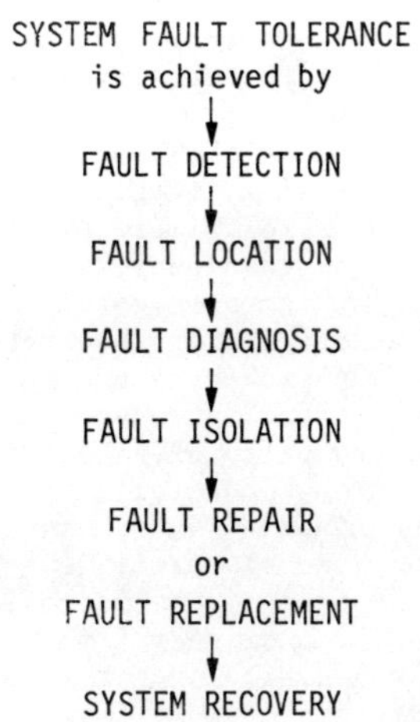

Fig. 1.38 The redundancy operation steps required for achieving fault-tolerance

remaining modules. Only after all these fault-tolerance procedures have been dealt with can the faulty module be replaced. The system recovery after the faulty module has been repaired may require the repetition of some of the operation steps which were affected by the fault.

The question could be asked why there is no provision for switching to the redundant module as soon as a fault is detected. This is in fact the ideal solution, but in practice it cannot always be provided. It is not always possible to pinpoint the source of the fault as soon as it is detected. Furthermore, the change over to the standby redundant module causes some disruption in the system operation which could be mistakenly taken as a malfunction. It is essential then that the change-over should be orderly for assurance of full recovery of the system operation. Finally, the fault tolerant techniques are generally embedded in the system circuits together with additional logic and coupled with special software programs.

1.11 EXAMPLE OF C^2 SYSTEM APPLICATION

It would undoubtedly help to clarify so highly a subject as Command and Control Systems if its application in a specific field were descriptively illustrated. The one selected for this purpose is that of a hospital medical care involving as it does most of the communication features discussed in this book.

A modern hospital is provided with both processing centres and separate Command and Control systems. For the example presented here, all the various computer installations and subsystems are regarded as part of a single comprehensive Command and Control system.

To start with, accounting in a hospital is highly complicated owing in part to graduation of charges for the various services the patient gets, as well as the cost of treatment and medicine. This however, is only a processing application, while the major command and control tasks are in actual medical care. Here real-time on-line systems are needed for the control of the patients medical condition as well as for functions in clinical diagnosis. The same large C^2 system can even incorporate subsystems for medical research studies and on a practical level, can assist in dispatching of ambulances as well as providing first-aid medical care at the place of an accident.

Bedside medical care can be regarded as being provided by two C^2 subsystems although both may be serviced by the same computer processing system. The basic subsystem is designed to alert the nurse as soon as there is a deterioration in the condition of one of the patients. The advantage of this subsystem is that it relieves the nurses of much of their load of routine and repetitive tasks and enables a single nurse to look after a number of patients concurrently. This subsystem operates in real-time and presents the monetary condition of the patients. For the physician, however, the patients' past case history is also essential. The second bedside subsystem is therefore intended to produce the complete medical history of each patient's ailments by merely pressing a set of buttons. It is assumed that the physician requires not only the latest reading of the patient's clinical measurments, but also graphs and tables of the patient's progress and reaction to treatments, as processed by the computer. This second subsystem, intended for the physician, can also include suggestions for diagnosis and for further treatment based on previous medical experience accumulated by the system.

The system operation is based on the continuous monitoring of the patient's reaction and this is accomplished by attaching to the patient various measuring instruments which can appraise the patient's rectal temperature, blood pressure, external weight, heart rate, urine output, inter-arterial pressure etc. This type of on-line system is essential for critically ill patients, but in parts, it could also be applicable for other less critical patients. Such a scheme, therefore, releases the nurses from recording and then charting the patient's progress, both of which are time-consuming tasks.

The measurement output is continuously sensed and then scanned at short fixed repetitive intervals. That is, the patient's health valuation is measured by sensors which are sensitive to small electrical currents or voltages. These sensors, after being scanned, provide D.C. amplitude output pulses which are a function of the momentary measurement readings. The data pulses collected are in either analog or state form and must first be converted into digital form (as described in Chapter 2) before they can be transmitted to the computer. The process of analog-to-digital conversion is performed near the patient as part of the measuring instrument. Once the data is available in digital form, it can be transferred over the communication lines, but because of the distances involved it must be modulated first (as described in Chapters 3 and 6) so as to ensure the safe duplication of the data in the computer. In this specific case the communication conveyance is twin copper wires, as described in Chapter 5.

Each patient has a number of sensors attached to him, while the data collected from each of the instruments must reach the computer. It is an impractical solution to have separate wires from each patient direct to the computer; instead, special data communication concentration equipment is essential (as described in detail in Chapter 7). In this case, as the sensors are scanned at a fixed rate and thereby the data rate that flows in the communication lines is constant, synchronous-time-division multiplexing (STDM) is suggested. Such equipment enables all the patient's data to flow on a single line with negligible delays between the time the measurement is made and the time it reaches the computer. This ensures that the patient is connected on-line to the computer and enables the computer to react in real-time.

The data collected from the patient is accordingly recorded in the computer, then correlated with both the patient's past clinical measurements and the predesign tables. The results can then be exhibited on a large display in the nurses centre command post. The display shows all the measurement readings of all the patients under care in the ward in the form of actual readings, tables or any other form required by the hospital. The display is programmed to give an alarm, ring bells, or flash lights whenever any of the measurement readings shows sudden changes and/or the readings are beyond specific precalculated boundary values. The command post is planned so that a single nurse can maintain supervision over all the patients in the ward.

The command post provides the means whereby the nurse is in command of the operation of the ward; such a system, however, is devoid of any feeling of intuition, and is useful directly only for alerting. Real medical treatment is based on direct contact between the physician and the patient where he can talk, feel or only look at him. Nevertheless, when visiting a patient, the physician must have access to the patient's medical file as processed by the computer. For this purpose each patient's bedside is provided with a small display monitor that can produce the patient's accumulated progress record. In fact there is no need for a separate display monitor for each patient for it can be wheeled round the ward and be plugged in as needed. These monitors are in the form of alpha-numerical displays which can supply means to permit a detailed dialogue between the physician and the system for the elicitation of any further information regarding the patient's history as well as suggestions for diagnoses and treatment. It should be added that whatever the system can provide, the final decision is always that of the physician.

Each monitor is provided with a code number which can thus allow a number of such monitors to operate in the same ward in parallel. The bedside monitor operates only on request, and its method of communication with the computer is completely different from that used for transmitting the measurement readings. Also, the information to and from the monitor is fully digital

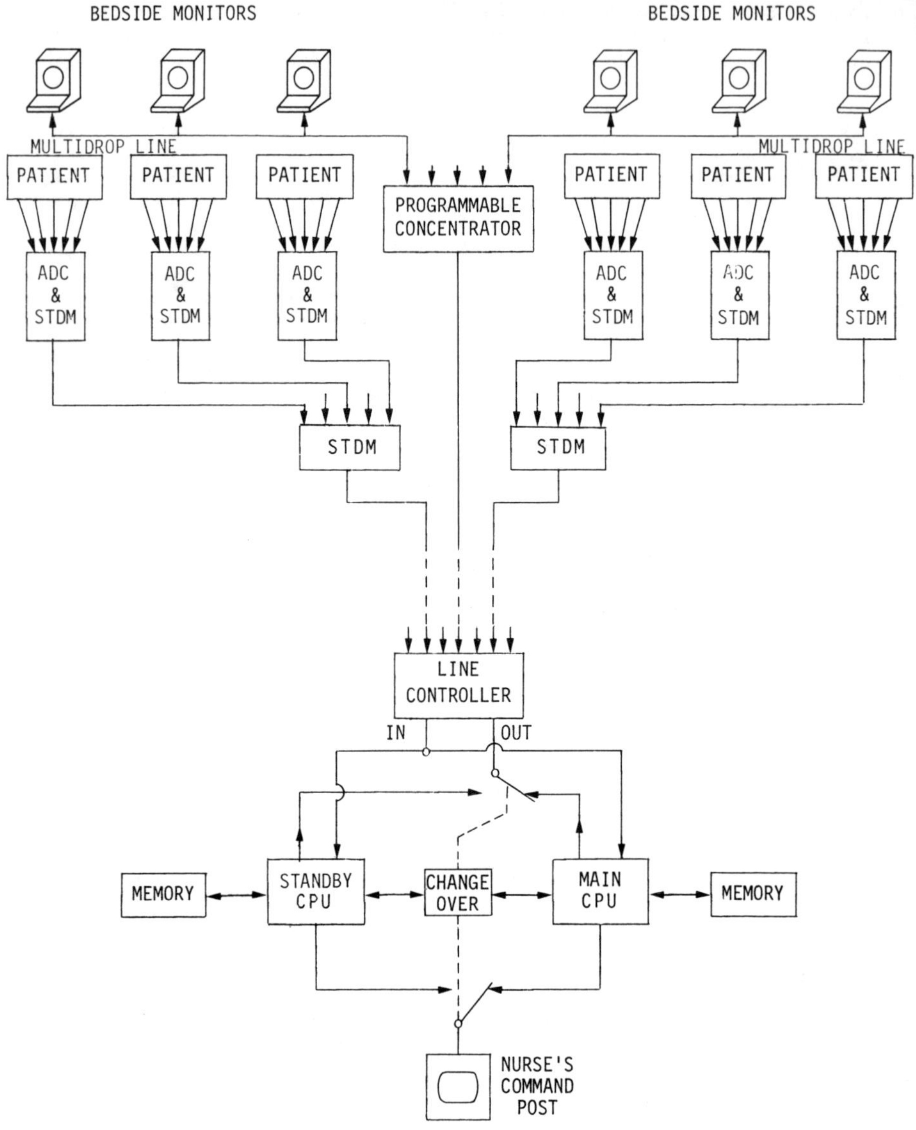

Fig. 1.39 Bedside patient care — hypothetical real-time system

and in alpha numerical representation, as distinct from the measurement reading, which is in analog or state form. Again, there is no need to provide each bedside monitor with direct lines to the computer. Here multidrop lines (see Chapter 4) together with programmable concentrators (see Chapter 7) are suggested for each ward, with the data flow controlled by polling procedures. The delays that may be caused by these procedures are negligible compared with the reaction time of the human physician.

It should be added here that the common lines to the patient's bedside of each system are separate from each other. The measurement readings flow in a single direction, i.e., in simplex form, while the bedside monitor requires the data to flow in both directions and are therefore in either half or full duplex form. The data that is collected from the patient is transmitted at a constant rate while the data to and from the monitor is communicated at completely random times, depending entirely on the physician's request.

The communication network construction and the computer installation of both bedside subsystems is presented in Fig. 1.39. It should be noted that the number of lines from the patient's instruments to the computer can be further reduced by adding in series an extra stage of time multiplexing. The computer in this system must be able to process the data in real-time so as to provide immediate alerts. In other words, the care of the patients is completely dependent on this computer operation and so the factor of reliability of the computer installation must be very high. This is achieved by using two parallel processing units (similar to those suggested in Fig. 1.33) both of which are active and perform identical tasks. In this way both processors operate in synchronization and produce identical results; nevertheless, the output results are taken from only one processor, while the other output is inhibited to avoid any conflict. Both processors are constantly checked and if a fault is located in the active one, the change-over unit switches to the standby processor. This type of processor system construction ensures that there are no delays affecting the real-time operation of the system when a fault is located in one of the units and a change over operation takes place.

It should be added here again that such a system, as described so far, is intended only for critically ill patients or for patients after an operation. In such cases the essential requirement is real-time alerts. Similar systems are already in operation in many hospitals which specialize in open heart operations. Patients who can walk about need not be confined to the bed and be connected permanently to the system. In that case, they may be attached to a measurement instrument only when a reading is necessary and the result is then automatically transmitted to the computer.

It is not advisable to have a single large computer installation, such as the one described above for the bedside care, serving the whole hospital or all the wards. Many wards may have specific requirements for patient care and this consequently makes it necessary to provide each ward or each group of wards with a separate system. In addition to the hospital's direct patient care systems, there is an essential need for a hospital computer centre for all the other tasks of the medical care, one of which is the availability of the diagnosis file needed to assist the physician in making a decision about possible treatment. This file, which does not relate to any particular patient, hence is for general use of the whole hospital and can be obtained from any of the bedside monitors in all the wards. To be effective, the file must be continuously updated following the experience gained from treatment of patients. The centre computer can also provide facilities for retaining a patient's permanent file and for the hospital accountancy both based on data supplied by the ward computers. It is obvious from the system description given above that all the ward computers must be able to communicate with each other and with the hospital's centre computer. This, however, does not mean that all the computers must be identical or even similar. All the direct patient care computers must operate in a real-time environment, with most of the functions on-line, while the hospital's centre computer can provide facilities at a much slower rate with many of them off-line.

The centre computer is mainly intended to furnish data processing facilities rather than ensure alerts; hence its reliability requirements can be slackened. The construction configuration of the centre computer can therefore be different from that of the ward computer, with fail-soft reliability being sufficient. In this case a multi-processor configuration, (as shown in Fig. 1.35) is suggested for this application. The centre computer can also have further functions such as assisting the hospital services by providing hard copy facilities.

As mentioned above, the centre and all the ward computers must be able to communicate directly with each other. The traffic load in the connecting lines is relatively high and the transmission speed fast. In a small hospital, the FEP of the centre computer can control the message flow on all the lines. If the hospital is large, with many computers spread over wide areas, this solution is inadequate and a separate stand alone communication processor is required. This computer can employ message or packed switching techniques, as discussed in Chapter 8. Since the distances involved are still limited, a much better approach for this application is a loop transmission network (see Chapters 4 and 7), where all the computers are connected in an endless ring, as shown in Fig. 1.40. The network configuration will ensure the required high speed communication and allow for growth of the system, all at a reasonable initial cost. A further suggestion for this network is that the communication lines should be in the form of fibre optical links (see Chapter 5). This will prevent any X-ray or other radiation affecting the communication.

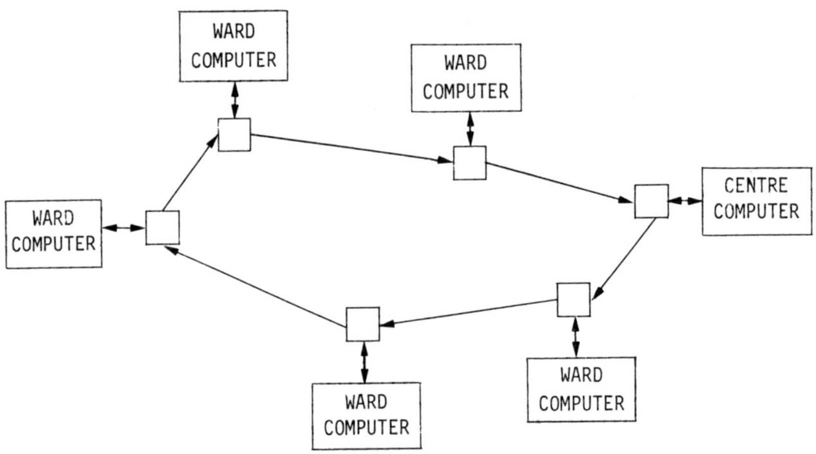

Fig. 1.40 Suggested network configuration for internal hospital communication

An additional function of the hospital's centre computer is to keep a record of the past history of the patients' illnesses and treatment which can easily be retrieved whenever a patient is admitted once more. Such a procedure is essential so as to provide continuation of treatment. It should be added that these files should be kept in the centre computer as this data is retained off-line and in many cases the patient may not return to his original ward. The most important task, however, of the centre computer is medical research where the measurement readings of each patient can be used as data necessary for statistical analysis of ailments. In this respect, all the measurement results may be collected in separate files for research analysis of a general nature. The data accumulated in these files will be recorded in accordance to illnesses and/or symptoms, with no reference to any particular patient, so as to assist future treatment of other patients. For this to operate successfully, the patients should be measured by sensors for data beyond that basically necessary for the patient's care. In fact each patient may serve as a guinea-pig providing data for research work.

Medical diagnosis and research analysis cannot be based only on the data adsorbed in one hospital. That is, for any statistical research, many medical centres must join together in creating a large data file. In many medical diagnosis processes a hospital may require the assistance of others. In other words, a single hospital or medical centre cannot rely only on the data collected locally and may need the cooperation of other research medical centres, both in the country and abroad, for exchange of information between them. Each medical centre usually specializes in a particular field of medicine and each updates and builds its files differently. For this reason, each centre must be able to operate independently but be capable also of exploiting information gathered in other centres. Such a requirement, defined as interoperability, calls for all the medical centres to be connected through a single communication network. For such a fully distributed network configuration (see Chapter 4) is suggested employing packed switching techniques (see Chapter 8). Since illnesses recognize no political boundaries, this communication network can be spread over the whole world. A typical example of such a long distance communication requirement could be the occurrence of a tropical illness in a cold northern country. In this case, the channels in the distributed network communication may be not only of copper wires or radio links, but also of satellite links as discussed in Chapter 5. Such a fully distributed network, as shown in Fig. 1.41, provides fast communication using packet switching techniques. Each centre, in such a configuration, is of equal importance and can contact any other centre either directly or via other centres.

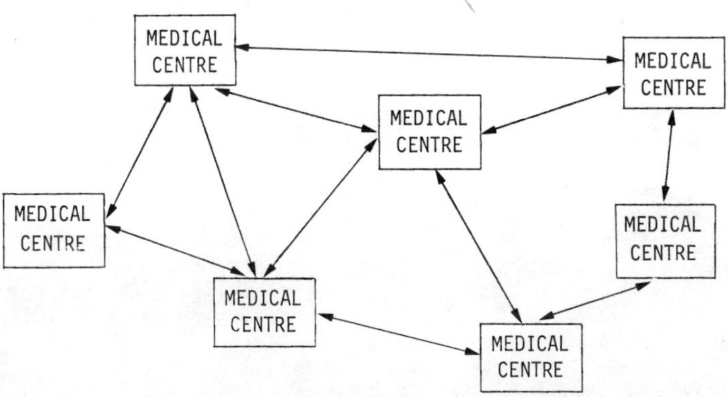

Fig. 1.41 Distributed network configuration providing communication between widely distributed medical centres

Another important function of the hospital centre is to provide real-time assistance to ambulances. Here the ambulance driver or nurse can ask the system for suggestions regarding ailments they have not encountered previously. In such a case the nurse types a message on the ambulance terminal which is subsequently radio transmitted to the centre (as seen in Fig. 1.42) and the computer then replies in the same way. The answer is

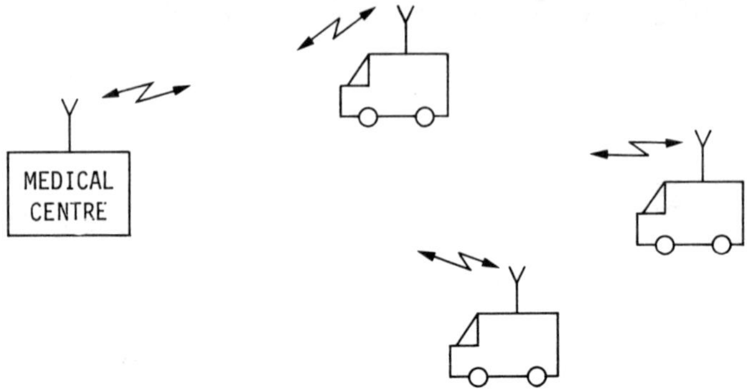

Fig. 1.42 Radio communication network for ambulance control

displayed in the ambulance, allowing the nurse to act accordingly. In this application, all the ambulances serving in one area will be operating on the same radio frequency band as the traffic to and from each ambulance is extremely small.

In the 'old days' when the nurses conducted speech communication with the centre over the radio channel, common sense reaction prevented two ambulance transmissions from disturbing each other. Even then, prevention of transmission collision could not be assured as the distances between the ambulances and the signal strength might have obstructed them from hearing each other. This problem is more acute with the communication performed solely by computers, which in this case can be regarded as stupid machines. In other words when using a single radio frequency band for a number of mobile stations, collision of the digital messages is apt to occur. To prevent such situations, special radio communication procedures have been introduced, as described in detail in Chapter 9. In this particular application, since the traffic load is small, CSMA procedure together with priority acknowledgement seems to be most appropriate.

Today it is possible to connect the patient in the ambulance to special measuring equipment which attaches him directly to the hospital's computer via the radio link. This scheme will enable the nurse to provide first-aid treatment to patients after a heart attack even before they reach the hospital. In this case, since the data has to flow from the patient continuously and be then transmitted, a dedicated radio bandwidth channel is necessary so as to avoid blocking the channel for use by the other ambulances. Nevertheless, since the information load content from the patient is relatively small, it is possible to apply time sharing techniques by using one of the request TDMA procedures (see Chapter 9).

The radio network communication as suggested above could be most suitable for most types of mobile communication. In the case of medical assistance, where life and death can depend on the success or failure of the communication link, a much more reliable communication system may be necessary. That is, the channel must be immune to all noise interference and must be proof against jamming from any other transmissions in the same bandwidth. For this application, spread spectrum communication is recommended (see Chapter 10) for providing the best means of external noise rejection. In this way the ambulances can receive and transmit messages even if the signal strength is near to or below the noise level.

1.12 REFERENCES

(a) TECHNICAL TERMS

1. Day, F. *IEEE Standard Dictionary of Electrical and Electronic Terms*, IEEE, December, 1977, 882 pages.

(b) COMMAND AND CONTROL SYSTEMS

2. Unold, R. *Communication Command and Control – A few Wroms from a Can of Worms*, Telecommunication, March 1968, pp.13 - 16.
3. Hilsman, W.J. *The Design and Operation of an Automated Command and Control System*, Military Review, February 1967, pp.22 - 29.
4. Steinkraus, L.N. *Command and Control*, Ordinance, March/April 1971, pp.440 - 442.
5. Davis, C.J. *Command Control and Cybernetics*, Army, January 1963, pp.51 - 55.
6. Gould, G.T. *Command, Control and Communication Systems*, Signal, May 1966, pp.70 - 71, 79 .
7. Mayer, A.S. *Some Notes on the Design of Command and Control Systems for the Ground Forces*, Mede, Germany, October 1979.
8. Wachs, R.E. and Corrigan, M.J. *Areas: In Support of C^3 Network Design and Evaluation,* Proc. Computer Network Symposium, 1978, pp.76 - 81.
9. Barruclough, J. *Reflection on C^3*, Natos Fifteen Nation, vol. 24, no.5, October/November 1979, p.12.
10. Mallorie, P.R. *Command and Control – Some Ace Aspects*, Natos Fifteen Nations, vol.24, no.5, October/November 1979, pp.34 - 38.
11. Wagner, K.H. *Interoperability in Command and Control by Networks*, Natos Fifteen nations, vol.24, no.5, October/November 1979, pp.50 - 55.
12. Diedrichsen, L.D. *Accomplishment in Communication Interoperability among Nato Tactical Forces*, Easion, 1978, pp.28 - 31.

(c) THE HUMAN POSITION IN C^2 SYSTEMS

13. Hartman, R. *The Human Equation: C^3 is for People*, Electronic Warforce/Defence Electronic, vol.10, no.6, June 1978, pp.61 - 64.
14. Schneider, R.H. *Human Factors in the Establishment of System Design Requirements*, IRE Wescon Convention Record, Pt.4, 1960, pp.127 - 129.
15. Freed, A.M. *Measuring Human Interaction in Man Machine Systems*, IRE Wescon Convention Record, Pt. 4, 1960, pp.189 - 201.

(d) COMPUTERS IN C^2 SYSTEMS

16. Reitman, J. *Design Decision Through Simulation*, IEEE Spectrum, March 1974, pp.76 - 79.
17. Thompson, J.F. *Information Acquisition, Transmission and Storage Considerations in Large Automated Transit Systems*, NTC , vol.3, 1978, pp.41.6.1 - 41.6.5.
18. Watson, R. *Time Sharing System Design Concepts*, McGraw Hill Book, 1970.
19. De Mercado, J. *Introduction to Time Sharing Systems*, International Seminar on Remote Data Processing, ILTAM, Israel, Pt. 1, 1972, pp.159 - 181. 279 - 314.
20. Fano, R.M. *On the Social Role of Computer Communications*, Proc. IEEE, vo.60, no.11, November 1972, pp.1249 - 1253.
21. Westerhouse, R.A. *Time Sharing and its Applications*, Computer, vol.2, no.7, January 1969, pp.3 - 7.

(e) COMMUNICATION CONSIDERATION IN C^2 SYSTEMS

22 Wachs, R.E. and Corrigan M.J. *In Support of C^2 Network Design and Evaluation*, Proc. Computer Networking Symposium 1978, pp.76 - 81.
23. Townsend, M.J. *Communication Control by Computer – An Introduction*, Telecommunication, May 1972, pp.33 - 38 and 60 - 62.

24. Newport, C.B. and Ryzlak, J. *Communication Processors*, Proc. IEEE, vol.68, no.11, November 1972, pp.1321 - 1332.

(f) SYSTEM RELIABILITY

25. Avizienis, A. *Design of Fault Tolerant Computer*, AFIPS, Fall Joint Computer Conference, vol.31, 1967, pp.733 - 743.
26. Mathur, E.P. and Avizienis, A. *Reliability Analysis and Architecture of Hybrid Redundant Digital System : Generalized Triple Modular Redundancy with Self Repair*, AFIPS, Spring Joint Computer Conference, vol.36, 1970, pp.375 - 383.
27. Chang, H.V. and Scanlon, J.M. *Design Principles for Processor Maintainability in Real-Time Systems*, AFIPS, Fall Joint Computer Conference, vol.35, 1969, pp.319 - 327.
28. Short, R.A. *The Attainment of Reliable Digital Systems through the use of Redundancy — A Survey*, Computer Group News, vol.2, no.2, March 1968, pp.2 - 17.
29. Dent, J.J. *Diagnosis Engineering Requirements*, AFIPS, Spring Joint Computer Conference, vol.32, 1968, pp.503 - 507.
30. Avizienis, A. *Fault-Tolerant Computing : An Overview*, Computer, vol.4, no.1, January/February 1971, pp.5 - 8.
31. Carter, W.C. and Bouricuis, W.G. *A survey of Fault Tolerant Computer Architecture and its Evaluation*, Computer, vol.4, no.1, January/ February 1971, pp.9 - 16.
32. Hansler, R., McAuliffe, G.K. and Kilkov, R.S. *Optimizing the Reliability in Centralized Computer Network*, IEE Trans. vol.COM-20, No.3, June 1972, pp.640 - 644.
33. Burrow, L.D. *The Fail-Soft*, Design of Complex International Conference On Distributed Computer Control, IEE Conf. Publication no.153, September 1977, pp.151 - 156.
34. Bennets, R.G. *Designing Reliable Computer Systems*, IEE Electronic and Power, November/December 1978, pp.846 - 851.
35. Morgan, D.E. and Taylor, D.J. *Computer Network Reliability and Availability : The State of the Art*, ICC, 1978, pp.3.3.1 - 3.3.5.
36. Beaudry, M.D. *Performance-Related Reliability Measures for Computing Systems*, IEEE Trans. vol.C-27, no.6, June 1978, pp.540 - 547.
37. Avizienis, A. *Fault-Tolerance : The Survival Attribute of Digitual Systems*, Proc. IEEE, vol.66, no.10, October 1978, pp.1109 - 1125.
38. Schaffer, E.J. and Williams, T.J. *Analysis of Fault Detection, Correction and Preventation in Industrial Computer Systems*, U.S. Dept. of Commerce, Report No. PB-278880, October 1977, 157 pages
39. Bennels, D.A. *Distributed Fault-Tolerant Computer Systems*, Computer, vol.13, no.3, March 1980, pp.55 - 65.

CHAPTER 2

Data Acquisition and Coding

2.1 INTRODUCTION

Command and Control systems function in a closed loop configuration, with one end of the loop serving as the Controllee and the other as the command post. The part of the system where the Controllee performs can be viewed as operating in the real-world as differentiated from the other parts of the loop which can be regarded as operating in the processing-world. The processing world deals only with coded digital data, in a precise and well defined language. This is not the case in the real-world, where all the operations are in a natural language; here the signal produced by it is referred to as analog. Analog data, in contrast to discrete data, imply continuity.

Decision making performed in the command post of C^2 systems is intended to ensure the efficient performance of a process or plant operating in the real-world environment under the control of the system. Correct decision making can only be exercised if it is based on real-time accurate and exhaustive information which can be displayed with sufficient time for these decisions to have an effect on the Controllee's operations.

The real-time information displayed at the control centre is based on data gathered throughout the C^2 system. Most of the data is collected at the Controllee's premises, indicating its present performance, while other data items need to be collected elsewhere. That is, the collected data which is based on the controllee's operation must be correlated with many other data sources before it can be computed into meaningful information. There is no C^2 system which does not have data collection as part of its loop cycle. In fact, in a single C^2 system there may be many data sources each in a different form. Before the data items collected from all the sources can be correlated into useful information, all the items must first be transformed into a compatible form. The ultimate purpose of data acquisition is to bring all the source data-symbols into a uniform category, this being the binary form which is the common mould that would be used for computers. Data acquisition can, therefore, be regarded as the front end of all C^2 systems and serves accordingly the subject of this chapter.

Source data could come in either analog or digital form. Analog form is derived from the outside world while the digital form represents the data which is correlated with it. A third source of data may be used which represents a relatively steady state. Although the division into three main types

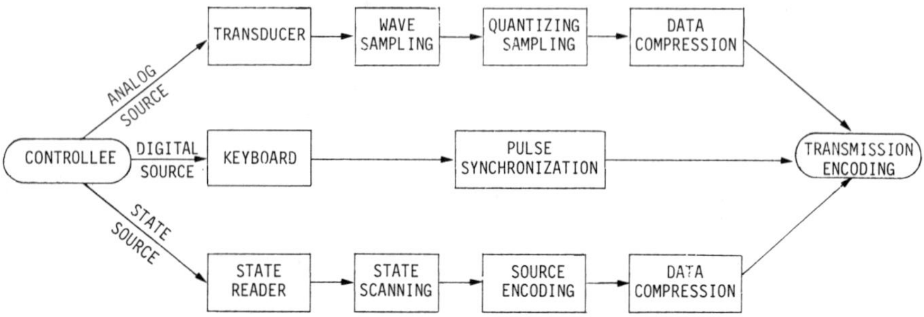

Fig. 2.1 Various forms of data sources and their processing stages

of data may seem artificial to some of the readers, the steady state is categorized as a separate class to mark its uniqueness. The steady state could be regarded as a form of digital representation although it still requires some processing stages before it can be used for correlating with other data sources. In fact, these processing stages are similar to those required for analog source data as seen in Fig. 2.1. These three data sources are defined as follows:

(a) Analog data source, where the signal obtained from the controllee is a direct result of its operational changes. This provides a continuous waveform signal which is generally produced as a voltage or current measurement where its magnitude variation represents the deviation in the performance of the controllee. An example in which a controllee produces such a signal could be the rotating speed of the rotor in a machine or the target movement received in a radar system.

(b) Digital data source, where the signal symbols obtained are already in binary form. That is, each symbol pertaining to a form of a discrete number of codes. Such a data source could be obtained as a direct result of a manual operation where the operator presses various buttons with which he can define the exact data symbols he is inserting into the system. Each time he presses a different button, a new set of digital symbols is transmitted. In this category one must include all the manual data entry devices, such as teletypewriters and keyboards.

(c) State source data, where the signal obtained from the controllee shows one of a number of steady states, each indicating a given performance condition. This signal could signify the three states of the controllee performance, one in which it is operated with no malfunction, another where it is over-performing and the last when it has completely ceased to operate. The signal could also represent an array of steady states such as those produced by magnetic or light scanners obtained from badge or imprint readers. An example of such a reader can be found in most supermarkets. The data form received from this source, as distinct from digitial source, is that it can generally be read time and time again without the contents changing. Furthermore, the signal need not necessarily produce a continuous waveform.

Before one can exploit the data signals in one system, all the three data forms must be processed into a common binary form. Both the analog and state source data need similar numbers of processing stages while the digital data requires very little processing. The data contained in analog and state signals is generally full of redundancies, enabling the data to be scanned and even compressed. This is not the case in digital data, which contains no redundancies. The sensors used for each of the three data forms also requires a different class of hardware devices. Analog data is received from transducers, digital data from keyboards, and state data from state readers. This chapter will only concentrate on analog data acquisition, since digital interface is simple.

2.2 ANALOG DATA ACQUISITION

Most, if not all, of the data sources in the natural physical world are in analog form and thus they must be conditioned, by suitable conversion, into the 'artificial-world' of computation. The actual analog waveform obtained from the sensor is continuous, while for any processing application it must be first sampled, as shown in Fig. 2.2, enabling each sample to be treated separately by the processing equipment. It should be pointed out that the processor regards the analog waveform not as a continuous waveform but as a set of amplitude reading. Once the analog waveform has been sampled, the representation is only an image of the original waveform and a copy of it. Nevertheless, a very close replica of the original waveform can be recreated with minor variations by charging up a capacitor with the samples and smoothing the results.

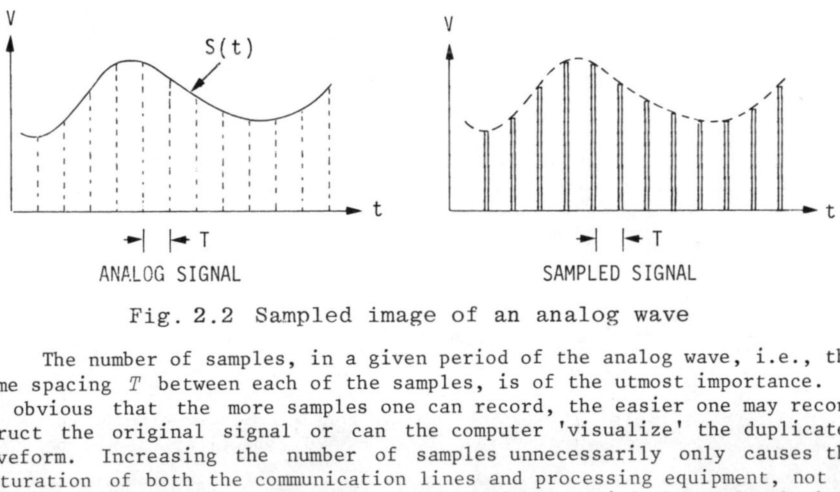

Fig. 2.2 Sampled image of an analog wave

The number of samples, in a given period of the analog wave, i.e., the time spacing T between each of the samples, is of the utmost importance. It is obvious that the more samples one can record, the easier one may reconstruct the original signal or can the computer 'visualize' the duplicated waveform. Increasing the number of samples unnecessarily only causes the saturation of both the communication lines and processing equipment, not to mention the wastage of computing time. It is essential that only the bare minimum number of samples be produced, just adequate enough to represent the original waveform.

According to Shannon's theorem, the sampling rate of a signal should be at least twice the highest frequency contained in the original waveform. If the signal waveform $S(t)$ is bandlimited by $B\,Hz$, it can be completely determined by giving the value of the waveforms as a series of amplitude pulses spaced $1/2B$ seconds apart. This sampling spacing is known as Nyquist sampling interval $T_{nyq} = 1/2B$. This means that if the original waveform shape is to be retained, the sampling spacing must be $T_s \leqslant 1/2B$, i.e., the sampling rate $R_s \geqslant 2B$.

The whole operation of periodically sampling an analog wave and then recording the corresponding instantaneous amplitude heights, is known as pulse-amplitude modulation (PAM). The advantage of PAM is that the spaces between the various samples can be used to transmit other signal data. That is, since in most of the time between samples no useful data is being produced, it is possible to transmit in this spare period data from other analog sources. The transmission of samples from several sources through a single communication channel is known as multiplexing. Staggering the samples in a fixed time pattern is called time-division multiplexing (TDM).

In the TDM system, as shown in Fig. 2.3, all the sensor outputs are continuously sampled at fixed intervals, with each sensor source producing a separate train of pulses of various amplitude levels. The sampling time of each sensor is staggered relative to the others, so that every sensor is scanned at different times. The TDM scans all the sensors in turn, providing each one with an identical operating time. After completing a set of scanning of all the sensors it starts once again with the first. The timing arrangement of the TDM is set so that each individual sensor is scanned every T seconds, where T is the Nyquist interval. This means that the clock timing is T/N, with N being the number of sensors being multiplexed. The pulse train of variable pulse amplitudes produced in each channel, is transmitted into the common link, each interleaved with the other PAM samples.

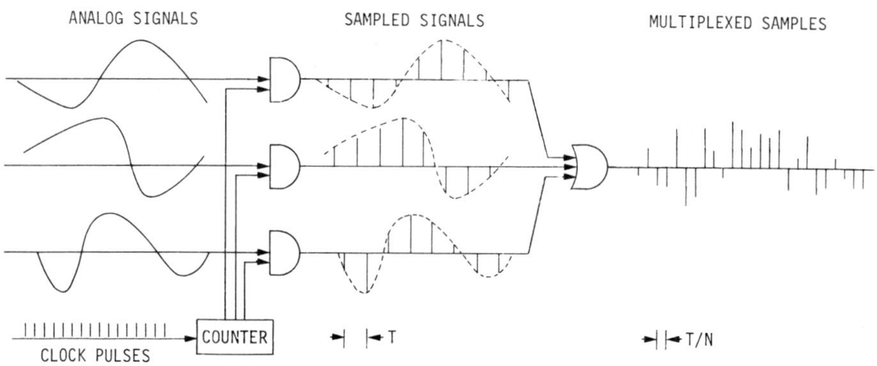

Fig. 2.3 Pulse-amplitude modulation in time division multiplex

The processes for achieving both pulse amplitude modulation and time division multiplexing are very simple. Each sensor output feeds a separate 2-input AND gate. The other gate input is fed in from a series of sequential clock pulses, produced by a decoding counter. The AND gates are amplitude sensitive, producing an output voltage with a level corresponding to the coincidence of both input pulses.

PAM is but one of several types of analog waveform sample representation. Some of these uncoded pulse modulated schemes used in practice are shown in Fig. 2.4. The disadvantage of PAM is that any noise 'riding' on the amplitude sampled pulse will introduce distortion. One solution used is pulse-duration modulation (PDM) in which the instantaneous samples vary the time duration of a pulse in accordance with the sample height. Such a scheme produces a series of pulses of variable duration with uniform height. In pulse-width modulation (PWM) the leading and trailing edges of the pulse are varied from the unmodulated position by samples of the modulated wave. This scheme is also known as pulse-length modulation (PLM). In pulse-position modulation (PPM) the samples of the modulation waves are used to vary the position of a pulse in time relative to its fixed sampled time. All these pulse modulation schemes are very similar and are thus known as pulse-time

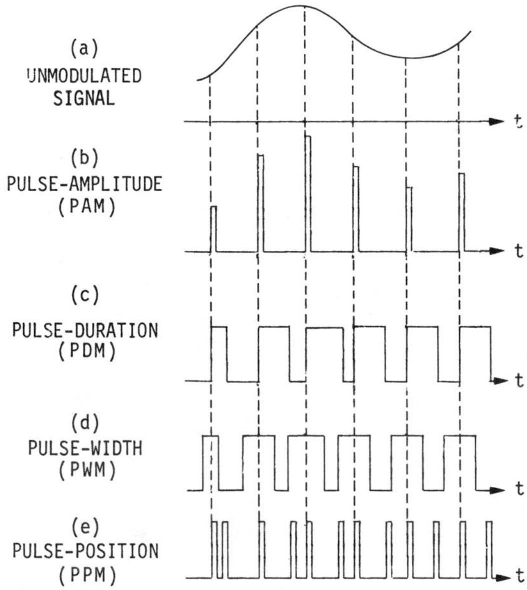

Fig. 2.4 Pulse-time modulation schemes

modulation (PTM). These PTM schemes have the common disadvantage that any distortion in the pulse amplitude or period may change the apparent information contents.

The advantage of time-pulse modulation schemes is that the original analog wave is represented by a series of pulse samples and the spacing between each sample can be exploited for other sampled pulses, using TDM techniques. Nevertheless, the application of the multiplexed samples is limited, since their amplitude or duration are very sensitive to noise. The aim in sampling the analog wave is to be able first to transmit them over a communication channel and then process them by a computer. Both these aims cannot be reached with the pulse modulated scheme so far described. To achieve these aims one must transform the pulse samples into digital form, pursuing the following steps.

The original analog sensor output is scanned vertically, producing a set of horizontal pulse amplitude samples. These, in turn, are also 'scanned', providing a train of binary pulses. The horizontal scan of the amplitude pulse is, as will be explained later, known as quantizing. The vertical sampling is performed by a periodical clock scan every T_s seconds, while the horizontal quantizing height measurement is represented by one of m discrete amplitude levels. These m levels are translated into binary form by converting them into a set of $k = (\log_2 m)$ bits. This means that every T_s seconds a new set of k bits is produced. The time interval between each clock pulse of the binary bits will thus be T_s/k.

The set of k bits can be regarded as binary-coded-analog (BCA), since each set represents an analog sample separately coded into binary. The process of deriving a digital representation of an analog voltage amplitude pulse is known as analog-to-digital converter (ADC or A/D). The method of transforming the analog wave into digital coded data, through the process of sampling, auantizing and coding is generally known as pulse-code modulation

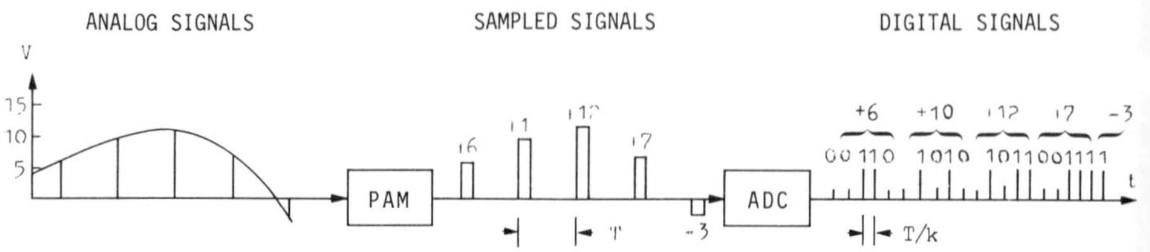

Fig. 2.5 PCM processing stages

(PCM). In other words, PCM comprises of both PAM and ADC processing schemese, as seen in Fig. 2.5.

When multiplexing PCM data, each BCA set of k bits corresponding to one analog data source sample is allocated a different time interval. It is essential to note that in this case of multiplexing, the staggering of the data is modified to that described before. Here, each group of k bits refers to a different sample and is then followed by another group of k bits from another sample, with no spacing between them. This means that the multiplexing staggering is not bit wise but set wise. In some cases a synchronization bit is added to each set to indicate its sequence start.

Multiplexing of PCM can be achieved in two schemes, as seen in Fig. 2.6 When considering individual sensor outputs, three successive operations are needed to transform the analog waveforms into a series of multiplexed digital pulses. Each individual waveform is sampled at the rate twice that of the highest frequency of all the waveforms. Then each individual sample amplitude level is quantized and the closest value is coded into binary form. The train of pulses obtained is time multiplexed with other sequences of coded pulses. The multiplexed data can then be transmitted on a common line to a computer centre. Such a multiplexing scheme is used only when the sensors are spread over a wide area.

When a number of sensors are situated in the same location, it is far more efficient to perform the multiplexing operation near the sensors, thus avoiding the need for a separate analog-to-digital converter for each sensor. Although the first scheme seems wasteful, there are many applications where it is unavoidable. The train of digital pulses that are obtained in both schems are identical, i.e., each set of k bits is associated with an instantaneous sample of a different sensor reading. After a cycle of all the sensor samples it returns to the first sensor for a second sample reading.

The pulse train of digital data obtained from the two schemes has been referred to in the literature both as multiplexing data and as PCM data. Neither is an accurate definition since it should be referred to as PCM time multiplexed data.

The timing rate in the pulse train of PCM in TDM is illustrated as shown in Fig. 2.7. Each analog waveform is sampled every T seconds where $T = 1/2B$, and B is the highest frequency. If there are N analog sources in the multiplexed system, the distance between each multiplexed level sample will thus be $t = T/N$. In this case the multiplexor cycle time will be T, so as to keep the sampling rate of each sensor as twice the highest frequency. This means that the pulses corresponding with a given source are repeated every T seconds, with all the samples corresponding to one source shifted by $it = iT/N$ seconds relative to the other sources, where i defines the relative

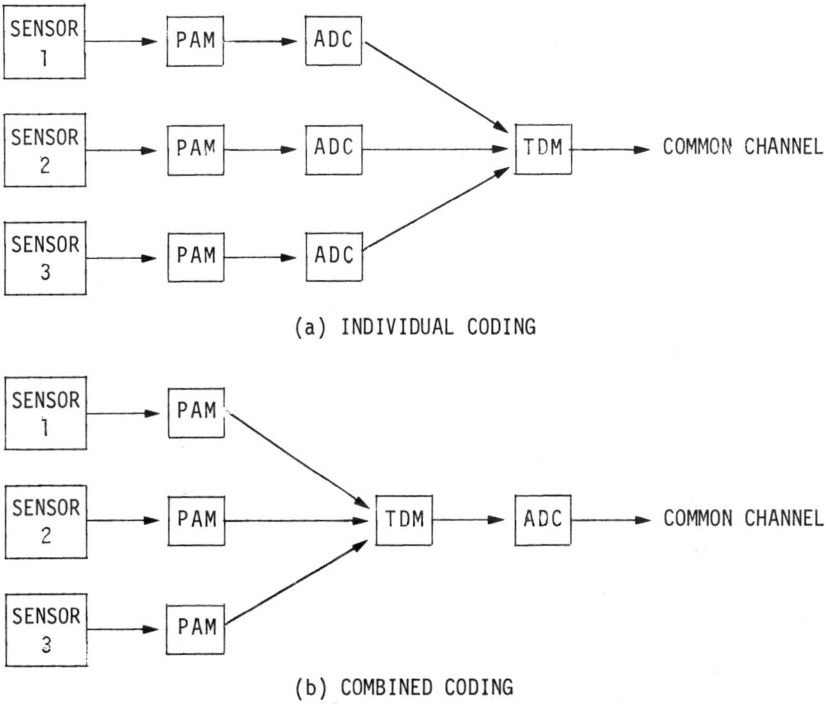

Fig, 2.6 Pulse-code modulation in time-division multiplexing

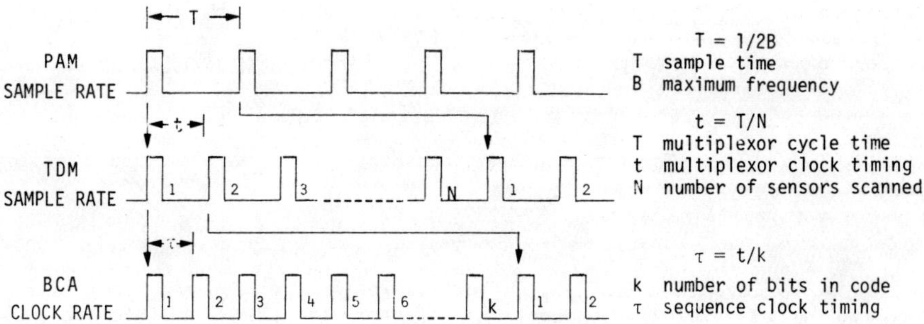

Fig. 2.7 Pulse data rate of PCM in a TDM system

source number, i.e., $1 < i \leqslant N$. The timing 't' therefore defines the scanning speed of the multiplexor. When converting each amplitude sample into binary, the set of k bits must fit into the spacing given by each sample, i.e. within the t seconds. In other words, when multiplexing PCM signals, the BCA sets of k bits must not be split, but handled as a unit. The timing period τ between each of the BCA pulses defines the data rate and corresponds to $\tau = t/k$ The final multiplexed data stream consists of k consecutive bits corresponding to one source, followed by $(N-1)k$ bits from the other sources before the same source data is produced again.

The multiplexed data rate, R, can be rather high. Take, for example, eight audio sources, each with a maximum of a 4 kHz frequency band, having the sample quantized into 8 binary bits. This gives the following rate in bits per second:

$$R = 1/\tau = k/t = kN/T = kN \cdot 2B$$
$$= 8 \cdot 8 \cdot 2 \cdot 4 \cdot 10^3 = 512 \cdot 10^3 \text{ bits/second.}$$

2.3 MODULATION AND MULTIPLEXING

The terms modulation and multiplexing have been used rather extensively in the previous section. Both these terms have been taken from the analog field while here they have been used in the pulse field. Before continuing it is essential to redefine them.

MODULATION: The process by which the amplitude, frequency or phase of one wave (often referred to as the unmodulated carrier) is modified in accordance with the characteristics of another wave (the modulating signal).

PULSE MODULATION: The process by which the amplitude, duration, time-of-occurrence or shape of a series of regular pulses (the unmodulated carrier or, as often referred to, the clock source) is modified in accordance with the characteristics of a wave or signal samples.

Figure 2.8 demonstrates the difference or the similarities of both forms of modulation.

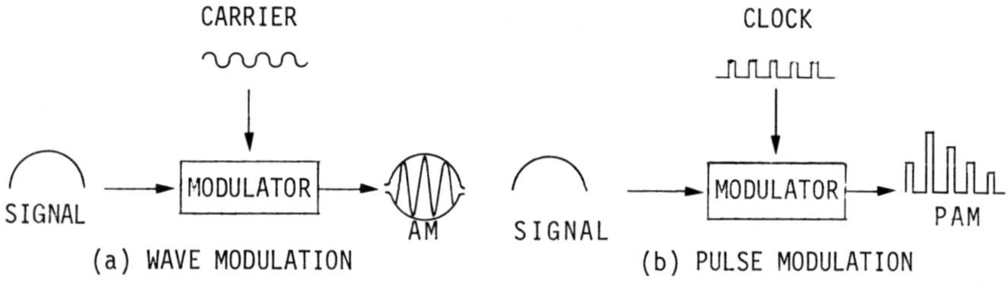

Fig. 2.8 Modulation techniques

MULTIPLEXING: The simultaneous transmission of two or more signals using a common carrier wave.

PULSE MULTIPLEXING: The means by which many data sources can share a common transmission facility, either by frequency, by time or by code.

TIME-DIVISION MULTIPLEXING: The division of the transmission facility into two or more exclusive channels by allotting the full available spectrum to each data source in succession. The time interval assigned to each source may be narrow enough to transmit a single pulse, or wide enough to transmit a whole set of pulses, provided the same time duration is assigned in turn to all the data sources.

2.4 QUANTIZING AND DIGITAL ENCODING

The basic feature of analog data acquisition is the conversion of the analog waveform into digital encoding. As already explained, this process is performed in two stages: first the waveform $S(t)$ is scanned vertically at fixed time intervals T and then the voltage pulse V_i obtained is sampled horizontally at fixed voltage levels. It is easy to scan a signal in the time domain, but it is nearly impossible in practice to sample in the voltage level domain. Instead, the pulse levels are quantized by comparing each individual sample level with a set of standard scales of voltage references. Each pulse sample is therefore assigned the reference level closest to the last comparison, as shown in Fig. 2.9. This process means, in effect, that when time sampling a waveform, the resulting pulse amplitude level is the exact replica of the waveform level at the instantaneous sampling time. However, when level quantizing, the resulting measured amplitude is only a close approximation value of the comparison. This approximation rounding off process is called quantizing.

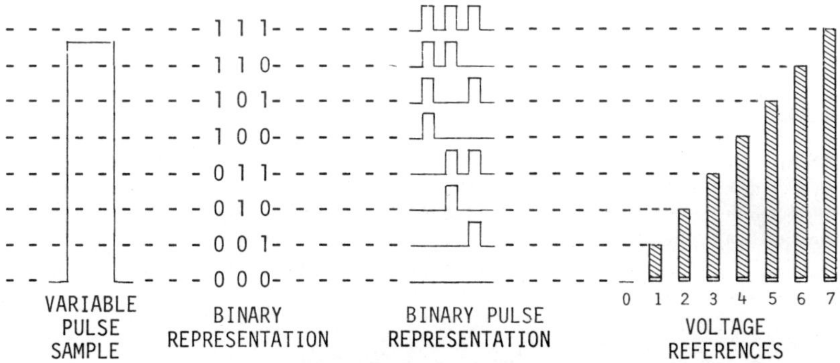

Fig. 2.9 Binary coded analog using fixed voltage references

The number of the voltage level references is selected so as to be equal to the scale of power of two. Therefore each fixed voltage reference can be coded into a set of binary codes. If $m=2^k$ discrete reference amplitude levels are selected, the binary scale is constructed by $k=\log_2 m$ bits. The factor k also defines the resolution of the quantizer, the factor m also defines the full-range scale (FRS) of the analog wave and each quantizing step is defined as (FRS)/m.

The relation between the input analog voltage level and the equivalent digital reference voltages can be plotted on a graph, as shown in Fig. 2.10. An ideal transfer curve is a linear one: however, in practice this is impossible. The binary scale is fixed, causing the corresponding output of the input voltage to change in steps. Although all the values of the analog input signal are presumed to exist, they must be quantized by partitioning them into $m=2^k$ discrete steps. All the analog values which fall within a given binary code range are assigned the same code. With the maximum analog amplitude signal range defined as 2^k the differences between each sampled level is 1 bit, as shown in Fig. 2.10(a). Therefore, the maximum deviation of the sampled amplitude pulse from the corresponding reference voltage level is 1 bit, which is defined as the amplitude uncertainty level ΔV. In order to reduce this uncertainty, the ADC transfer curve can be modified. Since the deviation affects only the smallest change that can be resolved by the k-bit scale, the quantization approximation should affect only the least significant bit (LSB) of the binary scale. Therefore, instead of rounding off to the closer lower digital value, a half offset is added to the transfer

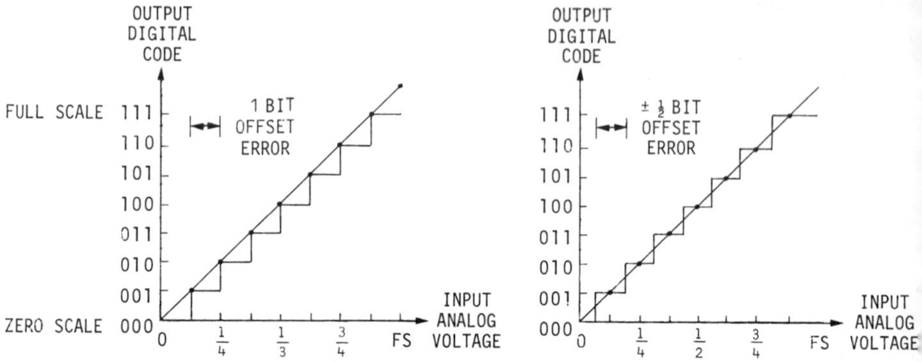

(a) ONE BIT UNCERTAINTY (b) PLUS/MINUS HALF BIT UNCERTAINTY

Fig. 2.10 Transfer curves of ADC

curve, as shown in Fig. 2.10(b). The amplitude level uncertainty ΔV is now $\pm\frac{1}{2}$ bit and the quantization approximation is associated with only the least significant bit of the code. This inherent uncertainty of $\pm$ LSB is also called the quantizing error.

The quantizing error is irreducible, causing a permanent deviation from the original analog wave. In digital-to-analog convertion (DAC) no other levels exist besides the discrete reference levels. Therefore, the plotted transfer curve for the DAC is ideally linear, as shown in Fig. 2.10. This consequently means that all the errors introduced in the ADC remain in the DAC.

The only method for reducing the quantizing error is to increase the number of discrete voltage levels m, since this is a direct function in recreation of the original waveform. In the example shown in Fig. 2.10, only 8 reference levels are used, i.e., $k = 3$ bit code. It is obvious that increasing the number of levels to 32, 64 or even 128 will reduce the quantizing error. The disparity between the original waveform and the quantized digital representation can be seen by the example in Fig. 2.11. This shows the reconstructed shape in the ADC of both the vertical and horizontal sampling, i.e., the close approximation received from the comparison with the fixed discrete levels.

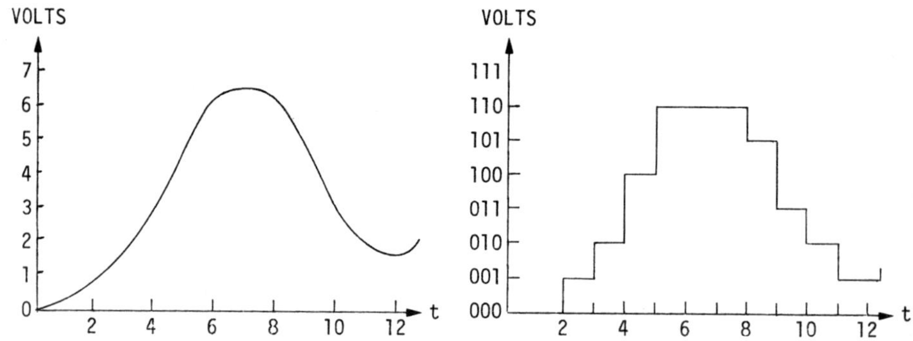

(a) ORIGINAL ANALOG SIGNAL (b) SAMPLED AND QUANTIZED VERSION

Fig. 2.11 Quantization errors in an ADC

In the absence of the sampling errors, the distortion introduced by the quantizer is the limiting factor in the performance of the ADC system. The distortion is commonly quantified by the signal-to-quantization error ratio (similar to signal-to-noise ratio in communication systems). This ratio is determined by the step sizes of the quantizer, its range and the stability accuracy of the input sampled pulse amplitude. A useful graph is shown in Fig. 2.12 which shows the quantizing as a function of the input sample values.

Up till now we have regarded the transfer function as an ideal transition. In the ADC circuit there are many more error contributions, besides the quantizing error. These are the linearity, drift and gain errors. Fig. 2.13 shows the bands of all those errors which express the accuracy of the ADC relative to the reading.

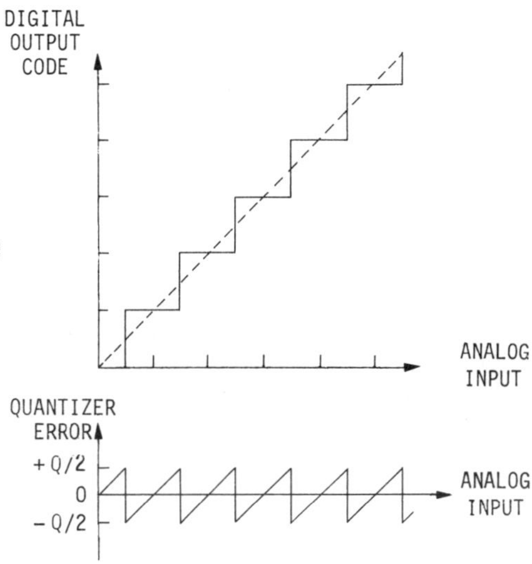

Fig. 2.12 Quantizing error graph

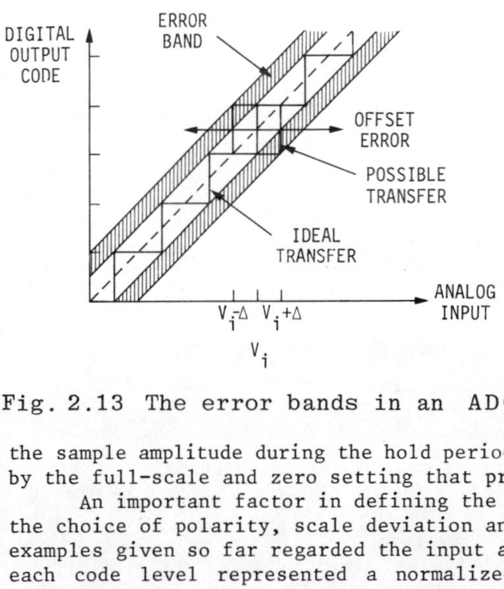

Fig. 2.13 The error bands in an ADC

The 'linearity' of the ADC is defined as the maximum deviation from the best straight line drawn through the transfer function of the converter. It is expressed as percentage of the full-scale range. The differential linearity is the measure of the linearity from one digit state to the next. If the differential linearity is specified as $\pm\frac{1}{2}$ LSB, the step size from one step to the next may be 1/2 to 3/2 of an ideal 1 LSB step. The 'drift' is caused by undersirable signal-coupling around the switches that are supposed to be switched off. This drift, therefore, caused by feed-through currents, usually affects the sample amplitude during the hold period. The 'gain' of the ADC is caused by the full-scale and zero setting that provides the conversion relationship.

An important factor in defining the conversion accuracy of an ADC, is the choice of polarity, scale deviation and full scale magnitude range. The examples given so far regarded the input analog signal as unipolar, whereby each code level represented a normalized magnitude of a voltage level without regard to its sign. Analog waveforms are usually regarded as bipolar signals which therefore require the addition of an extra bit to represent the sign bit, using the 0 to signify the plus, and the 1 to signify the minus, as shown in Fig. 2.14.

Most ADC systems use fixed voltage differences between each reference level, which means a uniform quantizer. This is not essentially so in all ADCs. If the measurement of the analog signal requires important differences near the zero level or at the full scale points, a logarithm scale may be used.

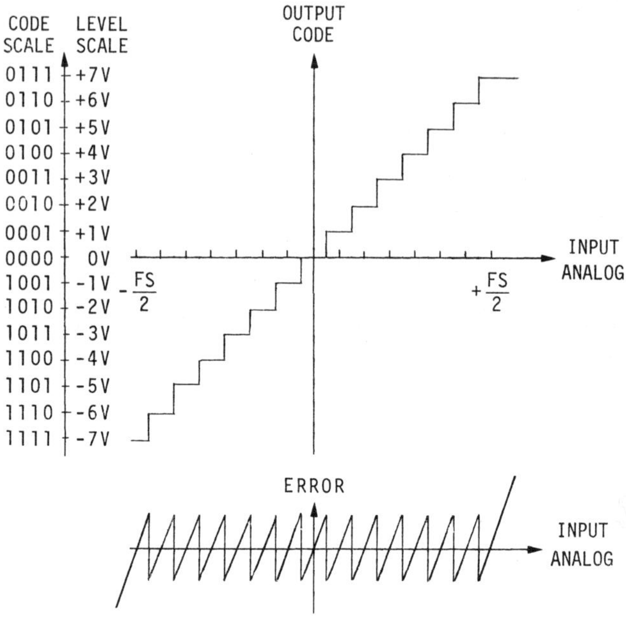

Fig. 2.14 ADC transfer curve of a bipolar analog input signal

2.5 ANALOG-TO-DIGITAL CONVERTERS

Analog-to-digital converters (ADC) are devices that translate the analog waveforms into computing information language. This operation, as shown in Fig. 2.15, ties together the combined processing of wave sampling, pulse amplitude quantizing and binary encoding. A low pass filter (LPF) is added at the analog end to get rid of the high frequency noise signal which may have been added to the input signal.

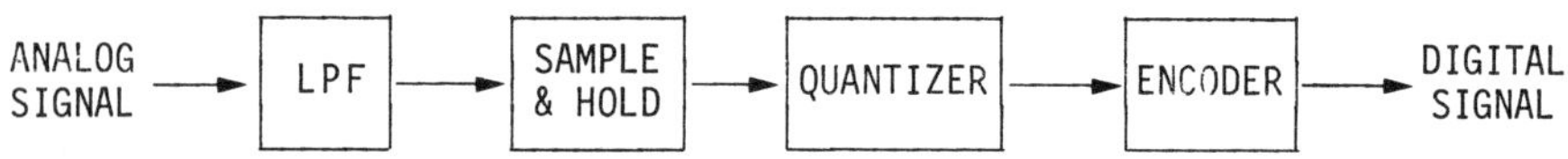

Fig. 2.15 Block diagram of an analog-to-digital converter

As already explained, the initial stage of the ADC process is sampling the analog wave, while the second stage is that of quantizing. This second stage, however, is not instantaneous during the sampling period, but a time consuming operation much after the sampling time. There is no cheap and simple method of comparing the sampled pulse level with the set of discrete voltage levels, selecting that which is the closest. Instead, a series of comparisons are conducted, till the closest can be found and retained. This calls for a method of retaining the original magnitude level throughout the quantizing comparison operation, well within the $\pm\frac{1}{2}$ code level change. Such an operation is performed by a circuit which is known as sample and hold (S & H), shown schematically in Fig. 2.16. S & H circuit usually contains an amplifier with a high input impedance and provides a low stable output drop. The amount of input voltage change that appears at the output, when the

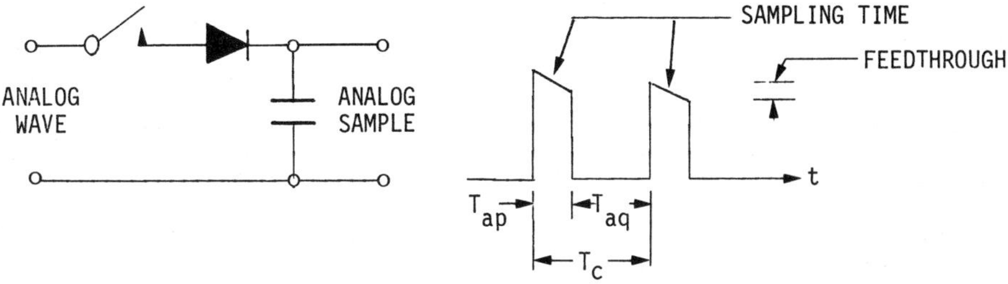

Fig. 2.16 Sample and hold schematic diagram

amplifier is held, is known as feedthrough. The time between the S & H common transition and the point at which the output ceases to follow the input is referred to as the aperture time, T_{ap}. This means that the maximum T_{ap} allowed is when the feedthrough is equal to the $\pm\frac{1}{2}$ level change. The quantization time, T_q, must therefore be less than the aperture time, $T_q < T_{ap}$. Another factor affecting the S & H operation is the time it takes to change from its previous value to the new value to be sampled. This referred to as acquisition time, T_{aq}. The conversion speed, T_c, of the ADC is defined as the minimum time required for a complete conversion process: hence it must include both the aperture and acquisition times, $T_c = T_{ap} + T_{aq}$.

The conversion speed is a most important factor in selecting an ADC. The faster the ADC, the more analog samples that can be multiplexed together. It has already been pointed out that the sampling spacing is limited by twice the highest frequency, i.e., $T_S = \frac{1}{2}B$. This means that the limiting factor in defining the number of sensors that could be multiplexed together, N, is the ratio of the sampling time to the conversion time, i.e., $N < T_S/T_c$.

There are several popular ADC systems, presenting different conversion speeds, all based on using a DAC in a feedback configuration. The most common are ramp, tracking, and successive approximation.

The most simple ADC is the ramp which is based on the full 2^k counts for each quantizing stage. In this system, as shown in Fig. 2.17, the counter steps up one position every clock pulse, and its decoded output selects one of the reference voltage levels. This reference voltage is compared with the analog input level. If there is an output from the comparator, the counter is stepped up one position, causing the next reference voltage level to be selected. This in turn is compared again with the input level. The procedure is continued till the analog input level is larger than the reference voltage level. The absence of an output from the comparator looks the gate which

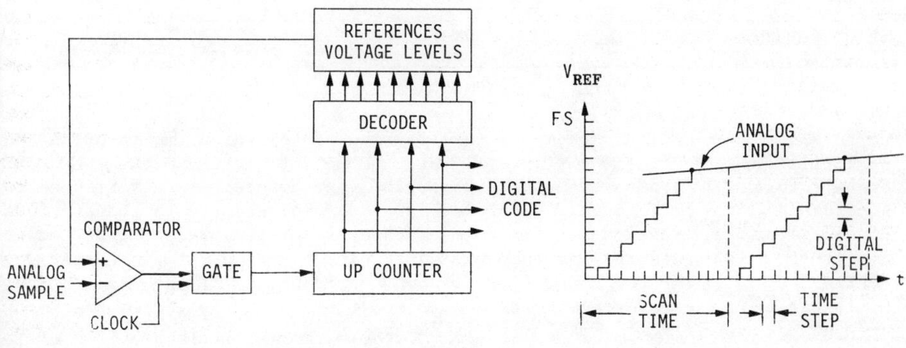

Fig. 2.17 The ramp ADC

inhibits the step operation of the counter. Each conversion operation must enable the counter to reach its full 2^k steps, so as to be able to measure the full voltage scale. The binary translation of the analog pulse is the counter position after it has stopped count. Such an ADC scheme is only good for very slow changing analog data.

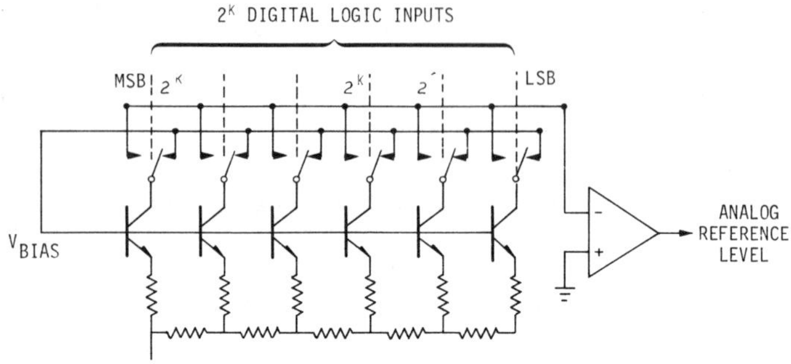

Fig. 2.18 Schematic drawing of an ADC

The operation of the set of voltage reference levels and their accompanying decoder is that of digital-to-analog converter (DAC). That is, the circuit receives a set of binary codes, translates them into a digital code and then selects the corresponding analog reference level. The actual DAC circuit incorporates current sources, decoding steering logic, resistor ladder network and a fast summing operational amplifier. A schematic drawing of such a DAC is shown in Fig. 2.18. Each digital decoding combination selects the appropriate current steering logic which defines a given output reference voltage level.

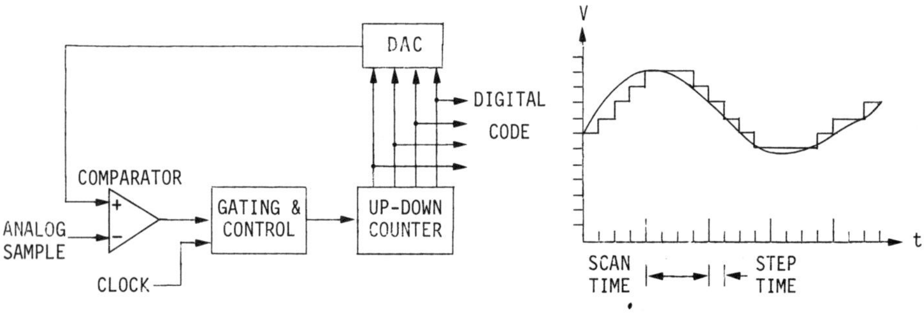

Fig. 2.19 The tracking ADC

A much better ADC scheme is the tracking system which is based on the same principles as the ramp system, but instead of counting the full 2^k steps through the complete conversion period, the counter is only stepped in accordance with the changes of the new input sample relative to the previous count. That is, the system tracks the changes of the present input pulse amplitude pulse relative to the previous pulse, whether it is larger or smaller, by stepping the counter up or down accordingly. If the input pulse is larger than the reference voltage obtained from the DAC, the counter is stepped up, and if it is smaller it is stepped down. This circuit is very simple, as seen in Fig. 2.19, and is very similar to that of the ramp circuit. The comparator continues to examine the input voltage sampled pulse in comparison with that of the DAC output till it 'locks' when the loop is balanced at $\pm$ LSB. In this state the counter output is the binary coded equivalent of the analog input pulse.

The tracking ADC can operate much faster than the ramp ADC, provided the relative amplitudes of the input pulse changes are small. Nevertheless, if these changes are large, such as occurs when the analog pulses are fed from a multiplexer, the full conversion period of 2^k steps may be required for each ADC scan.

High speed analog-to-digital conversion is achieved by using successive approximation, which requires only $k+1$ clock steps (where k is the number of the binary bits) instead of the 2^k clock steps required for the ramp and tracking ADC. In this scheme, as shown in Fig. 2.20, the conversion operation is performed by a series of trials in which the input pulse is compared with half scale range of a given measurement.

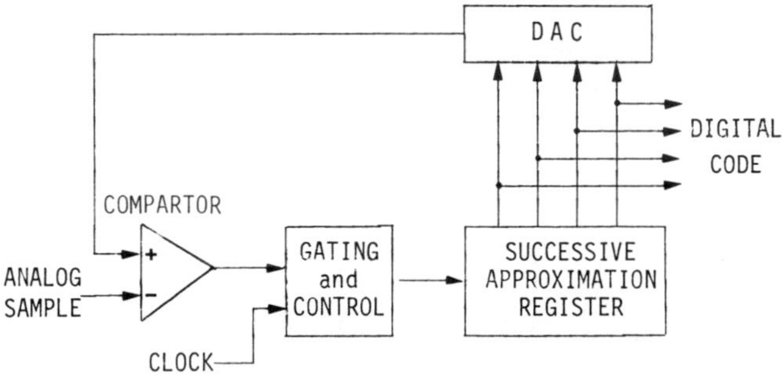

Fig. 2.20 Successive approximation ADC

Since the full input voltage scale (FS) is divided into a binary code, the most significant bit (MSB) defines whether comparison is in the upper or the lower half of the scale. That is, the MSB equals 1 if the trial shows the input pulse to be in the upper half of the scale, i.e. $V_{in} > V_{FS}/2$, and equals 0 if it is in the lower half, i.e., $V_{in} < V_{FS}/2$. Once the MSB is defined, the next significant bit is defined by comparing whether the input pulse is in the upper or the lower half of the remaining quarter full scale, i.e., $V_{in} \gtrless V_{FS}/4$ or $V_{in} \gtrless 3V_{FS}/4$. This progression of trials is continued for each successive bit until the smallest bit has been tried. After k such trials, the LSB is defined. This means that after k trials, an end-of-conversion signal indicates that the binary coded analog signal is ready to be used. Fig. 2.21 shows the flow diagram of the progressive of trails for a three-bit successive approximation ADC.

The successive approximation is performed with logic and the result retains in a register. If the full scale is matched with the binary scale of 111, the half scale is compared with 011. That is, it is compared with the voltage reference level 3 which is equivalent to that of 011. These values can easily be selected and they are used to define the next comparison reference level.

Successive approximation schemes can perform a conversion of an analog sample into 8 binary bits in less than 1 μs. For much faster speeds, such as may be needed in real-time radar application, different methods must be employed. The method to be used then is a simultaneous operation where each sample is compared in a single instance with 2^k-1 reference voltage levels, as shown in Fig. 2.22. This means that this scheme is 2^k times faster than the tracking ADC and k times faster than the successive approximation ADC. This, of course, is true provided that it is not limited by the internal components of the circuitry. Although this instantaneous ADC scheme sounds obvious, it is rather an expensive solution, since it requires the employment of $2^k - 1$ parallel analog comparators. All of these comparators have one of their inputs fed from a series of resistors producing a ladder of voltage reference levels spaced apart by $(FS)/m$ volts. The analog input signal is fed simultaneously to the input of all the other computers, thereby producing a digital code output in one-step operation. The least significant digit (LSD) is defined as large or smaller than $(FS)/m$ confined by $R/2$ ohms. In this way all the other values are defined. The digital code produced by all the comparators is proportional to the analog sample and it must be translated into a binary code using a decoder.

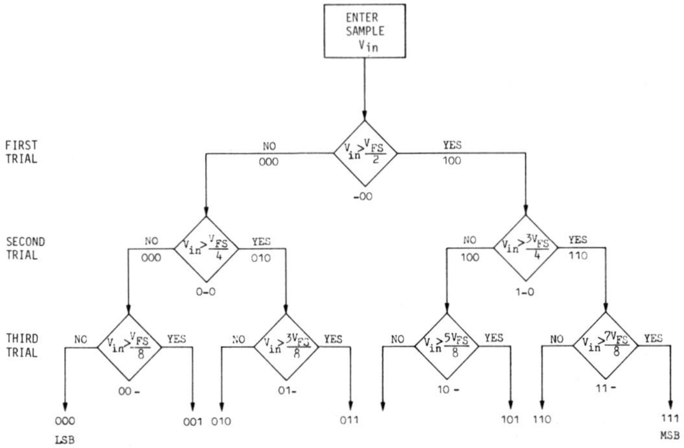

Fig. 2.21 Flow diagram of a three-bit successive approximation

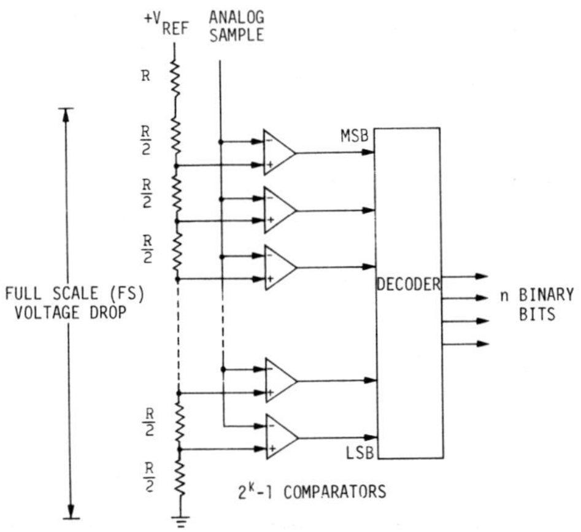

Fig. 2.22 Instantaneous analog-to-digital converter

For many Command and Control systems, the linearity and resolution of the quantization code is extremely important that it takes preference over the conversion speed. For these applications a different ADC is used. In this new ADC, the input analog sample of an unknown level is converted from a voltage level scale into a time duration scale. Then the resulting proportional time interval can be measured digitally. The principle of this ADC scheme is based on the operation of an integrator, in which the input sample is integrated for a predetermined time and then it is integrated again, although this time, in the opposite direction, for a measurable time, till it reaches the zero level. Measuring the variable time duration is performed with the aid of a counter as seen in Fig. 2.23.

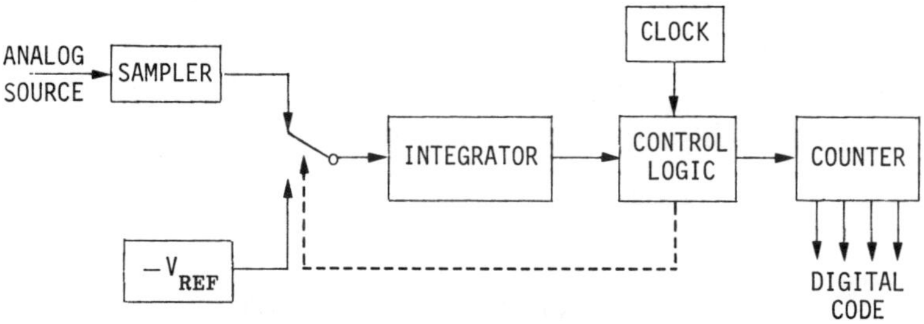

Fig. 2.23 Integrator type ADC

The input sample with the unknown amplitude sample is fed into the integrator for a fixed period of T_{IN} seconds (i.e., T_{IN} = constant). The output voltage of the integrator after period T_{IN} is proportional to the input amplitude sample level. At this instance of operation, the integrator is switched to a known reference voltage of negative polarity. The integrator is therefore integrated down from the level determined by the input until the zero level is reached. Since the reference voltage used for the down integration process is fixed and known, all the possible down voltage slopes of different analog input samples must be parallel to each other. For this reason, this ADC technique is usually referred to as a dual slope ADC, and each sample operation can be plotted by two slopes, as seen in Fig. 2.24. The time interval after the second integration operation, i.e., the time after switching in the negative fixed reference voltage is also proportional to the input sample amplitude level. This time interval, T_{REF}, is measured by a counter, where its output gives the desired binary code.

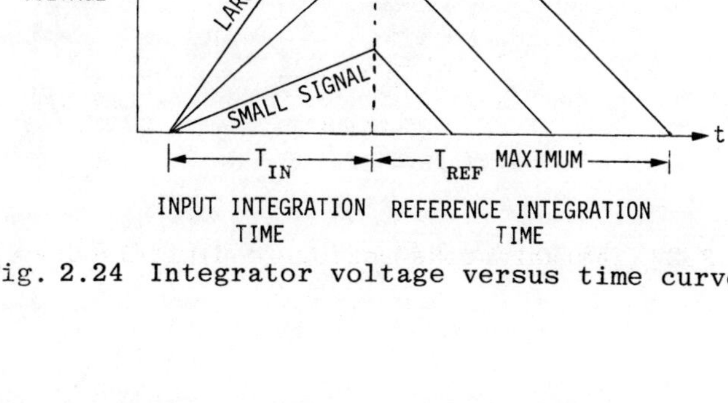

Fig. 2.24 Integrator voltage versus time curves

The input unknown amplitude sample level, V_{IN}, is related to the maximum full scale voltage range V_{FS} as the inverse ratio of the integration timing of the input and the reference sources, as seen by the following expression

$$V_{IN} = V_{FS} \times T_{REF}/T_{IN} = V_{REF} \times T_{REF}/T_{IN}$$

where T_{IN} is the pre-determined integration time and T_{REF} is the one time period measured by the internal counter.

2.6 ANALOG DATA COMPRESSION

The aim of data acquisition is to obtain meaningful information of the Controllee's performance. This information must be accurate and precise; however, most of the data collected is redundant, providing little useful information. It has been found that two adjacent samples of the same analog source are nearly of the same amplitude. The second value is clearly redundant. Information is drawn out of the data by evaluating the actual changes in the Controllee's actions, but little (if any) information can be drawn out of the data describing the Controllee's steady state actions. The more data acquired, the larger number of bits are transmitted to be processed. This makes the processing task tedious and wasteful, occupying much memory space. Above all, this also overloads the expensive communication channels, making them ineffective. Reducing the sampling rate or the quantization levels will only decrease the accuracy when it is really needed. What is wanted is a system which increases the number of bits when there are input changes but decreases them when there are little or no changes.

There are various methods of improving the data acquisition whereby less number of bits are needed to represent satisfactorily an analog signal while still retaining its reconstructed quality. Most of these methods have a major disadvantage, that if the ADC system should operate efficiently, i.e., producing less bits, a separate ADC encoder circuit is required for each analog source. This means in effect that the analog sources using any of the modified ADC systems are unsuitable to be time shared in a multiplexer as described for the PCM. Some of these modified ADC systems are described in this section. All of them are based on what is known as prediction coding (PC) where advantage is taken of the signal predictive values of the past samples.

Differential-pulse-code modulation (DPCM) produces data in accordance with the differential changes between the samples, rather than the samples themselves. Since the average changes between two adjacent samples are usually much less than the total magnitude of each sample, the number of bits transmitted in DPCM is less than that of PCM. The DPCM scheme holds (delays) the amplitude pulse of each sample for a full sample period and then compares it with the present amplitude sample, as seen in Fig. 2.25. The old amplitude sample, $S(t)_{n-1}$, is subtracted from the new amplitude pulse, $S(t)_n$, and the result $S(t)_n - S(t)_{n-1}$ is then quantized. The quantizer in this case is similar to any of the ones used in the ADC circuits described in the previous section.

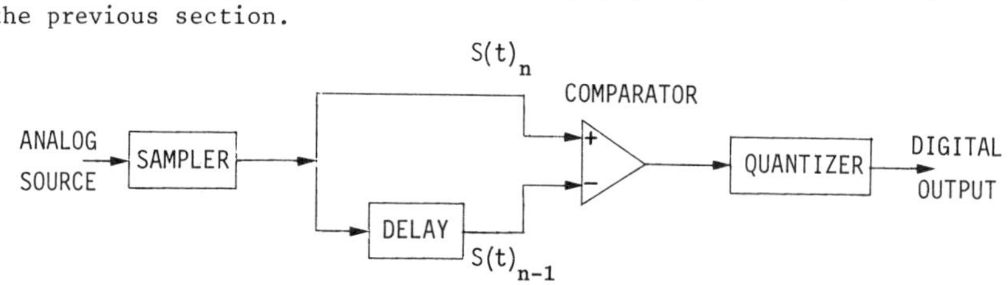

Fig. 2.25 The basic concept of coding the differential comparison of two adjacent analog samples

As already explained before, it is difficult to sample and hold an analog pulse amplitude and keep its magnitude stable for prolonged periods. Nevertheless, digital data, representing the pulse magnitude can be held indefinitely in appropriate digital registers. Since the data produced by the quantizer is already in digital form, this may be exploited by feeding the quantizer output back to the input and comparing it with the input sample. The feedback code will require a DAC to convert the code back to analog, as shown in Fig. 2.26. It is usual to regard the extra register and DAC as parts of the quantizer and not show it in the feedback loop. Crudely speaking, in the differential coding of analog sources, one takes sample differences of the input and internal signal and quantizes them. Then, from the resulting code, the approximation of the analog signal is recovered by integrating the quantized difference sample.

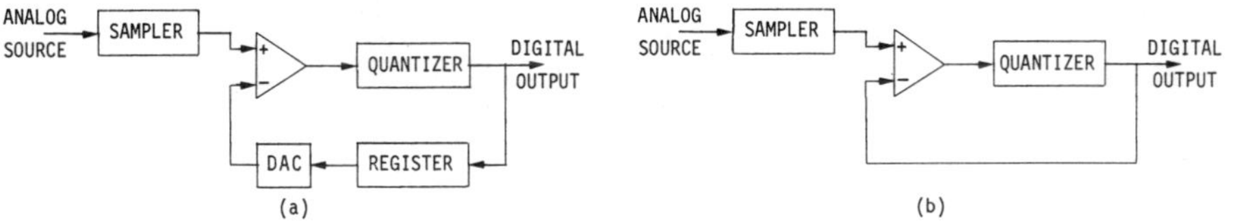

Fig. 2.26 Analog differential coding with the previous sample held digitally

The DPCM scheme described above suffers from the fact that the quantizer error is now increased as it goes through the extra DAC. That is, the comparison is not done with the directly past sample but with a reconstructed version of it. A solution to this problem is given by inserting a predictor circuit in the feedback loop, as shown in Fig. 2.27. The present analog input sample is not only compared with the directly previous sample, but also with a predicted value, and the difference between them is transmitted. It can be shown that in this case the quantizing errors do not accumulate with the successive differences.

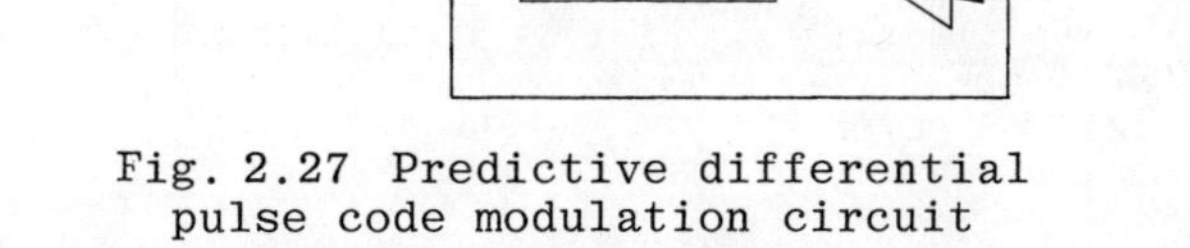

Fig. 2.27 Predictive differential pulse code modulation circuit

The predictive DPCM produces less bits per second than PCM, although the reduction in the bit rate is not very significant. It requires only one or two less binary bits per sample than the equivalent PCM coder to produce the same signal-to-quantized error rate. Furthermore, the DPCM is less efficient than the PCM, since the prediction is down with a quantized reconstructed signal. This is based on adjusting the statistical predictor to the amplitude values of the past three samples, using fixed weights in defining the predictor constants.

In DPCM, the prediction is based on fixed statistical estimation of the analog wave and has to be adjusted to the characteristics of each analog source. In another approach, the predictions of the statistical estimation of the reconstructed values are continuously updated to reflect the signal changes. This scheme, known as adaptive-differential pulse code modulation (ADPCM) further reduces the redundancy in the output data, therefore presenting a lower data bit rate. Here, too, the quantizing errors do not accumulate with successive differences.

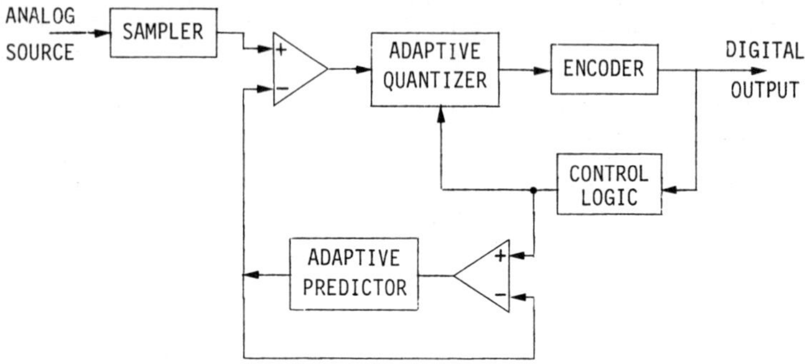

Fig. 2.28 Adaptive differential pulse code modulation circuit

A further improvement in the ADPCM operation can be obtained by using also an adaptive quantizer, as shown in Fig. 2.28, which increases or decreases the step size of the quantizer as a function of the quantizer load. When the load difference between the two samples is small, the quantizer load is also small and the steps may be decreased and, vice versa, when the difference is large. In this case, not only is the prediction modified, but also the quantizer steps are modified in accordance with the previous output pulses. However, the predictor depends on the statistical characteristics of the past few samples, while the quantizer adapts itself to the directly past sample.

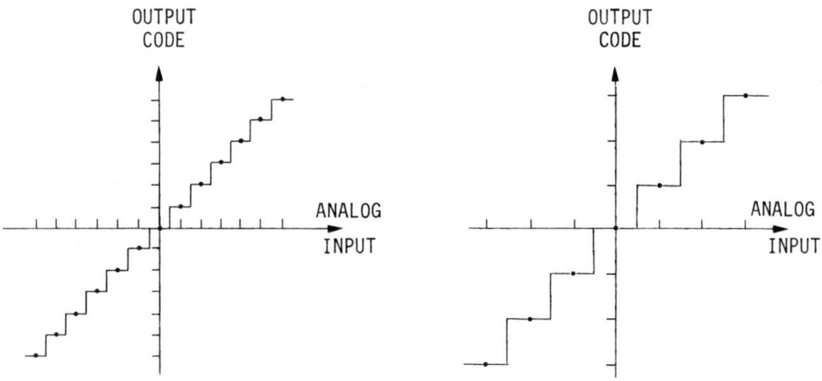

(a) LOW INPUT SIGNAL (b) HIGH INPUT SIGNAL

Fig. 2.29 The characteristic of an adaptive quantizer

Figure 2.29 shows the changes in the quantizer characteristic as it adapts itself in accordance with the input load. With low input signal power the steps are reduced and with high input signals the steps are expanded. In other words, it shrinks or expands the steps like an accordian. This means that the quantizer must follow the quantizer output rather than its input; it could also mean that the adaptation tracks the input rather slowly.

Another form of pulse code modulation is known as delta-modulation (DM), which can in certain respects be considered as a simple form of DPCM. In PCM, the instantaneous input voltage sample pulse is converted into a k bit coded word. The principles of delta modulation are quite different. There is no sample-and-hold operation, and the quantizer produces only one bit word. The spectrum of the transmitted digital signal for a DM also differs considerably from that of a PCM signal. With DM, the average number of bits per second is constant, while with PCM, there is no linear relationship between the instantaneous bits.

In delta modulation the input analog signal is compared with a predicted reconstructed analog signal. At given periodic instances the two analog signals are compared and the results are coded into one of two states. If the input signal is larger than the predicted level, a 1 is recorded, and if is is smaller a 0 is recorded. The quantizer now has only to define one bit and therefore it could be made much simpler, as shown in Fig. 2.30. In fact, it need only be a decision circuit which must define whether the input is larger or smaller than a given threshold. The predictor in the feedback path is also comparatively simple, since it need only to produce a continuous signal. It consists of an amplifier circuit and an integrator. The integrator output level is adjusted after each quantizer step in accordance with the single coded bit output pulse train.

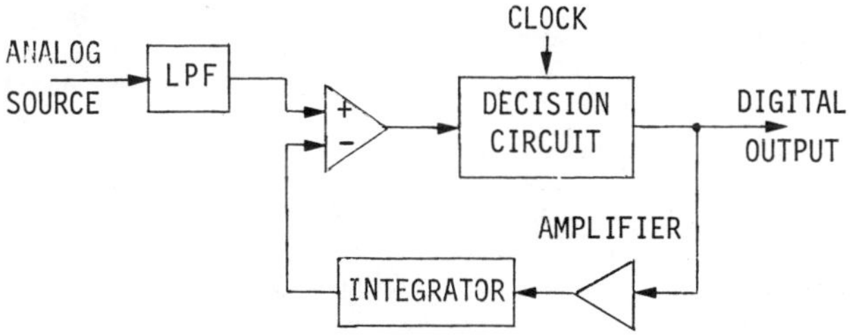

Fig. 2.30 Delta modulation basic circuit

In this delta modulator, the one bit quantizer defines two levels, either positive or negative. The prediction error is therefore quantized to one of only two values, a positive one or a negative one. This means that the integrator output is stepped up if the quantizer code is 1 and it is stepped down if it is 0. The stepped waveform is incremental representation of the previous 'sampled' input waveform. The comparison is, if so, between the present level of the analog level and the present integrator output. The present integrator level, represents in effect the approximation value of the previous comparison, performed one clock period back. In this manner a whole output bit train is produced, a bit for each comparison, as seen in Fig. 2.31. This bit pulse train consists of a positive going pulse for a 1 and a negative for a 0. These inverted pulses control two current sources which are fed through the feedback path to the integrator. The integrator output is therefore incremented in correspondence with the pulse train polarity.

The delta modulator circuit is simple and inexpensive: however, since the quantizer is adjusted by only bit at a time, the quantization errors are accumulative. This makes it unacceptable where d-c levels are important. This error can be reduced by using double integration to improve the signal-to-error ratio. Another problem that affects the quantization errors is that the reconstructed signal does not follow the input signal which occurs when

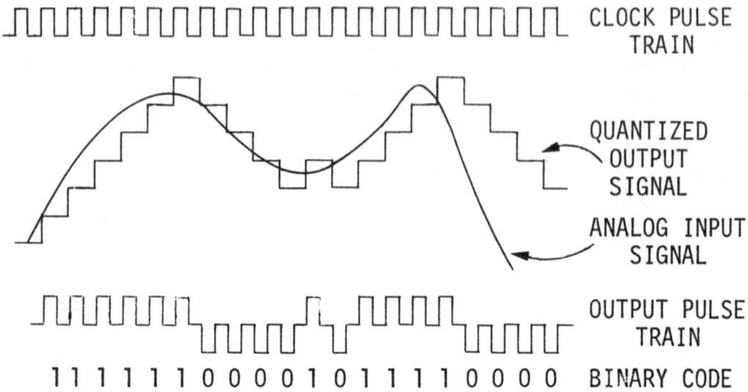

Fig. 2.31 Delta modulation waveforms

the analog exceeds the maximum step slope. Such a situation is referred to as the 'overload performance of the integrator'. It means in effect that the delta modulator is dependent on the frequency component of the signal being encoded. This can be avoided by increasing the sampling rate. Coders that adapt the step size to the load of the integrator are referred to as adaptive delta modulator (ADM), as shown in Fig. 2.32.

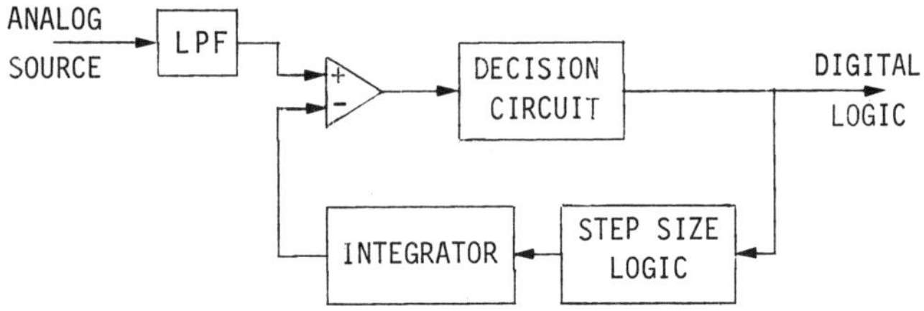

Fig. 2.32 Adaptive delta modulation circuit

In the ADM the step size is varied according to the instantaneous need, as shown in Fig. 2.33. If the last two bits transmitted are only 1's or only 0's, the step size is doubled. If the last two bits are an alternate 1 or 0's the step size is halved.

It should be added that there is no need to transmit the step size together with the data, as the decoding can follow the same logic of doubling or halving the step in accordance with the data contents.

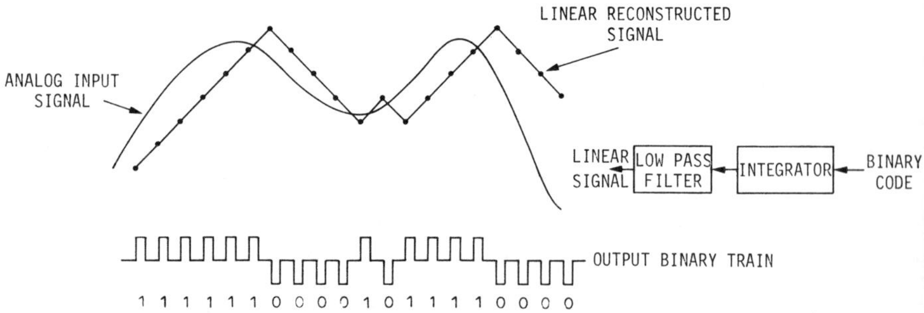

Fig. 2.33 Adaptive delta modulation waveforms

The conventional delta modulation, as distinct from adaptive delta modulation, is also referred to as linear delta modulation (LDM). In linear delta modulation the step size of the feedback path is constant, while in ADM the modulation is actually a nonlinear operation. Although this cannot easily be seen in Fig. 2.31, a simple low pass filter after the integrator produces a linear reconstructed signal, as seen in Fig. 2.34. This means that if the last bit was a 1, the reconstructed signal produced by the filtered integrator output will continue to rise till the next 0 bit, and then the reconstructed signal will start declining, and so on. Therefore, the reconstructed signal increases until, at some future sample time, it exceeds the analog input signal and then it starts to decline. In this example, the same input analog signal is used in both figures, 2.31 and 2.34, thus exhibiting the same code bit train.

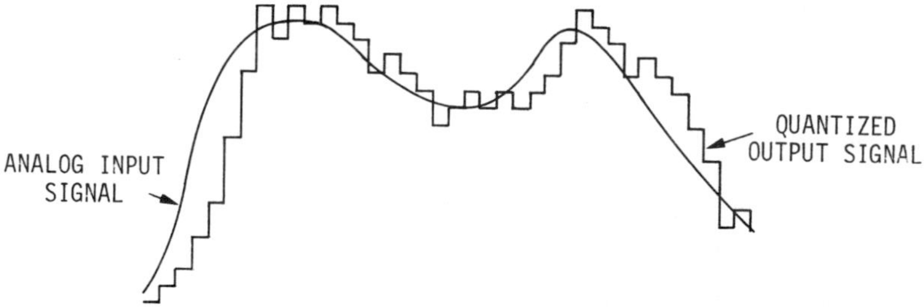

Fig. 2.34 Linear tracking waveform of the quantized signal

In conclusion of the presentation of the various ADC compression systems, the following tables give a quality comparison between them. For this comparison, a voice analog signal which has the frequency band of 4 kHz, is used as an example. This means that the sample rate (PAM clock rate) is essentially 8 *K* bits per second. If the number of bits quantized for each sample is 7 or 8, the PCM bit rate will be between 56 *K* bits/sec to 64 *K* bits/sec.

TYPE OF PULSE MODULATION	BITS/SEC REQUIRED TO REPRESENT THE ANALOG SIGNAL
PCM	56,000 to 64,000
DPCM	40,000 to 48,000
ADPCM	30,000 to 40,000
LDM	10,000 to 32,000
ADM	8,000 to 16,000

It is important to add that ADCs are relatively very expensive and so it is suggested that the designer should employ multiplexing systems with PCM techniques. This means that the designer must decide whether the compression of the data, which reduces the communication channel cost, is more economical than using a single ADC with time division multiplexing per channel.

2.7 DIGITAL CODE FORMATS

The output from all the ADC systems described are essentially fixed voltage levels referred to the ground potential. At any given clock period this voltage is determined according to the coding scheme used. Although all the outputs may have different voltage levels, each individual output is usually switched only between two fixed levels, either V_i or 0 volts, where the same V_i is usually used for all outputs. The changes in the voltage levels as a function of time of each output, produces a pulse or series of pulses. These pulses represent digital coded data. By digital one means data in the form of one of a discrete number of codes. Each individual pulse or each individual output terminal, at any given time, represents a single bit. A 'bit' is the abbreviation of binary digit. In this respect it means a single unit or symbol employing exactly one of two kinds of states. It is the logarithm to the base two of the number of possible states of each individual output.

There are many possible codes used in A/D converters: some of them will be described here. Before one can analyse them, a general background of numbering is needed.

The number of digital symbols or states in each system is called the 'radix' of the system. The decimal system has a radix of 10 while the binary has a radix of 2. Any number N can be described by any numerical code system using the following polynomial expression:

$$N = F_0R^0 + F_1R^1 + F_2R^2 + \dots + F_iR^i + \dots + F_nR^n = \sum_{i=0}^{i=n} F_iR^i$$

R represents the radix which the number system uses, and the integer F_i represents the symbols or states of the system. The value of the integer in each system must be smaller than the radix itself.

$$0 \leqslant F_i \leqslant (R-1) \ .$$

In practice it is usual to write only the integers, one after the other, leaving out the radix. The order in which the integer appears is from right to left, which is the reverse of that shown above

$$N = F_n \dots F_i \dots F_2F_1F_0 \quad \text{with} \quad 0 \leqslant i \leqslant n \ .$$

The least significant bit (LSB) is expressed by F_0, and the most significant bit (MSB) by F_n. This means that the maximum number of digits represented in this number is $n+1$. This factor n must not be confused with m or k, mentioned before as representing the number of levels or bits in a given code.

Although, each individual output signal from the ADC is in binary form, i.e., either 1 or 0, the group of all the ADC output signals need not necessarily be in a binary code. In fact, there are many code formats that may be used. In this respect a code is a plan for representing each of a finite number of values or symbols as a particular arrangement or sequence of discrete conditions or events. The parallel output from all the ADC terminals usually refer to a single integer F_i while only the serial output refers to N. That is, F_i represents one of the m numerical positions in a scale of N reference voltages. To receive the binary equivalent of F_i, it must first be decoded (as shown in Fig. 2.35) a process which will be described later.

When quantizing by comparing m discrete reference levels with the analog input sample, only one of the m parallel terminals will produce an output, i.e., only one output will have the potential V_i corresponding to the 1 state, while all the other outputs will have the ground (zero) potential corresponding to the 0 state. The code integer F_i representing this format has m levels. The factor m describes the full scale range of the ADC and is also the number of digits required to describe the integer F_i. This means in effect that m also represents the radix of the system, and its integer value can be expressed as follows:

$$0 \leqslant F_i \leqslant (m-1)$$

The m set of parallel outputs can be described as a polynomial in a dummy variable of X. The figure X in this case describes the numerical position of the output terminal. Since the number of outputs for this type of ADC terminal points is m, the polynomial will be of the order $(m-1)$ and is expressed as follows:

$$F_i = a_{m-1}X^{m-1} + \ldots + a_j X^j + \ldots + a_p X^p + \ldots + a_2 X^2 + a_1 X^1 + a_0$$

where the coefficients a_j and a_p may have the values of either '0' or '1' only, provided that only one coefficient, say, a_p, may have the value 1 while all the other coefficients a_j must have the value 0, with $0 \leqslant j \leqslant (m-1)$, $0 \leqslant p \leqslant (m-1)$ and $a_j \neq a_p$.

A train of m pulses, representing this integer code, with only one pulse indicating a '1' and $(m-1)$ pulses representing a '0' is rather wasteful. Instead, this format and most of the other ADC formats are converted (decoded) into the binary code format. This binary format is the most economical method of representing a number where the coefficients have the value of 1 or 0 only. In this type of ADC, the integer F_i of the original code, which had m output terminals, is decoded and therefore producing only k output terminals. This factor k has already been defined as the resolution of the ADC. The relationship between the binary radix system with k bits per code and the previous level radix system with m parallel bits per code is $k = \log_2 m$ or $m = 2^k$.

In the instantaneous ADC, a different code format is obtained although having the same number of m-bits. With each instaneous output, representing one integer F_i, there are p output terminals with the potential V_i representing the value 1 and all the other $(m-p)$ output terminals at the ground potential representing the value 0. Since the same number of quantization levels are used in both ADC systems, the same polynomial will express the integer, although the coefficients are different:

$$F_i = c_{m-1}X^{m-1} + \ldots + c_j X^j + \ldots + c_p X^p + \ldots + c_2 X^2 + c_1 X^1 + c_0$$

where all the coefficientcs c_p have the value 1 with $0 \leqslant p \leqslant (m-1)$ and all the coefficients c_j have the value 0 with $(p_{max}+1) \leqslant j \leqslant (m-1)$. Here too, the integer F_i is expressed by m pulses which is again a rather wasteful means of representation and should therefore be converted into a binary code of k bits with $k = \log_2 m$.

The binary code format notation, also known as 'natural binary' or 'straight binary' is the most efficient means of expressing other radis numbers where the output terminals may have one of two possible levels. In the binary scheme the radix equals 2 and each integer can have the value of 1 or 0 only. Each integer in the binary number therefore represents a different power of 2. The binary number can also be expressed by a polynomial to the power of 2 with k representing the maximum number of bits in each code word, as follows:

$$N = b_{k-1}2^{k-1} + b_{k-2}2^{k-2} + \ldots + b_q 2^q + \ldots + b_2 2^2 + b_1 2^1 + b_0$$

where b_q may have the value of 1 or 0 only with $0 \leqslant q \leqslant (k-1)$. For example, an integer F_i representing 16 levels with only level number 14 conducting may therefore be represented by a 4 bit code, as follows:

$$F_i = 0 \cdot X^{15} + 1 \cdot X^{14} + 0 \cdot X^{12} + 0 \cdot X^{11} + \ldots + 0 \cdot X^2 + 0 \cdot X^1 + 0$$

$$N_i = 1 \cdot 2^3 + 1 \cdot 2^2 + 1 \cdot 2^1 + 0$$

$$N_i = 14_{16} = 1110_2 \,.$$

A useful summary of the binary code, defining the resolution of the

ADC (i.e. the number of k bits in the binary representation), and the maximum number of quantization levels specifying the number of quantization levels m is given in Table 2.1. Added to this table is the weight of each step, i.e., the full range scale divided by m, (FSR)/m which equals $1/m = 1/2^k = 2^{-k}$.

Table 2.1 Decimal equivalents of power of 2

RESOLUTION BITS k	QUANTIZATION LEVELS $m = 2^k$	LSB WEIGHT $1/m = 2^{-k}$
0	1	1
1	2	0.5
2	4	0.25
3	8	0.125
4	16	0.0625
5	32	0.03125
6	64	0.015625
7	128	0.0078125
8	256	0.00390625
9	512	0.001953125
10	1024	0.0009765625
11	2048	0.00048828125
12	4096	0.000244140625

In the ADC, the number of levels is taken to be exactly equal to the power of two of the codes. This is not always the case, as in some applications the number of levels m is not equal to a power of two. In these applications the number of binary bits k is selected as the next highest power of two.

$$m \leqslant 2^k .$$

The k binary bits representing the m levels are all produced in parallel. To transmit these codes from the Controllee's ADC to the computer, the data must be in serial form. To obtain a serial pulse sequence of these codes, the k parallel binary digits produced by the decoder are temporarily stored in a k stage shift register, as seen in Fig. 2.35, and then pulsed out serially. The same applies if the k bits are produced directly from a k stage binary counter. That is, the parallel outputs are obtained directly from the counter, while from the shift register it is necessary to produce the serial output. The time period between each two parallel output pulses has already been defined as t and the period between each two serial output pulses as $\tau = t/k$.

The binary code format is the most commonly used, although there are a number of other popular codes which are also used, adopted to assist in various applications. Some of these codes will be described here indicating their particular applications.

The output from some converters may be required to drive integrated circuits with negative going logic. For these applications all the bits are represented by their complements. Such codes are called complementary binary codes. In these codes each 1 is represented by a 0 and each 0 by a 1, which is simply the one's complement of straight binary.

In many ADC applications it is essential to reduce the conversion

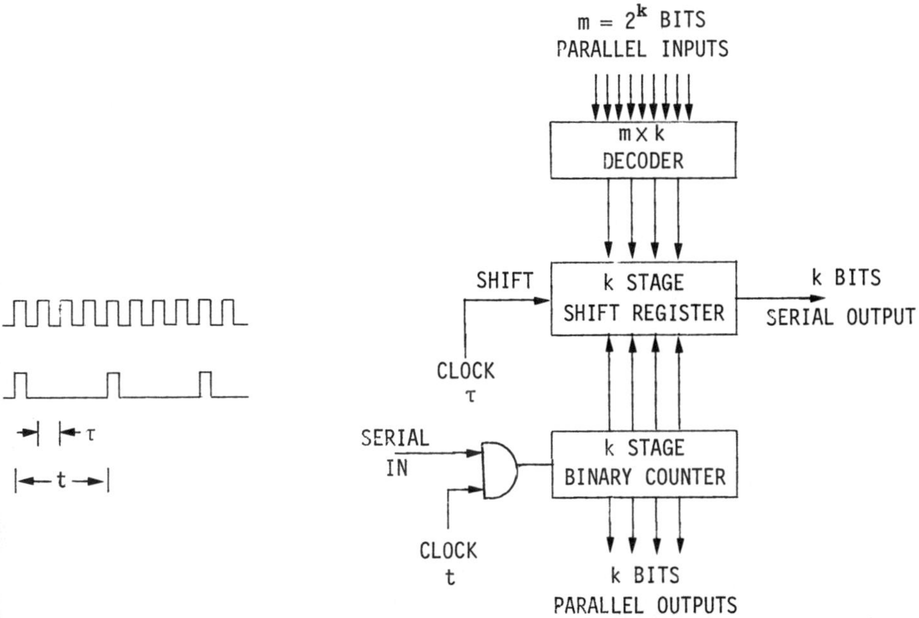

Fig. 2.35 ADC serial and parallel pulse output circuits

error probability. In the straight binary code, an ambiguity of one digit in the conversion result can cause an utterly false error in the reading. Such situations can occur in fast converters, where the intermediate result is retained in a register, and therefore many bits may have to switch in a single transfer transaction. Take, for example, the possible transition between code words (7)0111 and (8)1000 which calls for a change in all four digits. A particularly important code format, in which successive code words differ is in only one digit, is the Gray code. The Gray code is a binary code in which there is only a one bit change between each level transition. As the quantization scans through the various levels, using the Gray code representation, an ambiguity between two adjacent levels will result in an error by only a small fraction of the LSB. In straight binary, each bit position signifies the numerical weighting of the analog sample integer. This is not the case with the Gray codes, where the bit position signifies no numerical weighting; nevertheless, it is easily translatable into straight binary, and vice versa.

If the binary code is denoted by the coefficients $b_{k-1} \dots b_i \dots b_2 b_1 b_0$ and the Gray code by $g_{k-1} \dots g_i \dots g_2 g_1 g_0$, then $g_i = b_i \oplus b_{i+1}$ and $g_n = b_n$ with $0 \leqslant i < k-1$. To convert from the Gray code back to straight binary, one starts from MSB where $b_n = g_n$ then proceeds to the LSB, making $b_i = g_i \oplus b_{i+1}$. The relation between the binary and Gray code is given in Table 2.2, together with complementary binary.

There are other forms of binary codes which have the same characteristics as the Gray code, i.e., only one bit change for each transition. Nevertheless, the translation to and from straight binary is more difficult. All the codes with this transition characteristics are known as 'cyclic codes'.

Although the binary code format is the most efficient method of representing analog data, in many cases it is convenient to represent the sample magnitude in a decimal code format. It has already been shown that using 10 levels is inefficient, while the straight binary code is inconvenient

Table 2.2
Straight binary and Gray code formats

DECIMAL EQUIVALENT	STRAIGHT BINARY	COMPLEMENTARY BINARY	GRAY CODE
0	0000	1111	0000
1	0001	1110	0001
2	0010	1101	0011
3	0011	1100	0010
4	0100	1011	0110
5	0101	1010	0111
6	0110	1001	0101
7	0111	1000	0100
8	1000	0111	1100
9	1001	0110	1101
10	1010	0101	1111
11	1011	0100	1110
12	1100	0011	1010
13	1101	0010	1011
14	1110	0001	1001
15	1111	0000	1000

when referred to the decimal weighting. A group of binary digits, arranged so that each decimal number is represented by unique arrangement of the minimum binary digits is called binary-coded-decimal (BCD). This means that the code remains binary although it represents decimal quantities.

In order to represent 10 decimal digits, 0,1,...,9 in the binary code, a group of four bits are needed. Nevertheless, a group of four bits can represent 16 possible levels, since the relationship of the binary code bits k to the maximum describable levels is $m = 2^k$. Therefore, there are six possible combinations of these $k = 4$ binary bits which are not used in describing the decimal number. This makes the BCD code rather wasteful, since 6/16 (i.e. 36.5%) of the possible combinations of the binary bits are not used in the code. Nevertheless, the BCD code enables the ADC results to be displayed direct in their familiar decimal numbers, with no need of translators. Many processors today have adopted the BCD code in preference to straight binary, because they are easily converted to pure decimal.

Manipulation with the BCD is more complicated than with straight binary, since a group of four bits, representing values in excess of 9 are not permitted. Nevertheless, BCD arithmetic is simple and easy to implement. Adding two BCD numbers which result in excess of 9 will call for a carry over of a digit to the next power of four digits. This is done be adding a six to the original BCD result. Table 2.3 shows the relationship between straight binary and BCD showing various ADC readings. For this example the maximum input sample pulse to the ADC is taken as 10 volts, which means that its full scale range is displaying 0 to + 10 V, and an accuracy of 0.01 volt for the LSB. To avoid overflow, the maximum possible display range is:

$$(FS) - (LSB) = 10 - 0.01 = 9.99\,V.$$

Table 2.3
Straight binary and BCD coding of unipolar samples

FRACTION OF FS	IDEAL FS + 10 V	STRAIGHT BINARY CODE	BINARY CODED DECIMAL (BCD)
+ FS − 1 LSB	9.99 V	1111100111	1001 1001 1001
+ 3/4 FS	7.50 V	1011101110	0111 0101 0000
+ 1/2 FS	5.00 V	0111110100	0101 0000 0000
+ 1/4 FS	2.50 V	0011111010	0010 0101 0000
+ 1 LSB	0.01 V	0000000001	0000 0000 0001
ZERO	0.00 V	0000000000	0000 0000 0000

Note that in the Table only 9 straight binary bits are required to display the full-scale range (less 1 LSB) while 12 bits are required for the BCD code. There are a number of BCD code systems, the one presented here is known as 8-4-2-1. This is based on the relative weights of the binary digits and their corresponding digit values.

The binary notation described so far regarded all the input analog samples as uinpolar, i.e., only positively going pulses. In practical applications, however, the input sample can either be positive or negative going. Therefore, a sign must be introduced into the binary code representation which can distinguish between the positive and negative pulses. There are four basic code representations of the bipolar pulse samples: offset binary, signed magnitude, one's complement and two's complement. Table 2.4 demonstrates these codes, with each code expressed by 4 bits. Note that in most of the code formats the positive pulses are expressed by the same notation.

Table 2.4
Level representation of bipolar pulses employing commonly used codes

LEVEL NUMBER	OFFSET BINARY	POSITIVE-SIGN MAGNITUDE	NEGATIVE-SIGN MAGNITUDE	ONE's COMPLEMENT	TWO's COMPLEMENT
+7	1111	1111	0111	0111	0111
+6	1110	1110	0110	0110	0110
+5	1101	1101	0101	0101	0101
+4	1100	1100	0100	0100	0100
+3	1011	1011	0011	0011	0011
+2	1010	1010	0010	0010	0010
+1	1001	1001	0001	0001	0001
+0	–	1000	0000	0000	–
ZERO	1000	–	–	–	0000
-0	–	0000	1000	1111	–
-1	0111	0001	1001	1110	1111
-2	0110	0010	1010	1101	1110
-3	0101	0011	1011	1100	1101
-4	0100	0100	1100	1011	1100
-5	0011	0101	1101	1010	1011
-6	0010	0110	1110	1001	1010
-7	0001	0111	1111	1000	1001
-8	0000	–	–	–	1000

Offset binary, also known as bipolar offset binary (BOB) or offset straight binary (OSB), is the easiest binary to implement. It simply ignores the fact that the signal is bipolar and regards the most negative point of the full scale as the zero code notation. In other words, the maximum negative point is given the all zero code, i.e., 0000, while a different code is necessary for the ground (zero volt) level. The level of all zeros is only used as a reference and is not used for processing. The same applies for the most positive point of the full scale range, as this will call for code overflow, i.e. a carry bit to the next power of two. This means that the actual scale will be from 0001 to 1111, both representing the two extreme measurable points of the full scale, which are both less 1 LSB. The ground level voltage will therefore be represented by 1000 binary code.

Sign — magnitude code is a form of straight binary with the addition of an extra digit representing the polarity of the number. There are two forms of signed magnitude codes: in one the positive number is labelled with a 1 and the negative with a 0, and with the other form the notation is reversed. Both forms are commonly used although negative-sign magnitude is preferred. This enables all the positive operations to ignore the sign. Sign-magnitude would appear to be the best means of expressing analog samples. Unfortunately these codes have the drawback that they require two codes for expressing the ground (zero) voltage level, one for the positive going pulses and the other for the negative. This short-coming of the sign-magnitude codes makes them difficult for processing.

One's complement code has the characteristic that the sum of the positive and negative code expressions will always produce all ones. In other words, the positive code is a direct binary complement of the negative code, when both codes express the same magnitude. This code may seem ideal, although it again calls for two different codes to express the ground (zero) voltage level. That is, 0000 for the positive going pulses and its complement, i.e., 1111, for the negative going pulses. This disadvantage of the one's complement, means that besides the zero level ambiguity it is also difficult to implement.

The two's complement code is similar to the one's complement although the ground ambiguity is removed. This code has the characteristic that the sum of the positive and negative codes for the same analog sample magnitude of the polar pulses will always produce an all zero code together with a carry forward bit to the next power of two. In other words, the positive codes is the two's complement of the negative code when both express the identical magnitude. The ground (zero) voltage is expressed by an all zero code, with the positive going codes equal to that of unipolar straight binary. The negative codes are simply the two's complement of each of the positive codes, that is, the binary complement of the positive going code with the addition of one bit. If letter 'A' represents the positive going pulses, then the negative going pulses of identical magnitude will be $\bar{A}+1$. For example, take level +5 expressed by 0101, then level −5 will be expressed by 1010 + 1 = 1011. The two's complement is very useful for processing applications since most computers use this notation for their arithmetical operation

2.8 BINARY PULSE FORMAT

The task of data acquisition is to interface between the real world of analog parameters and the processing world of digital parameters. This transition from the analog format to the digital format is performed with the aid of the A-to-D converters, as already described. Once the analog signal has been digitized, the data can be transmitted to a computing centre for processing. The data obtained from the ADCs is in a form compatible with that of the computer, which means in effect, that the data is now in a standard code format of binary bits of either 1's or 0's only.

One would expect that the voltage pulse waveform representation of the binary data of 1's and 0's, will also have a standard format. This assumption is based on the fact that the data contents have already been constrained by the selection of the ADC scheme, the definition of the number of quantization levels and the rationalization of the digital code format. This assumption is not correct. In fact, there are many forms and shapes of the electrical pulse waveforms that may be used to represent the same binary bits of 1 and 0. The designer is presented with a large number of formats from which he can select; nevertheless, some of the formats are preferable than others. All these formats may be used to transfer the data from one location to another, all conveying the identical digital data contents, although each format has its distinctive pulse shape as required for the particular application.

All the pulse formats can be divided into three main classes, depending on the criterions used. Within each of the main categories, there may be a number of similar formats, to be described later.

(a) Number of binary states in the transmitted format of the pulse sequence:

1. Full binary sequence, where both the '0' and the '1' bits are part of the pulse patterns;
2. Half binary sequence, where only the 1 bits form the pulse pattern, having the 0's recognized by the absence of a pulse at the time of the clock;
3. Pseudo-ternary sequence, where three voltage levels may be transmitted, with each level representing a bit state, although still retaining the binary notation.
4. Multiple binary sequence, where each pattern level represents a number of binary bits.

(b) The number of possible pulse directions in the pulse stream pattern that can be considered or treated as a single bit and are adjusted in time:

1. Unipolar, where the pulses are only in a single direction.
2. Bipolar, where the pulses are in both directions relative to the centre level.

(c) The adjustment of the pulse pattern between two consecutive adjacent bits:

1. Return-to-zero (RZ), where there is a return to the zero level after transmission of each data bit. This format has a third steady state or rest condition in which the pulse returns during each pulse period;
2. Non-return-to-zero (NRZ), where there is no voltage level change between two consecutive digits in the bit stream if they possess the same binary state, although there is a level change when there is a change in the state pattern.

As already stated each of these pulse patterns is intended for a particular application. The points the designer has to consider when selecting a certain code pattern for his application are as follows:

(a) Simplicity in the construction of the pulse pattern from the bit stream;

(b) Error probability at the receiver side, in the discrimination between the two transmitted states;

(c) The stability of the pulse pattern of the code if the bits are in a continuous given state, such as when a stream of all 0's is transmitted.

That is, has the code pattern got a dc path and does it drift when a continuous single code group is transmitted. This point is extremely important when the transmitted stream passes through a transformer on a capacitor coupling.

(d) The probability of recovering the clock timing out of the transmitted pulse pattern. It should be noted that in most data transmission systems, the clock is not transmitted with the data stream. In these applications the clock contents are recovered by the transitions from one pattern level to the other.

(e) The band required to transmit the data stream which is a function of the ratio of the data contents and the number of transmitted pattern levels.

Each of the various pulse patterns for the respective code formats, which will be illustrated in the figures of this section, represents the same binary data of 0001 0111 0100 1101. (The formats are explained in this section, where the prefix corresponds to those shown in the illustrations.)

The most commonly used pulse formats are those in whic the full binary of both the '1' and '0' states are easily recognizable, as shown in Fig. 2.36, and are known as RZ bipolar and NRZ unipolar.

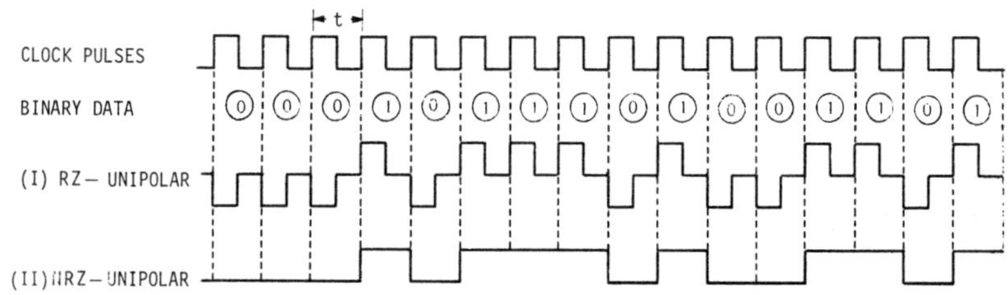

Fig. 2.36 Full binary formats pulse waveforms

(I) Return-to-zero, bipolar, where opposite polarity pulses are used to transmit the '1' and '0' bits. Between the pulses no power is transmitted, i.e. there is a space between each two adjacent pulses. This coding method is most popular for slow speed transmission and is used mainly for FSK of speeds up to 600 bits/second. It is most reliable, since mutilations of the pulses during transmission are unlikely, owing to the fact that the change in polarity is improbable without other causes. The clock timing can be easily recovered since there must be a pattern transition for each clock pulse, although this is at the expense of the transmitted band. That is, for each bit transmitted there must be two steps, which means the ratio of data contents to the number of transmitted pattern levels equals 1/2.

(II) Non-return-to-zero, unipolar, where the pulses are spread out in time so that they occupy a full time slot, thus permitting acceleration of transmission. The term NRZ is applicable here because the pulses lose their individuality in the process of transmission and do not return to zero between successive pulses of '1's and '0's. The data can be recovered from the pulse stream by the voltage level of the signal at the time of the clock pulse; a '1' is recorded if the level is high and a '0' if the level is low. This NRZ format is the most popular data-transmission format and is accepted as a standard in transmission speeds of 600/1200/2400 bits/second. The duration for the transmission of each bit is considerably reduced on average, thus resulting in higher transmission speeds than with the RZ format. This code format exploits the transmission bandwidth efficiently, as the entire bit period contains signal information. In other words the ratio of data contents to the number of transmitted pattern levels equals one. Since the number of level transitions is a function of the data contents and not of the number of bits, this NRZ format is not ideal for clock recovery. Its major limitation is the lack of a D-C centre path when consecutive pulses of all 1's or all 0's are transmitted.

The NRZ unipolar sequence is one of the most extensive pulse code formats employed. Although there are many non-return-to-zero patterns, the technical literature refers to NRZ-unipolar as simply NRZ. The reason for this popularity of the unipolar format is that most, if not all, of the computer waveforms are already in the NRZ-unipolar pattern. The serial ouput from any shift register produces this format, with no further manipulation required. In this case, the ground level represents the '0' state and the positive (or negative) voltage level represents the '1' state.

(III) One of the accepted methods for preventing errors in the transmitted sequence is the use of differential coding. In this method each transmitted bit is a result of the comparison of the present data bit with the previous transmitted bit, as shown in Fig. 2.37. This comparison is based on the formula that a '1' is transmitted if the present and previous bits are of opposite bit notations and '0' is transmitted if they are in the same bit notation. This formula is known both as exclusive-OR and modulo-two addition, with the operation given the sign of $\oplus$, i.e., $1 \oplus 0 = 0 \oplus 1 = 1$ and $1 \oplus 1 = 0 \oplus 0 = 0$. This method of coding can be used in any code format, although it is mainly employed with full-binary family of NRZ code formats. The pulse format obtained by coding the NRZ sequence is known as differential-non-return-to-zero (D-NRZ). It is important to note that the D-NRZ can also be regarded as a form of unipolar-NRZ representing the identical data bits as with the standard NRZ although in a coded form, and therefore it must be decoded again at the receiver.

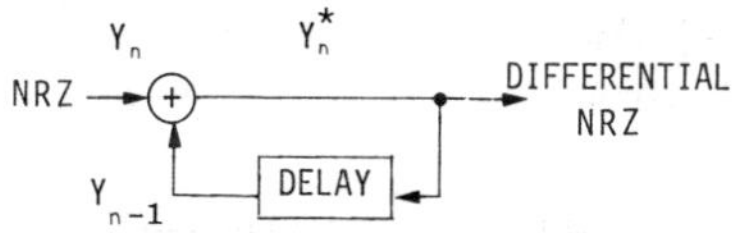

(a) DIFFERENTIAL CODING CIRCUIT

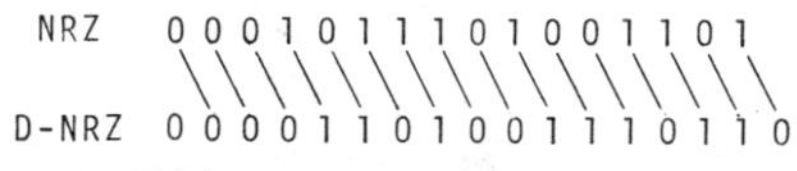

(b) DATA CODING

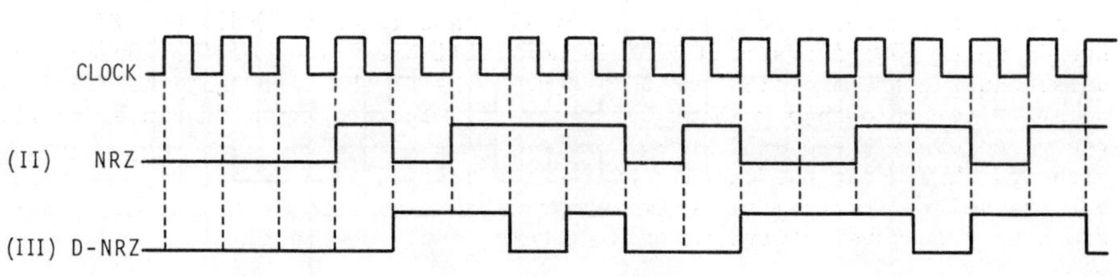

(c) PULSE FORMATS

Fig. 2.37 Differential–NRZ coding circuit and waveforms

When transmitting NRZ data pulses over long distances, the pulse stream must pass through some transformer coupling. This makes the state pattern no longer defined in relation to two fixed voltage levels, but in relation to the dc path, as seen in Fig. 2.38. That is, the '1' is defined

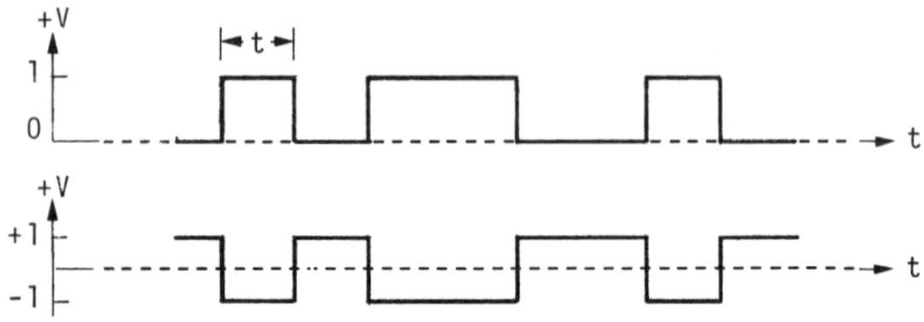

Fig. 2.38 NRZ format as DC level deviation

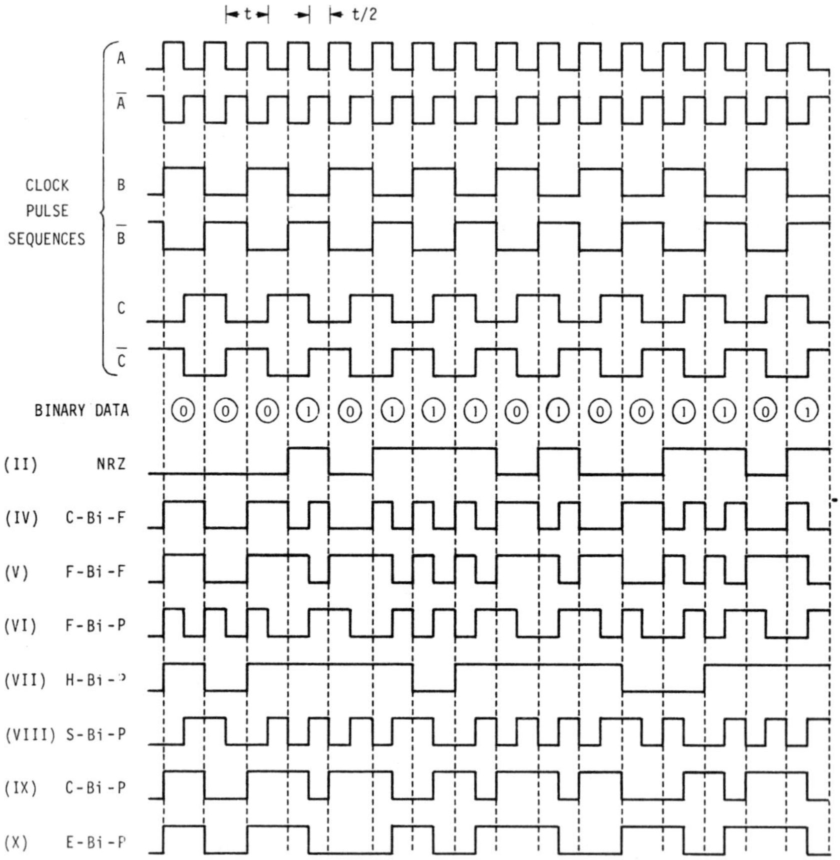

Fig. 2.39 Pulse format waveforms which prevent DC drift

as negative to the dc level and the '0' is defined as positive to the dc levels. These two new levels are specified as a unit distant from the dc level, i.e., they equal +1 or −1. It is not important which binary state is defined as the negative or positive level, although it is accepted in most disciplines as defining the 'one' as the negative level. The reason for this is that when passing the signal through an exclusive −OR circuit the operation can be regarded as a multiplication process, i.e., instead of $1 \oplus 0 = 0 \oplus 1 = 1$ and $0 \oplus 0 = 1 \oplus 1 = 0$ one can insert:

$$(-1) \times (+1) = (+1) \times (-1) = (-1) \quad \text{and} \quad (-1) \times (-1) = (+1) \times (+1) = (+1).$$

When transmitting data over long distances it is essential to ensure that the net average dc level will be zero, i.e., the transmitted waveform around the zero crossing is symmetrical. This imposes no problems when employing the NRZ format and transmitting the data at high speeds when it is constructed of both 1's and 0's, as seen in Fig. 2.38. Nevertheless, this requirement does present difficulties when only all 1's or only all 0's are transmitted in a sequence. When such a pattern is transmitted, it causes dc drift, bit slippage and prevents the clock recovery which consequently causes loss of receiver synchronization. Such a situation can occur during the idle time between message transmissions. If the synchronization is then lost, the heading of the message may not be decodable. In order to cope with this requirement, it is essential there should be at least one transition for each bit transmitted. There are many pulse formats that may be employed for these applications, some of which are shown in Fig. 2.39. It should be noted, however, that adding a transition for each transmitted bit only doubles the required transmission bandwidth.

The clock rate for all the pulse patterns shown in Fig. 2.39 is double the clock rate required for the NRZ format. That is, two clock pulses are required to code each binary bit, one at the leading edge of the bit period and the other at the centre of the period. These two basic clock pulses are used to form three different time pulses sequences, labelled A, B and C, as shown by the circuit given in Fig. 2.40(a), and used as means of coding the various formats. That is, for each pulse format, two time generated sequences are used, one to represent the 1 bit state and the other the '0' state bit. The actual encoder circuit, shown in Fig. 2.40(b), produces the new pulse format by combining the clock time sequences with that of the standard NRZ data sequence. Pulse formats produced by the combination of NRZ with two clock sequences, each with a different duration, are referred to as 'bi-frequency', while the combination with two clock sequences of equal duration (although different timing) is referred to as 'bi-phase'.

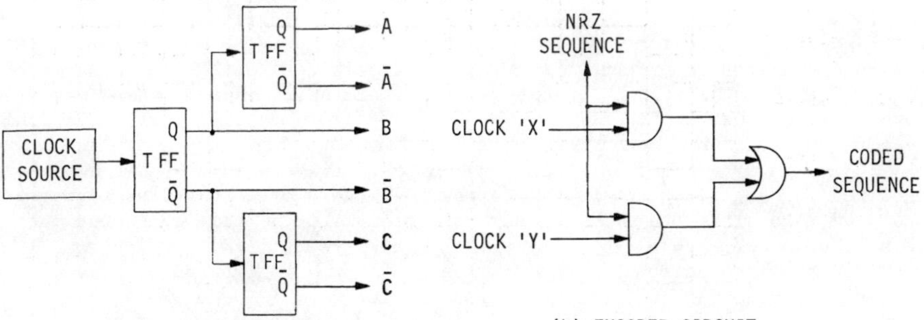

(a) CLOCK PULSE GENERATOR (b) ENCODER CIRCUIT

Fig. 2.40 Pulse generator and encoder circuit for drift free formats

(IV) Continuous-bi-frequency (C-Bi-F), is formed by having the data digits shifted between two frequencies, A and B, one for transmitting the '0' bit and the other for the '1' bit. The format has the characteristic that frequency phase must be continuous during the data transition: that is, the first 0 after 1 must follow the phase dictated by the 1, and vice-versa. In other words, each data bit, whether '1' or '0', must have a level transition at the leading edge of each bit period. While it is not important what frequency is allocated to each bit state, it is preferable that if the idle time between messages is 0, then this state will be allocated with the lower frequency, i.e., having one of the B clock pulses for the '0' state, leaving the A clock pulses for the 1 bit. This means, in effect, that each data bit which equals 1 is recognized by a full A frequency cycle with a level transition at the centre of the 1 bit duration. On the other hand there is no centre bit transition when transmitting a '0'. In this C-Bi-F format, the worst pattern for high speed transmission (i.e. one which requires the widest bandwidth) is produced when all 1's are transmitted, and the best pattern when all 0's are transmitted. In other words, this format does not suffer from any dc drift, but is limited by the requirement of a transmission bandwidth double that of the NRZ format.

(V) Fixed bi-frequency (F-Bi-F), has the same features as C-Bi-F, with a centre bit transition for all data bits which equal 1, and with no centre bit transition for the '0' bits, although the restriction that the phase must be continuous is removed. This enables the pattern to be formed by having the data bits keyed between two fixed frequencies A and B. In this modified format, the phase of the two frequencies remains the same throughout the coded pulse sequence, irrespective of the phase of the previous bit. It means, in effect, that it differs from the C-Bi-F format by not necessarily having a level transition at the leading edge of each bit duration. In this format the best pattern is achieved when alternate 1's and 0's are transmitted, while the worst pattern is produced when all 1's are transmitted.

(VI) Full-bi-phase (F-Bi-P). It follows that if it is possible to key the data bits between two frequencies, it should also be possible to key the data between two phases of the same frequency. Bi-phase format is formed by keying the data bits between the two clock pulses A and $\bar{A}$ in accordance with the bit state. In this coding system each bit exploits a full frequency cycle for each bit duration, with 180° phase shift for each data state change, either from 1 to 0 or from 0 to 1, while there is no phase shift when either all 1's or all 0's are transmitted. It is possible, therefore, to define the coded data pattern according to its phase, since it can be in only one of two phases, either 0° or 180°. This format is one of the most popular of the dc drift free pulse patterns and it is also referred to as the Manchester or split-phase format. This format assumes a zero level transition at the centre of each bit duration, regardless whether the bit state is a 1 or 0. The transition occurs during the split period, where the '1' reverses from positive to negative and the '0' from negative to positve. This feature of at least one level crossing for each bit duration simplifies clock recovery, although at the expense of increasing the total bandwidth required in comparison to the NRZ format. The best pattern in the format is produced when alternate 1's and 0's are transmitted, while the worst pattern is produced when all 1's or all 0's are transmitted. Owing to the reliable synchronization characteristics, this Bi-phase format is used for airborne to ground slow-speed data transmission.

(VII) Half-bi-phase (H-Bi-P). It follows that if a code format can be formed by the full cycle of clock A, it could also be formed by keying the data bits between the two clock pulses B and $\bar{B}$ in accordance with the bit state. These two clock pulse sources are each 180° phase shifted relative to each other, but have half the frequency rate of the A clock pulses. It is logical to assume that this format, too, will be dc drift free. However, contrary to the expectations this is not true. In this format, there is a 180° phase shift for each data state change, irrespective of the state of the data. Since the leading edge of each data bit duration coincides with the zero crossing of the NRZ pattern, these two transitions cancel out, making its characteristics similar to that of NRZ. That is, when alternate 1's and 0's are transmitted, there is no level transition.

(VIII) Shifted-bi-phase (S-Bi-P), is an improvement on the H-Bi-P format by preventing the two level transitions to cancel out. In this format the data bits are keyed between the two clock pulses C and $\overline{C}$ in accordance with the bit state. Selecting the C clock pulses instead of the B pulses presents a format which is dc drift free, but it still requires a wider transmitted bandwidth than that of the NRZ format. In this format the worst pattern is produced when alternate 1's and 0's are transmitted, and the best when all 1's or all 0's are transmitted.

(IX) Combined-bi-phase (C-Bi-P), is formed by keying the data bits between the two clock pulses B and C in accordance with the bit state. This format is a combination of both the H-Bi-P and S-Bi-P formats, although here the two clock pulses used for keying are only 90° phase shifted from each other. It has some of the wide bandwidth characteristics of H-Bi-P, although the danger of level transition cancellation is removed as in S-Bi-P. The worst pattern, however, occurs with the transmission of the last bit which equals to 1 coming before a 0 bit. This is the best pattern presented so far, although it still required double bandwidth when the worst case occurred.

A close study of the C-Bi-P pattern shows that it is characterized by having a level transition at the leading edge of the bit period for all those bit states which equal '0' and a level transition in the centre (mid) period for all those bit states which equal '1'. When recovering (decoding) the data back to the NRZ format, it is possible to look only at the leading edge of the centre of the bit period to define the state of the bit. In other words, if one only looks at the centre of the period, it is possible to neglect the transitions at the leading edge and still decode the data reliably. This means, in effect, that if the leading edge of the first 0 after 1 is omitted by some means while keeping all the other format characteristics, the data contents can still be retained, as explained in the following format.

(X) Extracted-bi-phase (E-Bi-P), is produced by extracting from the C-Bi-P format all those level transitions which are important for the data-state contents or for the dc drift free performance, while omitting all those level transitions which limit the transmission bandwidth. The C-Bi-P sequence is, therefore, first differentiated and then passed through a full-wave rectifier. this produces a sequence of narrow pulses, corresponding to all the level transitions. By delaying the NRZ sequence by half a period and adding it to the transition sequence, it is possible to exclude the leading edge of the first 0 after 1 while retaining all the other transitions. The remaining sequence can then be fed into a *T* flip flop resulting with a new format pattern which has the same bandwidth as the NRZ format but is completely dc drift free. It is possible to produce other E-Bi-P formats extracted for the C-Bi-P format by forming the latter from clocks B and $\overline{C}$, $\overline{B}$ and C or $\overline{B}$ and $\overline{C}$. This 'ideal' format is also known as the 'Miller format.'

In a summary of the various dc drift free formats, Table 2.5 presents the condensed results. The transmission bandwidth is limited by the inverse bit duration, whereas the dc drift danger occurs when the period between two level transitions is long. These two conflicting conditions can be dealt with only by compromise.

Table 2.5
DC drift characteristics of frequency and phase combination formats

CODE FORMAT	CLOCK CONTENTS	DATA BIT DURATION		PATTERNS TO ACHIEVE HIGH SPEED	
		MIN	MAX	WORST	BEST
NRZ	A	t	∞	Alternate 1's & 0's	All 1's or all 0's
C–Bi–F	A* & B*	t/2	t	All 1's	All 0's
F–Bi–F	A & B	t/2	3t/2	All 1's	A 1 after a 0
F–Bi–P	A & $\overline{A}$	t/2	t	All 1's or all 0's	Alternate 1's & 0's
H–Bi–P	B & $\overline{B}$	t	∞	All 1's or all 0's	Alternate 1's & 0's
S–Bi–P	C & $\overline{C}$	t/2	t	Alternate 1's & 0's	All 1's or all 0's
C–Bi–P	B & C	t/2	3t/2	The last 1 before 0	The last 0 before 1
E–Bi–P	B & C	t	2t	All 1's or all 0's	A 0 after a 1

In half binary formats only the bit states which equal 1 are represente by a pulse or by polarity transitions, whilst the 0 bits are seen as spaces. Nevertheless, the 0 bits are recognized by the absence of a pulse at the tim of the clock period. The philosophy behind this format is based on the stati tical conception that since the number of 1's in a long pulse sequence is equal to the number of 0's, therefore the reduction of the number of pulse results in the reduction of the transmitted power with the possible increase in transmission speed. That is, the restriction of the pulse format to repre sent only the 1's limits the minimum pulse density which ensures sufficien energy to operate the receiver extraction circuit. The omission of the 0 bit transmission, however, enhances the clock timing recovery from the transmitted pattern. Therefore, these half binary formats are generally only used where

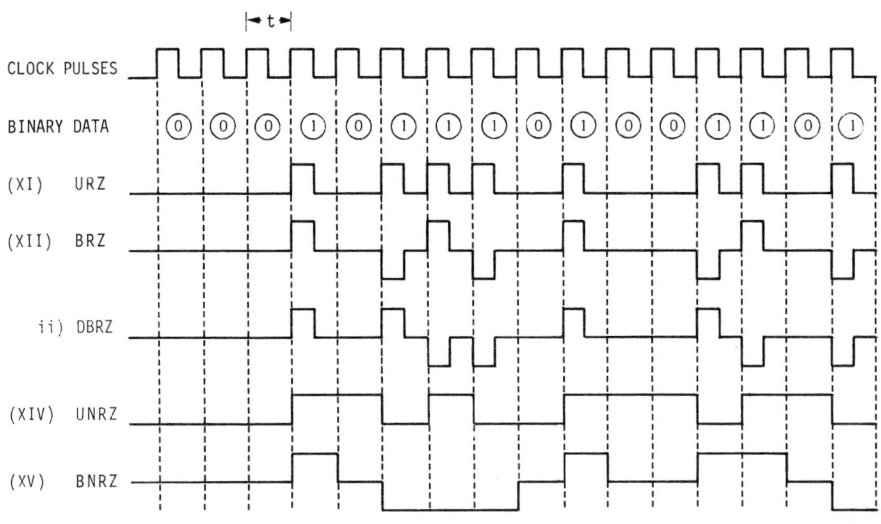

Fig. 2.41 Half binary and pseudo ternary formats pulse waveforms

the clock timings are transmitted separately, such as in a parallel PCM audio transmission using frequency-division-multiplex techniques. For this reason, these formats are also known as PCM formats.

In pseudo-ternary sequences three voltage levels may be transmitted, +1, 0 or −1, each representing a different bit state. It is called pseudo-ternary since these formats do not transmit three-state-bit information but pure binary information, with the +1 and −1 levels representing all the bits which equal 1 and the centre level representing the 0 bits.

Figure 2.41 presents a number of half binary and pseudo-ternary formats, Some of the formats presented could come under both headings, while others are distinctive, belonging to one category only.

(XI) Unipolar, return-to-zero (URZ), which is the same format as the bipolar return-to-zero pattern, except that the 0 pulses have been eliminated. In other words, the pulse pattern presents a pulse for each bit which equals a '1' and no pulse if the bit equals a '0'. This code format modifies the frequency spectrum by reducing the high frequency signals which cause cross-talk. However, it also produces an unstable dc level which is difficult to transmit. This format method also wastes 50% of its bandwidth, since only half of its bit period contains bit data. It is not recommended to employ this format for transmission purposes, although the format itself is extremely useful in obtaining other half binary and pseudo ternary formats.

(XII) Bipolar, return-to-zero (BRZ), is based on the NRZ format, except that the polarity of each consecutive pulse is reversed. By inverting the alternate '1' pulses, the dc level is well established and the inter symbol interference is reduced. This code also facilitates error detection, since all that is required is to remember the polarity of the last pulse received. In other words, any single error will result in two pulses of the same polarity without an inverting pulse of opposite polarity between them. However, with the '0's not transmitted, it gives no indication of the location of the error. This code format is most popular for PCM audio transmission at high speeds and is known as alternate-mark-inversion (AMI). This format which is based on half binary can be regarded as pseudo-ternary.

Although the BRZ format reduces the danger of dc drift when all 0's are transmitted, the lack of any transmission pulse will restrict the low frequency contents of the transmitting signal, which consequently causes the loss of the receiver synchronization. If no parallel timing is transmitted with the BRZ signal, then it is necessary to introduce preventive measures. If in a stream of data bits there are four consecutive bits which equal 0, then a substitution pattern is inserted instead. This substitution is arranged so that it is easily recognized for decoding purposes, though with a pattern that does not violate the alternative polarity pulse. This is achieved by employing four substitute bits, two of which are positive and two are negative. The first pulse in the substitutive group must always be of the same polarity as the last data bit. If the last data bit was +1, then the zero substitution for four bits will be +1, −1, −1, +1 (or alternatively +1, +1, −1, −1) and if the last data bit was −1, then the zero substitution is −1, +1, +1, −1 (or −1, −1, +1, +1). These two possible zero substitutions for the half binary BRZ format are shown in Fig. 2.42. It is important to note that each set of substitution bits is complementary of each other and it assists in the decoder with the zero substitution extraction. In the format described and illustrated, only $n = 4$ zero bits were substituted. This leaves, in any long sequence of only all 0's, room after substitution of 3 adjacent 0 bits. There are other zero substitution code formats which have the common name of 'bi-polar n-zeros substitution' formats (BnZS). The format just presented is consequently known as B4ZS.

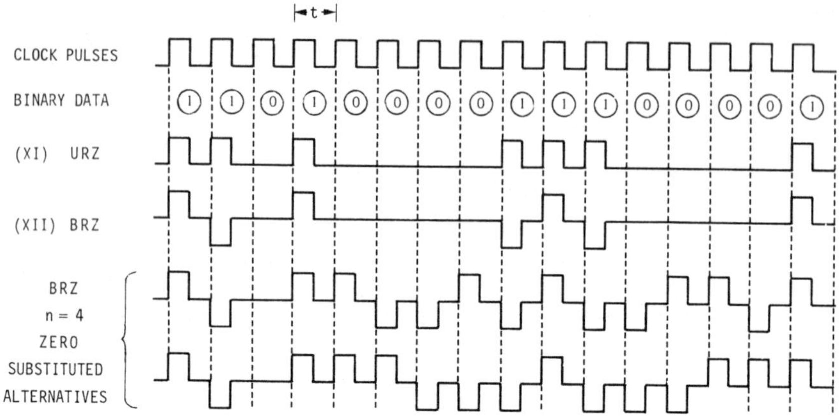

Fig. 2.42 Zero substitution format pulse waveforms

(XIII) Double bipolar, return-to-zero (DBRZ), is another form of the BRZ format. This code, suggested by M. Karnaugh, extends the basic BRZ pattern, with the difference that the polarities are inverted only after two consecutive '1' bits. The format pattern further reduces the inter-symbol interference, thus promoting better utilization of the bandwidth.

(XIV) Unipolar, non-return-to-zero (UNRZ), is a half binary format which efficiently utilizes the bandwidth. It is formed by reversing the signal polarity each time a '1' bit is transmitted. No polarity level transition takes place when a '0' bit is transmitted, regardless of the transition direction prevailing as the last '1' bit. This means in effect that the format presentes difficulties in clock timing recovery and therefore requires external timing synchronization. This format which is also known as NRZI (non-return-to-zero invert) format, is widely used for application of tape recording, but it may also be used for data transmission applications.

(XV) Bipolar, non-return-to-zero (BNRZ) is a pseudo-ternary format with the three levels representing binary information and is known as 'duobinary' format code. This code format has the advantage of a narrower spectrum and thus is intended for increasing the transmission speed. It is used with three-level-frequency shift keying and is explained in detail in section 6.4. In this format, the bits which equal the '0' bit-state are represented by the zero level, and the bits which equal '1' are represented by a bipolar form of +1 or −1. The polarity of the levels representing the 1's is reversed only if the number of consecutive bits which have the value '0' between any two bits with the value '1', is odd. If the number of consecutive '0' bits total up to an even number, the succession of the two spaced '1' bits remains the same polarity.

Multi-level sequences are a means of better utilization of the transmission bandwidth. The allowable bandwidth controls the number of symbols per second which can be transmitted. In most data transmission systems hitherto discussed each bit corresponds to one transmitted symbol. By having one symbol signal to represent a number of bits, the speed of transmission could be increased. However, a penalty must be paid for higher data rate, as the data is now more vulnerable to noise and other distortions. This is because individual states are more difficult to distinguish in a multiple level method than in a two level method.

There are many multi-level code formats; for example, a ternary code which replaces the binary code. This code could be transmitted by a three level as in an NRZ code. This method is not widely used because the circuitry involved is rather complicated. The popular method is to have the number of levels equal to a binary scale, i.e., two, four, eight, sixteen...

(XVI) Non-return-to-zero, quadrature binary (QNRZ), is a four level code format where each level represents two binary bits, that is 00, 01, 10, and 11 respectively. The two bit pair is known as a 'dibit'. As illustrated by the pulse waveform in Fig. 2.43, the pulse sequence duration of each dibit can be transmitted at the same times as a one bit pulse. In practice, however, the dibit is transmitted at speeds less than double that of a bit, since the speed is limited by spectrum disturbances which reduce the received threshold.

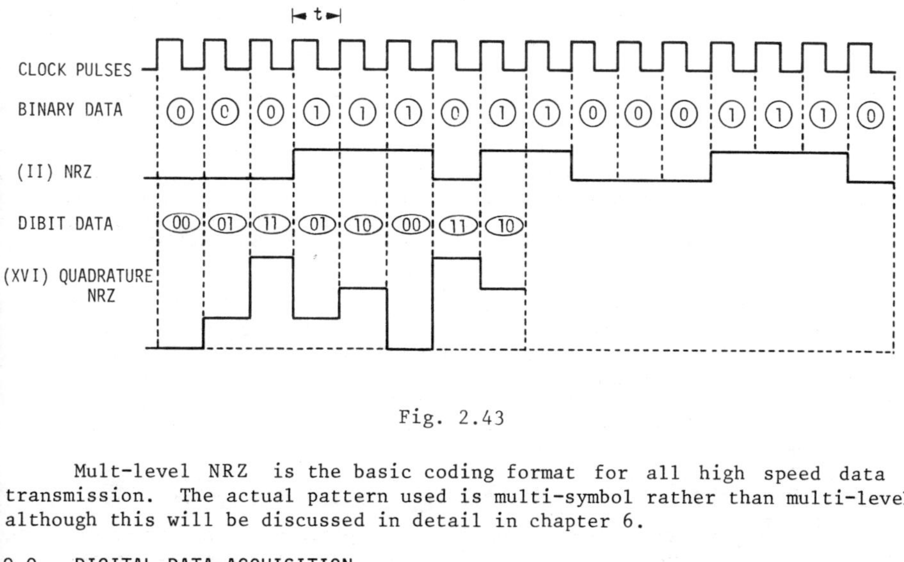

Fig. 2.43

Mult-level NRZ is the basic coding format for all high speed data transmission. The actual pattern used is multi-symbol rather than multi-level although this will be discussed in detail in chapter 6.

2.9 DIGITAL DATA ACQUISITION

Whereas analog data originates from the natural physical world, digital data can be regarded as part of the 'artificial world'. Digital data acquisition is usually executed with the aid of manually operated terminals, which could be either in the form of teletypewriters, keyboards or CRT terminals. The data produced by these terminals is already in digital pattern and in the format ready for transmission, with no further processing needed. The digital format may be in any of the binary pulse formats described in the previous section, depending on the requirements of the communication channel. The data acquired either from analog sources or manual digital sources is in a form that is available for processing in a computer.

As stated above, the digital acquired data from the manually operated termanals are already in a binary pattern and therefore no further conversion processing is necessary. The digital data can, if so, be fed directly to the transmitter, be conveyed to the computer. Nevertheless, the sequence format used for digital data is different from that used for analog data. While binary-coded-analog (BCA) is produced in a continuous form of binary bits, digital source data is produced in discrete groups of binary bits. In

BCA sequence pattern, each group of bits represents one sample and is immediately followed by a second group representing an adjacent sample. The difference between any two adjacent samples is very small, a characteristic which enables the binary-coded-analog (BCA) data to be compressed. In digital source data, each group of bits represents a unique character of information, with one group unrelated to the other group and therefore this data cannot be compressed.

The digital data source originates from manually operated terminals which are usually slow and where each time the operator presses a keyboard a single character is transmitted. A character is a set of binary bits which represent elementary symbols which normally include both alpha and numerical codes plus punctuation marks and any other symbol which may be read, stored or written. The size of the character could be 5, 6, 7 or 8 bits depending on the code symbol convention used. It is not the intention here to present all the various conventions available, but only to point out their existence. Each keyboard or teletypewriter uses a particular accepted code. Fig. 2.44 presents an example of a code format known as ASCII (American Standard Code for Information Interchange). This code consists of 7 bits labelled b_1 to b_7, with each symbol character defined by the coordination of four by three bits. When pressing any of the 128 possible keys or their combination, a 7 bit digital group is transmitted representing either figures or letters. Other character symbol conventions worth mentioning are the Baudot telegraph code of 5 bits per symbol and the EBCDIC (Extended Binary Decimal Interchange Code) which is a form of BCD code format extended to 8 bits.

BITS												
b_7 →					0	0	0	0	1	1	1	1
b_6 →					0	0	1	1	0	0	1	1
b_5 →					0	1	0	1	0	1	0	1
b_4 ↓	b_3 ↓	b_2 ↓	b_1 ↓	Column → / Row ↓	0	1	2	3	4	5	6	7
0	0	0	0	0	NUL	DLE	SP	0	o	P	`	p
0	0	0	1	1	SOH	DC1	!	1	A	Q	a	q
0	0	1	0	2	STX	DC2	"	2	B	R	b	r
0	0	1	1	3	ETX	DC3	//	3	C	S	c	s
0	1	0	0	4	EDT	DC4	δ	4	D	T	d	t
0	1	0	1	5	ENQ	NAK	%	5	E	U	e	u
0	1	1	0	6	ACK	SYN	&	6	F	V	f	v
0	1	1	1	7	BEL	ETB	´	7	G	W	g	w
1	0	0	0	8	BS	CAN	(	8	H	X	h	x
1	0	0	1	9	HT	EN	)	9	I	Y	i	y
1	0	1	0	10	LF	SUB	*	:	J	Z	j	z
1	0	1	1	11	VT	ESC	+	;	K	[	k	{
1	1	0	0	12	FF	FS	,	<	L	\	|	¦
1	1	0	1	13	CR	GS	—	=	M	]	m	}
1	1	1	0	14	SO	RS	.	>	N	^	n	~
1	1	1	1	15	S1	US	/	?	O	——	o	DEL

Fig. 2.44 ASCII code table matrix

There are two main types of manually operated terminals, those which transmit a character at a time and those which transmit a block of characters. The data transmitted by character terminals are extremely slow, with each transmitted character preceded by a start bit and ended by one or two stop bits, as seen in Fig. 2.45. For this reason these terminals are commonly known as start-stop terminals. The receiver can detect the beginning of the character and determine the character symbols by counting the rest of the bits. The character bits can then be stored in a buffer which is checked when the stop bit is detected.

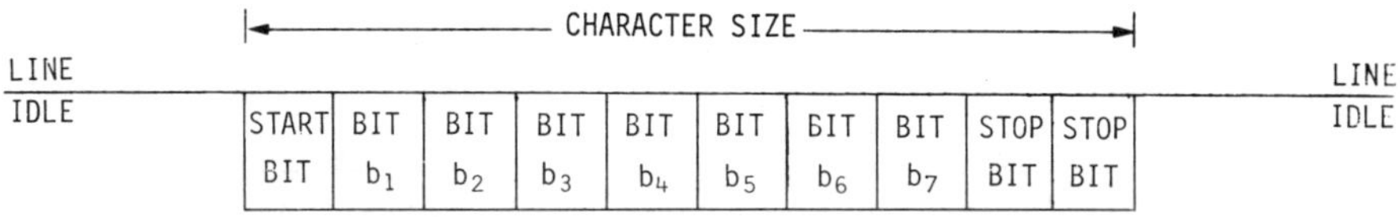

Fig. 2.45 Digital format of a transmitted character symbol

Most of the terminals today need not transmit their data a character at a time, but can transmit large bulks of data, that is, a whole block of characters handled and transmitted as a single unit. Although basically these terminals also have keyboards and are operated in the same manner as teletypewriters, they have the advantage of memory capacity and processing ability. The operator may still type its message a character at a time although the contents are retained till the whole message is complete and only then is it transmitted in one block. Most of these terminals also contain a visual display (CRT) whereby the operator can visually edit the message before transmission and change items accordingly. The speed of transmission of these terminals is no longer limited by the manual, typist operation but may be as fast as the communication channel allows it. Such terminals therefore need not be limited to the character start-stop format but may employ much more efficient techniques. All the characters now form a message and are transmitted in one block, as seen in Fig. 2.46, with no gaps between the characters. Here too, the message is headed by a start flag signal and ended with a stop flag. The flags are intended to be used in the receiver to assist in identifying each of the characters. Although the flag's characters could be compared with the start-stop bits described before, the channel utilization here is high, since the overhead of the flag bits relative to the message bits is small.

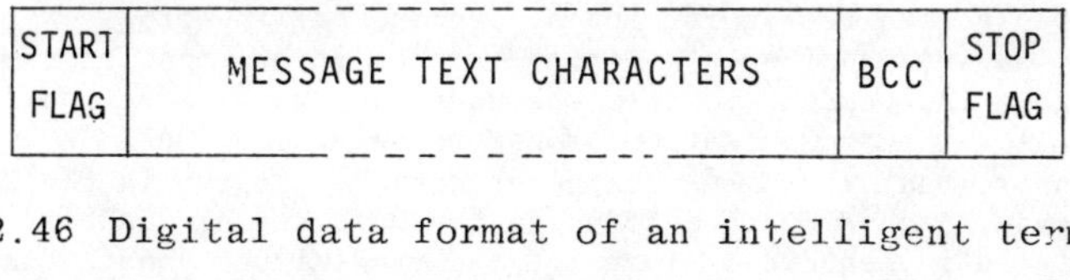

Fig. 2.46 Digital data format of an intelligent terminal

A message is defined here as an arbitrary amount of characters whose beginning and end are defined or implied. In character transmission the size of the character is known and fixed; this is not the case in block transmission. To ensure that no characters or bits are lost while being transmitted, a BCC (Block Count Character) code is added before the stop flag which specifies the number of characters in the message block.

With message transmission the whole unit is transmitted in series as a single block leaving no gaps between adjacent bits or characters, although leaving intervals between adjacent messages. This calls for, as already stated, special control characters to be added to each message, not only to identify its beginning and its end but also to ensure its efficient handling in the equipment along the communication path. This consequently makes it necessary to insert into the transmitted message means of distinguishing the data bits from the control bits. Such a provision is known as data transparency and it is achieved by special data link control (DLC) protocols.

There are two general DLC protocols, character oriented protocols and bit oriented protocols, as seen in Fig. 2.47. Both these classes of protocols achieve transparency by having the source terminals adding special characters or bits at the header and trailer of the message, which the receiver terminal removes.

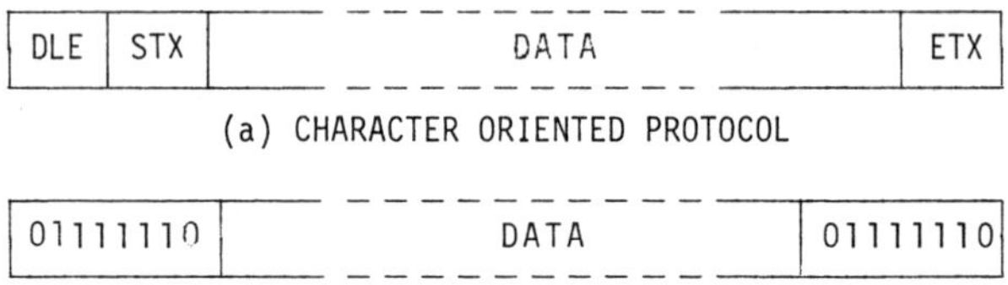

(a) CHARACTER ORIENTED PROTOCOL

(b) BIT ORIENTED PROTOCOL

Fig. 2.47 Data link formats for achieving transparency

In character oriented protocols (COP) a definite set of control characters effect the orderly operations of transmission of the messages over the communication channels. Formats which may be classified in this category are the ASII and EBCIDIC already mentioned as well as the BSC (binary synchronous communication protocol) and DDCMP (digital data communication message protocol). Examples of these code characters (were given in Fig.2.44) are DLE (data link escape), STX (start of text) and ETX (end of text). The message text itself consists only of characters of fixed size and this type of message format is used in the majority of data links in operation today. Such a format places a burden on hardware and software necessary to distinguish between the data and control characters within the transmitted block,

Bit oriented protocols (BOP) are more flexible and their implementation are more straightforward, which makes them efficient formats for computer type communication. Among the BOP formats are SDLC (synchronous data link control), ADCCP (advanced data communication control protocol) and HDLC (high-level data link control). With BOP only two characters are used for control operation of the channel, leaving the data text to be bit-oriented rather than character-oriented. These two characters of eight bits sequence 01111110 , are used to signify the header and tailing flags of the message frame. The advantage of such a format is that it provides transparency operation with text independence, where the user can choose a bit pattern without any concern for the communication procedure being employed.

To ensure that the flag character does not appear in the data portion of the message, a zero must be added to the data stream whenever five ones are transmitted in sequence. That is, the transmitter counts the data bits as they are dispatched and if it detects a five ones in a row, i.e. 011111..., it automatically adds a zero, i.e. 0111110... The receiver also counts the bits and automatically removes the added bit after each sequence of consecutive five ones within the message text. Such an operation is known as 'bit-stuffing'. This advantage of BOP formats makes it, however, unsuitable for use in asynchronous data transmission since the procedure there is based on fixed length characters which are destroyed with bit-stuffing. BOP are intended for high speed synchronous data transmission facilities.

2.10 REFERENCES

(a) ANALOG DATA ACQUISITION AND TELEMETRY

1. Foster, L.E. *Telemetry Systems*, John Wiley and Sons, 1965.
2. Martin, J. *Telecommunications and the Computer*, Prentice-Hall Inc., 1965, pp.265-292.
3. Zuch, E.L. *Principles of Data Acquisition and Conversion*, Digital Design, Pt.1, May 1979, pp.60-68, Pt.2, July 1979, pp.30-36. Pt.3, September 1979, pp.72-83.

(b) ANALOG TO DATA CONVERTERS

4. Sheingold, D.H. *Analog-to-Digital and Digital-to-Analog Converters*, Digital News, December 1972, pp.50-56.
5. Analog Devices, Inc., Publication *Data Acquisition Products Catalog* 1978, pp.381-470.
6. Sheingold, D.H. and Ferrero, R.A. *Understanding A/D and D/A Converters*, IEEE Spectrum, September 1972, pp.42-56.
7. Gersho, A. *Quantization*, IEEE Communication Society Magazine, September 1977, vol.15, no.5, pp.20-29.
8. Milojkovic, D. *Data Conversion with Companding DAC Devices*, Advanced Micro Devices, Inc., Publication, February 1978.
9. Cuccia, C.L. *The Handbook of Digital Communication*, Microwave System News, pp.30-33.
10. Talambiras, R.P. *Digital-to-Analog Converters: Some Problems in Producing High-Fidelity Signals*, Computer Design, January, 1976, pp.63-69.
11. Soderquist, D. *A low Cost, Easy-to-Build Successive Approximation Analog-to-Digital Converter*, Application Report AN-11, Precision Monolithics Inc. Publication.
12. Soderquist, D. and Schoeff, J. *Low Cost, High Speed Analog-to-Digital Converter with DAC-08*, Application Report AN-16, Precision Monolithics Inc. Publication.

(c) PULSE MODULATION

13. Bayless, J.W., Campahilla, S.J. and Goldberg, A.J. *Voice Signals: Bit by Bit*, IEEE Spectrum, October 1973, pp.28-34.
14. Beards, P.H. *Pulse Code Modulation for Broadcast Signal Distribution*, Electronics and Power, June 1979, pp.425-430.
15. Steele, R. *Partial Removal of Transmission Errors in Digitally Encoded Speech*, Electronics and Power, June 1978, pp.451-454.
16. Rosenthal, L.H., Rabiner, L.R., Schafer, R.N., Cummiskey, P. and Flanagan, J.L. *Automatic Voice Response: Interfacing Man with Machine*, IEEE Spectrum, July 1974, pp.61-68.
17. Schindler, H.R. *Delta Modulation*, IEEE Spectrum, October 1970, pp.69-78.
18. Maschoff, R.H., *Delta Modulation*, Electro-Technology, January 1974.
19. Flanagan, J.L. Schroeder, M.R., Atal, B.S., Crochier, R.E., Jayant, N.S. and Tribolet, J.M. *Speech Coding*, IEEE Trans., vol. COM-27, no.4, April 1979, pp.710-736.
20. Chakravarthy, C.V. and Faruqui, M.N. *Amplitude-Controlled Two-Loop Adaptive Delta Modulator*, Proc. IEE, vol.126, no.4, April 1979. pp.285-290.
21. Steele, R. *Chip Delta Modulators Revive Designers Interest*, Electronics, October 13, 1977, pp.86-93.
22. Gosh, S. *Code Conversion Techniques for Digital Translation*, Computer Design, August 1978, pp.103-111.
23. Schwartz, M. *Information Transmission, Modulation and Noise*, McGraw Hill Book Co., 1959, pp.116-117.

24. Ristenbatt, M.P. *Alternatives in Digital Communication*, Proc. IEEE, vol.61, no.6, June 1973, pp.703-721.

(d) DIGITAL CODE FORMATS

25. Morris, D.J. and Alexander, W. *An Introduction to the Ternary Code Number System*, Electronic Engineerings, September 1960, pp.554-557.

26. Horton, A.W. *An Introduction to Computer Binary Arithmetic*, Bell Telephones System, Monograph 3050, 1956, 29 pages.

(e) PULSE CODE FORMATS

27. Croisier, A. *Compatible High-Density Bipolar Codes: An Unrestricted Transmission Plan for PCM Carriers*, IEEE Trans., vol.COM-18, no.3, June 1970, pp.265-268.

28. Johannes, V.I. *Comments on Compatible High Density Bipolar Codes: An Unrestricted Transmission Plan for PCM Carriers*, IEEE Trans. vol.COM-20, no.1, February 1972, pp.78-79.

(f) DIGITAL DATA ACQUISITION

33. Hobbs, L.C. *Terminals*, Proc. IEEE, vol.60, no.11, November 1972, pp.1273-1284.

34. *USA Standard Code for Information Interchange* published jointly as X3.4 - 1968 and Federal Information Standard 1 (FIPS-1).

35. Murphy, D. *Digital Techniques in Data Communication*, Telecommunications, February 1969, pp.19-25.

36. Weissberger, A.J.W. *Data-Link Control Chips: Bringing Order to Data Protocols*, Electrics, June 8, 1978, pp.104-112.

37. Conard, J.W. *Character-Oriented Data Link Control Protocols*, IEEE Trans. Vol.COM-28, no.4, April 1980, pp.445-454.

38. Carlson, D.E. *Bit-Oriented Data Link Control Procedures*, IEEE Trans., vol.COM-28, no.4, April 1980, pp.455-467.

(g) TECHNICAL TERMS

39. Jay, F. (Editor-in-Chief), *IEEE Standard Dictionary of Electrical and Electronic Terms*, IEEE and Wiley Interscience, 1977, 882 pages.

CHAPTER 3

Data Communication

3.1 INTRODUCTION

In the previous chapter a description was given of the means of acquiring the data and coding them into digital form. This chapter will deal with the means required to enable this data to be transferred to the computer and the problems that must be overcome to enable the coded digital data acquired to reach its desired destination without distortion.

The two extreme poles of the command and control operation loop are data acquisition and data processing. For the data collected in the field stations to be processed, it must be transported to the processing site, where it is regenerated and fed into a computer. The connection between the two poles of the C^2 system is via a labyrinth of communication networks, as seen in Fig. 3.1, The communication medium, whether wire, cable or radio is simply a highway over which the data travels. Nevertheless, the data communication systems are possibly the most important feature of the command and control system. The data, at the acquisition site, is already in digital form ready for processing into useful information needed for decision making. The communication channel, however, can distort this data while it is transported by introducing errors into its coded contents whereby a transmitted 1 may be received as a 0 or a 0 as a 1.

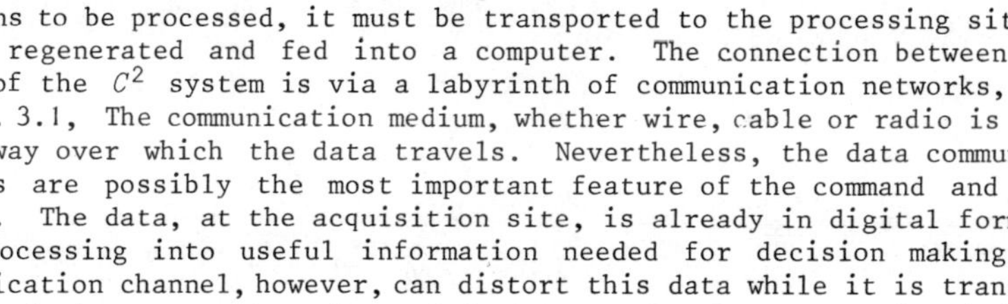

Fig. 3.1 The position of data communication in the C^2 system loop

The command and control system requires expensive computers and other sophisticated hardware equipment. Yet, it has been found that one of the most expensive items required in running a command and control system (and also the most unreliable) is the complex of communication channels which is not actually part of the data processing equipment.

Data communication involves the transmission of discrete coded pulses over a communication channel. For this purpose the communication channel serves to establish the communication between two points over which digital coded messages can be forwarded in a determined direction, as shown in Fig. 3.2. The communication channel may present either a physical or only a virtual link connection between the two points. In both cases a transmitter and a receiver are required at both ends of the communication channel. The data transmitted over a single communication channel must always be in a serial format, whereas, if parallel data is to be transported, separate transmitter-receivers are needed for each channel termination. The terms transmitter and receivers are used here in their broad sense, that is, as devices required in assisting in transferring the data from one location to another.

Fig. 3.2 The data communication basic formation

Data communication is usually in both directions between the two points of the channel, and then a transmitter and a receiver are necessary at both ends of the channel. The data transmitter and the data receiver are generally referred to by the general term of data-communication (or circuit-terminating) equipment (DCE). The data source, where the data is collected, and the data sink, where the data is transported to, are known by the general term of data-terminal-equipment (DTE). The communication channel with its associated terminating equipment is known as data link. These definitions are illustrated in Fig. 3.3. The design goal of a data communication system is the conveyance of data from one DTE to another DTE with the data link only acting as a transparent media.

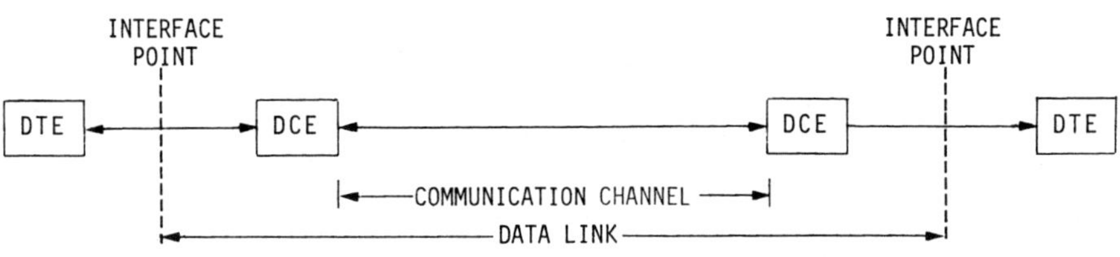

Fig. 3.3 The general units required in a data communication system

3.2 THE PROBLEMS OF MOVING DATA

Data acquisition, whether the data originated from an analog, digital or state source, produces a series of pulse patterns of voltage levels changing as a function of time. When transmitting these pulses over short distances, the voltage pulses are simply transmitted over twisted copper wires with a transistor driver as the transmitter and a transistor amplifier as the receiver, as shown in Fig. 3.4(a). These pulses are, in fact, transmitted by simply turning the dc voltage level on or off; nevertheless this may require considerable power when the voltage levels change at high speeds. As the distances over which the pulses are transferred increase, the pulse shape is distorted and other means are necessary. One of these means that may be employed is the use of current pulses, with the receiver amplifier closing the current loop, as seen in Fig. 3.4(b). Other means may employ the use of

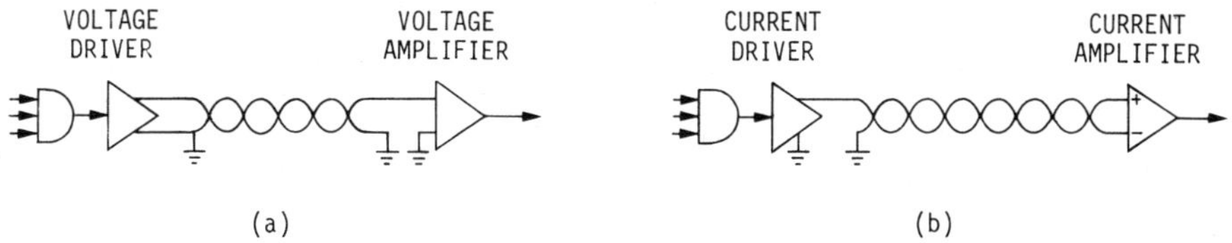

Fig. 3.4 Voltage and current pulse transmission

differential pulse transmissions, where each of the two balanced wires convey the same data contents, although complemented. At the receiving end, the amplifier responds only to the differential signal.

With all the practical techniques it is impossible to transmit narrow pulses (i.e. fast changing pulse patterns) over long distances and reconstruct them again at the receiving end. Rectangular pulses travelling through a communication channel get distorted by the channel capacitance, inductance and leakage. The twisted copper wires, with either current loop or differential pulse transmission, have limited distance beyond which the pulses are completely 'smeared-out'. The changes in the pulse waveform shape, as seen by the example given in Fig. 3.5, may be received in such a form that the reconstructed data is totally different from that which was transmitted.

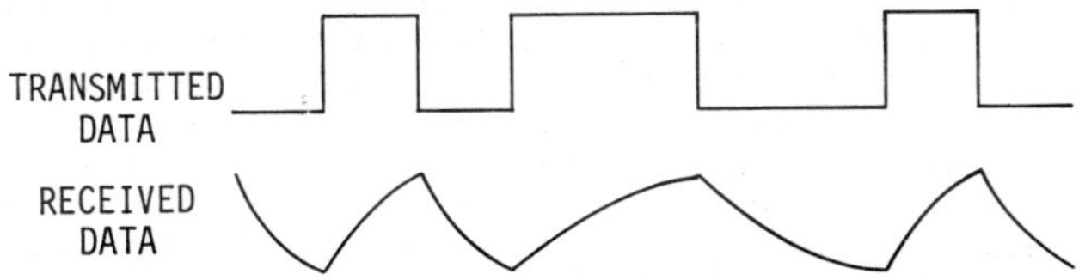

Fig. 3.5 Distortion of pulse data in a transmission channel

Instead of transmitting the data in the form of electrical level pulses, the data is converted into analog form. Analog signals are superior to pulse signals, since analog signals carry the data in the form of waves which are continuously changing as a function of time, whilst the pulse signals assume various fixed voltage levels which may change instantaneously. The frequency spectrum of an analog wave may be relatively narrow while the steady levels with the instantaneous changes requires an extremely wide frequency spectrum. For these reasons, the pulse stream is converted into analog by having the transmitter modulating the digital data into a continuous analog wave with the receiver demodulating it again into pulse form. In spite of the fact that data in the communication link is analog, the data link is still transparent, since the data at both link terminals is in the same pulse form.

As already stated, most data links may transmit and receive in both directions over the communication channel. In other words, each end of the communication channel is terminated both by transmitter and receiver equipment, The link termination equipment, which can both MOdulate and DEModulate the data is therefore known by its abbreviated form of 'modem' (Fig. 3.6). Although modem is the most popular name for the data communication equipment (DCE), they are also called data-set, data-termination and data terminal. (The latter must not be confused with data terminal equipment, DTE, which is the equipment comprising the data source, the data sink, or both.)

Although the long communication channels transport analog wave signals they do present a potential source of distortion of the data, whereby a

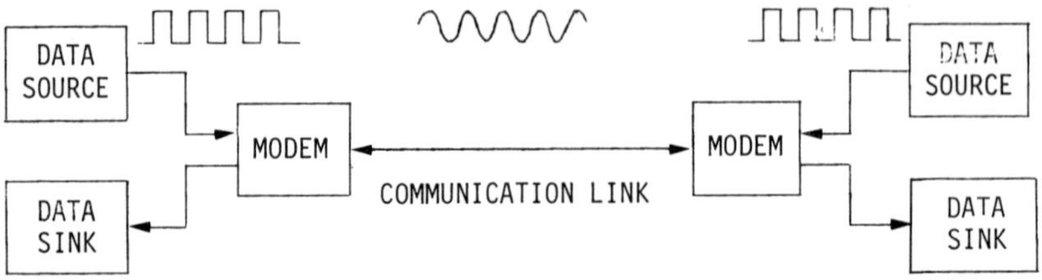

Fig. 3.6 The MODEM communication system configuration

transmitted 1 can be regenerated as a 0 and vice-versa. The inevitable interference source is primarily due to noise, attentuations and delays, as seen in Fig. 3.7. The noise is the worst cause of external disturbances and may be both of background (or white) noise which is characterized by random Gaussian type impulses that have a flat frequency spectrum and impulse noise which is characterized by sudden high peaks transient disturbances separated in time. While it is possible to visualize the performance of white Gaussian noise it is impossible to predict impulse noise. Both delay and attenuation are not uniform over the frequency spectrum and must be compensated accordingly. These channel interferences impose severe restrictions both on the speed and the distances that the data may be transported over the communication link.

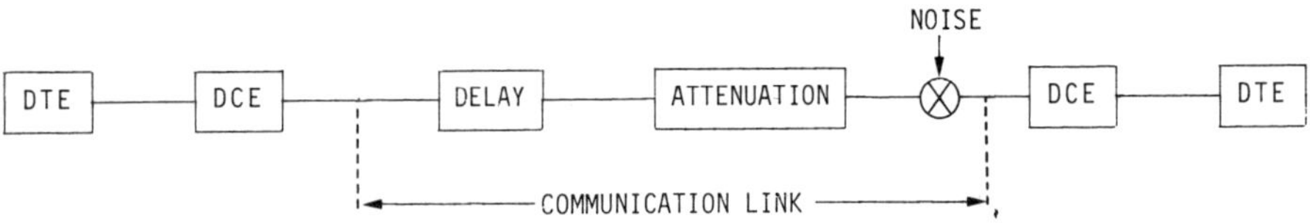

Fig. 3.7 The disturbance threats in the communication links

It is the principal objective of the data link design to transfer the maximum amount of data with the least possible number of errors. In other words, the intended purpose of data communication design is to improve the data transmission quality and efficiency between two termination points by overcoming the inevitable link interferences.

3.3 DATA TRANSMISSION MODULATION

The fact that a series of rectangular looking pulses is made into continuous type waveforms does not make it analog, though seemingly so. Analog data implies continuity, in contrast to digital data which is concerned with discrete states, The information contents of an analog signal is conveyed by the magnitude of the signals. This is not the case here since the continuous analog wave is concerned with discrete states. For this reason the analog waves can be regarded as analog-coded-binary (ACB). This means that the transmitted signal in the link has analog characteristics useful for communication although it is still a form of binary data. This also means that the ACB signal does not necessarily require an ADC (as described in the previous chapter) to convert it back into its original pulse form but may employ much simpler techniques.

The purpose of the modem is to exploit the analog communication channel for digital transmission, that is, its function is to perform the conversion from binary pulses to ACB and from ACB back to binary. In principle, the essential elements of the modem follow the block diagram shown in Fig. 3.8. The binary data signals are fed into the modem serially and they are first encoded into a pulse form for efficient transmission over the

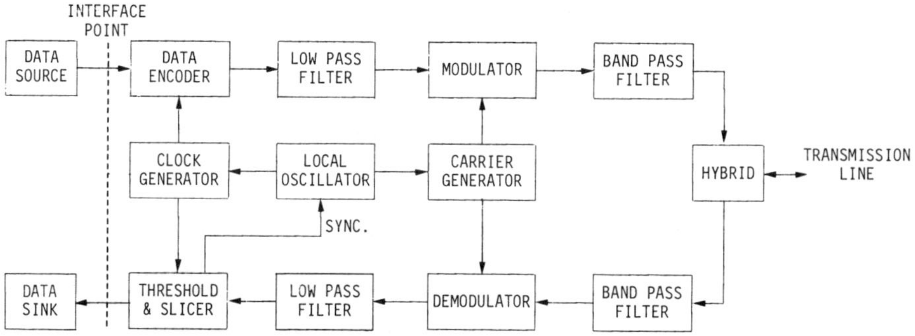

Fig. 3.8 The basic block elements of the MODEM

communication channel. The sharp instantaneous edges of the pulse level transition are then smoothed out by means of a low pass filter (LPF). The encoded pulses are modulated by a carrier signal having the resultant waveband limited by means of a band-pass filter which keeps the transmitted signal within the boundaries of the link frequency spectrum. The binary data is restored by a complementary process. After demodulating the signal it is sliced and reshaped back into pulse form.

There are three basic types of modulation which are used for data communication: Amplitude Modulation (AM), Frequency Modulation (FM) and Phase Modulation (PM). A pure sine wave, $\sin \omega t$, known as the carrier, is used as a modulation signal, whereby some of its characteristics are varied in accordance with the binary data stream. The pure sine wave can be made to carry data by varying one of its three parameters, amplitude A, frequency f or phase ϕ. In the pure sine wave carrier these three parameters are constant and can therefore be represented by

$$S(t) = A \sin(2\pi f_c t + \phi)$$

where $S(t)$ is the instantaneous value of the carrier voltage at time t, $A = 1$ the maximum amplitude of the carrier, $f_c = \omega/2\pi$ is the carrier frequency, and $\phi = 0°$ is the phase angle.

The coded binary pulse stream acts as a modulating wave which varies either the amplitude, the frequency or the phase of the carrier signal, as shown in Fig. 3.9. In amplitude modulation, the amplitude A of the sine wave carrier signal gets the values of $A = 0$ for the binary 0 state, and $A = 1$ for the binary 1 state, in accordance with the data stream contents. In frequency modulation, the frequency f_c of the carrier signal gets the values of f_1 for the 0 state and f_2 for the 1 state in accordance with the data. The two frequencies f_1 and f_2 are obtained by shifting the carrier frequency up or down by a given frequency deviation ΔF, i.e., $f_1 = (f_c - \Delta f)$ and $f_2 = (f_c + \Delta f)$. In phase modulation, the phase angle θ of the carrier signals gets the values of $\phi = 0°$ for the 0 state and $\phi = 180°$ for the 1 state in accordance with the data contents. A more detailed analysis of these modulation techniques in given in chapter 6.

Signal modulation involves the process by which the carrier characteristics vary in accordance with a modulating wave. The pulse data stream has the quality of being normally at constant level values for a period of time and then being subject to an instantaneous transition to another level and remaining there for a period. For this reason data modulation is better known today as data shift keying (DSK). Keying is a form of modulation process which involves the selection from a finite set of discrete states.

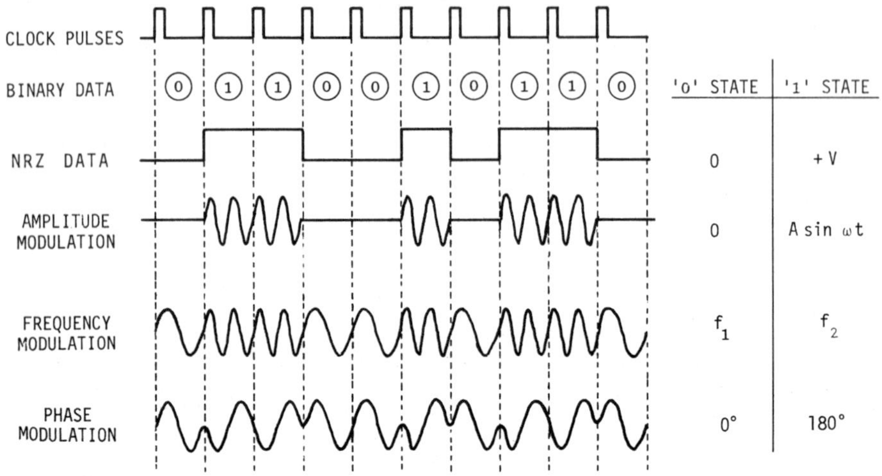

Fig. 3.9 The three basic methods of modulating a sine wave by a pulse data stream

Shift keying the carrier signal between two amplitude levels in accordance with the binary states is known as amplitude shift keying (ASK). Shift keying the carrier signal between two frequencies is known as frequency shift keying (FSK) and between two phase angles as phase shift keying (PSK).

ASK is the simplest scheme to implement, while PSK is the most difficult to realize. ASK, on the other hand, is very susceptible to noise impulses which can change the amplitude level, and thereby changing the data contents. Both FSK and PSK transmit the modulated signals with constant amplitude. For this reason FSK is the most popular scheme for most modems operating at medium transmission speeds, while PSK is used for faster transmission speeds.

When the amount of data transmitted in the communication channel is relatively small, the modulated sine wave carrier signal can carry the binary data contents by defining the two binary states according to the two characteristical values of the shifted carrier. For example, an ASK wave may be of high or low magnitude in accordance with the binary data of 0 or 1. The capacity of data communication link is limited by the frequency spectrum of the link. In data communication systems operating up to about 1200 bit/second over voice grade telephone lines, the economy and simplicity of the modem operation are generally of greater concern than the maximum frequency bandwidth utilization. Voice telephone lines having a nominal 3 kHz bandwidth could theoretically transmit up to a capacity of 3000 pulses per second. In practice, however, the available bandwidth is significantly reduced by the line characteristics which limit the capacity of the lines to a maximum of approximately 2400 bits/second. Increasing the transmission speeds is achieved by using non-binary coding techniques, i.e., each transmission pulse may have the data contents of more than one bit (binary digit).

As already been shown in the previous chapter (Fig. 2.43) each pulse in the data stream can be made to convey more than one bit. With a dibit, each pulse level conveys two bits while with a tribit each pulse load conveys three bits. In the case of dibit, each pulse can be in one of four possible levels, while with a tribit, each pulse can be in one of eight possible levels.

Grouping the bits in the data stream into dibits, tribits or even four-bits allows the capacity of the communication channel (i.e. the transmission speed) to be theoretically doubled, tripled or even redoubled again. Each of these groups can then be used to modulate one of the carrier parameters. In this way ASK may be transmitted by selecting the carrier amplitude as one of a number of possible levels to serve as a function of binary-group contents, and PSK may be transmitted by shifting the carrier signal between a large number of phases angles, say every 90° or even 45°. These forms of multi-array modulations are known as MASK and MPSK. Another method of packing more digits in a single modulation period is to vary more than one of the carrier signal parameters, for example, the variation of two amplitude levels and two phase angles. This means that for each transmitted modulation signal there are four possible symbols formed by the combination of the two amplitude levels and two phase angles. In other words, these combination modulation symbols can represent two dibit contents. These forms of modulations are known as APSK.

When transmitting with multi-array modulation schemes the receiver must also be designed to detect the transmitted signal. Reliable techniques are known for the detection of MASK, MPSK and even APSK. With frequency keying it is theoretically possible to shift the data between three discrete frequencies by using the carrier frequency signal and its two deviation frequencies to transmit the three tribit states. Nevertheless, shifting the frequency between more frequency positions creates technical problems in detecting them. The are modulation techniques which combine FSK and ASK but then further restrictions are imposed on the frequency shift characteristics. Such a modulation scheme is known as MAMSK, and will be described further in chapter 6.

Most of the modems operating above 2400 bits/second use multi-level techniques. The most common method of achieving this is by the employment of quadrature-phase carrier signal. That is, two carrier signals are used which ar orthogonal to each other (i.e they are 90° apart). If one of the carriers is $\sin 2\pi ft$ then the other is $\cos 2\pi ft$. The principle block diagram of such a modem is given in Fig. 3.10. The data is fed serially into the modem, where it is converted to parallel, enabling two bits to be modulated simultaneously.

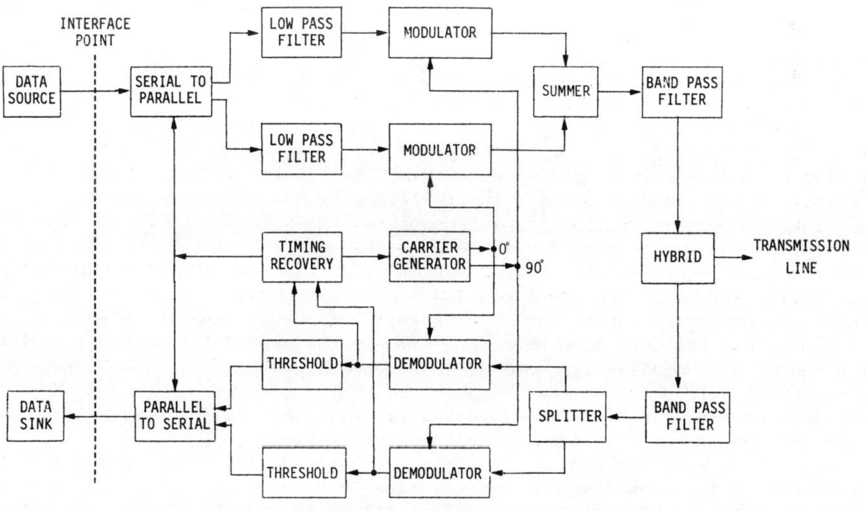

Fig. 3.10 Schematic diagram of a quadrature MODEM

3.4 COMMUNICATION SIGNAL RECEPTION EVALUATIONS

The use of modems with multilevel transmission symbols calls for sophisticated means of receiving the transmitted signals and above all, regenerating them back to their pulse binary data format with no errors. The communication channel, as already stated, is susceptible to noise, delay and attenuation, which cause the transmitted pulse shape to be distorted with a transmitted '1' detected as '0' and a transmitted '0' as '1', as seen in Fig. 3.11. The decision whether the regenerated pulse after modulation is a '1' or a '0' is performed by threshold level detection. That is, if the regenerated signal at the time of the clock signal is positive to the centre level the signal is defined as a '1' and if it is negative it is defined as a zero. Such a process is known as 'hard decisions'. Two important factors are therefore necessary for the decisions: to define when is the right timing for looking and what is the allowed threshold.

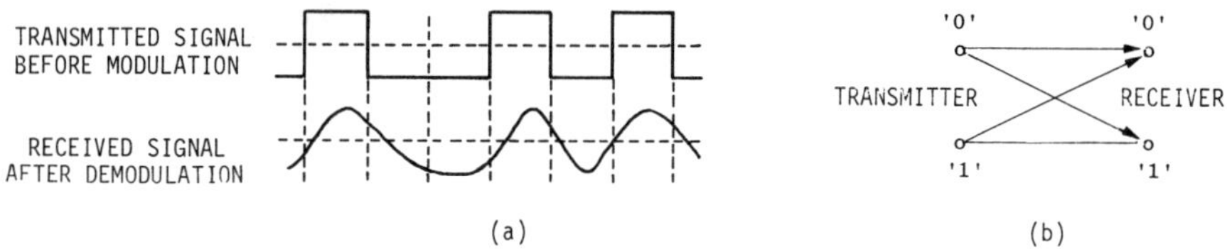

Fig. 3.11 Received distorted signals may be decoded wrongly

A useful concept in evaluating the performance of a communication channel and hence the quality of the received demodulated signal is 'eye pattern'. The eye patterm, shown in Fig. 3.12, is formed by superimposing the various demodulated signal waveforms of random data. Such an eye pattern can be obtained experimentally with an oscillator displaying the demodulated signal of a random sequence stream, synchronized to the clock timing source. The upper sequence shown in Fig. 3.12 is based on ideal conditions while in most cases the received signal is distorted, causing the eye pattern to 'close', as shown by the lower sequence.

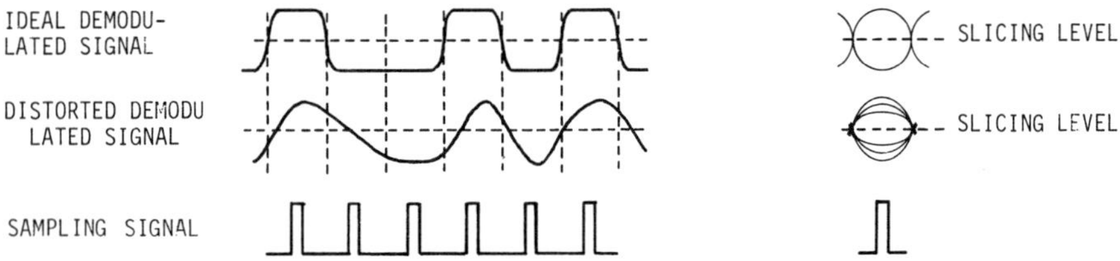

Fig. 3.12 The concept of the eye pattern in evaluating the demodulated signal

The vertical eye opening at the time of the sampling pulse defines the available threshold levels, while the horizontal eye opening defines the available timing tolerances for sampling the results. The slicing level defines the demarcation line between the result being decoded as a 1 or as a 0. The sampling time is at the centre of the clock intervals. These points are better illustrated in Fig. 3.13. It is obvious that as the sample period is widened, the available threshold decreases.

When transmitting a multi-level signal, such as a four level AM scheme, the eye pattern consists of three vertical eye openings, as seen in Fig.3.14. In this case, each transmitted symbol represents more binary bits and hence

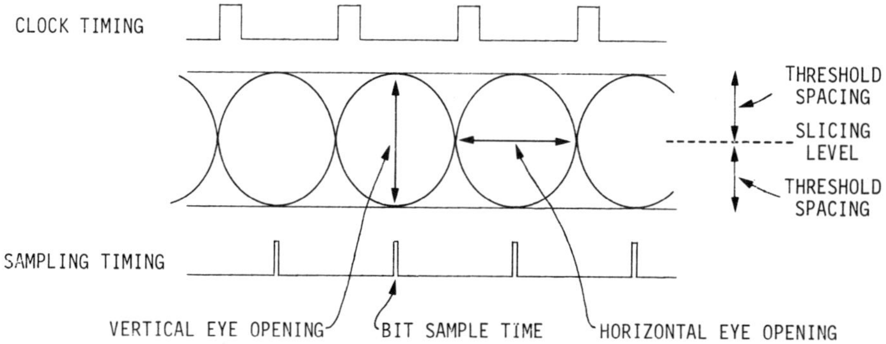

Fig. 3.13 Eye pattern of an ideal demodulated signal

more than one error may result when the symbol is distorted. As the number of levels increases, both the horizontal and vertical eye opening are compressed, thereby increasing the probability of errors. With the presence of external channel disturbances, such as noise, delay and attenuation, the eye pattern opening could be much worse. For this reason, multiple symbol transmission is generally limited to a maximum of 4 or 8 symbol levels.

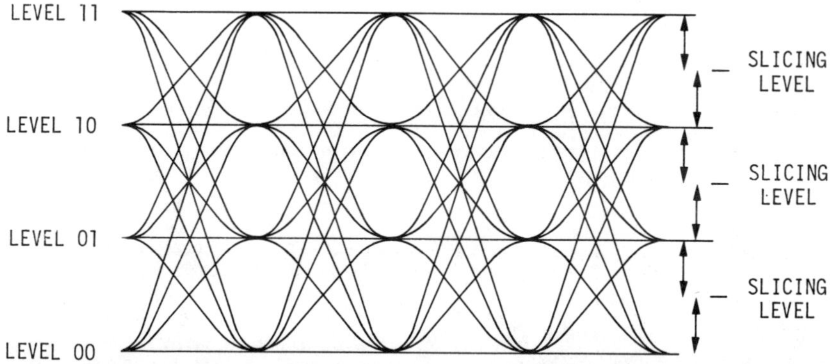

Fig. 3.14 An ideal eye pattern for four level transmission

One means of improving the transmitted speed is achieved by the employment of mixed modulation techniques. In such a case, each technique may have only four levels, hence the combination of two techniques can produce sixteen different symbol possibilities. This means that in each transmitted time period four binary bits may be transmitted in a group. This combination has increased the transmitted speed although now each error can affect 4 bits. It is of advantage then to use the Gray code instead of the pure binary code, so as to reduce the symbol error probability to only one bit.

When mixed or even multiple phase modulation techniques are employed, the eye-pattern concept is not of much assistance. The reason for this is that it is most difficult to represent both amplitude and phase variation of each transmitted symbol state. Instead, it is preferable to have a vector-space diagram concept which graphically represents the demodulated signal peaks as points on a Cartesian or polar diagram. In this type of modified pattern only the maximum peak points of the demodulated signal at the instantaneous sampling time are exhibited. The eye pattern, as already explained, is formed by superimposing the instantaneous amplitude values of the received demodulated signal and observing them on a full clock time period. The maximum eye opening points which provide the maximum threshold spacing are obtained at the centre of the clock period. The space pattern exploits only these extreme points since they exhibit most of the required information. For example, if a four level amplitude modulation technique is employed, four horizontal points are formed at the sampling time, as seen in Fig. 3.15. That is, each point represents the maximum magnitude of A at the sampling time, with the transmitted signal being $S(t) = A \sin 2\pi f t$.

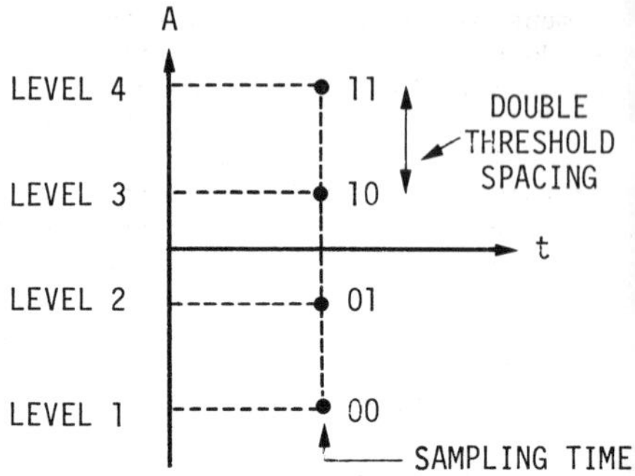

Fig. 3.15 Maximum eye opening points in a 4-level ASK

In the case of orthogonal carrier modulation techniques, as used in multiple phase and in mixed modulation, a two dimensional diagram is employed with the magnitude of one carrier signal used for the diagram abscissa and the other carrier signal for the diagram ordinates (i.e., the horizontal and vertical coordinations of the diagram). The geometrical interpretation of the two demodulated signal maximum peaks are therefore plotted as points on a two dimensional space diagram. That is, each of the diagram coordinates exhibits the instantaneous magnitudes A and B of the two orthogonal carrier signals, with the transmitted signal being $S(t) = A \sin 2\pi f t + B \cos 2\pi f t$. Each of the space points on the diagram corresponds to all the pairs of the maximum demodulated signals at the sampling time. If, for example, A and B get the values of ± 1, then four space points are formed, as shown by the pattern in Fig. 3.16.

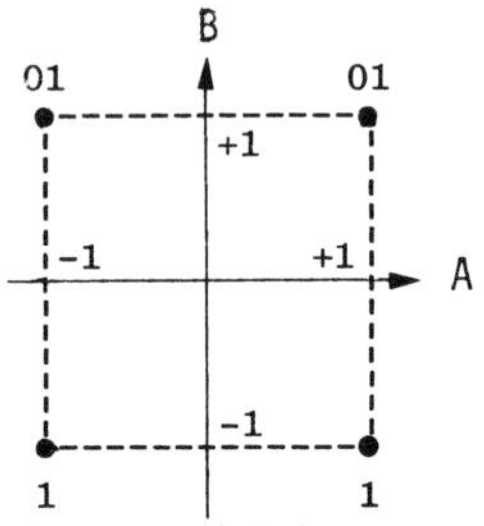

Fig. 3.16 Space pattern for a four symbol state modulation

Other space patterns can be formed by varying both the amplitude and phase of the transmitted quadrature carrier signal (as will be explained in detail in chapter 6). In the example illustrated in Fig. 3.17, a 16 symbol-state modulation technique is exhibited, formed by 4 level variations of each of the two orthogonal carrier signals. The modified eye pattern for mixed modulation technique represents the maximum eye opening points of each symbol and is exhibited on a two

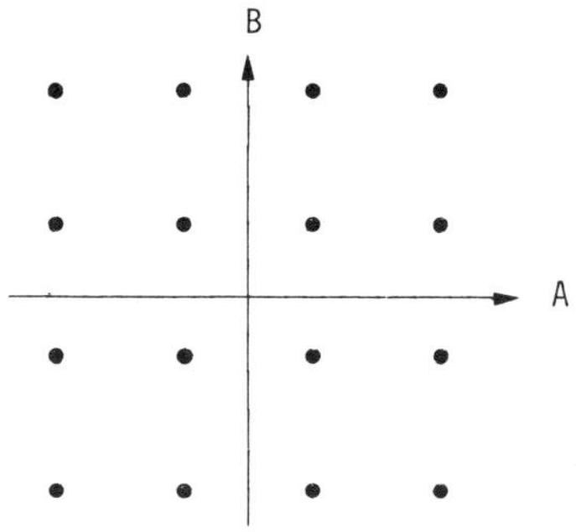

Fig. 3.17 Constellation pattern of a 16 symbol transmission

dimension space signal diagram. The space pattern of the demodulated signal is known as a 'constellation' pattern. The constellation points may be defined either by the coordination on a Cartesian diagram or by the phasor shifts on a polar diagram, or as a combination of both.

The constellation pattern concept serves the designer with the same usefulness as the eye pattern concept. In the eye pattern, the horizontal eye opening representes the maximum available threshold for specifying whether the symbol is in one binary group state or in another. In the constellation, the vector created by the coordination of the two orthogonal signals, defines the same threshold spacing. In general, it can be broadly said that the maximum spacing between the closest two constellation points defines the maximum available threshold spacing for specifying the binary group state of the symbol.

The constellation pattern represents the best achievable limits of the demodulated signal, with all the other communication impairments neglected. Additive noise to the received signal will cause the random displacement of the constellation points from their predicted position. Any amplitude variations in the received signal will be exhibited on the constellation pattern as changes in the vectors' magnitude. That is, the vector changes are relative to its centre of gravity and not to its position in relation to its closest other constellation points which are not on the same vector axis. Any phase jitter variations in the receiver demodulated signal will be exhibited as circular rotation changes about the original in the position of the vector. That is, the vector changes are relative to its phase angle and not to its closest constellation points, which are not on the same circle. These phase and amplitude variations are illustrated in Fig. 3.18. There are some communication systems which are sensitive to amplitude distortions, while others are more sensitive to phase distortions.

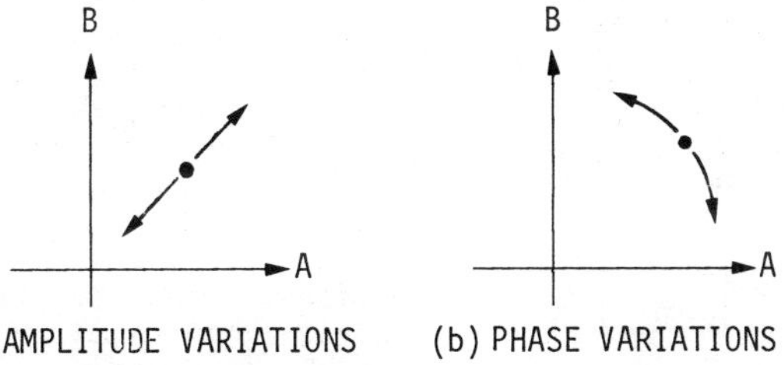

Fig. 3.18 Constellation point change due to additive noise

With very high speed transmission, means are inserted into the modems to assist in defining whether demodulated constellation points are in one position or another. It is obvious that the essential parameter to define is the slicing axis between any two constellation pattern points. This slicing axis can generally be presented to lie in the centre spacing between the two constellation points which are being considered. This is not always the case, as the probability of distortion may affect one constellation point more than the other. (Take for example, a constellation pattern, two points at which are on the same vector line. Transmitting the larger amplitude signal which is associated with the outer vector point, requires more signal power and is more susceptible to noise than the inner point. There is less probability that a demodulated signal which should be associated with the inner vector point, i.e., near the centre of gravity, will be assumed as being associated with the outer point. There is more probability that the opposite would occur, that is, that the demodulated signal would be assumed as the inner point although it is associated with the outer point). In this case we are assisted by software decisions which take into account the statistical probability of error.

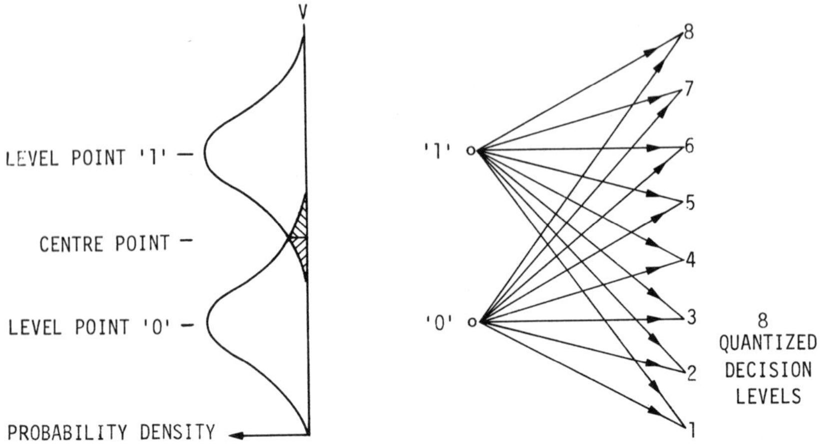

Fig. 3.19 Software decisions based on quantizing the probability density

It is not enough only to define the axis and the allowable threshold spacing, since in practical conditions the centre axis is not well defined and the probability density of the results being in one constellation state point rather than the other may be overlapped, as seen in Fig.3.19. In each constellation spacing the error probability density is the highest near the centre axis than near the pattern points.

The threshold spacing between the two constellation points is divided into a discrete range scale with each range given a weight corresponding to the probability of its being in one of the two points. Each demodulated result at the sampling time is therefore associated with this weight. It is obvious that the most probability of errors lies near the centre slicing axis and decreases as the distance from the centre increases. Smaller weights are therefore given to the results near the centre. The software decisions are then performed with quantizing techniques coupled with error correction techniques. The transmitted data includes redundant data which assists in checking the received data stream. If an error is detected, the weighting principle can locate the low and high probability of the demodulated signal.

3.5 COMMUNICATION SIGNAL SHAPE CONSIDERATION

The error probability of the received signal is not due only to external disturbances in the communication channel, such as noise, but also to internal disturbances in the transmitter. These internal disturbances caused by the signal shape could reduce the eye opening, and are brought forward by the bandpass limits of the communication channel. An ideal rectangular binary signal has a frequency spectrum extending to infinity, as seen in Fig. 3.20. Perfect reception of this pulse requires therefore a channel with infinite bandwidth. The main problem in data communication is to define how much bandwidth restriction is allowed without introducing excessive impairments, and thereby producing unacceptable errors at the receiver.

The communication channel places the data pulse energy completely within the frequency band of the channel by means of some form of efficient modulation technique. When transmitting an incidental data pulse stream it requires a baseband signal having a rectangular spectral shape and a linear phase characteristic, as shown in Fig. 3.21. The amplitude response is unity at 0 frequency and it stays unity up to frequency f_b where it has a sharp

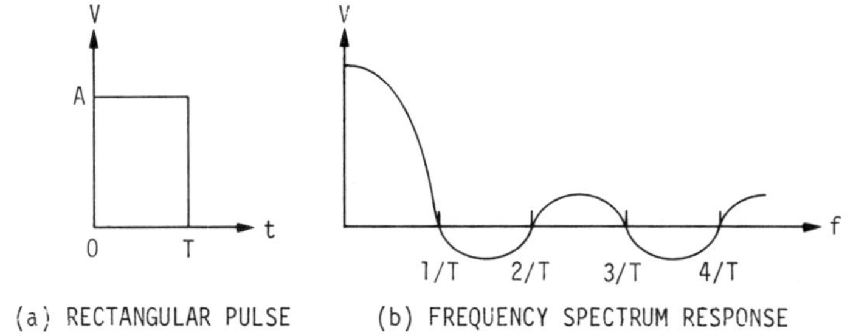

Fig. 3.20 Frequency response of the transmitted rectangular data pulse

cut-off. The cut-off frequency is equal to the baseband, i.e. is equal to $f_b = 1/2T$ where T is the internal clock pulse duration. The phase response is linear over the same range. The amplitude modulation frequency band is double that of the baseband.

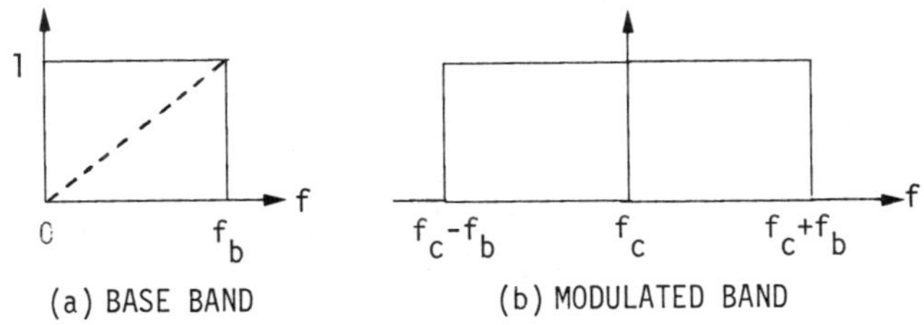

Fig. 3.21 Frequency spectral shape of a modulated signal

The limiting condition of the communication channel as described here was first discussed in 1928 by Nyquist, who showed the means of digital transmission without internal interferences. This theory is best introduced by considering the baseband system, as shown in Fig. 3.21(a). In this case the modulation employs a frequency spectrum from 0 to f_b. This could also be regarded as having the communication channel acting as a low pass filter with a cut-off frequency of f_b.

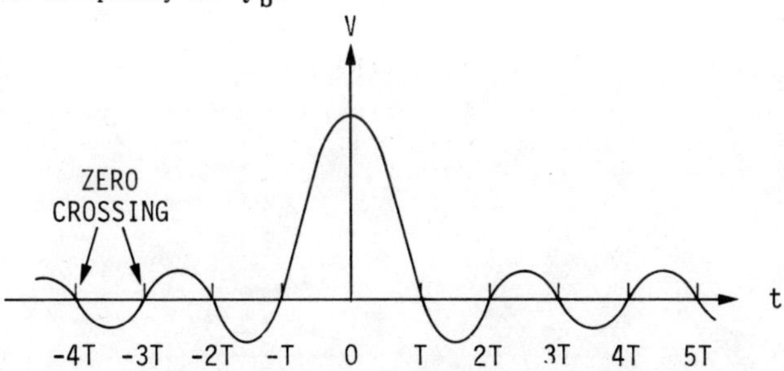

Fig. 3.22 Ideal time response of a low pass filter

The corresponding time response of a transmitted rectangular frequency spectrum has the known characteristics of $\sin 2\pi f_b / 2\pi f_b$, as shown in Fig. 3.22. This is formed by the Fourier analysis of the impulse response of an ideal low-pass filter. That is, it is just the $\sin X/X$ response of a rectangular pulse. Note that the signal shape has a symmetrical impulse response on both sides of the pulse which gradually 'rolls off' to zero. The time response of the signal pattern crosses through the zero level at intervals of $T = 1/2f_b$, where T refers to the clock interval and f_b to the cut-off frequency of the low pass filter.

The transmitted data stream consists of consecutive pulses every T seconds which produce the impulse response as shown in Fig. 3.23. It can be seen that it is possible to transmit at a clock rate of $1/2T$ without the peaks of received pulses interfering with each other. This means that it is theoretically possible to transmit with a baseband of $f_b = B$ Hz frequency a width data stream of $2B$ bits/second without interference. This may be expressed as the theoretical maximum capacity of a channel with a cut-off frequency B Hz, as equal to $2B$ bits/second. The cut-off frequency $B = f_b = 1/2T$ of the base-band is referred to as Nyquist frequency. In general, the Nyquist conclusion states that to transmit N symbol elements per second a bandwidth of $N/2$ Hz is theoretically sufficient.

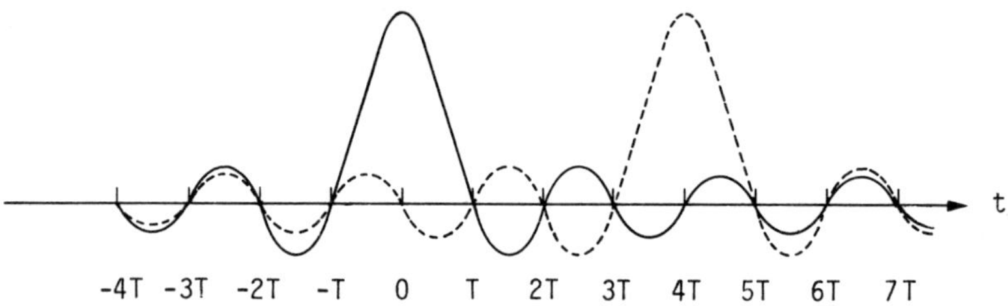

Fig. 3.23 The ideal impulse response of a data stream

Although the data pulses can be transmitted T seconds apart, the overshot impulses of the past pulses into the time slot pulse being currently transmitted could cause considerable interference. These interference impulses are therefore referred to as 'intersymbol interference'. The ideal impulse response is characterized by having the zero crossing every nT seconds. This corresponds ideally with the sampling instances of the proceeding pulses. The communication channel, however, does not present the ideal impulse response of a low-pass filter since it can cause both group delay and attenuation distortions. In this case, the impulse response of the received signal will not always be equal to zero at the sampling instances, as seen in Fig. 3.24. That

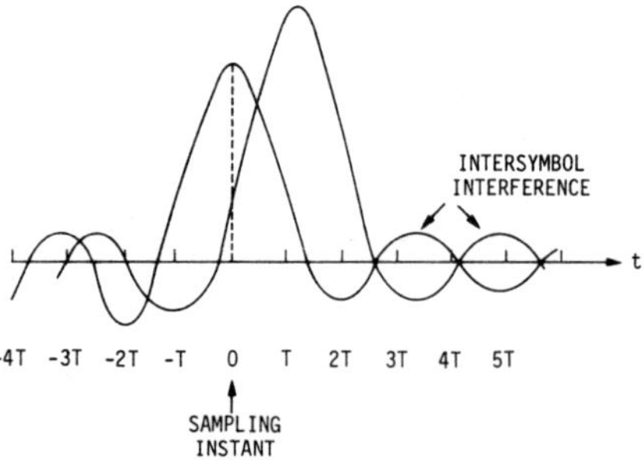

Fig. 3.24 Intersymbol interference in a communication channel with group delay and attenuation

is, the intersymbol interference of a communication channel can be quite considerable and consequently reduces the capacity of the channel.

Even without external interferences, the intersymbol interference from previous transmitted pulses can reduce the available threshold at the sampling instances. These disturbances are caused by the sharp cut-off frequency in the transmitted spectrum. These sharp cut-offs, when realized in practice, could cause drastic amplitude and phase distortions reflected back into the frequency band. Nyquist also showed that the amplitude of the roll off impulse, i.e. the intersymbol interferences, could be reduced by modifying the amplitude shape of the frequency spectrum without disturbing the zero crossing points. That is, more gradual and careful controlled cut-off of the spectrum will reduce both the intersymbol interference and other disturbances. The modified frequency shape, shown in Fig. 3.25, results in additive bandwidth which is expressed as a percentage of the basic baseband. The effect is achieved by means of a low pass filter which shapes the signal before modulation (see modem block diagram Fig. 3.8) with a cosine characteristic. The additive bandwidth is achieved at the expense of some reduction in the transmission symbol rate (i.e., in the transmission speed). Excess bandwidths of 10% to 50% are of general use, depending on the modem ability to compensate for the departure from the ideal shaping.

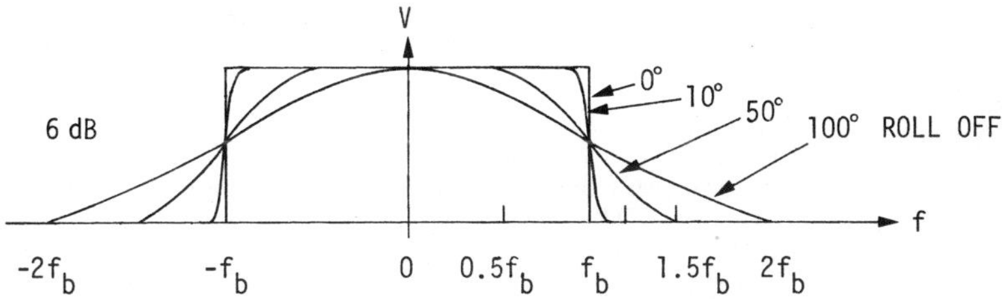

Fig. 3.25 Roll off cosine shaping effects on the frequency spectrum

A useful concept in evaluating the intersymbol interference of a modem is the eye pattern. The modem is wired by linking the transmitter with the receiver and thus avoiding all the communication channel disturbances. The superimposed results of the demodulated signal, with the transmission of a long random train of pulses, presents the characteristical performance of the modem.

3.6 COMMUNICATION MODULATION SYSTEM EVALUATION

The communication system designer may use the eye pattern to evaluate the modem performance after it has been selected. Nevertheless, these eye patterns do not assist the designer in selecting the best modem scheme for his particular application. The following chart presents the basic trade off points that should be considered when selecting a specific modem scheme. If the application calls for higher speeds, then it consequently requires a more expensive modem. The designer should then decide whether the price for transporting each bit is justified.

Higher Speed ⟷	Lower Speed
↓	↓
Wider Bandwidth	Narrower Bandwidth
↓	↓
More Levels	Less Levels
↓	↓
Lower S/N	Higher S/N
↓	↓
More Errors	Less Errors
↓	↓
Sophisticated Designed Modem	Simple Designed Modem

The trade-off points give only a superficial indication of the conse quences in selecting a specific transmission speed but it does not assist the designer in selecting the modulation technique which could be most suitable for the particular application. What is required are means by which the designer can best measure the modem characteristics to be used for the comparison between the various modulation techniques. There are four major measures that could be used when comparing different modulation techniques:

a. Nominal speed of transmission in bits per second per cycle.

b. The required bandwidth necessary to transmit the data.

c. The performance of the signal in the presence of the noise, i.e., the received signal-to-noise ratio.

d. The received signal error probability.

These measures are related and must be used in conjunction with each other. Before this can be done, some fundamental parameters must be defined. All these measures take into consideration not only the modulation system performance but also the communication channel characteristics. The behaviour of the modem, under different operation conditions is of importance here. For this the known theoretical performance of the modulation techniques is used as reference. The communication channel is considered only with background noise which is characterized as being close to additive white and Gaussian noise (AWGN).

(i) TRANSMISSION SPEED (RATE)

By the speed of transmission one refers to the instantaneous rate at which the information is transferred over the communication facility. This quantity is usually expressed in number of elements per unit of time. The term 'speed', or more precisely the 'rate', must be better determined than the general statement given above since it could very easily be interpreted wrongly.

The term 'rate' is specified as the reciprocal of the interval time unit measure in seconds, i.e. $R = 1/T$. It is possible, however, that two different rates of unequal domina may be used to express the speed in the same communication link. The term rate or speed must therefore be associated with a qualifying secondary term.

Modulation rate is defined as the number of discrete conditions, or symbol elements, or signal events that can be transmitted in a unit of time. The modulation rate is also referred to as the signalling rate or the symbol rate. It is expressed in Bauds or unit pulses.

A 'BAUD' is a unit of signalling speed derived from the duration of the shortest signalling pulse per second. It is named after its inventor Baudot who introduced it as a unit for telegraph transmission. The Baud is strictly a statement of modulation signalling speed, and does not refer to the data flow. The Baud is equivalent to the number of transmitted pulses (or symbols) per second (PPS), i.e., 1 Baud = 1 PPS.

Data rate is defined as the number of information bits that can be transmitted in a unit of time. This is also referred to as the source rate or the bit rate. The data rate is expressed in bits per second (BPS or B/S).

The unit BPS refers to the number of binary bits that are transmitted and not to the actual number of symbols that are transmitted in the communication channel. The difference between BPS and Bauds can be realized in two planes; with the signal either in pulse form or in analog form. In the case of 2-level NRZ pulse formats, the Baud rate equals the bit rate; but this is not the case with RZ pulse formats, as seen in Fig. 3.26. When transmitting an RZ signal pulse stream, each level is give a different modulation symbol; therefore three different symbols are used for this format. The

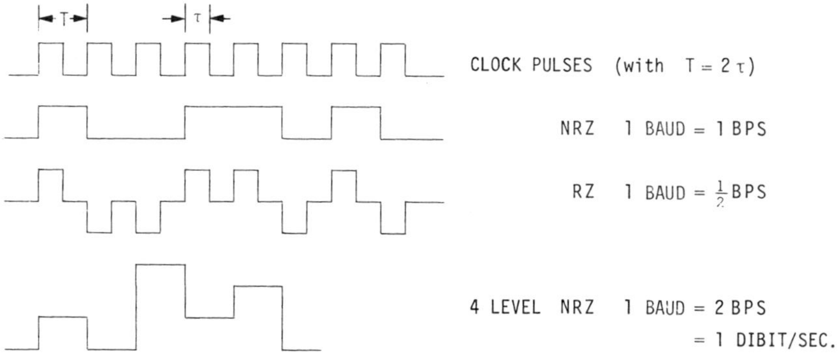

Fig. 3.26 Definitions of pulse signalling rates

difference between the two rates is more emphasized in multi-level pulses, where a number of bits may be transmitted in each level. In all the examples presented, the Baud is always defined by the minimum pulse duration, i.e. $1/\tau$.

In data communication systems both rate units may be used in the same system, with BPS for the data stream and Bauds for the transmission channel. The difference may be better understood with reference to Fig. 3.27, which shows an NRZ pulse stream produced by the data source and its equivalent modulated ACB signal produced by a QPSK modem. The quadrature phase shift keying modem modulates two bits in parallel, i.e. a dibit, and can therefore transmit one out of four possible symbols in each transmitted time interval.

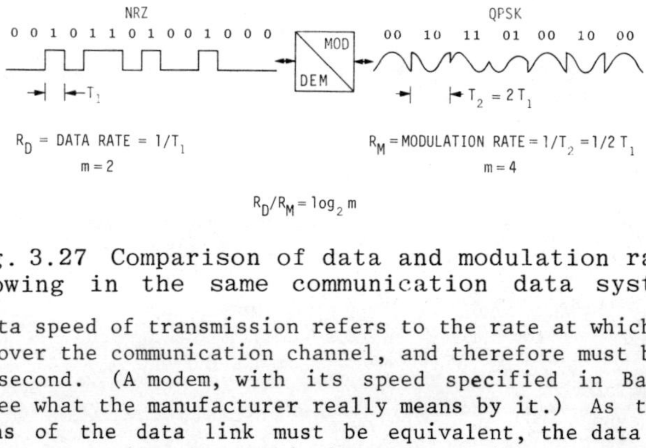

Fig. 3.27 Comparison of data and modulation rates flowing in the same communication data system

The data speed of transmission refers to the rate at which the data is transferred over the communication channel, and therefore must be expressed in bits-per-second. (A modem, with its speed specified in Bauds must be checked to see what the manufacturer really means by it.) As the speed in both portions of the data link must be equivalent, the data rate must be related to the modulation rate. This is expressed in the following equation having the data rate equal the modulation rate times the number of bits/symbol.

$$\text{Data Rate} = (\text{Symbols/Second}) \times (\text{Bits/Symbols})$$

$$R_D = (1/T) \log_2 m = k/T$$

where T is the shortest elementary time interval, m is the number of possible different modulation symbols and $k = \log_2 m$ is the number of bits per symbol. In the case of two-level NRZ there are only two symbol states (0 or 1), i.e., $m = 2$ and $k = 1$, therefore $R_D = 1/T_1$. In QPSK, however, there are four modulation symbols, i.e. $m = 4$ and $k = 2$, therefore $R_D = 1/T_2 \log_2 4 = 2/T_2$. Nevertheless, since $T_2 = 2T_1$, the two data rates are equal.

It should be noted that the above equation is correct only if the data pulse format and the modulation schemes are based on non-return-to-zero binary formats. Return-to-zero formats introduce time redundancy which must be compensated for in the equation.

(ii) CHANNEL CAPACITY

The communication channel capacity is defined as the maximum possible data rate that can be transmitted through a channel and subjected to the constraints of the channel. The channel capacity is expressed in bits per-second or symbols per-second

$$C = \max\ (R_D).$$

Nyquist defined the maximum number of symbols that may be transmitted in a bandwidth of B Hz and then be received without intersymbol disturbances as $2B$ pulses per second. This refers to the modulation rate but not to the data rate. Furthermore, it refers only to NRZ type data. In a multi-level transmission system the theoretical capacity of a channel with a band limit of B Hz is

$$C = 2B \log_2 m\ .$$

Shannon extended Nyquist's theory by regarding also the noise in the communication channel. While Nyquist dealt only with the modulation rate and its maximum capacity due to intersymbol disturbances, Shannon deals with the maximum data rate that can be transmitted through a band limited communication channel. Shannon has shown that the channel capacity is not only limited by the bandwidth but also by the ratio of the mean received signal power S to the mean noise power N. In any communication channel with a bandwidth of B Hz, the maximum data that can be transmitted and received with no error was found by Shannon to be given by

$$C = B \log_2(1 + S/N)$$

where C is measured in bits per second and S/N is the band limited signal-to-noise ratio. The white noise of a random type with a flat frequency spectrum is at the frequency range of interest, and therefore it is usual to present the white noise as n_0, which is the mean noise power per unit bandwidth, i.e. $n_0 = N/W$. The capacity can then be rewritten

$$C = B \log_2(1 + S/n_0 W).$$

Any limiting of the bandwidth below its minimum value will only introduce errors in the demodulated signal. That is, if we try to push too many bits per second through the band limited channel, the errors will begin to mount up. Shannon also showed that by coding the data before transmission into m levels or symbols, the capacity of the channel is still limited by the above equation.

In any band limited communication channel, the additive white and Gaussian noise (AWGN) in the spectrum limits the transmission. Coding of the data may increase the transmission data rate above the Nyquist limit and enable it to approach the Shannon capacity limit, although never above it. Whenever the data rate R_D is below the channel capacity C, there are several coding and modulation techniques that can be employed. Whenever $R_D > C$ the error probability cannot approach zero and then with the introduction of data coding the effect may be worse.

(iii) AMPLITUDE-FREQUENCY SPECTRUM

The frequency bandwidth that is required for data transmission is a function of both the modulation rate $R_M = 1/T$ and the modulation technique employed. It is important, therefore, to consider the transmitted signal not as an amplitude-time function, but as an amplitude-frequency function.

When one deals with data in the form of pulses, it is usual only to consider them as a function of time. When transmitting the data over a communication channel, the time element is still considered, although the limited factor is now the available bandwidth. It is therefore necessary to convert the time function signal $S(t)$ into a frequency function $S(f)$. Each modulation technique requires a different bandwidth to transmit the same data rate, which makes this an important factor in selecting a particular modem scheme.

Every time function has a unique frequency spectrum associated with it. The explicit relation between the time-domain and the frequency-domain is formulated by Fourier series. Figure 3.28 shows the transform of a pure sine wave and a series of periodic rectangular pulses into frequency spectrum. In the case of $S(t) = A_m \sin 2\pi ft$, the frequency function is $S(f) = A_m f$, that is, a single amplitude at frequency f. In the case of the rectangular pulse, however, with $S(t) = A_m$ for $0 \leqslant t < it$ and $S(t) = 0$ for $it \leqslant t < 2it$, the frequency function contains frequency components of all integral multiples of the fundamental frequency. Nevertheless, the envelope of the amplitude decreases at high frequencies. The transform of the time function of a rectangular pulse produces the well-known $(\sin 2\pi f)/2\pi f$ frequency spectrum.

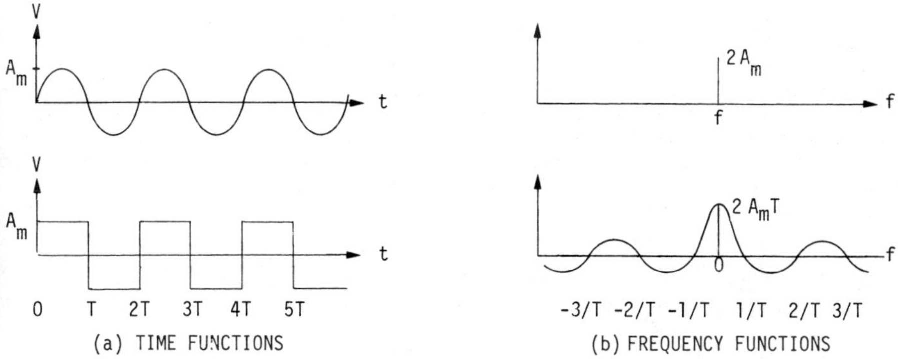

Fig. 3.28 The frequency functions of a sine wave and a rectangular pulse

In amplitude modulation the amplitude of a pure sine wave carrier is varied in accordance with the amplitude of a modulating signal. The modulation signal time function is

$$S(t) = [A_c + A_m(t)] \cos 2\pi f_c t$$

where $S(t)$ is the modulated signal during interval T, having A_c the carrier signal amplitude and $A_m(t) = A_m \cos 2\pi f_m t$ the modulating signal amplitude during the same interval T. As will be explained in chapter 6, the time function can be expanded, giving

$$S(t) = A_c \cos 2\pi f_c t + \tfrac{1}{2} A_m \cos 2\pi (f_c - f_m)t + \tfrac{1}{2} A_m \cos 2\pi (f_c + f_m)t.$$

Taking the Fourier transform of this amplitude-modulation signal gives the frequency function

$$S(f) = A_c F(f_c) + \tfrac{1}{2} A_m [F(f_c - f_m) + F(f_c + f_m)].$$

The amplitude modulation process has the effect of shifting the basic base-band frequency f_m up to the carrier frequency f_c, as shown in Fig. 3.29. It produces a frequency spectrum around the centre carrier frequency f_c. This frequency spectrum consists of three sections: the centre frequencey f_c,

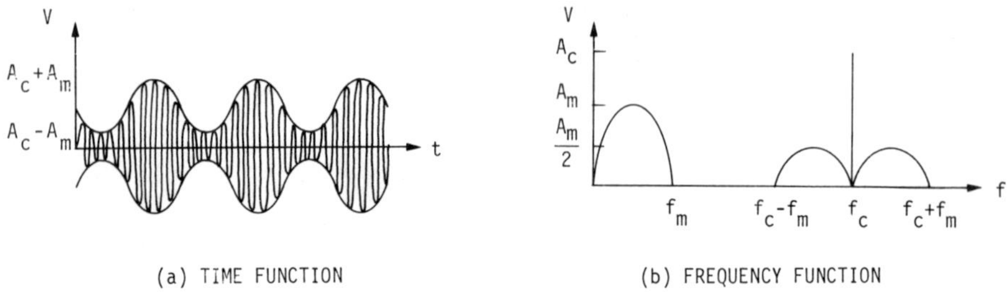

Fig. 3.29 Time and frequency functions of amplitude modulation

upper side band $f_c + f_m$ and lower side band $f_c - f_m$. The centre portion $A_c \cos 2\pi f_c t$ contains no modulating components and therefore can be omitted in the transmitted signal. It is used only as a marker for shifting the base-band. The two side bands are equal; with both containing the modulating signal. That is, they are both of the same spectrum as the base-band, although they are a mirror image of each other. In other words, the Fourier transform of the amplitude modulation time signal is a stable signal at the centre frequency with amplitude A_m and two side bands which are symmetrically distributed about the centre frequency.

In a similar process it is possible to transfor all the time functions of the various modulation techniques to frequency functions.

(iv) POWER SPECTRAL DENSITY

The spectrum analysis in the previous section dealt with the time function of a deterministic signal which was transformed into an amplitude-frequency function using a Fourier series process. In the case of random signals, as is the situation in digital data transmission, it is the power spectral density that plays the equivalent role, using a Fourier integral process.

The analysis of this transform process is beyond the scope of this book, although some basic background is presented. If $S(t)$ is the random transmitted signal over interval T and $S(f)$ is its Fourier transform, then

$$S(f) = \int_{-\infty}^{+\infty} S(t) e^{-j\omega t}\, dt.$$

The signals finite average energy

$$E_s = \lim_{T\to\infty} \frac{1}{2T} \int_{-T}^{T} |S(t)|^2\, dt$$

with the inclusion of random function

$$R_n(\tau) = \lim_{T\to\infty} \frac{1}{2T} \int_{-T}^{T} S(t+\tau)\, S(t)\, dt$$

The power spectral density, $G_n(f)$, is the Fourier transform of random power spectrum over an interval τ, and may be written as

$$G_n(f) = \int_{-\infty}^{\infty} R_n(\tau)\, e^{j\omega\tau}\, dt.$$

In the previous section, where the frequency band for each modulation technique was considered, the information was given on the frequency band required for each transmission; this however, is not sufficient when considering the frequency band required for many adjacent transmissions. The amplitude-frequency function present the main spectrum lobe, which is sufficient for analog communication. Nevertheless, in the case of digital random data communications there are further points to consider. If there was only one communication link with a transmitter and receiver at both ends of the link, as is the case in voice telephone lines, the problems might be simple. However, when a number of links are crammed together in a wide-band cable or in a radio communication channel, the problems could be critical. In these cases, each communication is allocated a different carrier frequency with sufficient band spacing between each allocated band. Since spectrum conservation is highly important these days, it is important to design the communication link so as to make the most efficient use of the available spectrum. Therefore it is not enough to know what spectrum is required for each modulation technique, it is essential to know what spectrum guard is required for each shift keying technique. The main design consideration is to add as many links as possible in the same transmission spectrum. Even with side lobes being 30 or 40 dB down their additive effect or nearby links may be disastrous.

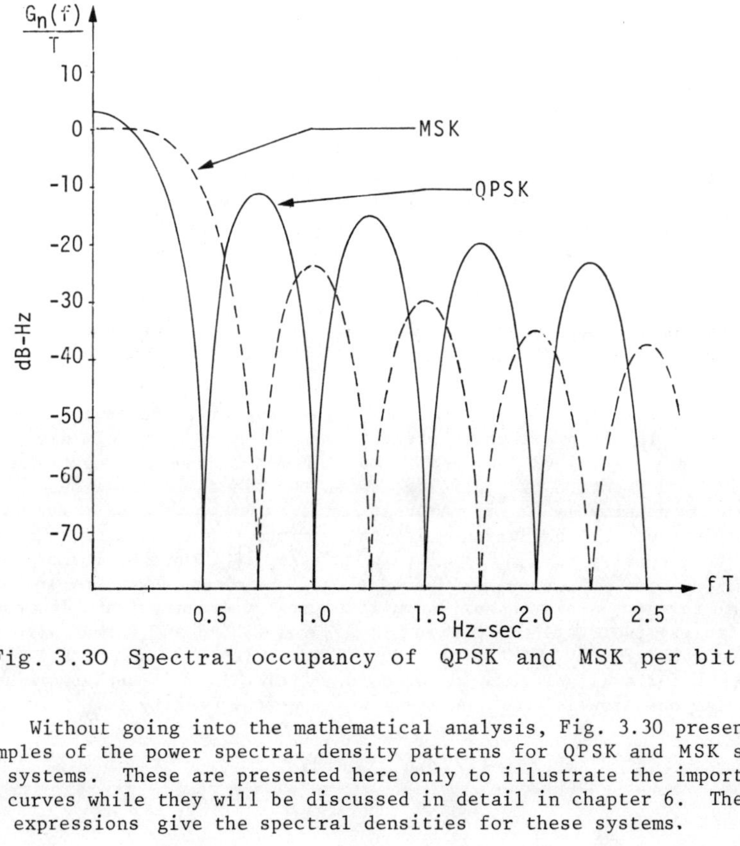

Fig. 3.30 Spectral occupancy of QPSK and MSK per bit rate

Without going into the mathematical analysis, Fig. 3.30 presents two examples of the power spectral density patterns for QPSK and MSK shift keying systems. These are presented here only to illustrate the importance of the curves while they will be discussed in detail in chapter 6. The following expressions give the spectral densities for these systems.

$$G_n(f) = 2T\left(\frac{\sin 2\pi ft}{2\pi ft}\right)^2 \quad \text{for QPSK}$$

$$G_n(f) = \frac{16T}{\pi^2}\left(\frac{\cos 2\pi ft}{1-(4ft)^2}\right)^2 \quad \text{for MSK}$$

The power spectral density, also known as power spectrum, is plotted in the terms of magnitude of spectral components, i.e. all frequency components are plotted as positive quantities regardless of their actual polarity. Furthermore, when comparing the various shift keying systems, we need only be interested in the relative magnitude of the frequency components. For this reason we plot the power spectrum curves per bit rate, i.e., $G_n(f)/T$, as seen in Fig. 3.30. The curves are plotted on a decible scale, as this facilitates comparison of the side lobes as well as the main lobe.

(v) BASE-BAND SPECTRA

The modulation power spectral density is a function of the instantaneous modulation frequency about the centre frequency, i.e. by the amount proportional to the baseband frequency. Up till now it was assumed that the baseband is formed by NRZ (non-return-to-zero) format. That is, there are only two binary states with the level transition only at the clock timing (i.e. the minimum distance between two level transitions is T seconds). This is also the format used by Nyquist for his intersymbol disturbance interferences. With any other pulse formats there is a different data-base spectrum which could consequently affect the modulated power spectral density.

The baseband is the frequency band of frequencies occupied by the signal before it modulates the carrier frequency to form the transmitted line or radio signal. The signal in the baseband is usually distinguished from the transmitted signal by having a frequency spectrum ranging over the lower frequencies. The lower range of the spectrum may include dc (zero frequency) which may have an effect on the power spectrum. The power spectral density for the baseband is usually referred to as the baseband spectra.

Pulse formats which have a clock duration different than T, or formats with more than two voltage levels, will have a different data-base spectra. Examples of some data-base spectra for several binary pulse formats are presented in Fig. 3.31. All these baseband spectra curves are plotted on the same relative coordination for comparison purposes. All the signals are normalized to peak levels of ± 1 and to the same clock interval T. Four formats are give, NRZ, URZ, BRZ amd F-Bi-P (Manchester), and have already been introduced in the previous chapter.

As already explained the NRZ format uses the baseband most efficiently since the entire bit period contails the signal information. Its power spectrum density is $G_n(f) = T[(\sin 2\pi fT)/2\pi ft]^2$. The unipolar RZ wastes 50% of the total bandwidth, with the minimum pulse duration $T/2$. Its power spectrum density is $G_n(f) = T[(\sin \pi fT)/\pi fT]^2$. The BRZ format has a pulse duration of T; however, it does not have a dc component, as in most three level formats. This causes a null spectral value at $f=0$. Its power spectral density is $G_n(f) = 4T[(\sin \pi fT)/\pi f]^2$. The bi-phase format also has a null spectral value at $f=0$ since there is no level transition at the clock period for a '1' after a '0' or a '0' after a '1'. Furthermore its minimum pulse duration is $T/2$ and its power spectral density is:

$$G_n(f) = 4T[(\sin^2 \pi fT/2)/\pi fT]^2 .$$

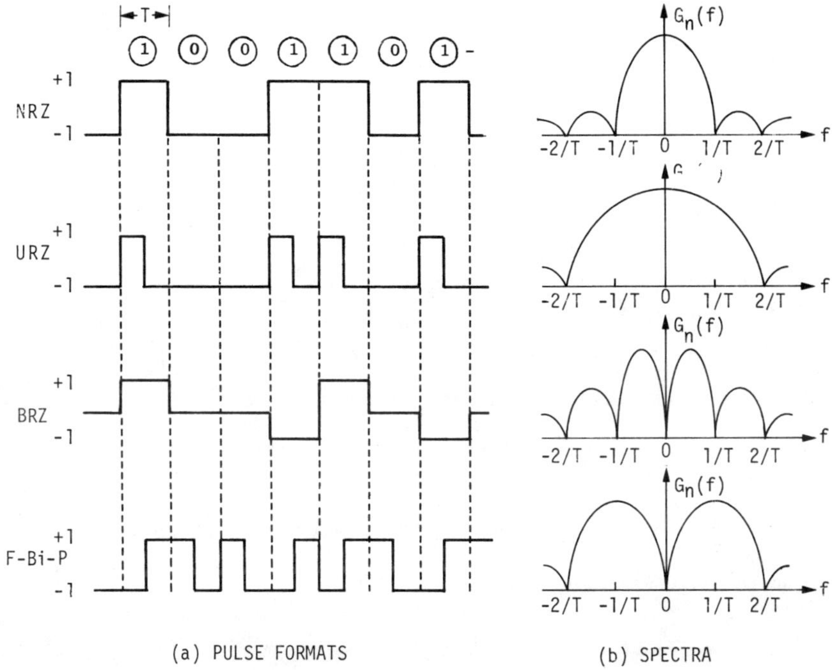

Fig. 3.31 Examples of baseband spectra for different pulse formats

(vi) BIT ERROR PROBABILITY (BEP)

An important criterion in comparing the performance of different modu-ation techniques is the signal-to-noise ratio (S/N) required to achieve a given probability of error.

$$P = f(S/N).$$

A symbol error occurs when the receiver makes an incorrect interpretation of the signal that was transmitted. The signal error rate is the probaility that such a mistake can occur during a particular signal interval T. Noise, in itself, is not the main reason for errors in the decoded data. What is important is the ratio of the signal energy to the noise energy.

Each modulation technique has a different signal-to-noise ratio and therefore it introduces into the system a different probability of errors. When comparing different modulation techniques it is necessary to normalize their values relative to the data bit rather than to the transmitted symbols. In this case one defines the bit error probability P_B to the ratio of the average received signal energy E_b to the mean noise energy per unit bandwidth n_0. That is, the signal-to-noise ratio per data bit would then be E_b/n_0 and this is a function of the bit error probability,

$$P_B = f(E_b/n_0).$$

The theoretical error probability for a type of modulation technique can be calculated and plotted as points on a graph. An example of some of these curves is presented in Fig. 3.32 for three types of modulation techniques. These are phase-shift-keying, on-off amplitude keying and frequency shift keying. It can be seen that better results are achieved with PSK than with FSK, and the worst results with OOK.

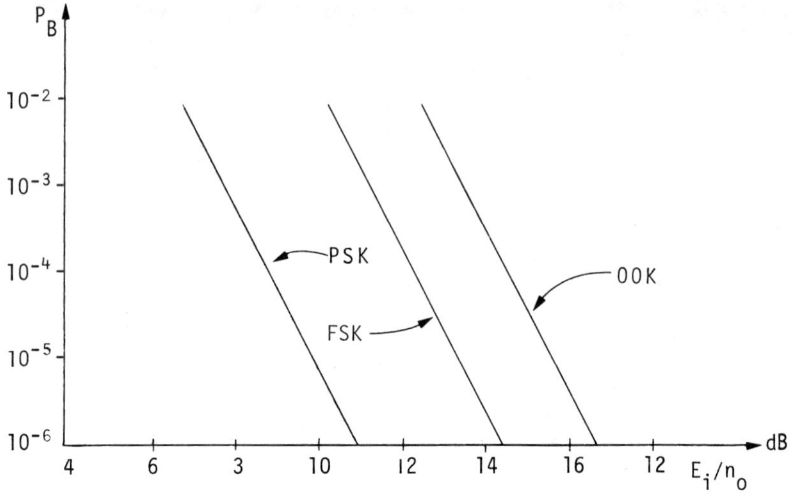

Fig. 3.32 Relative error probability performance of three types of modulation techniques

The received signal energy per data bit is given by:

$$E_b = \frac{1}{k}\int_0^T S^2(t)\,dt$$

where $S(t)$ is the signal waveform during interval T (which is unique for each modulation technique) and k is the number of bits per symbol.

The noise sources are also readily available and are regarded as white and Gaussian. That is, it is assumed that the statistics of the noise waveform has a flat frequency spectrum at the frequency range of interest and may be regarded as a deterministic waveform. The noise per unit bandwidth can therefore be given as the near square spectrum noise power divided by unit bandwidth, i.e., $n_0 = N/B$.

3.7 DATA TRANSMISSION DEMODULATION

Digital demodulation is the process of regenerating the data from the received signal. This digital demodulation process is often referred to as 'detection', in the same way that digital modulation process is referred to as shift-keying. The process may be regarded as consisting of four stages, as seen in Fig. 3.33. The band-pass filter (BPF) separates the received signal from the whole frequency spectrum, and so limits the noise components and other interferences to the frequency spectrum band of interest only. The detector recovers the baseband wave together with other signal harmonics. The low-pass filter (LPF) separates the baseband from the high frequency components, having the resulting signal sliced and reshaped in order to reconstruct the rectangular pulses.

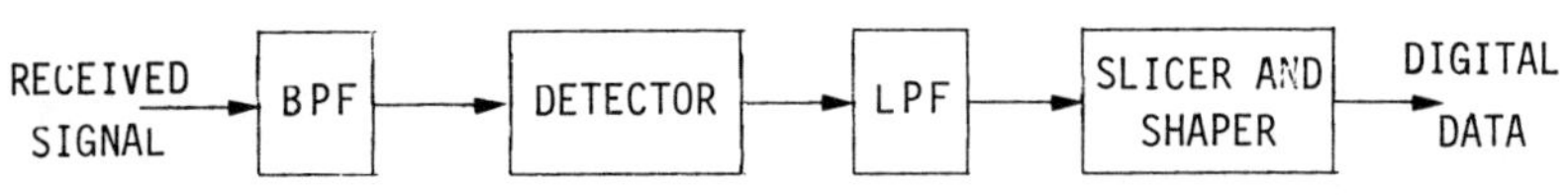

Fig. 3.33 The stages of detections of binary signals

In the previous section we considered how to evaluate the different modulation systems. The problem, however, does not lie within the modulation techniques but in the possibility of demodulating the signal and regenerating the original data without any errors. This process of detection is therefore the one that should be considered since this is what ensures the required wide eye-opening and consequently defines the error rate performance of the technique.

There are essentially two common modes of detection: coherent and non-coherent detection. Coherent detection is also referred to as synchronous or homodyne detection, while non-coherent detection is also referred to as non-synchronous or envelope detection. By synchronous one refers to the relation of the frequency and phase of the received signal with the frequency and phase of the receiver's local oscillator. It is important at this stage to distinguish between synchronous modulation (or detection) from synchronous transmission. (The latter refers to the relation of the clock source of the receiver, to be discussed at a later stage.) For this reason synchronous detection will be referred to as coherent detection.

Coherent receivers require the exact knowledge of the incoming received signal structure, whereas such information is not required in non-coherent receivers. For this reason the first operation in coherent receivers is that of carrier synchronization.

Coherent detection involves the multiplication of the received signal $S^*(t)$ with the locally generated signal source $S(t)$, as seen in Fig. 3.34, having both the same frequency and phase. The detection is essentially that of mixing the two signals and then filtering out the high frequencies.

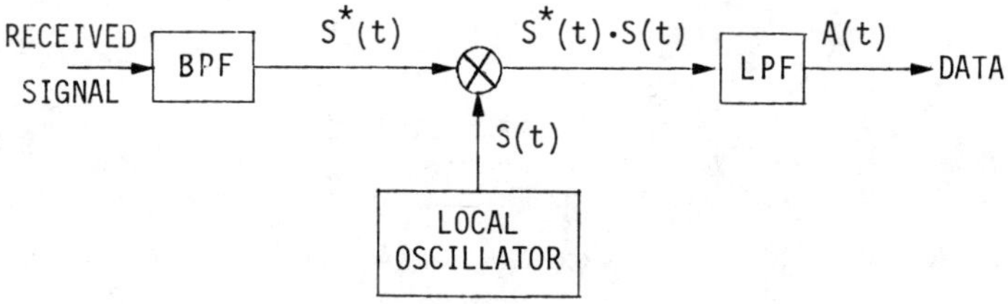

Fig. 3.34 Coherent detection circuit

The coherent process may be better understood by following the mathematical operation of the receiver, taking ASK formula as an example: the received signal,

$$S^*(t) = A(t)\cos\omega_c t\ ,$$

and the local generated signal

$$S(t) = \cos\omega_c t,$$

where $S^*(t)$ is the instantaneous value of the carrier signal at time t, $f_c = \omega_c/2\pi$ the carrier frequency and $A(t)$ the data baseband modulating signal. Thus the mixing process will produce

$$S^*(t)S(t) = A(t)\cos^2\omega_c t$$

$$= \tfrac{1}{2}A(t)\,[1+\cos 2\omega_c t]\ .$$

The low-pass filter rejects the second harmonic of the mixer output, $\frac{1}{2}A(t)\cos^2\omega_c t$, with the remainder, $\frac{1}{2}A(t)$, producing the desired baseband of the binary sequence. The same applies with quadraphase signal modulation, as seen in Fig. 3.35. In this case there are two internal signals. each orthogonal to the other, multiplying the input signals in two separate mixer channels.

The received signal

$$S^{*}(t) = A(t)\cos\omega_c t + B(t)\sin\omega_c t\,,$$

where $A(t)$ and $B(t)$ are the two data-bases of the two bits of the same dibit signal.

Channel 1 mixer output

$$S^{*}(t)S_1(t) = A(t)\cos^2\omega_c t + B(t)\sin\omega_c t\cos\omega_c t$$

$$= \tfrac{1}{2}A(t)\,(1+\cos 2\omega_c t) + \tfrac{1}{2}B(t)\sin 2\omega_c t$$

after LPF $\tfrac{1}{2}A(t)$.

Channel 2 mixer output

$$S^{*}(t)S_2(t) = A(t)\cos\omega_c t\,\sin\omega_c t + B(t)\sin^2\omega_c t$$

$$= \tfrac{1}{2}A(t)\sin^2\omega_c t + \tfrac{1}{2}B(t)(1-\cos 2\omega_c t)$$

after LPF $\tfrac{1}{2}B(t)$.

After filtering each channel separately, the two parallel data-base signals are converted back to serial data. (The whole process will be described again in detail in chapter 6.)

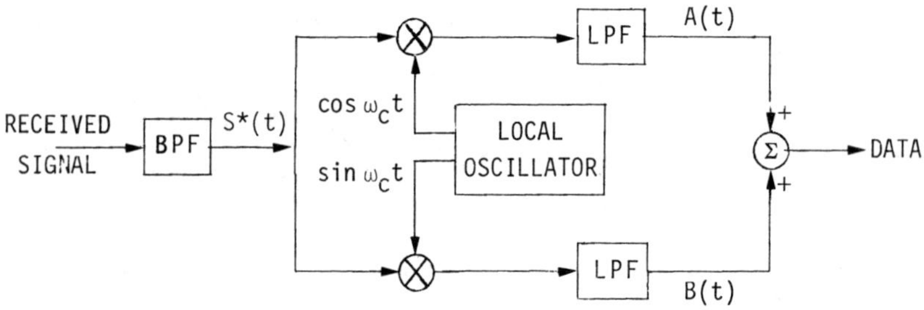

Fig. 3.35 Coherent detection of quadrature phase signal

In the case of coherent detection of FSK signals, the situation is more complicated. FSK modulation involves the shifting of the carrier signal between two separate frequencies, f_1 and f_2, as a function of the data. This means in effect that for coherent detection of the two FSK signals two separate synchronization processes are required in the receiver, one for each frequency. That is, both f_1 and f_2 frequencies must be precisely defined in the receiver, both in phase and frequency. Each locally generated signal will then be multiplied by the received signal, as seen in Fig. 3.36. Furthermore, two separate band-pass filters are required, one for f_1 band and the other for f_2 band. In this decoder the two channels produce the same data bit, with one channel producing the '1' state and the other the '0' state. The output of the two channels is subtracted to produce the desired bit.

It is obvious from the above description that it is difficult to implement, in practice, an FSK coherent detection. The difficulty is amplified since each of the two frequency signals f_1 and f_2 is transmitted on the average only half the time. If a string of ones or zeros are transmitted, then the receiver will soon lose synchronization. If however, both f_1 and f_2 could be produced as harmonics of the same source, then coherent detection may be possible, although this is somewhat difficult practically. For FSK detection, non-coherent processes are required.

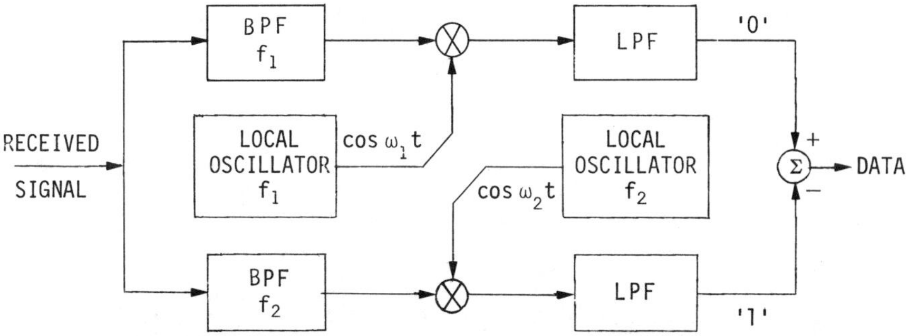

Fig. 3.36 Coherent detection of FSK signals

It is important to add here that the probability of errors in noncoherent detection systems is higher than in coherent detection systems, that is, when considering the system performance with additive Gaussian noise. For noncoherent detection a higher signal-to-noise ratio power is required. In other words, for the same probability of errors, noncoherent detection requires more SNR.

Noncoherent detection is based on the receiver following the envelope of the received signal. In the case of ASK, it recognizes the presence of the signal on the bases of the amplitude of the carrier envelope. Envelope detection is often accomplished by simply rectifying the received signal. In the case of FSK, if one uses envelope detection, the received constant amplitude signal must first be converted to amplitude variations. This is achieved by differentiating the received signal, followed by envelope detection, as seen in Fig. 3.37. To ensure that the received FSK signal is of constant amplitude, it is necessary to limit the signal prior to differentiation, (i.e. *d/dt*). The signal output is then sampled at the clock times, which had previously been retrieved by the peak times of the envelope.

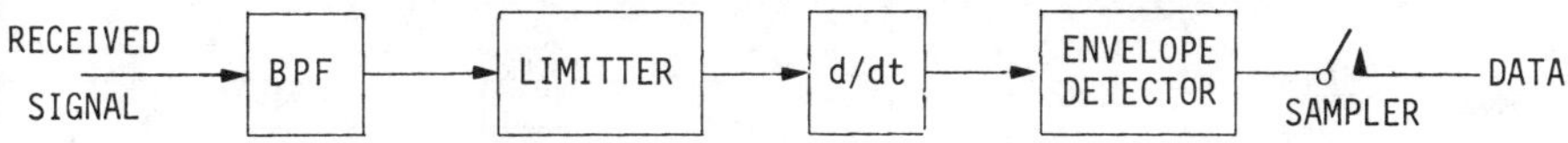

Fig. 3.37 Noncoherent envelope detection principles

To perform the necessary differentiation, a tuned circuit can be used. In the case of FSK two tuned circuits are required, one tuned to frequency f_1 and the other to frequency f_2. In fact, the FSK refeiver, as seen in Fig. 3.38, consists of two detector channels, one for each frequency. The outputs of the two detectors are then compared to determine whether the transmitted binary bit is '1' or a '0'. The comparator ensures better accuracy since it does not rely on the base-band results of one channel only.

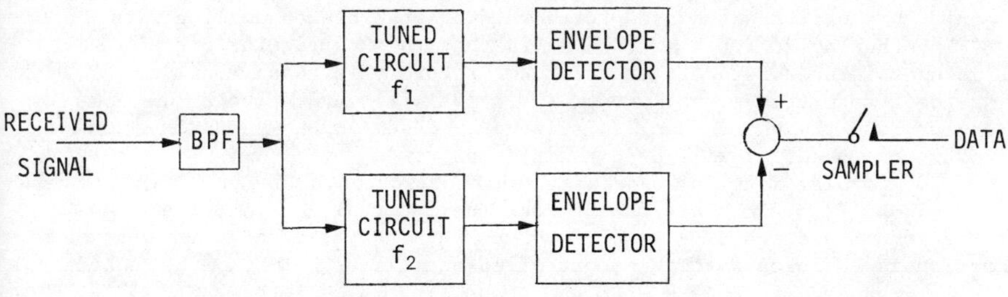

Fig. 3.38 Noncoherent envelope detection of FSK signals

Another form of noncoherent detection used for FSK is known as zero crossing detectors. In this type of detector circuit, the number of times the signal crosses the zero level are counted. If the number of crossings in a given time interval is larger than a given threshold figure the result is a '0' and if it is smaller the result is a '1'.

With multiple frequency shift keying it is not a practical solution to use multiple tuned circuits. For this reason coherent detection is preferable. It follows that for high speed data transmissions, multiple amplitude and phase techniques coupled with coherent detection produce better results than with FSK.

As already stated, coherent detection can only be implemented if the receiver's local oscillator is synchronized with that of the transmitters. In other words, the receiver must reconstruct the carrier frequency and phase, and this must be performed on the distorted received signal and in the presence of the received noise. Two problems are involved here, first to extract the carrier frequency by separating it from the data base components and then to synchronize the resulting signal with the locally generated signal. There are various methods available for obtaining the required synchronization, some of which are presented here.

The received carried signal can be derived by doubling the input signal (i.e., using what is known as a square-law modulator) and then filtering out the high frequencies only, as seen in Fig. 3.39. The received signal, as already stated, is

$$S^{\star}(t) = A(t) \cos \omega_c t.$$

After the frequency doubler, it obtains the value of

$$[S^{\star}(t)]^2 = A^2(t) \cos^2 \omega_c t$$

$$= \tfrac{1}{2}A^2(t) + \tfrac{1}{2}A^2(t) \cos 2\omega_c t.$$

After the high pass filter, only the second harmonic of the carrier signal remains, i.e., $\cos 2\omega_c t$. This result is then divided into two, to give the required signal of $\pm \cos \omega_c t$.

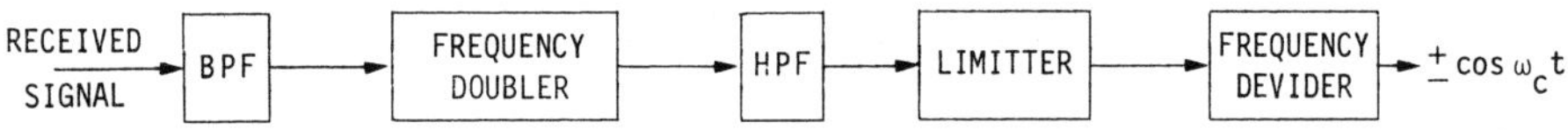

Fig. 3.39 Extraction of the carrier reference out of the received signal

It is important to note the ambiguity in the phase of the resultant signal due to the frequency divider. This is an important point to consider since the 180° phase ambiguity could mean also an ambiguity in the meaning of the decoded data. One means of overcoming this point in PSK is by using differential coding, where each transmitted bit is referred to the previous transmitted symbol phase and not to the carrier phase. The data base symbols can then be decoded correctly, irrespective of the polarity of the carrier signal This will be discussed further in chapter 6 when dealing with PSK.

Another point worth mentioning is that this particular circuit is not of much assistance for carrier reference recovery in phase modulation. Doubling the transmitted signal $\cos[\omega_c t + \phi(t)]$ will double both the carrier frequency and the phase changes. In such cases other means of carrier recovery are used, as will be explained later.

One popular means of obtaining synchronization, is by employing a phase-locked-loop (PLL) circuit which locks the internal local frequency on that of the incoming signal. A phase-locked-loop is a feedback circuit whose function is to force a voltage-controlled-oscillator (VCO) to be coherent both in frequency and phase with the input signal, as seen in Fig. 3.40. The

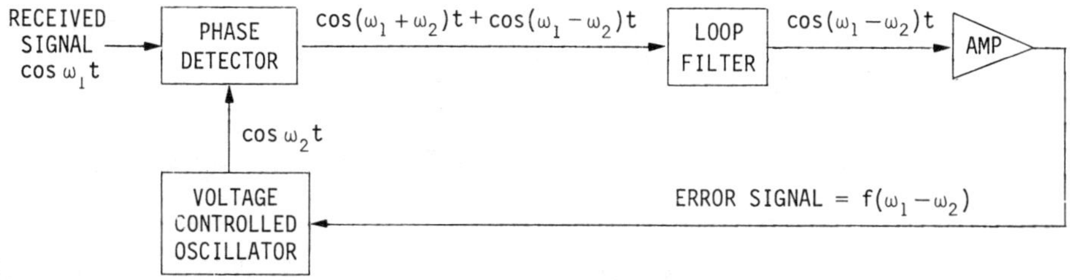

Fig. 3.40 Phase locked loop circuit

phase detector compares the instantaneous phase differences of the two signals. This comparison process is performed in fact by a mixer operation. If the input signal is $\cos \omega_1 t$ and the internal signal is $\cos \omega_2 t$, the mixing operation will result in $\cos(\omega_1 + \omega_2) + \cos(\omega_1 - \omega_2)$. The loop-filter filters out the high frequency, leaving $\cos(\omega_1 - \omega_2)$, which can then be put out as a voltage signal proportional to the error between the two signals. The signal is then amplified by the loop filter to force the VCO to reduce the error. It should be added that the PLL tracks the frequency of the input signal and not its phase.

In the case of phase modulation, the phase depends on the quadrature signal and the carrier recovery is derived accordingly. The carrier recovery circuit is based on a Costas method whereby the incoming signal is multiplied by a quadraphase locally generated signal. That is, the incoming signal is multiplied in one channel by $\cos \omega_c t$ and in the other channel by $\sin \omega_c t$, as seen in Fig. 3.41. The results of the two channel mixers are each passed through low-pass filters to obtain only the low frequency components. The results of the two channels are again mixed to produce the error voltage required to adjust the local oscillator.

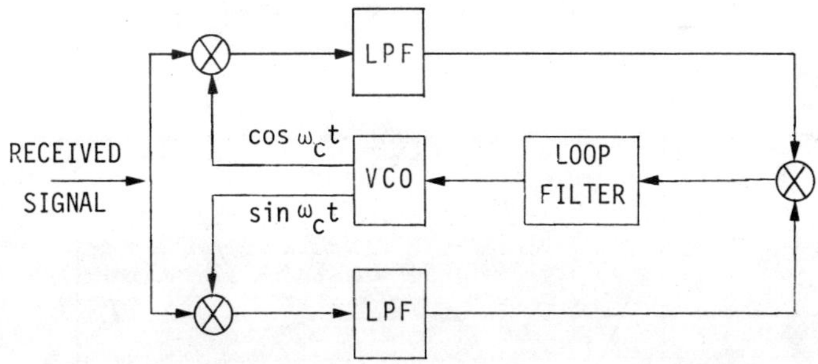

Fig. 3.41 Costas method of receiver synchronization

The Costas loop consists essentially of two phase locked loops which are used in-phase and quadraphase to generate the error signal. The error signal is filtered to produce a voltage to reduce the error of the oscillator so that the internal signals are in phase and quadraphase with the received signal.

3.8 COMPENSATION FOR COMMUNICATION DISTORTION

The intersymbol interference discussed before regarded the communication channel as having an ideal low-pass filter pulse response. Such a response produces eye-patterns with zero crossing at the clock timing. In practice, however, the communication channel is far from ideal. The departure of the received demodulated signal from the ideal waveform shape reduces the eye opening, thereby increasing the error probability.

Distortions of the received signal result from the non-uniform characteristics of the communication channel, that is, through unequal variations in the three parameters of frequency, amplitude and phase which cause both delay and attenuation of the transmitted signal. Even over short communication lines the transmitted signals become quickly degraded and they are progressively distorted as the distances increase. By the time the signal reaches the receiver, the demodulated pulse may be completely deformed both in shape and timing. This distortion causes the tails of preceding and succeeding pulses to increase the intersymbol interference, having them mounting up into the time slot of the pulse which is being currently sampled. These distortions may be either tolerated, accepting the speed limitations, or equalized to allow higher speeds, a process which will add to the cost of the transmission of each bit.

It is impossible to improve the communication channel characteristics to a point where there are no delays or attenuations. Nevertheless, it is possible to compensate for both the delay and attenuation, although this too is limited. At low transmission speeds (up to 2400 BPS) it is possible to reduce the distortions by adding fixed delays and attenuations to selected portions of the transmission frequency spectrum, thereby providing an ideal relatively uniform delay and attenuation throughout the frequency band. This can be seen in Fig. 3.42 which illustrates the envelope delay through a voice telephone line and the effect of a compensating network. That is, the receiver adjusts the communication characteristics by compensating controlled portions of the spectrum. Such networks may compensate for amplitude and frequency distortions but are inadequate in the case of phase delays which are important in higher speeds. At higher transmission speeds, therefore, fixed delay networks are not sufficient to compensate for the demodulated pulse shape, and more efficient techniques are required whereby the modem is precisely adjusted to the particular communication channel performance.

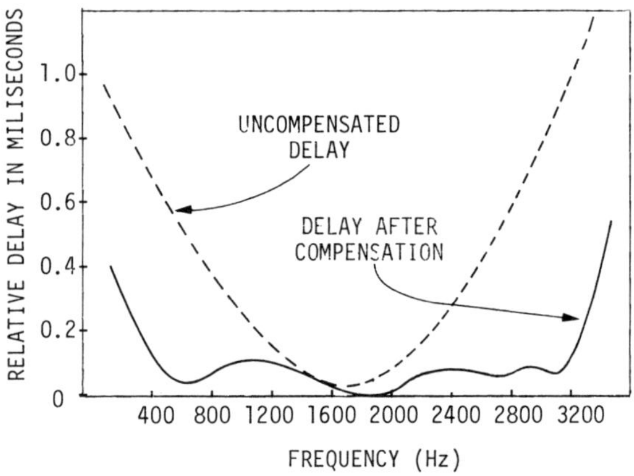

Fig. 3.42 Compensation for the communication channel non-ideal characteristics

The device designed to compensate for the undesired characteristics of the communication channel is known as an equalizer. The equalization process reduces the frequency and phase distortions of the channel by the introduction of a circuit to compensate for the difference in attenuation and delays at the various frequencies in the transmission band. That is, the aim of the equalizer is to reduce the excess intersymbol interference back to zero. This is made possible by taking statistical samplings of the signal performance over the communication channel to determine its characteristics, thereby making it possible to design equalizers which compensate for the average line performance. International standards define the communication line characteristics used for data communication as a function of the frequency band so that the user can decide the maximum speeds that can be achieved on these lines and the type of equalizer required for this. There are three types of equalizers which may be used: fixed, manual adjustable, and automatic adjustable

Fixed equalizers employ matching network based on the known characteristics of the particular communication line. This will improve the theoretical line at the critical points of the frequency band. It centres the compensated equalization of variation of the distortion at about zero. The design is based on defining the ideal line characteristics and then theoretically compensating it. This has many practical limitations and may not be suitable for all lines. For modems operating with speeds up to 1200 BPS it is usually adequate to use fixed equalizers. The fixed compensated equalizers provide nominal improvements of very bad lines; however, their use will sometimes impair, rather than improve an existing good transmission line since it adds attenuation to the line.

Manual adjustable equalizers are based on a set of fixed equalizers which permits the modem user to match the available communication line to a specific equalizer network. The manual adjustable equalizer can compensate for a wide range of characteristics, leaving the user the choice to select the best network suited for his particular line. Its main advantage is that the same modem could be matched to various types of lines. The equalizer installation, however, requires skilled technical personnel and special test equipment to match each communication line. Admittedly the process of installation is slow and far from perfect. Experience shows that the equalizers are installed by the technician's intuition rather than by his skill. Each time the modem is switched to a new line it ought to be rematched to the line, a practice which is seldom observed. The manual adjustable equalizers are used in modems with speeds up to 2400 BPS. In these modems there is a set of fixed compensating equalizers which can be chosen to match a range of expected line distortions.

Both fixed and manually adjustable equalizers are limited in use, since they are restricted to the ideal communication line characteristics. They may compensate for known amplitude and frequency distortions in particular lines and so allow the modem to operate over a given range of lines. It should be added that not all the communication channels have the ideal stable characteristics as the physical lines, whereas with radio lines the characteristic changes could be unforeseen. Furthermore, fixed equalizers fail to correct unexpected phase distortions which may cause large intersymbol interference. This may become critical in *M*-ary modulation transmission and in particular those with phase parmetens. The present trend is towards automatically adjustable equalizers which adapt themselves to the specific line being used, thus eliminating the need for any manual adjustments. For higher speeds, where multi-level codes may be used, the fixed and manual adjustables cannot provide rapid correction of the distortion. Where the number of states is more than two, the intersymbol interference could be so severe that it would limit the speed. The equalizer line matching must therefore be rapid and maintain continuous correction.

Although some fixed and automatic adjustable equalizers may use similar techniques, the design philosophy behind the automatic adjustable equalizers is completely different. While fixed equalizers are intended to compensate the communication channel characteristics, the automatic adjustable equalizers are intended for improving the distorted pulse waveform shape. That is,

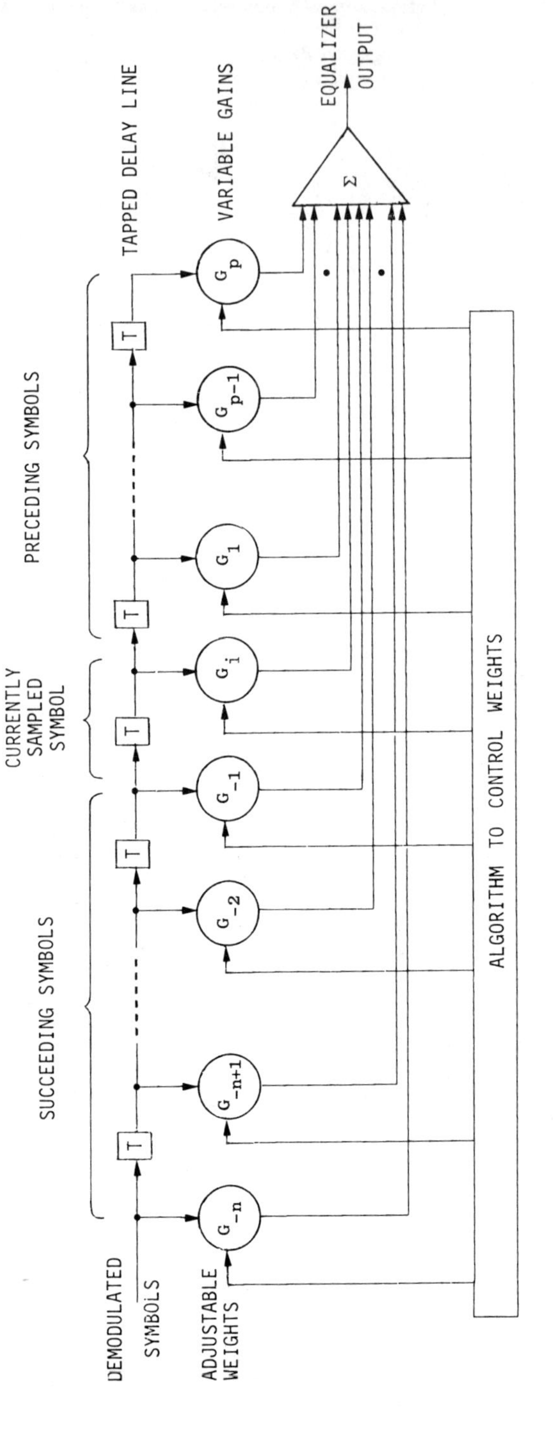

Fig. 3.43 The tapped delay line traversal filter

instead of regarding the whole frequency spectrum of the transmission channel by improving its delay and attenuation pattern, each received symbol is considered individually by separating it from the intersymbol interference caused by the preceding and succeeding symbols. In other words, the automatic adjustable equalizers are intended to control precisely the intersymbol interference rather than the spectrum distortions. This is accomplished by passing the demodulated signal through a delay line having taps at the symbol interval T, as shown in Fig. 3.43. The centre delay unit then holds the symbol currently being sampled, while the fixed delay units hold the preceding symbols and the last delay units hold the succeeding symbols. In other words, each individual symbol is adjusted with the disturbances that are caused by the symbol coming immediately before and by the symbols which immediately follow the particular symbol. The output of all the delay units are summed to provide the qualizer output. Such a network is known as a traversal filter.

The tapped delay line transversal filter is conceptually a simple device which is ideally suited for the purpose of adaptive equalization. It can be used for both analog and digital signals, although in this application only the digital form is of interest. There may be infinite number of taps (number of delay unit) in the line, although in practice the number need only be sufficient to ensure the residual involved values of intersymbol interference to be reduced to a satisfactory level.

The adjacent symbols of the centre symbol, may have a different bit state and their polarity may then have to be reversed before they can be summed. What is of interest here is the sign of the pulses. Each impulse must accordintly be provided with a variable gain which can be positively or negatively adjusted. This means in effect multiplying its contribution by any value between +1 and -1.

There are two types of automatic adjustable equalizers, depending on the means used to adjust the tap weights, preset equalization in which the error components are measured and inspected at the sampling time only, and adaptive equalization which continuously adjusts the demodulated data during transmission. In both automatic equalizer types, the tap gains are adjusted by feedback control circuit. The equalizer output impulse is passed through a decision device, such as a quantizer, to define the discrete digital levels. The results are then delayed in relation to the other impulses by shift registers. The output of the shift registers undergo binary multiplication in exclusive-OR gates. The binary output of +1 or -1 are then fed into binary averaging circuits before being used to control the variable-gain circuits.

In preset equalizers, commonly known as 'zero forcing' equalizers, the variable gain settings of each symbol are so adjusted that they force the disturbances to zero at the sample time. That is, the departure of the received demodulated symbol is used to adjust the taps of the traversal filter in an iterative fashion until the intersymbol interference has been forced to zero. The output of the traversal filter is sampled at selected times by a peak timing detector, as seen in Fig. 3.44. The mid-point of the centre symbol is sliced at +1, while the other samples are sliced at zero level. The voltage impulses produced are the error voltages. These error results of the slicer (equalizer) are fed back through a shift register and are later used to command the tap gains of the traversal filter to count up or down according to the values obtained.

Preset equalizers are used mainly for fixed equalizers where the weights are adjusted by a training sequence which simulates the line characteristics. For automatic adjustable equalizers it is usual to use adaptive type equalizers.

Adaptive equalizers are based on 'decision directed' operation in which the iterative corrections are derived from the equalizer output. The errors are continuously estimated during the course of the transmission and corrections to the gains are continuously effected. That is, the mean square error

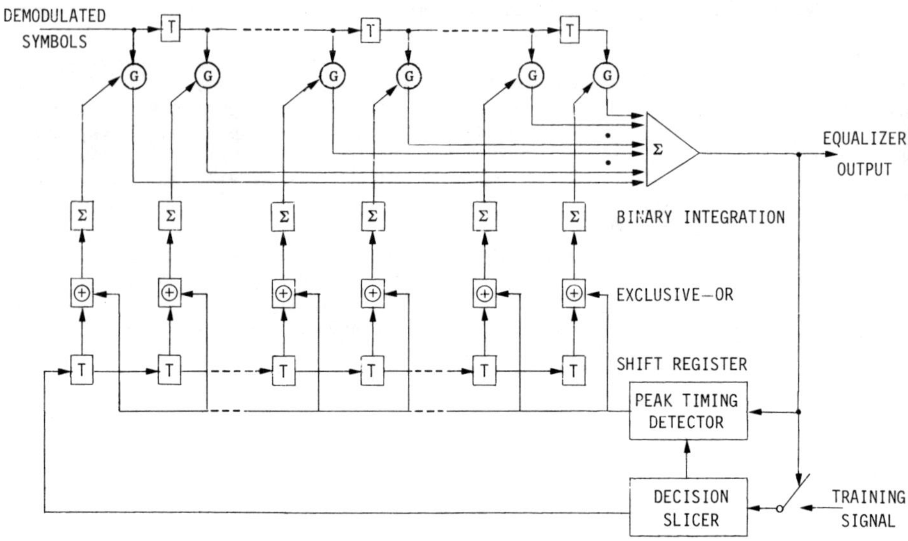

Fig. 3.44 Zero forcing adjustable equalizer

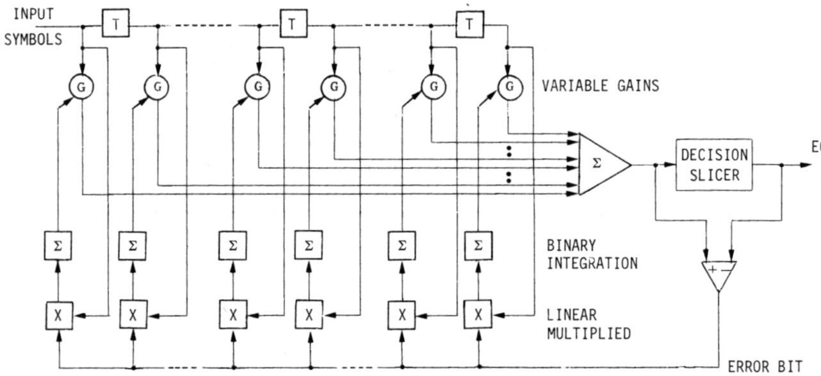

Fig. 3.45 Mean square error adaptive equalizer

is used to adjust the gain rather than the values at the sampling points, as seen by the circuit in Fig. 3.45. The advantage of this is the ability to adapt to any changes in the channel characteristics during transmission.

While zero-forcing equalizers use binary multiplication (i.e. exclusive-OR gates), the mean-square error equalizers require more complex linear multipliers. Nevertheless, the properties of the zero forcing is inferior since the sampling at the mid symbol timing means that some information is discarded.

Another form of automatic adjustable equalizers is shown in Fig. 3.46. It is a hybrid between the zero forcing and the mean square error. In this circuit the error signal is clipped as in zero forcing while the tap signal is as in mean square error.

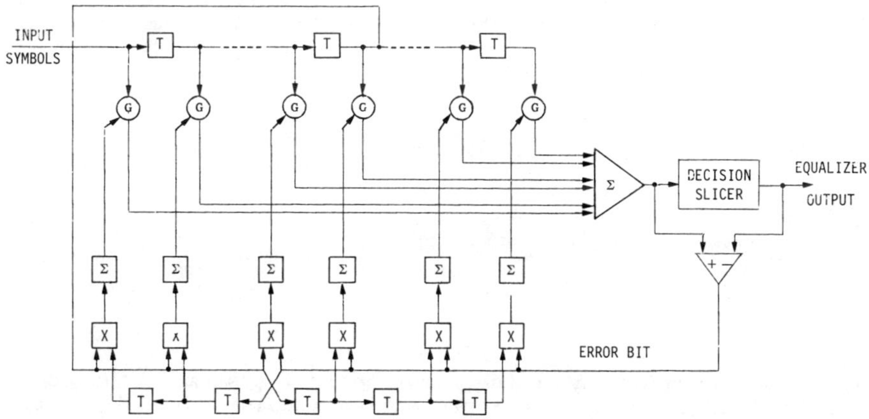

Fig. 3.46 General configuration of a hybrid automatic adjustable equalizer

3.9 COMPENSATING FOR COMMUNICATION IMPULSE NOISE

It has already been shown that additive white Gaussian noise (AWGN) distorts the signal in the communication channel. Nevertheless, since AWGN has a flat frequency spectrum at the frequency range of interest, it is possible to compensate for it and prevent it from distorting the regenerated data in the receiver. White noise, however, is not the only source of external noise disturbances which affect the signal in the communication channel. A major source of external disturbance is caused by impulse noise, which is characterized by transient disturbances separated in time by quiescent intervals.

Impulse noise can occur in any type of communication channel, whether it is a physical wire channel or a radio channel. The impulse noise may be caused by nearby voltage switching, nearby power lines, cross talk coupling, lightning, fading, magnetic sun storms or other noise sources. The noise impulses generally come in bursts where a number of disturbing pulses may be clustered together, although they could also be random impulses with single disturbing pulses. Whatever the form of these noise impulses, each single disturbing pulse can cause considerable damage to the transmitted data (with a transmitted '1' received as a '0' and vice versa). It has been found that on the average one impulse can alter three to four bits in a sequence. While it is the modem that compensates for the white noise, other means are required to compensate for the errors caused by impulse noise.

Since impulse noise is caused by unexpected sources, the communication system has four options: it can either ignore the errors, cancel the message containing the errors, ask for the retransmission of the message containing the errors or correct the errors in the receiver. Ignoring the errors or cancelling the message is only possible in continuous measurement data, where the absence of part of the data will not greatly affect the overall measurement picture. Nevertheless, in most communication systems, no part of the transmitted data is allowed to be destroyed or even damaged. In other words, means have to be inserted into the communication system which will prevent any of the transmitted data bits from changing their original state. This is achieved by introducing an error preventing control system into the communication overall systems, as seen in Fig. 3.47. That is, the error control system is inserted as a separate unit between the data-communication-equipment (i.e. modem) and the data-terminal-equipment (i.e. computer or data entry device). The error control system can either be designed into the DCE or into the DTE. In either case it is an integral and essential part of most data communication systems. In early communication systems it was designed as part of the DCE but today the tendency is to transfer it to the DTE. Whatever its position in the communication systems, there is no message transmission without an error control system attached to it.

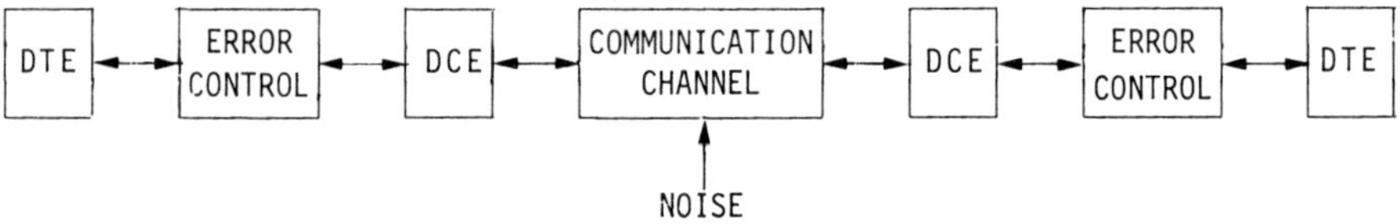

Fig. 3.47 The position of the error control system in the communication systems as means of preventing impulse noise distortions

The error control systems generally consists of three operation stages:

(a) Error detection is the stage in which the presence of an error in the data is detected.

(b) Error location is the stage in which the exact position of the error in the data is located.

(c) Error correction is the stage in which means are introduced to correct the errors.

Error control can only be made possible by adding redundant data bits into the transmitted data stream, which contain no useful information beside introducing means for error detection and correction. The amount of redundancy inserted into the transmitted stream depends on the degree by which the individual error must be located and corrected. In some communication systems only the presence of an error in the data message is needed and then the correction can be performed either by cancelling the message or by the retransmission of the whole data message. In other systems the exact location of the error in the data is essential so as to correct the error in the receiver. Error correction can therefore be classed in three main categories:

(a) Forward error correction (FEC) in which the individual errors are corrected in the receiver after they have been detected and located.

(b) Automatic repeat request (ARQ) in which the mere presence of the error calls for retransmission.

(c) Hybrid error correction (HEC) in which minor random errors are corrected in the receiver while a burst of errors will require retransmission.

ARQ requires that the data message size will be precisely defined in a fixed unit referred to as a block. A block is usually coded as a single portion for the purpose of error control, with its size being as large as 1000 - 2000 bits or as small as a single character. In FEC, the data stream may be continuous or be of fixed size. FEC, however, requires much higher redundancy inserted into the transmitted stream, than that inserted with ARQ. While with ARQ the overhead redundancy may be only in the range of a few percents, FEC may require a few hundred of percents of overhead redundancy. Despite the high redundancy, FEC has limited capabilities and it just cannot be designed to correct all the errors. It should be pointed out that the error control transmitted bits, in both ARQ and FEC systems, are usually staggered relative to the transmitted data bits. This is in order to try and prevent the same noise burst from damaging both the data bits and the error control bits.

FEC can correct the data up to a given number of bits, beyond which it starts introducing new errors into the system. After correcting B bits it must wait for a period of G bits before it can start properly correcting the errors. The guard period G, as seen in Fig. 3.48, is usually much larger than B, i.e., $G \gg B$. This is caused by the performance nature of the error correction used. In other words, FEC is effective as long as the number of errors in the system is less than B and the interval between errors is less than G.

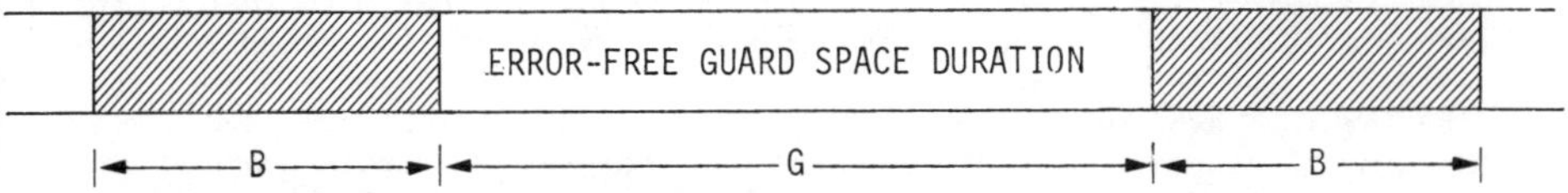

Fig. 3.48 The position of the error free periods in the FEC data systems

ARQ requires that every block transmitted will be confirmed by the receiver; either positively or negatively. If the block in the receiver is found to be error free, a positive acknowledgement (ACK) is returned to the transmitter. If, however, an error is detected in the received block, a negative acknowledgement (NCK) is returned to the transmitter. The block confirmation can be returned immediately after receiving each block or after receiving a number of blocks. In the case of a single block retransmission, the process is as described by the flow chart given in Fig. 3.49. That is, no new block can be transmitted before an ACK is returned to the transmitter confirming the proper reception of the first block; while, if a NAK is returned to the transmitter, the same block is retransmitted. The process of ARQ could introduce delays into the system and hence some systems transmit a number of blocks before confirmation for all the blocks is returned. In this case the confirmation states which blocks are error free and in which errors have been detected.

Relying only on FEC error control means that very high overhead redundancy is required, and even then it is not enough to guarantee that the data received is error free. By relying only on ARQ error control means, retransmission may be required even if the number of errors is small. Most communication systems today use hybrid error control systems which allow random errors to be corrected in the receiver, while if there are bursts of errors then retransmission is requested.

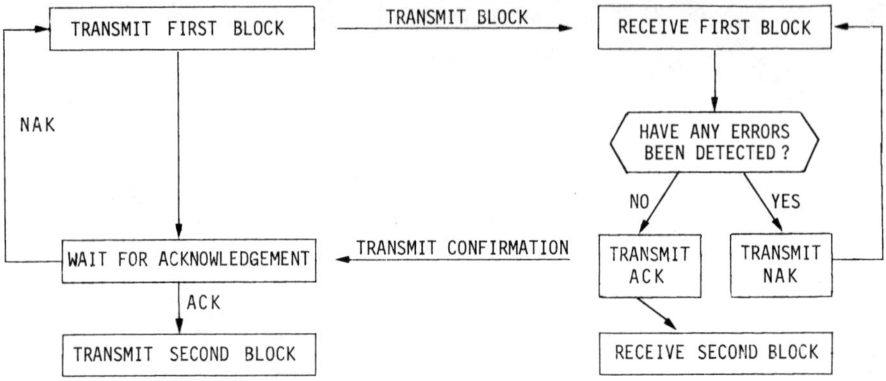

Fig. 3.49 Flow chart of the main processing stages in ARQ

3.10 SYNCHRONIZATION CONSIDERATIONS

One of the main problems in data communication is the ability of the receiver to reconstruct accurately the transmitted signal back to its original data form. This can only be accomplished if the receiver has some means of coordinating its internal oscillator and clock sources with that of the transmitter. That is, the receiver must have means of recovering both symbol and bit timing for use within the data-communication equipment (DCE) and the data terminal equipment (DTE). The symbol and bit timing are used to sample the demodulated signal at the centres of the reconstructed symbols which enable the data pulse to be regenerated. Reconstruction of the symbols and regeneration of the data can only be performed if the receiver knows 'when to look'. In other words the receiver must have some means of synchronizing its operations with that of the transmitter. This problem is critical for the performance of the communication system and is not as simple as it might appear. It is not only a problem of recognizing the symbols and converting them to rectangular pulses, it is also a problem of regenerating them back to their original data format. The latter is more complicated where each symbol may represent a number of bits. It is important not only to reproduce the data stream but also to restore the data back into its original message format. The problem of 'when to look' applies to all the three data reproduction requirements: pulse, data stream, and message format.

In data communication, where the data is transmitted serially over a transmission link, there are three synchronization aspects where the receiver must be interfaced with the transmitter: a) carrier, b) time, c) frame. There is a lot of confusion in the literature whereby the same terminology of 'synchronous transmission' is used for all three types of synchronization. In fact, the three synchronization requirements, shown in Fig. 3.50, are not necessarily directly related, although they may employ similar techniques in achieving synchronization.

Regenerating of the data can be regarded as a three level operation. The carrier and clock synchronizations are usually controlled within the data communication equipment, while the frame synchronization can be controlled either in the data communication equipment or in the data termination equipment.

The synchronization requirement for demodulation of the carrier signal has already been discussed. Synchronization requirements for the time and

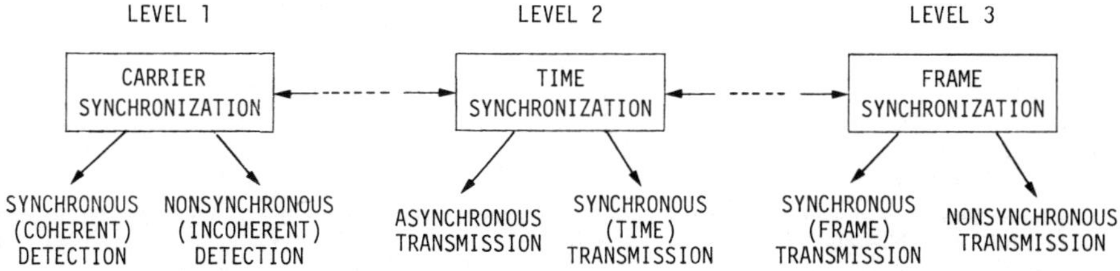

Fig. 3.50 The relationship between the various synchronization transmission operations

frame still need to be clarified. The definition of the two remaining synchronization aspects are as follows:

Asynchronous transmission: A mode of data transmission such that the time occurrence of the bits within each character or block-of-characters relates to a fixed time scale defined by the start of the character or block. The start of each character or block, however, does not relate to this time scale. That is, the time interval between the transmission of each character or block may be of unequal length and unrelated to any time scale.

Synchronous (time) transmission: A mode of data transmission in which the transmitter and the receiver units are operating at substantially the same rate. The receiver clock synchronization is continuously maintained by means of corrections, if necessary, in a fixed relationship.

Synchronous (frame) transmission: A transmission process such that between any two significant instances there are always an integral number of unit intervals. (Note, frame synchronization can only be effective if the timing is also synchronized.)

Nonsynchronous transmission: A transmission process such that between any two significant instances in the same character or block there are always an integral number of unit intervals. Between two significant instances in different characters or blocks there are not always an integral number of unit intervals. (Note, nonsynchronous transmission also requires that the timing is fully synchronized.)

Asynchronous transmission is favoured for low speed transmission while synchronous transmission for high speeds. Asynchronous transmission means in effect that the receiver clock operation is unrelated to that of the transmitter; its timing is always adjusted by the start of each receiver character or block. Synchronous transmission requires the essence of the clock to be transmitted with the data in order to mark the location of the clock. This marker is then used to adjust the receiver internal clock so that it follows the transmitted clock. Most digital communication today employ nonsynchronous transmission techniques as a synchronous time communication system. That is, the timings of the receiver is synchronized with that of the transmitter while the messages are only transmitted when there is data available to transmit. These messages are then recognized by a special start and stop code transmitted at the two ends of the message. When message transmission is continuously generated, i.e. one after the other with no spacing, synchronous time and frame are employed, although even here the messages are recognized by a start of message (SOM) code and end of message (EOM) code, depending on the message format. Continuous transmission is generated in special concentrating equipment and in analog coded data such as in digitized voice.

The term nonsynchronous transmission is a relatively new term introduced to distinguish non related character or message transmitted in a synchronous communication system from asynchronous transmission. This means that despite the interpretation of the term, nonsynchronous transmission refers to time synchronous transmission. While asynchronous transmission is associated with slow speed communication a new term was required to define fast speed communication where is is essential for the transmitter and receiver to be fully time synchronized, although the data flow may be in intermittent bursts. With nonsynchronous transmission the clock essence is transmitted with the data as in synchronous transmission, while with asynchronous transmission the internal timing is reproduced in the receiver each time a new character or block is received.

All these synchronization techniques require special equipment to perform the coordination between the sending and the receiving termination units.

(i) *ASYNCHRONOUS TRANSMISSION*

Asynchronous transmission refers to the oldest form of data communication where synchronization must be re-established with each received character It originated for telegraph and teleprinter type communication where each character is typed and transmitted separately. Today the most common example of asynchronous transmission source is the teletypewriter, where the data is transmitted only as fast as the typist's fingers operate the keyboard. This means that asynchronous transmission is characterized by the irregular source of the characters and the slow speed of transmission. Special start and stop bits are automatically added to the beginning and the end of each character to enable the receiver to know 'when to look'. For this reason asynchronous transmission is often referred to as start-stop transmission. The receiver uses these bits to initiaite a count process to provide its internal clock pointers as to 'when to look'. These coordination start and stop bits add extra overhead on every character and thereby reduce the actual bit rate.

There are various pulse formats used for start-stop transmission. While the earlier telegraph systems employed RZ bipolar pulses, the pulse format described here is, however, that of NRZ since this is the most common pulse format used with teletypewriters.

As already stated, asynchronous transmission is generally associated with electromechanical equipment which is mainly operated manually. The maximum speed is usually specified and the equipment can then operate randomly at any speed up to this point. Asynchronous transmission is also used for somewhat 'faster' equipment such as paper tape readers and punchers, card readers and line printers. It is customary to refer to the logical levels of the electromechanical equipment as 'mark' and 'space'. With the continuous current referred to as 'mark', representing the '1' bit state and the break in the current as 'space' representing the '0' state. Each character has a modest size of 5, 7, or 8 bits depending on the data format used, which is initiated by a start signal of a single space bit. At the end of each character there is generally a single parity bit followed by one or two mark bits, as seen in Fig. 3.51. This means that each character is generally of the size of $(N+3)$ bits where N is the number of bits in the character including the parity. The stop 'mark' signal level is maintained until the next character is transmitted: therefore there is no limit to the stop signal duration, although there is a minimum duration to its lower range, depending on the format used.

Once a chatacter is transmitted, its speed is controlled by the maximum transmission speed. This means that the nominal bit interval T is defined by this speed. No clock pulse is transmitted with the data, but the individual data bit state could be reconstructed in the receiver by claculating the location of the bit centre, i.e., an interval of $T/2$ seconds after the start of the bit interval. This is accomplished by dividing the bit interval into

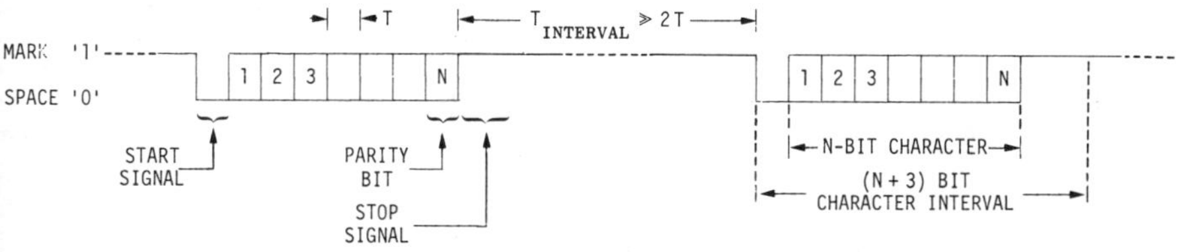

Fig. 3.51 Asynchronous character data sequence

n clock durations with t seconds between each clock pulse, as seen in Fig. 3.52. The ideal bit duration is equal to $T = nt$ and therefore it is sampled after $nt/2$ seconds. A new count is initiated at the beginning of each bit which is recognized by a voltage or current level transition. If a transition occurs before the end of the counter, the counter is reset to zero and the counting is freshly resumed irrespective of the previous count. If the counter reaches its maximum count of n, the counter is reset and simply resumes its count once more. If the transition is only delayed, then this will cause the counter to be reset a second time and then it will resume its count once more. If there is no transition between two bits, as in the case when both bits have the same data state, the counter is reset after n clock pulses and starts counting the next n clock pulses. In this case there could be a timing error displacement of Δt added to each bit.

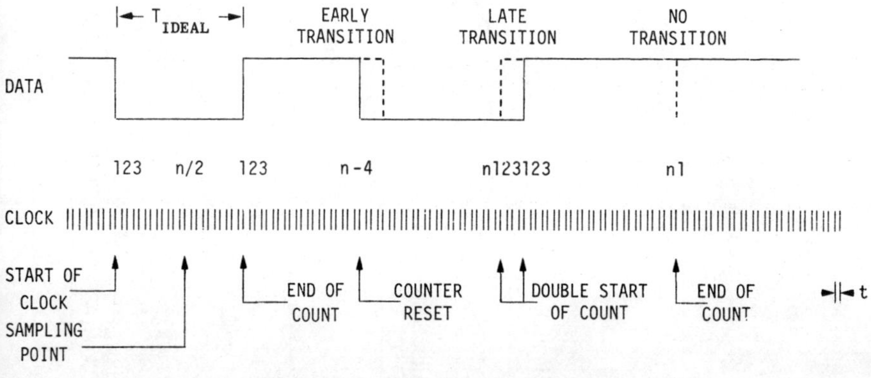

Fig. 3.52 Sampling the asynchronous data signal

When, in a character there are only ones or only zeros, then the cumulative timing error displacement of the character will be $N\Delta t$. The maximum accumulative bit displacement must ensure that the last bit can still be sampled in its presumed duration. This means that the error distortion is limited by $N\Delta t < T/2$. The character size N is therefore limited by the transmission line distortion whereby in the period of NT the delay is smaller than $N\Delta t$. Any distortion in the sequence above this figure will affect the reliability with which the character is assembled.

To demonstrate the operation of the asynchronous receiver, a simplified diagram of the circuits involved is given in Fig. 3.53. The start bit at the

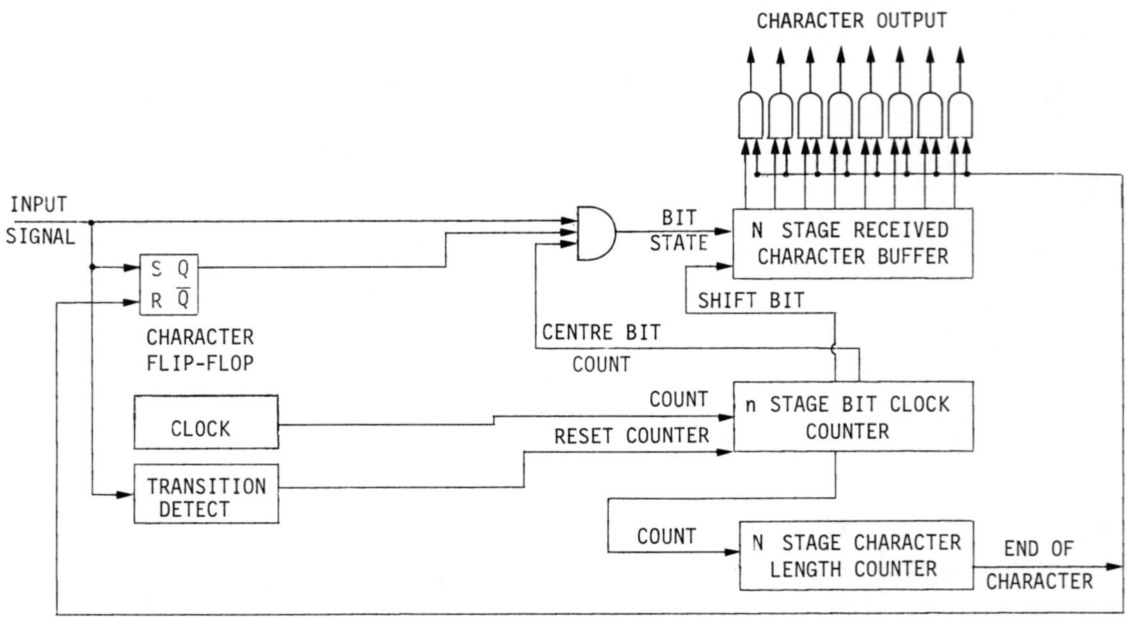

Fig. 3.53 Schematic diagram of an asychronous receiver

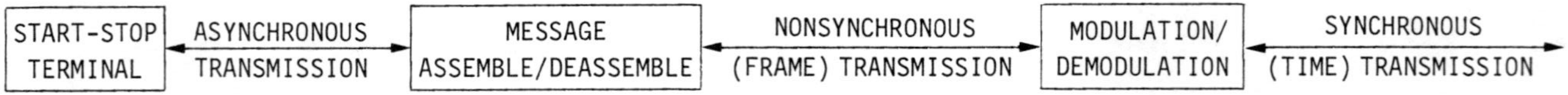

Fig. 3.54 Additional equipment for asynchronous terminal to adapt its transmission to synchronous communication

leading of the character is well defined and is used to initiate the full operation by setting the character flip flop. The n bit clock counter is reset by each transition, as explained above. When the counter reaches its centre position a pulse is used to sample the input signal, with the result inserted into a buffer. A count instance slightly before it reaches the centre is used to shift the buffer and adjust the count of the bits in the character length counter. This pulse cannot be taken at the beginning of the n bit clock count in case there is a late transition causing the bit clock counter to be reset twice. Nor can it be taken at the end of the count in case there is an early transition and the counter never reaches its maximum count. When the character length counter reaches its Nth stage, an end of character pulse is used to clear the buffer and transfer all the N bits of the character. The end of the character pulse is also used to reset the character flip-flop in preparation for the reception of the next character. Before the full character is transferred to other locations the data is usually checked for errors, although this is not shown in the schematic receiver diagram.

Asynchronous equipment is used over a wide range of data rates and organized for diverse applications. However, it must be noted that asynchronous transmission is slow, limited in accuracy and inefficient. It requires separate timing circuits both at the transmitting and receiving modems, with about 25% of the data stream being redundant for the start and stop signals, apart from the extra bits required for the parity and the spacing between characters. Asynchronous transmission equipment is available up to 1200 BPS, although most applications restrict the use of this technique for speeds much below 600 BPS.

Asynchronous transmission is inexpensive to instal and maintain although these advantages are offset by their slow speed and limited accuracy. Asynchronous transmission is a most ineffective means of sending data over a communication channel. For this reason the tendency today is to use synchronous transmission. Nevertheless, the number of start-stop terminals in operation these days is still great. Instead of replacing all these 'old' terminals means have been adopted so that they can operate in synchronous transmission mode. That is, new equipment has been designed for these start-stop terminals, known as message assemble/deassemble (MAD), as seen in Fig. 3.54, which enable the data to be received in slow speed asynchronous form and transmitted in high speed nonsynchronous message form. The modems can then use fully synchronous transmission. The MAD is designed to also do the function of concentrating, whereby it assembles the transmissions from a number of slow speed terminals. This, however, will be discussed in chapter 7.

(ii) *SYNCHRONOUS (TIME) TRANSMISSION*

Synchronous transmission is many times faster than asynchronous transmission since there is no need for gaps between transmission of the characters and there is no wastage of time for resynchronization of the receiver after each character. This type of transmission permits sophisticated handling of the data, although it is initially more expensive to install than asynchronous transmission. Nevertheless, synchronous transmission efficiently exploits the communication channel and soon compensates for the high cost of the equipment. What is important in data communication is not only the cost of the equipment and that of the lines but mainly the cost of transferring the data, i.e. the price per bit transmitted. (For real-time systems, which are the major concern in C^2 systems, it is the price per bit transmitted in a given time.)

Synchronous transmission is characterized by its precision, where the data is transmitted at a constant speed. This means that all the data communication equipment (DCE) must be able to transmit and receive accurately at this high speed and that their clocks could be adjusted to coordinate with the DCE operating at the other end of the communication channel. While in

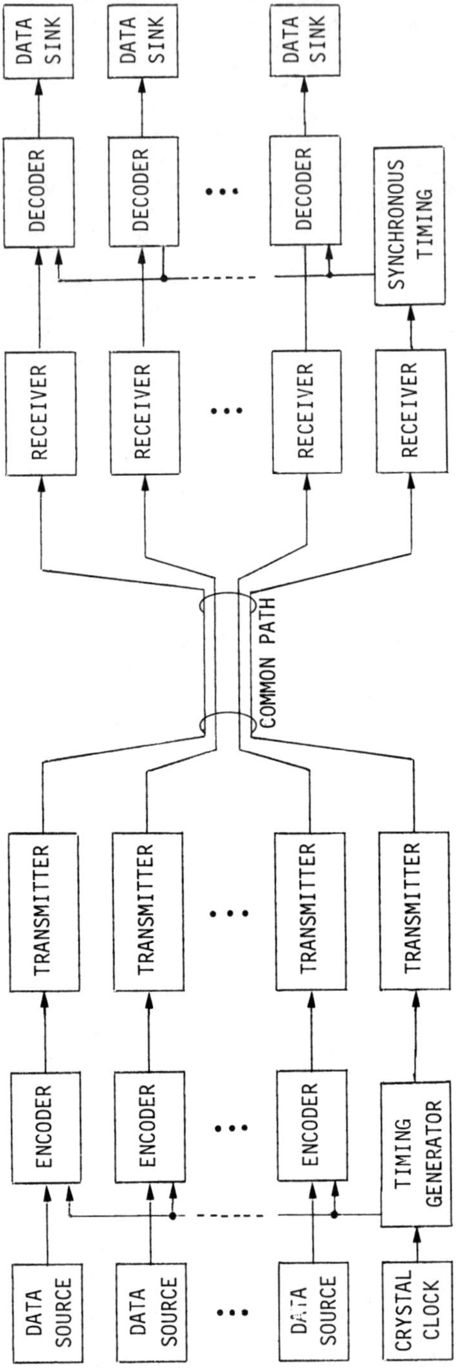

Fig. 3.55 Transmission of timing data over a parallel separate channel

asynchronous transmission timing distortion is allowed since each character is synchronized separately, this however, is not allowed in synchronous transmission.

A most important aspect of synchronous transmission is that the essence of the clock timing must be sent with the data from the transmitter to the receiver. It is then up to the receiver to continuously adjust its operations in accordance with that of the receiver. The term 'essence' is used here since it is not the actual clock timing information that is transmitted by only the digital data stream. The receiver recognizes the clock timing by extracting all the bit state transition in the data stream. It is thus the function of the receiver to maintain synchronization even when no data is transmitted, i.e., even when there are gaps between the messages. This is therefore the most critical design aspect of the receiver, and various techniques have been proposed to solve it.

The simplest synchronization method is to transmit all the timing synchronous data on a separate communication path, as seen in Fig. 3.54. This could be accomplished either on a separate transmission line or a separate frequency channel in an FDM system. Such a technique is most popular with PCM transmissions, where a number of channels are operating in parallel. In this case one of the channels is reserved exclusively to transmit the clock timing.

In the early synchronous transmission systems, which employed RZ data formats, a signalling bit was transmitted once every 7-8 data bits, as shown in Fig. 3.56. In such a system, synchronization is achieved by first transmitting a number of the signalling bits, then, once synchronization is acquired, the system can continuously track for this signalling bit. Similar systems to this one are employed in many PCM transmissions.

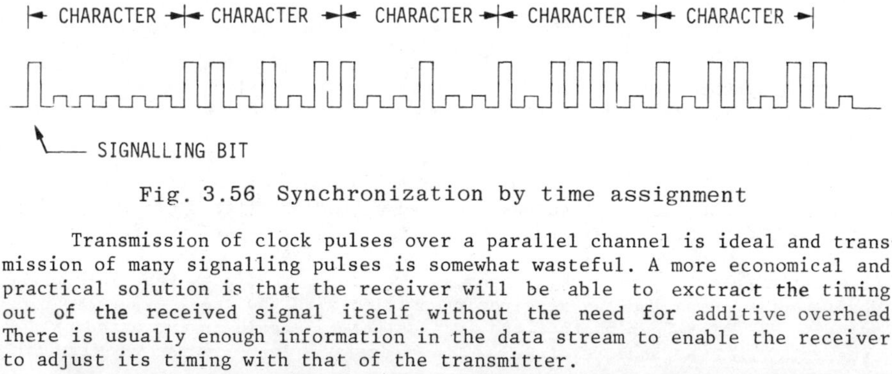

Fig. 3.56 Synchronization by time assignment

Transmission of clock pulses over a parallel channel is ideal and transmission of many signalling pulses is somewhat wasteful. A more economical and practical solution is that the receiver will be able to exctract the timing out of the received signal itself without the need for additive overhead. There is usually enough information in the data stream to enable the receiver to adjust its timing with that of the transmitter.

The timing extraction is very similar to the carrier extraction although their frequencies are different. With coherent synchronous detection, the symbol shaping is restored but not the timing. That is a separate process is necessary to coordinate the clock timing, nevertheless, they may use corresponding techniques to those used in synchronous detection, as seen in Fig. 3.57. In the schematic diagram shown, the clock timings are restored with a phase locked loop, although other techniques are also available.

In most synchronous systems the receiver consists of a free running local oscillator which is pulled into synchronization by the transition pulses, as shown in Fig. 3.58. In other words, it is essential for the oscillator output to be fed back as in the demodulation circuit. Reconstruction of the timing enables the receiver to sample the regenerated symbols and accurately reproduce each bit in the stream. When switching on the receiver, it must first search for proper synchronization by adjusting the local clock pulses

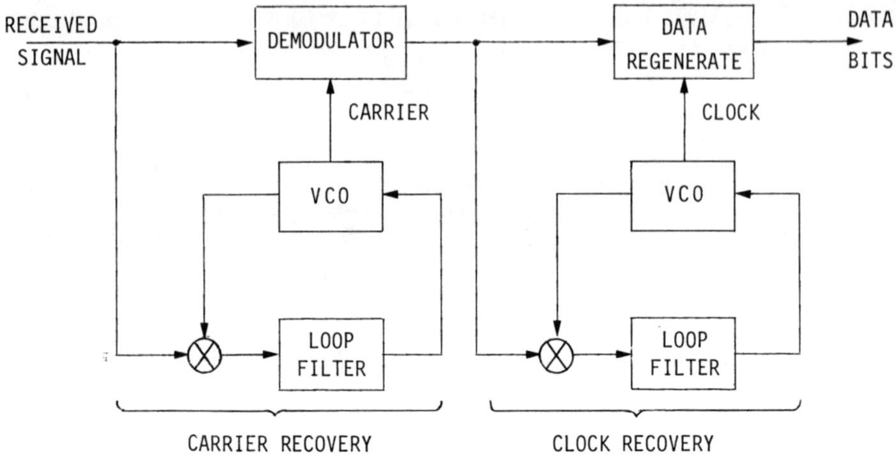

Fig. 3.57 Carrier and time recovery in synchronous transmission

with those it received, and then it must find means of maintaining the receiver in this synchronized position. In other words, the receiver must remain in synchronization with the transmitter even during intervals when no bit state transition occurs in the demodulated signal. The efficiency of a synchronous transmission can therefore be measured by two factors:

(a) Acquisition time, which is the process time for establishing synchronization.

(b) Tracking time, which is the process time for correcting the synchronous drift.

The synchronization process requires that initially the coordination must be acquired rapidly and it must then continue to track the data transition over the entire transmission period.

It is usual to have the receiver's local oscillator operating at a much higher frequency than that required for the clock timing, T. If the oscillator is N times faster, the N pulses produced by the oscillator can then be used to compare with the transmitter timing. Acquisition of synchronization is achieved by recycling the N internal pulses till T/N coincides with the next transition pulse. The tracking process continuously adjusts the clock phase, persistently comparing the received transition with the local generated clock pulses. It determines whether the internal clock should be increased or decreased by $1/N$ seconds relative to the received transitions. By limiting the steps to only $1/N$ increments in the oscillator stepdown chain, the resultant change in clock timing is small, thus providing smooth continuous tracking. This is only a general description of the synchronization process and there are several techniques that may be used for gaining and maintaining synchronization.

The first method, shown in Fig. 3.59, is based on a free-running multivibrator which locks its timing when it coincides with the extracted timing information. To maintain constant accuracy, the receiver clock timing must be frequently corrected, although the multivibrator remains locked even if it misses a transition. The disadvantage of this system is that it requires a relatively long acquisition time before synchronization is gained. This can only be achieved if dummy data of a number of '1' or '0' state bits (i.e., transitions) is transmitted before the receiver can establish synchronization with the transmitter. Another disadvantage of the system is that once it

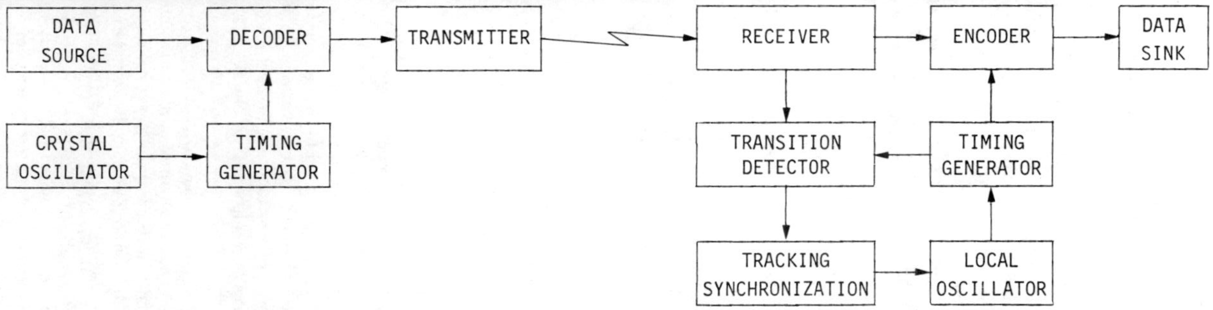

Fig. 3.58 Schematic general block diagram of synchronous data communication

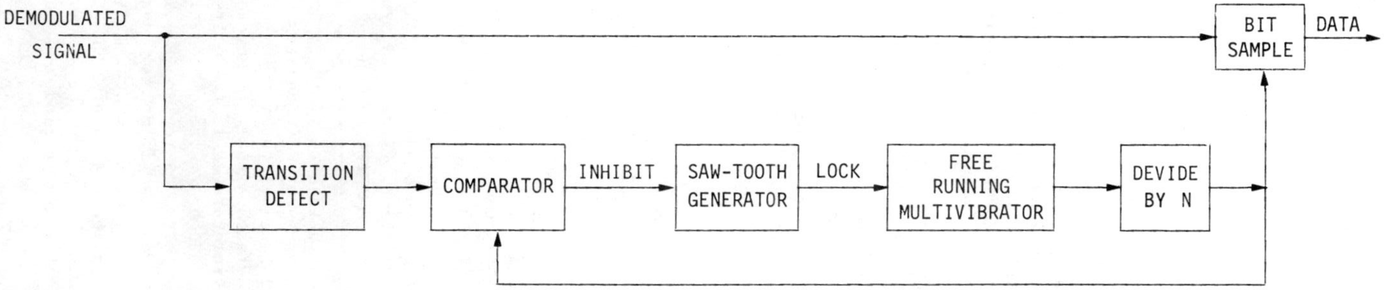

Fig. 3.59 Simplified diagram of a continuous bit synchronizer

loses synchronization it takes a long time to re-establish it. For this reason this technique is seldom used today although it was used in early modems.

A better solution is where the correction is performed slowly so that after synchronization is acquired, it maintains it also in the absence of a large number of bit transitions. An example of such a scheme is given in Fig. 3.60, where the clock timing is adjusted in steps of a small fraction of the symbol interval by adding or deleting count pulses at higher frequency points in the chain. The bit period is sampled once every N internal clock pulses. That is, the clock timing is counted down to produce the accurate timing of each bit, and the internal bit of the receiver is then compared with the transmitted data bit transition. If the internal clock-timing pulse arrives before the data bit, then the internal pulse is advanced by one sample pulse; if it arrives after the data bit, then it is held for one sample pulse period. In the absence of an input data bit, it first advances by one step at the beginning of the internal pulse and then deletes it at the end of the pulse, resulting in no change of the internal timing. This way the receiver can remain in the original timing even in the absence of data transitions.

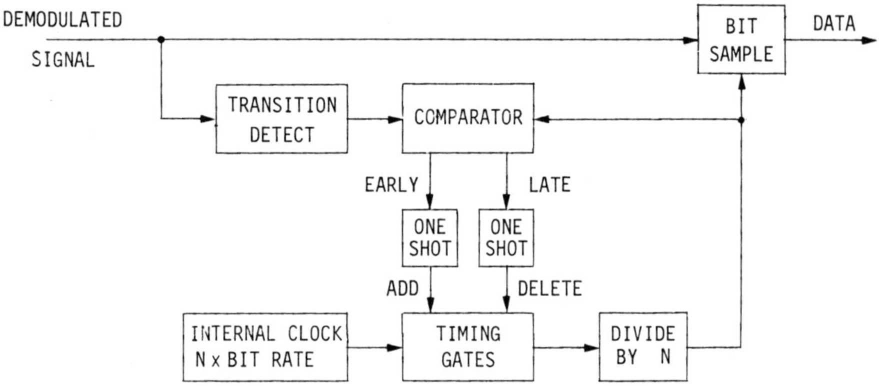

Fig. 3.60 Bit synchronization by adding and deleting 1/N increments

As already mentioned, the timing information can be extracted from the reconstructed received signal, using the same techniques employed for coherent demodulation, such as phase locked loops. In this case the timing is carried in the modulation of the data symbol itself. That is, there will be a component of the bit rate frequency in the received signal; for example, in amplitude shift keying the received signal $S(t) = A(t) \cos \omega_c t$ with $A(t)$ containing the data base timing. Some of the systems which employ these techniques are presented in chapter 6.

The receiver obtains its clock timing information from transitions in the demodulated signal. Therefore, synchronous transmissions can only be properly interpreted if there are transitions in the non-return-to-zero data waveform. Such a strict condition cannot always be enforced, and for this reason many systems replace the NRZ formats with other formats which have at least one transition for every one or two consecutive bits, such as the Manchester or the Miller code formats. Another popular technique is using differential coding, where the transmitted bit represents the transition of the present bit with the previous bit in succession. Each bit is held (delayed) for one period and then compared with the present bit, as seen in Fig. 3.61. This comparison process is performed by modulo-2 addition (exclusive-OR operation).

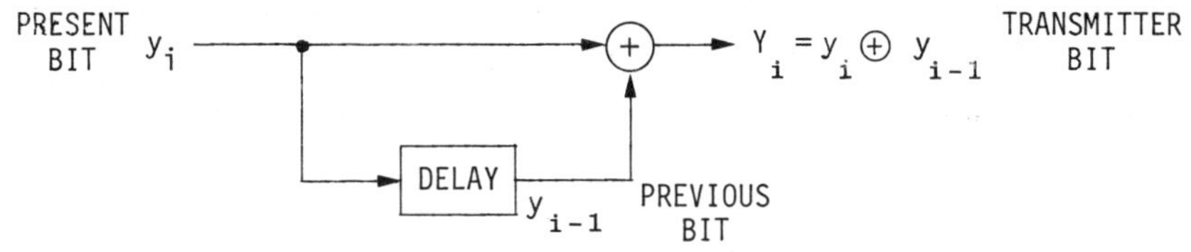

Fig. 3.61 Differential data encoder

In the case of quadrature phase shift keying, differential encoding can be taken one step further. Each transmitted phase, is not compared with a fixed phase reference but with the directly previously transmitted phase. In this case, if in a sequence, only the bits which equal 1 are transmitted then there is a phase shift of 45° between each successive transmitted symbol, as shown in Fig. 3.62. This technique ensures that there is a transition for each dibit, no matter what are the data contents. Similar techniques of forcing a transition for each transmitted bit are used also for minimum frequency shift keying, as will be explained in chapter 6.

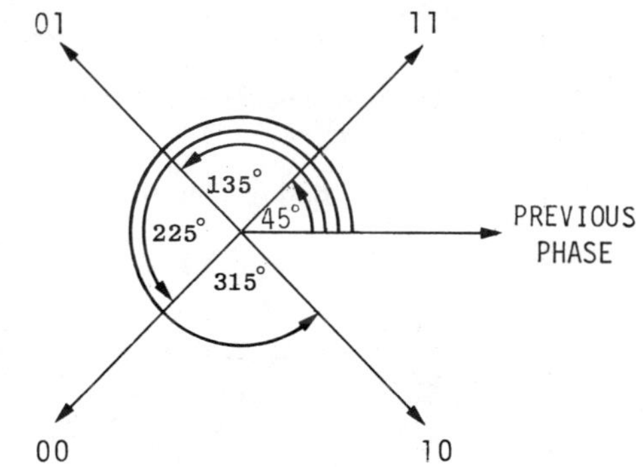

Fig. 3.62 Differential phase coding

(iii) *SYNCHRONOUS/NONSYNCHRONOUS FRAME TRANSMISSION*

Synchronous data transmission systems is defined as one that continuously transmits the data over the communication channel at a constant speed. Only analog or state source data can be regarded as forming a continuous infinite stream of data, while most of digital source data generally is a finite data stream. That is, data messages originating from terminals or computers may be short or long, even though they are always bounded by a given (although possibly flexible) length size. With short messages of single characters, asynchronous (time) transmission may be used. Time synchronization is essential, however, for message transmission, since the receiver is not expected to acquire synchronization with each message transmission. Furthermore, with fast speed transmission, the receiver is not expected to acquire synchronization before each message, since this will reduce the transmitted data rate. In other words, with message transmission the receiver must always be fully synchronized with the transmitter, irrespective of the length of the message or the gap between the messages. That is, when each message is decoded, the receiver must be in its time tracking mode and not in any time acquiring mode. Time synchronous transmission if so, solves the question of 'when to look' for the bits, while it is the frame synchronous transmission that solves the problem of 'where to look' for the messages. Although the channel uses synchronous time transmission, the messages are nonsynchronous.

The term nonsynchronous transmission refers to intermittent flow of data while it does not imply that the timing coordination between transmitter and receiver are not available. Nonsynchronous (frame) transmission usually refers to systems which produce the data in bursts of large number of bits with the data flow achieved only if the receiver is fully time synchronized with the transmitter. The reason for using the two terms, asynchronous and nonsynchronous transmission is to enable one to distinguish between the two transmission systems where both produce the data in intermittent flow, although one system requires synchronization before each character while the other system requires the receiver and transmitter to be fully synchronized. Asynnchronous transmission, therefore, refers to slow speed start-stop transmission, while nonsynchronous transmission refers to high speed message burst transmission. Both asynchronous and nonsynchronous require their transmission to have initiating and ending symbols, although with asynchronous transmission the start and stop symbols are provided for each character while with nonsynchronous transmission the initiating and ending symbols are provided for each message. As a result the overhead data with nonsynchronous transmission is considerably less than with asynchronous transmission. The main differences between the two transmission synchronization systems, however is not in the transmitted block size or in the overhead required, but in utilization of the communication channel. That is, in the channel performance during the spacing between the block transmission. In asynchronous transmission, as seen in Fig. 3.63, the interval period is completely random, while with nonsynchronous transmission the interval period always has an integral number of clock perios T, since the receiver must be fully synchronized with the transmitter. For this reason, nonsynchronous transmission efficiently utilizes the communication channel and therefore allows high speed transmission of the messages.

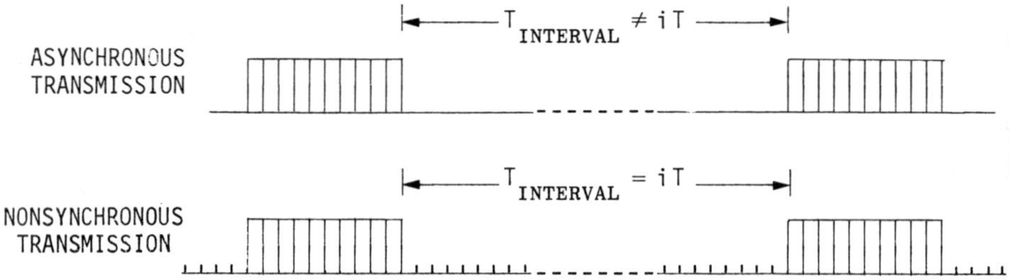

Fig. 3.63 Character transmissions in asynchronous and nonsynchronous systems

In general, frame synchronous transmission is arranged to transmit the data in fixed formats. The message block is typically referred to as containing characters rather than referring it to containing bits. Each of these characters refers to an alpha numerical code and to any other type of symbol which may be required for processing. Some of these characters are also reserved for communication purposes.

In asynchronous transmission, the start and stop signals enable the receiver to define the sampling timing references for 'when to look'. In nonsynchronous transmission the timing information is already established, with the initiating and ending symbols defining the framing references for 'where to look'. In any particular message communication system, the block frame format is known, and hence the initiating and ending symbols are also known. This knowledge is built into the data communication equipment (DCE) or data terminal equipment (DTE) and is defined by accepted communication protocols. The receiver recognizes the initiating symbols as an instruction to prepare the DTE to accept the transmitted message. The message length may be variable and therefore the ending symbol may be used as an instruction that the message transmission is completed. Nevertheless, even in fixed message length transmission the ending symbol is essential, as this is recognized as an instruction to test for transmission error. In variable message length it is essential to add a symbol giving the number of characters in the transmitted block.

There are a large number of message formats that may be used to construct the transmitted block in a communication system. These formats are defined in accepted national or internation standards, although each communication system is left the freedom to invent its own format. The only world wide accepted format is that used for packet switching (see chapter 8) since there it involves sending messages between countries. An example of this is shown in Fig. 3.64 which has a frame beginning and ending symbol of 8 bits referred to as flags. That is, the message frames are delimited by a flag byte which has the value of 01111110. The flag has an accepted character size of 8 bits, referred to as a byte or octet. This format, however, uses bit stuffing rather than character stuffing as described in the previous chapter.

FLAG 01111110	MESSAGE DATA BITS	CYCLIC-REDUNDANCY	FLAG 01111110

Fig. 3.64 Message frame in an accepted format

Just before the ending of the message frame a double octet is inserted for error detection and this is based on a 16 bit cyclic redundancy code. The rest of the message frame is reserved for the message text, which is completely divorced from the communication format process of 'where to look'.

3.11 COMMUNICATION CHANNEL FAULT DIAGNOSTICS

Communication data transmission operations cannot be measured by the performance collection of the individual equipment, but by the performance of a complete system involved in the transaction of data transfer between two points. The successful system operation depends not only on the correct functioning of each of the units, individually and collectively, but also on the system performance in the presence of the external disturbances, such as noise, delays, attenuation and fading which could cause damage to the data. With ordinary equipment it is usually sufficient to test to see if it is operative or faulty; this, however, is not enough with a communication system. That is, in a communication system most of the faults are due to intermittent faults rather than permanent faults, a situation that can only be tested with the whole system fully operating.

The communication system is intended to provide a service of transferring the data from the collecting point to the processing centre or between processing centres. Such a service can be guaranteed only if the system can operate reliably and at a low running cost. It is therefore expected that these systems would be able to perform without any manual intervention. When a fault may occur in the system, means must be provided for speeding up recovery from system degradation by transferring the data to alternative transmission paths or from system failure by transferring the malfunction unit to standby equipment.

The sophisticated data communication equipment is usually designed these days in such a way that it can monitor its own communication link and locate the fault by built-in diagnostic tests. Upon instruction, the equipment can allow the diagnostic operation to occur locally or remotely at both ends of the channel without any human intervention. The diagnostic test involves looping the signals back towards their origin and measuring the results. The full advantage of the remote diagnostic tests can only be gained if the system is transferring the data in both directions simultaneously. This, however, is not feasible in all communication systems since the equipment is designed to transmit and receive the data at both ends of the channel, although both do not function at the same time.

With a loopback circuit one feeds into the system a test signal and loops it back at given points in the system. This diagnostic process assumes that there are no permanent system failures but only system degradation. (System failures are continuously monitored and their results are displayed by special status reporting lamps.) This means that the diagbostic signal transmitted through a loopback circuit will always produce a received signal. What is tested is the quality of this received signal. By comparing the received diagnostic regenerated data with the transmitted data it is possible to check the error rate. The loopback is extended a stage at a time by adding new units to be tested in accordance with a routine test program. If the test conforms to the error rate specification, then the degradation is not caused by the unit under tests. These routines are as follows:

(a) Checking the communication terminal equipment (DTE) by looping it back on itself (as seen in Fig. 3.65(a)). This loopback is performed at the output of the DTE or at the input of the DCE.

(b) Checking the digital coding and decoding performance of the DTE and DCE by looping it back after all the digital circuiting (as seen in Fig. 3.65(b)).

(c) Checking the transmitting and receiving performance of one end of the communication channel (CC) by looping it back on the communication line termination (as seen in Fig. 3.65(c). This test controls all the digital and analog circuitry of one side of the communication data link.

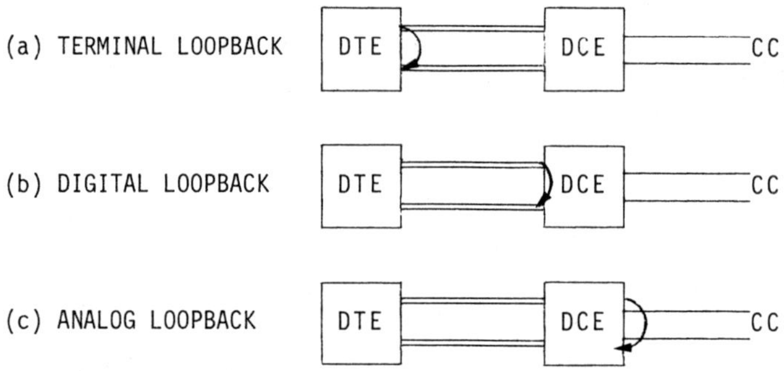

Fig. 3.65 Diagnostic test routine in the local installation by loopback

If the communication channel consists of two wires it enables the data to flow in a single direction at any given time. This means that only the local equipment can be diagnostically tested, while the rest of the communication system equipment must be tested remotely. That is, the same test routine must be repeated at the other end, as seen in Fig. 3.66. The instructions for performing these tests and their results must be transmitted to and from the centre over a difference communication link.

A much easier diagnostic routine test exists in a four wire link where each couple of wires transmits its data in the opposite direction. That is, the loopback circuit can be further extended to include also the remote equipment of the communication channel :

(d) Checking the performance of the communication link (as seen in Fig. 3.67(d) after the assurance that there are no faults in the local equipment. In this particular test, the result showing that the error rate is higher than the specification is an indication for the system to transfer to an alternative communication path.

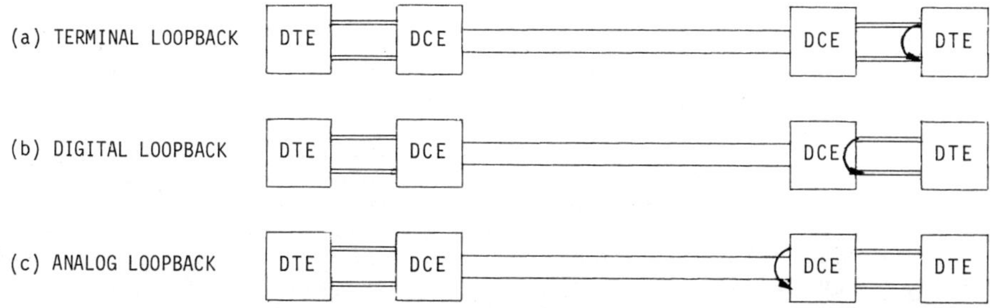

Fig. 3.66 Diagnostic test routine in the remote installation in a two wire link

(e) Checking the performance of the remote DCE, as seen in Fig. 3.67(e), which includes all the remote regeneration equipment.

(f) Checking the performance of the remote DTE, as seen in Fig. 3.67(f). This final diagnostic loopback tests the performance of the complete communication system.

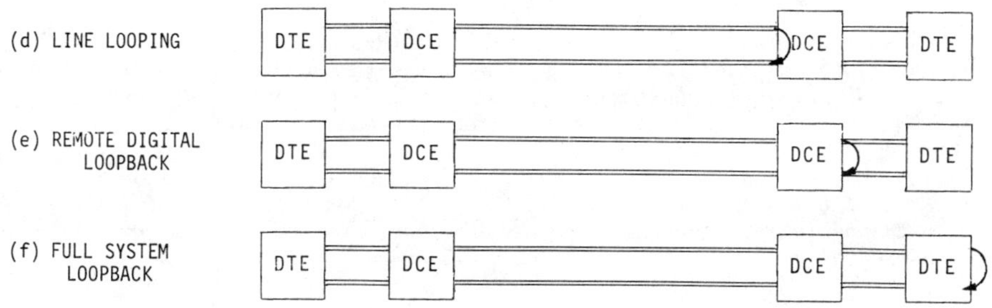

Fig. 3.67 Diagnostic test routine in the remote installation in a four wire link

In some applications, where no alternative paths are available, it may be neccessary to test the communication line performance to locate the reasons for the disturbances which cause the system to degrade its operation. In these tests only the analog circuits are tested, which requires taking the digital equipment off-line, as seen in Fig. 3.68. In this case each wire pair must be tested separately at both ends of the communication link. These tests can provide details of the quality of the link and can assist the designer in deciding what extra preventing equipment are needed, such as equalizers and error control.

Fig. 3.68 Diagnostic testing of the analog line performance

These tests usually measure the error rate as performed before, although they may be extended to inlucde extra facilities which can examine the level and loss performance over the link. That is, these tests can include accurate attenuation distortion and envelope delay distortion measurements, both as a function of the frequency. Other tests can measure the white and impulse noise as a function of time.

The bit error rate (BER) test is the basic test for all the diagnostic loopback circuits. The result is usually an indication of the predicted errors which are in the range of a number of errors per million bits transmitted. This, however, is not enough since it does not provide details of their inter-relations. That is, the errors may be randomly spread over in time or may be clustered together caused by a single impulse noise. The data rate should therefore give an indication both of the absolute number of errors and the number of block errors. The block errors give the predicted number of actual retransmissions required. That is, a cluster of errors will be recognized as a single error since it causes the cancellation of a single block. It should be added that all these error rate diagnostic tests are performed with the error correction techniques fully operative. In communication systems which measure the block error rate, the block size is defined by the DTE and not the DCE.

3.12 CONCLUDING REMARKS

The major operation of Command and Control systems is transferring the data collected in the field from remote geographical locations to central points where it is processed into meaningful information for decision making. The decisions are then transferred back to the remote locations for implementation. Nevertheless, the major task of Command and Control systems is not data transfer but decision making, while the communication operations are intended only as a service. This means that the designer must always keep in mind that the data transfer is not a target on its own, but is a means of assisting the implementation of the Command and Control tasks.

The design of the data communication equipment is a matter of trade-off where the following points are to be considered:

(a) Narrow excess bandwidth enables more links to be transmitted in a single channel, while wide excess bandwidth provides better signal and time recovery.

(b) Small number of signal levels are more immune against noise while large numbers of levels provide means for high data rates. The penalty for extra signal levels is a decrease in the signal-to-noise ratio of about 10 dB for each extra bit per level.

(c) Large overhead error coding ensures better data transfer reliability but reduces the data transfer efficiency.

(d) Synchronous transmission requires expensive time tracking circuits although it provides means for high speed transmission.

(e) Sophisticated modulation techniques may increase the speed but also increase the equipment cost.

When designing data communication equipment it is not the cost of only the equipment to be considered but the cost of the whole communication system. The costs of the DCE and DTE are possibly the most complex equipments in the data transfer operation, but it accounts for only a small part of the overall cost of the communication system. The cost of the equipment should be judged not by its price alone but mainly by its performance in exploiting the communication channel. The communication system is only intended as a service for the Command and Control system and therefore it must be regarded in this context only. That is, it is not the cost of the communication system that must be considered but its cost effectiveness for Command and Control application. The designer must determine his trade-off considerations by taking into account the price of transmitting a bit of information in a given time.

In communication systems the designer usually specifies the average data rate required, where the statistical average figure is generally measured over a period, say 24, 12, or 6 hours. The equipment system can then be guaranteed to transfer all the data accumulated in this period. In Command and Control systems there may be a need to transmit only a few messages per day, but these messages are required to be transferred in real-time. In this case the communication system must be judged by the cost of transferring N bits with a delay of T seconds, where T is defined by the real-time specifications. This means that it is not the average volume of data that is to be considered in defining the equipment speed, but the volume of data that must be transferred in the peak periods only. The cost effectiveness of the communication must be based on the price the designer is ready to pay to achieve the so-called 'real-time'. This probably is the main trade-off consideration where he must compromise between the price of transmitting a message at the peak period and the delay time he allows for this process.

The designer is confronted with a vast range of communication system speeds. These can be categorized into three main classes depending on the communication channel employed.

(i) *SUBVOICE TRANSMISSION LINES*

These communication channels are used for very low transmission speeds where the data source is usually a result of start-stop terminals. Since the speed is extremely slow, the bandwidth required is less than that available for telephone voice lines. As a result the transmission required is only part of the voice line and consequently many data transmissions can be implemented in parallel or the data can be transmitted in parallel with the voice. The latter is known as data-under-voice and is made possible since not all the voice bandwidth is really necessary for voice coherent transmissions. Table 3.1 gives some of the accepted speeds in this class.

Table 3.1

SPEED	MODULATION	TIME SYNC.
50 75 110 150 200 300 600	OOK or FSK	ASY

(ii) *VOICE TRANSMISSION LINES*

These communication lines are the most popular means of transferring data where the voice telephone lines or their radio equivalents are employed. The data source for these transmissions is from sophisticated DTEs, which have means of storing the data prior to transmission. The data sources can also be from computers, or even analog sources in the form of PCM. The data transferred in this category exploits the full voice communication line. In this respect the line is considered to be of only two wires which allows the data to flow in one direction only at any given time. Table 3.2 gives some of the accepted speeds in this class.

Table 3.2

SPEED	SUB CLASS	TRANSMISSION LINE	MODULATION	TIME SYNC.
600 1200	Low Speeds	Dialled Lines 3002	OOK FSK	ASY/SYN
1200 1800 2400	Medium Speeds	Leased Lines C1	FSK Doubinary DPSK	SYN
2400 3600 4800	High Speeds	Leased Lines C2	VSM QPSK QASK	SYN
4800 5400 7200 9600	Very High Speeds	Conditioned Leased Lines C4	VSM APSK	SYN

(iii) *WIDE BAND TRANSMISSION LINES*

These communication lines are used to transmit large volumes of data as required for PCM or computer to computer communications with ultra high speeds of 50 K to 20 M bits per second. The channels employed here are specially wideband such as can be achieved with cables or radio. These speeds are employed with satellite data communications and are becoming the present means of data transfer over long paths.

3.13 REFERENCES

(a) DATA TRANSMISSION PRINCIPLES

1. Bennett, W.R. and Davey, J.R. *Data Transmission*, McGraw Hill Inc. 1965
2. Lucky, R.W., Salz, J. and Weldon, E.J. *Principles of Data Transmission* McGraw Hill Inc. 1968.
3. Martin, J. *Telecommunication and the Computer*, Prentice Hall Inc.. 196
4. James, R.T. *Data Transmission – The Art of Moving Information*, IEEE Spectrum, January 1965, pp.56-83.
5. Franklin, R.H. and Law, H.B. *Trends in Digital Communication*, IEEE Spectrum, November 1966, pp.52-58.
6. Hersch, P. *Data Communication*, IEEE Spectrum, February 1971.
7. Gregg, W.D. *Analog and Digital Communication*, John Wiley and Son, 1977 603 pages.

(b) MODEMS

8. Davey J.R. *Modems*, Proc. IEEE , vol.66, no.11, November 1972, pp.1282 - 1292.
9. Edwardsson, K. and Nyman H. *Modems for Data Transmission on Voice Grad Lines*, ITT Electrical Communication, vol.48, no.162, 1973, pp.110-1
10. Andrews, H.R. *Let's Take the Mystery out of Modems*, The Electronic Engineer, July 1972, pp.DC-5 – DC-10.
11. Roiz, E.F. *Considerations in the Application of Modems to Data Processing Systems*, Telecommunication, December 1968, pp.19-23.

12. Lyon, D. *How to Evaluate a High-Speed Modem*, Telecommunication, October 1975, pp.35 - 39.

(c) EQUALIZERS

13. Westcott, R.J. *An Experimental Adaptive Equalized Modem for Data Transmission over Switched Telephone Network*, The Radio and Electronic Engineer, vol.42, no.11, November 1972, pp.499 - 507.
14. Hirsch, D. and Wolf, W.J. *A Simple Adaptive Equalizer for Efficient Data Transmission*, IEEE Trans. vol.COM-18, no.1, February 1970, pp.5 - 12.
15. Kretzmer, E.R. *The Evolution of Techniques for Data Communication over Voiceband Channels*, IEEE Communication Society Nagazine, January 1978, pp.10 - 14.

(d) TRANSMISSION ENCODING

16. Nyquist, H. *Certain Topics in Telegraph Transmission Theory*, AIEE Trans., vol.47, April 1928, pp.617 - 644.
17. Shannon, C.E. *Communication in the Presence of Noise*, Proc. IRE, vol.37, no.1, January 1949, pp.10 - 21.
18. Pierce, J.R. *Some Practical Aspects of Digital Modulation*, IEEE, Spectrum, November 1968, pp.63 - 70.
19. Lender, A. *Correlative Level Coding for Binary-Data Transmission*, IEEE Spectrum, February 1966, pp.105 - 115.
20 Gallager, R.G. *Information Theory and Reliable Communication*, J. Wiley, 1968.
21. Forney, G.D. *Coding and its Application in Space Communication*, IEEE Spectrum, June 1970, pp.47 - 58.
22. Schmidt, K.H. *Data Transmission Using Controlled Intersymbol Interference*, ITT Electrical Communication, vol.48, no.162, 1973, pp.121 - 133.
23. Ristenbatt, M.P. *Alternatives in Digital Communication*, Proc. IEEE, vol.61, no.6, June 1973, pp.703 - 721.

(e) MODULATION TECHNIQUES

24. Swartz, M. *Information Transmission, Modulation and Noise*, McGraw Hill Book Company, 1970.
25. Smith, J.G. *Spectrally Efficient Modulation*, ICC 1977, pt.1, pp.3.1-37 to 3.1-40.
26. Pasupathy, S. *Minimum Shift Keying: A Spectrally Efficient Modulation*, IEEE Communication Magazine, July 1979, pp.14 - 22.
27. Fielding, R.M., Berger, H.L. and Lockhead, D.L. *Performance Characterization of a High Rate MSK and QPSK Channel*, ICC, pt.1, 1977, pp.3.2-42 to 3.2-46.
28. Foschini, G.J., Gitlin, R.D. and Weinstein, S.B. *Optimization of Two-Dimensional Signal Constellation in the Presence of Gaussian Noise*, IEEE Trans., vol. Com-22, no.1, Jaunary 1974, pp.28 - 37.
29. Amoroso, F. *Bandwidth-Efficient Modulation for Voice Band*

30. Thomas C.M., Weidner, M.Y. and Durrani, S.H. *Digital Amplitude-Phase Keying with M-ary Alphabets*, IEEE Trans., vol. COM-22, no.2, February 1974, pp.168 - 180.

(f) PHASE LOCKED LOOPS

31. Wetenkamp, S.F., Wong, K.J. *Transportation Lags in Phase-Locked Loops*, Watkins Johnson Tech-notes, vol.5, no.3, May/June 1978.
32. Palmer, L.C., Klein, S.A. *Phase Slipping in Phase-Locked Loop Configurations that Track Biphase or Quadriphase Modulated Carrier*, IEEE Trans., vol. COM-20, no.5, October 1972, pp.984 - 991.

33. Messerschmitt, D.G. *Frequency Detectors for PLL Acquisition in Timing and Carrier Recovery*, IEEE Trans., vol. COM-27, no.9, September 1979, pp.1288 - 1295.

(g) TRANSMISSION SYNCHRONIZATION

34. Booth, R.W.D. *Carrier Phase and Bit Sync. Regeneration for the Coherent Demodulation of MSK*, NTC, 1978, pp.6.1.1 - 6.1.5.

35. Murphy, D. *Digital Techniques in Data Communication*, Telecommunication February 1969, pp.19 - 25.

36. Stuart, S.B. *Asynchronous Communication*, Data Communication Design, January-February 1973, pp.13 - 24.

37. Schunneman, R.F. *Serial Sampling of Asynchronous Signals*, Data Communication, November 1971, pp.28 - 30.

(h) TECHNICAL TERMS

38. Jay, F., (Editor in Chief), *IEEE Standard Dictionary of Electrical and Electronic Terms*, IEEE, 1977.

CHAPTER 4

Data Communication Networks

4.1 INTRODUCTION

Having dealt with the main problems involved in transferring data over a single communication channel, it is necessary to consider now the means for transferring data over a number of channels, serially or simultaneously. This calls for a prior expansion of the subject to present the various communication networks applicable to C^2 systems. This chapter will merely introduce the various available networks, leaving their detailed treatment for subsequent attention in the second half of the book.

The successful operation of command and control systems depends on its ability to transfer data between remote geographical locations, speedily and correctly. The data located in one location may be required to be transmitted to one particular central location or to any one of a number of distant location points. The data may also be required to be transmitted to a number of locations either simultaneously or according to the information content. The communication channel provides the means of connecting two distinct location points, while the communication network provides the means of selecting distant location points and the channel for communication over it.

A data network is a series of points interconnected by communication lines. The network itself can connect innumerable points, with the advantage that any two points, when connected, regard the communication channel as exclusively theirs, as seen in Fig. 4.1. When these two remote points are connected, the characteristics of the communication channel between them are defined. While the communication channel performance is specified in terms of frequency band with its limitation effects of envelope delay and attenuation, the communication network performance must be specified also in terms of availability and throughput with its limitation effects of acquisition delays.

All the analog communication networks are organized on the principles of the point-to-point connection. That is, each two communicating points must have a direct connection link between them. This may be permanent or fixed only for the period of transmission, with the communication path set up by a switching centre.

With data communication networks, some connections may still be organized on the principles of point-to-point (PTP) connection, even though other network structures are available, as appears in Fig. 4.2. Since the data is in digital form, it may be stored and processed en route by computers in a technique known as store-and-forward (SAF). While point-to-point is defined

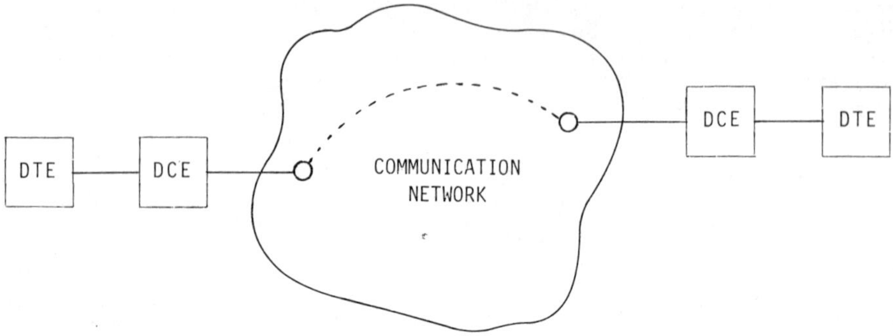

Fig. 4.1 Point-to-point communication connection via a network

as the method of direct communication between two points, store-and-forward is defined as a method of communication in which the messages may be received at intermediate routing points and stored. The messages are then transferred (when a route is available) directly or with further network points with interim storage, each in accordance with its own address. In the case of SAF, there is no direct connection between the distant points when they are so-called communicating, but the messages are stored in the network nodes and are retransmitted in the channels only when such a channel has free time to accept the messages. The advantage of the store-and-forward method is that the transmitting DTE need not acquire a transmission path nor need it ascertain that the receiving DTE is free to accept its messages, as is required in point-to-point systems. Furthermore, store-and-forward systems can provide means of efficiently exploiting the communication channels by ensuring smooth and continuous flow of data.

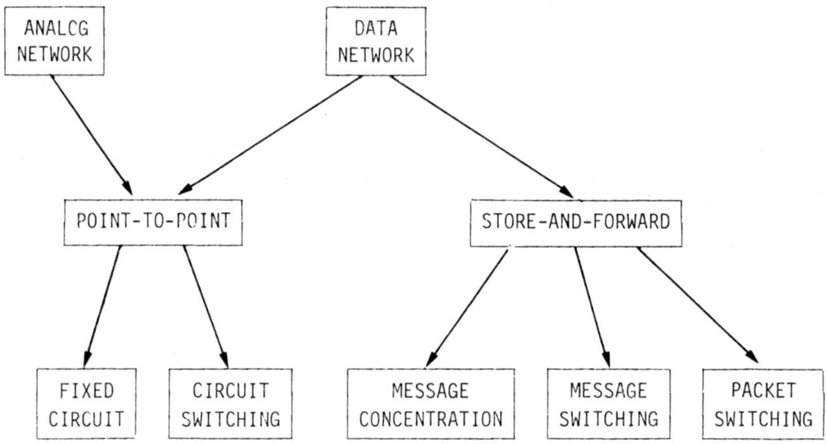

Fig. 4.2 The various structures of communication networks

One usually refers to point-to-point communication as the permanent circuit connection between two fixed points, while semi-permanent connection is generally referred to as circuit switching. That is, the transmission and reception of data between clearly defined termination points is specified as a permanent connection, while a semi-permanent connection is specified as a connection which is set-up for a short duration only. Circuit switching is defined as a method of communication where there is an electrical connection between the calling and the called stations for the exclusive use of the circuit until the connection is released. In this respect, PTP connections are also referred to as circuit connections since the connection provides a complete electrical path between two points and therefore furnishes a two way communication channel. It is important to add that time division multiplex, with the communication channels shared between a number of transmission channels, also provides a momentary point-to-point connection and therefore can be regarded as a circuit-switching network.

Store-and-forward is becoming the main means of data communication, since it can provide the most efficient resources of exploiting the network when short messages are transmitted. With long transaction times, point-to-point circuit-switching may be preferable. Nevertheless, even there, if there are gaps between the transactions, store-and-forward may be the best solution. Store-and-forward principles may be used in three types of communication techniques: message concentration, message switching and packet switching. These will be discussed in detail in chapters 7 and 8.

The communication channels used in data networks can be categorized in three groups, as seen in Fig. 4.3. The channels may be dedicated to a single user, shared by a number of users or switched through a centre or node. It is the communication network which provides these facilities to the communication channel and not a characteristic of the communication channel itself. Therefore, the three types of communication channels may employ both PTP and SAF techniques.

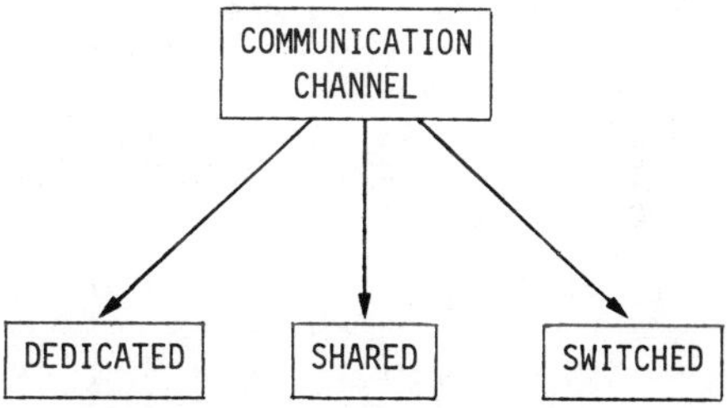

Fig. 4.3 Communication channel facilities as provided by the network

This chapter will survey the various forms of data communication networks and the types of communication channels which have an influence on the network design. The network will first be introduced by means of point-to-point dedicated channels as required for terminal-to-commuter communication. The subject will then be expanded to shared channels as required for both terminal-to-computer and computer-to-computer communication, using both PTP and SAF techniques. Finally, the highlights of computer-to-computer networks will be introduced, employing centre switching technologies.

4.2 TRANSMISSION MEDIA

Data communication can be implemented over various forms of transmission media. The early type of data transmission exploited the already available telephone network, and therefore most of todays DCE's are geared for this type of media. Telephone lines are limited in their frequency band, and consequentially the data communication over these lines is confined to a transmission speed of maximum 9600 BPS, while in most cases the speed is much less. These telephone lines are designed to transport voice communication, and all the equipment attached to these lines is intended to improve the voice transmission. This added equipment, which may be ideal for analog transmission, may be disastrous for data transmission. Examples of such equipment are the repeaters and amplifiers which are intended for speech coordination but may cause phase and frequency jitter and thereby distort the data. Many circuits in the DCE are therefore introduced not for data the transfer process but exclusively to enable the data communication to be piggybacked on the existing telephone media.

There are other various forms of media, although many of them still conform to the same voice frequency band, using the same types of modems with the same transmission speeds as in telephone lines. An example of such a medium is the radio channels. The reason for this is that most radio transmitter-receiver equipments are also intended for voice communication. When these are used for data transmission, they usually only exploit the readily available equipment.

Although the largest portion of data communication equipment is geared for voice channels, there are other channels that could be provided by the so-called telephone companies. The variety of communication channels that are available from the common carriers for data communication can be divided into three classes (as seen in Fig. 4.4) depending on whether the channel spectrum available is a larger or smaller than the voice spectrum. The voice band has a frequency spectrum of 300-3400 Hz, which can facilitate data transmission at speeds up to 9600 BPS. The narrowband channels provide data transmission facilities at speeds up to only 300 BPS. The broadband channels may use extremely wide frequency spectrums and thereby can provide facilities for extremely high speed transmissions: with today's equipment technology it is possible to transmit on these channels up to several million BPS.

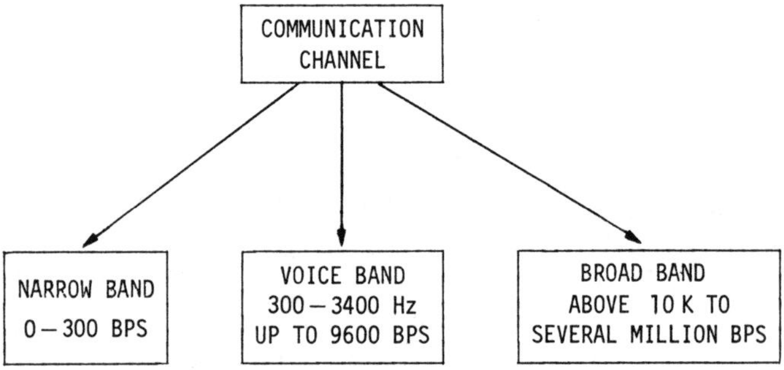

Fig. 4.4 Categories of the available communication channels

There are several forms of transmission media which offer frequency bands other than voiceband and thereby provide different communication facilities. In these cases different DCE modems which are designed to match both the speed and band are necessary. In the physical line communication one must mention twisted wire pairs, which have the advantage over telephone lines by presenting less noise disturbances. Such lines may transmit the actual rectangular pulses without distortions over short distances. The use of twisted pairs allows the transmission of data even without modems, although with restriction in distance. Twisted pairs can also be used with special short hall modems for moderate distances. Such a facility is used on private lines in a local installation.

Narrowband channels are usually supplied by the common carrier and telephone companies by sharing the voiceband channels. Each channel can then be regarded as an independent transmission facility. Such narrowband channels are used for telex and other forms of slow data rate transmission.

Another form of physical transmission media is the coaxial cable which can transmit over a much higher frequency spectrum than a pair of telephone wires. Such a communication link can provide the means of transmitting multi-megabits of data and is relatively immune to noise disturbances. A similar solution to the wideband cable is the optical fibre which provides many attractive properties. While the coaxial cable is intended primarily for analog transmission, the optical fibre can be used for direct digital transmission. Nevertheless, optical fibres still require modems to modulate/

demodulate the digital electrical pulses to light wave pulses. The attraction of optical fibres is that extremely high speed data communication can be achieved over them without any danger of external interferences. The optical fibre cable is very small in diameter and light in weight, and therefore cheap to install. Its main drawback is that its transmission distances are limited and hence it requires repeaters after a few miles.

All the physical transmission media suffer from the disadvantage that a direct connection is necessary between the two communicating points. This disadvantage is overcome by the use of radio transmission which can provide long range communication between points which have no physical wires between them. Such a transmission medium offers a connection for many users in a single network, while some of the points may be mobile stations. Radio transmission facilities are not in principle limited in bandwidth and therefore could be designed specially to accommodate data communication and provide high speed transmission.

The frequency spectrum of radio transmission is wide stretching, up to the gigahertz range. HF band presents the means of very long distance communication, using sky waves to bounce the signals, although the mode of transmission is susceptible to fading caused by changes in the height of the sky waves. VHF and UHF offer better reliable communication, but they are limited in distance to only the line-of-sight. In this case, the high speed transmission over long distances requires relay stations (as seen in Fig. 4.5) to enable the communication to take place. Such systems are used in microwave systems with relay stations spaced about 30 miles apart. Today the same principles are used for satellite communication, where the satellite acts as a single relay station between points hundreds of miles apart.

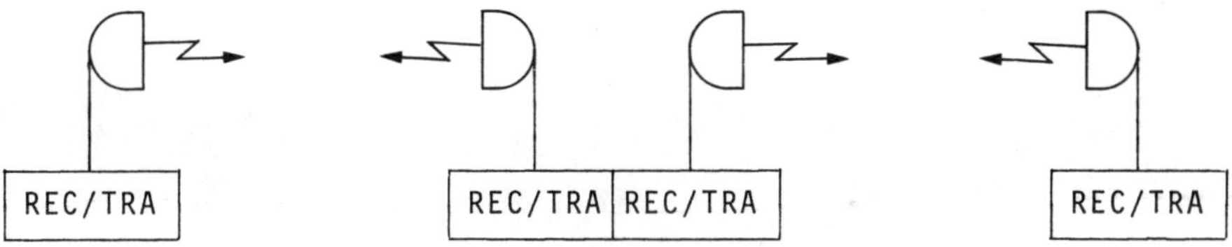

Fig. 4.5 The principle structure of a communication network with relay stations

A more detailed presentation of the various transmission media is given in the following chapter. The subject is introduced here only for assistance in the understanding of the effect of the transmission media on the design of the network.

4.3 TRANSMISSION FACILITIES

When a user needs to set up a data communication network, he is confronted with an extremely wide range of similar communication equipment from which to choose. This, however, does not apply to the communication channel facilities. In most contries, all the communication channels, whether wire or radio, are controlled by government bodies or by monopolized common carriers. It is generally impossible to install a dedicated communication channel which crosses over public areas without getting government approval. This does not mean that there are no privately owned dedicated communication channels, but these can only be set up for organizations with specialized private applications. Many organizations, such as banks and electricity suppliers, have established complex networks for their exclusive use whereby the equipment and channels are under their prime control. Nevertheless, most communication networks are designed with communication channels owned and controlled by government or public bodies.

The basic choice of transmission facilities is therefore that provided by the telephone companies which are regulated either by the government or by the common carrier companies. These facilities are not restricted only to voice communication channels but could also provide broadband communication channels. Today it is possible to lease from the common carrier companies long distance channels via satellite relays or even optical fibres where these exist.

The variety of network facilities is extremely wide, as shown in Fig. 4.6. Nevertheless, the choice of which facility to select is rather simple, since each type of facility provides both definite advantages and particular drawbacks. The basic consideration is that of speed, a determining factor where a public or dedicated channel is required. One must also consider here the effect of real-time and the time one is ready to wait before a channel is acquired. The second consideration is that of the application for which the channel is intended, whether it is exclusive or may be associated with other users. In most cases it is not advisable to set up an exclusive private network (as will be explained later) because of cost and services.

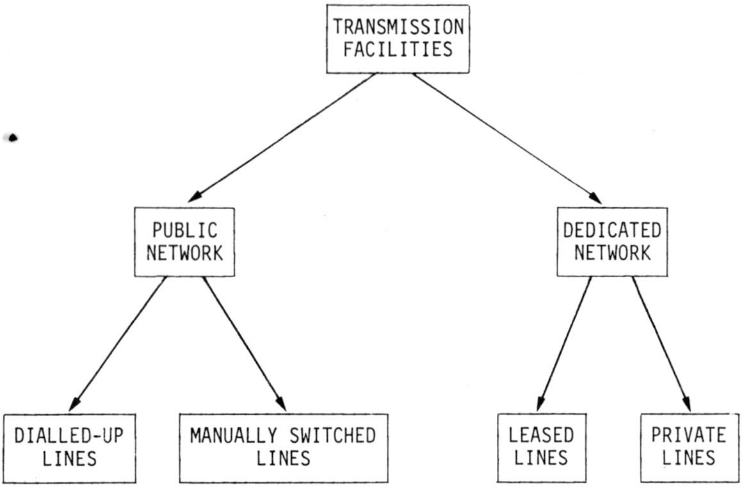

Fig. 4.6 The variety of the available transmission facilities

The public network is designed primarily to provide transmission whereby the subscribers throughout the world can communicate between themselves. Most of the channels in the public networks today are intended for analog transmission, while the data transmission is only piggy-backed on the existing facility. This being the case, the transmission speeds over these channels are limited. While humans can interpret words and missing vowels from the received noisy voice signal, data signal reception is very sensitive also to the line characteristics. Nevertheless, communication over analog switching circuits provides the most easily available facility and is for this reason the most commonly used.

The common carriers and telephone companies also can provide communication facilities which are not intended for voice communication, with channels suitable for either specially low or specially high transmission rates. The low rate channels are mainly used for the slow start-up terminal communication, with the telex world wide system being a typical example. High rate channels are also offered by the common carriers, both for data and video communications. These channels, such as coaxial cables, can provide megabit data transmission as well as the transmission of television programmes.

The tendency today is to replace many of the analog switching circuits in the exchange with pure digital circuits. The digital exchange can provide much wider facilities both for voice and data communication. Furthermore, digital switching introduces no errors into the system since it is less susceptible to external noise or crosstalk.

The simplest means of exploiting the public network is through the dialled-up lines. Dialled-up lines are in the form of two copper wires supplied by the common carrier on the public telephone network. These lines are intended and guaranteed to carry voice communication but they could also be used for data transmission. The subscriber simply attaches to the line his modem equipment instead of his telephone set, either permanently or through a switch. The subscriber can thereby dial up directly to the location of the computer or terminal he wants to communicate with. The computer can be permanently connected to the telephone line or can be switched in by the operator. Today the computer can automatically accept calls, with the calling subscriber inserting a particular code identifying him to the computer. In fact, there are available special automatic-call and automatic-answer units intended for unattended operations which are specially designed to recognize the telephone tones.

Dialled-up lines through the public network are also refrred to as direct distant dialling (DDD). The advantage of the DDD facility is that lines are available all over the world and anyone with the suitable data communication equipment can just dial up his computer and communicate over the public network. There is no special need to get permission for each connection nor is there any restriction on the location of the telephone. That is, the dialled up call can even take place from a public call box. There are special DCE units which enable one to communicate by special acoustic adapters where there is no need to physically connect the modem to the line as it can be attached only to the handset. Another important advantage of the DDD is that the same channel can be used both for speech and data (as seen in Fig. 4.7) which makes the channel most cost effective if the amount of data is small. In this case the channel is used for normal office work and is converted to data only when required. If the DCE is connected to the local office exchange, such an arrangement can provide alternative channels if one of them is faulty.

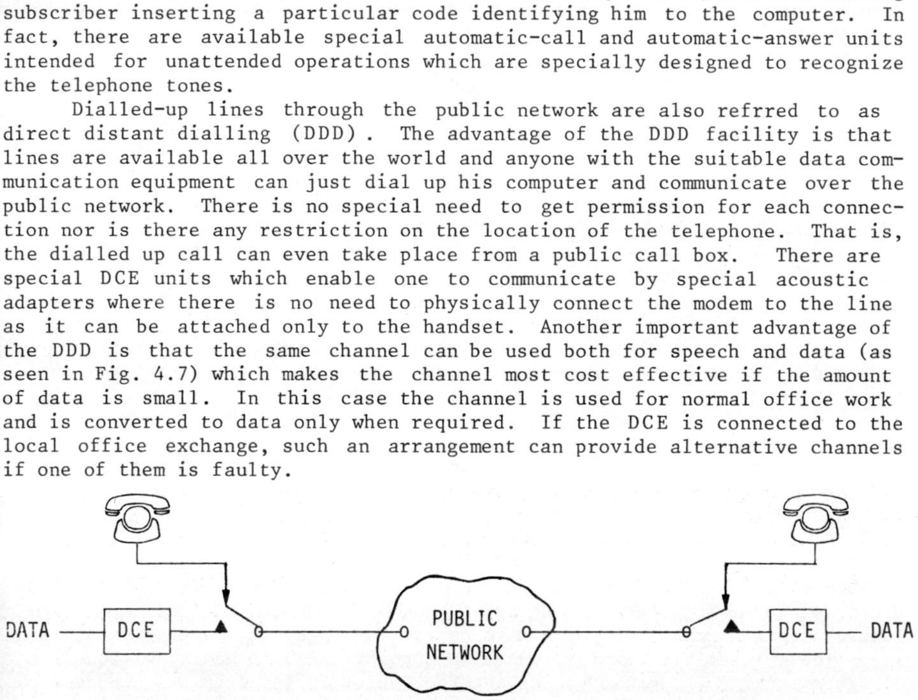

Fig. 4.7 Normal communication through the DDD facility

The disadvantage of the DDD facility is that the channels one obtains when dialling could be of inferior grade, for most DDD lines are noisy with considerable delay distortion. When one dials through the public network, one can encounter any type of channel performance suitable for voice communication, although some may not be suitable for data communication. When the operator talks one cannot analyse the line performance by ear, except by testing the line with data. The user therefore, has no choice of line performance and must be satisfied with whatever is made available. Sometimes the line obtained is relatively good, but normally it is noisy with impaired characteristics. It is essential to regard each line with caution, checking whether its characteristics are suitable. It may have equipment attached to it which would not only disturb the transmission but may even prevent it. The dialled lines usually have control exchange signals within the bandwidth and/or other equipment required in some voice communication systems, such as filters, repeaters, echo suppressors, etc. In this case, one can only try to make the transmission after making the connection, and if the number of errors mount up, the only solution is to close down and redial. Even if the encountered lines are relatively good, their characteristics still prevent high data transmission above 2400 BPS. It is therefore not recommended to use the public telephone network for speeds above 1800 BPS. With acoustic adapters, if good transmission is required, it is suggested that the speed should remain limited to 600-1200 BPS, so as to keep the error rate low.

The main drawback of the DDD facility comes from the fact that all the lines must pass through the switched exchange, which still employ electro-mechanical selectors. As such, the DDD lines not only present high background noise but are also extremely susceptible to impulse noise a feature which is disastrous to the data. Nevertheless, an important advantage of DDD lines is that there is no need to prepare for alternative lines, as these are always provided by the exchange. It is advisable to employ these DDD lines whenever the data rate is low, since the cost is also relatively low. In one's calculation of data rate, it is important to take into account the long acquisition time and the danger of getting access to impaired lines. Despite the many advantages of DDD lines it is not advisable to employ this facility for real-time applications even if the data rate is low, because the time it takes to set up a good connection may be too prolonged and the slow speeds one can achieve on these lines. The trade-off of selecting dialled up lines or dedicated lines is reached by comparing the overall amount of data to be conveyed with the time available for their transmission, after the subtraction of the delays in obtaining good lines.

Manually switched lines are seldom used today, although there are still many places throughout the world where an operator assists in selecting the required connection. These lines have the same advantages and disadvantages as the dialled up lines, although the mere fact that an operator makes the connection adds to the drawback of the system. An operator cannot help in selecting good lines, a fact which adds to the acquisition time. From personnel experience, it is common practice for the operator to listen in to the conversation, a practice which could be disastrous to the data communication. The operator, after hearing the strange transmission signals, may simply ask whether one is still talking!

For most command and control system applications, where real-time is essential, it is advisable to utilize dedicated lines; either leased or private. Dedicated lines are channels which are reserved for the personal use of a single proprietor. Leased lines are dedicated lines which are leased on full terms from the common carrier at specific degrees of conditioning. These lines are usually leased on a point-to-point basis, without interchange switching arrangement, leaving the leasor to build up his own network. Private lines are also dedicated lines which are set up by the proprietor to specific characteristics and are installed to encounter specific applications. Leased lines, although they are not switched, pass through the main distribution frame (MDF) of the switching exchange, which still leaves them susceptible to cross talk and impulse noise. Private lines, on the other hand, are completely cut off from any external switching equipment. With the use of leased or private lines, a high performance channel is ensured, because the lines are specially selected. Furthermore, the dedicated lines are always available, as they are permanently connected to the location site and their characteristic performance is known. Once the line is reserved, it may be individually compensated for distortion.

Dedicated lines are generally of high quality and present a low error rate. Leased lines however, may, without the user's knowledge, be channelled over carrier cables or microwave radio links. In addition, although theoretically permanently connected, changes are sometimes made to the routing without this being reported to the user. Private lines, on the other hand, have stable characteristics which are specially designed to the specification of a particular application. However, private lines are extremely expensive, leaving the maintenance to the user's responsibility, and provide no means of alternative routes.

Leased lines cost much more than dialled lines and the price doubles if alternative lines are also made available. When a leased line breaks down it takes time to locate the fault or to obtain an alternative line if one was not provided for originally. Leased lines are thus much more expensive than dialled lines, although the cost effectiveness is reversed when there is a large amount of data to transfer. The break-even point depends on the actual charges for the various lines and the amount of data transmitted.

4.4 TRANSMISSION MODE

The facility of the point-to-point communication data in either one or both directions is an important design factor of the network structure. In this respect, the data links many be categorized into three classes (as shown in Fig. 4.8) according to the mode of transmission that can be transported over them:

(a) Simplex transmission, where the link carries the data in one direction only, while prohibiting the data flow in the other directions.

(b) Half duplex transmission, where the link can carry the data in either direction, but only in one direction at a time, i.e. it prohibits the data flow in both directions simultaneously.

(c) Duplex (or full duplex) transmission, where the link can carry the data in both directions simultaneously.

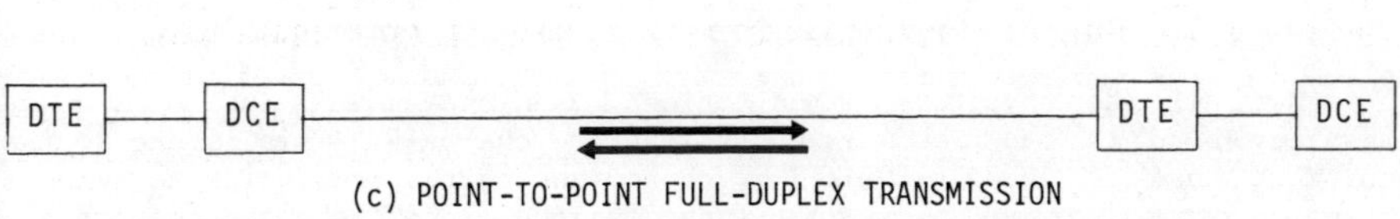

Fig. 4.8 The three categories of transmission modes over a point-to-point communication link

These three modes of transmission in the data lines are usually selected in accordance with the system's needs. Simplex transmission is used for data collection, half duplex transmission is used in most terminal-to-computer communication, while full duplex transmission is used for computer-to-computer communication. Although full duplex transmission seems to be the most efficient means of channel utilization, this is a completely wrong concept; a communication channel can provide, at any given time, a transmission only in a single direction. When the channel is used for voice communication this is not felt by the users, since a person cannot speak and keep listening at the same time. In data communication, the transmission of signals through the same frequency band in both directions simultaneously will only distort the data regeneration at both ends. For this reason, the most common transmission facility is to operate the communication channels in a half duplex mode.

Duplex transmission is only possible if two separate communication channels are used, one in each direction, as seen in Fig. 4.9. In other words, duplex transmission can be achieved with two simplex channels.

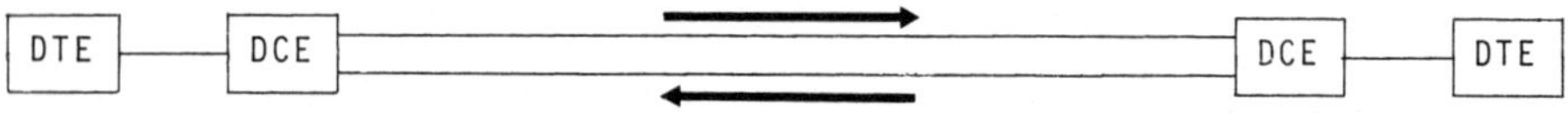

Fig. 4.9 Duplex transmission made possible by double simplex transmission

With radio communication, duplex transmission can be achieved by employing two separate frequency bands, one for each direction. In this respect the two radio channels act as double simplex channels, similar to the physical line transmission which requires two separate lines, each of two wires.

It is possible to achieve duplex transmission over a single voice line (two wires or their equivalent) by sharing the available frequency band of the line between the two channels, as seen in Fig. 4.10. That is, each channel is assigned only part of the frequency band available in the two wires. Contrary to expectation, the efficiency of the line is reduced, since a guard band must be inserted between the two channels so as to ensure that they are separated and do not disturb each other. This means in effect that the data transmission speed in each direction is much less than half the transmission speed which can be transmitted in a single direction.

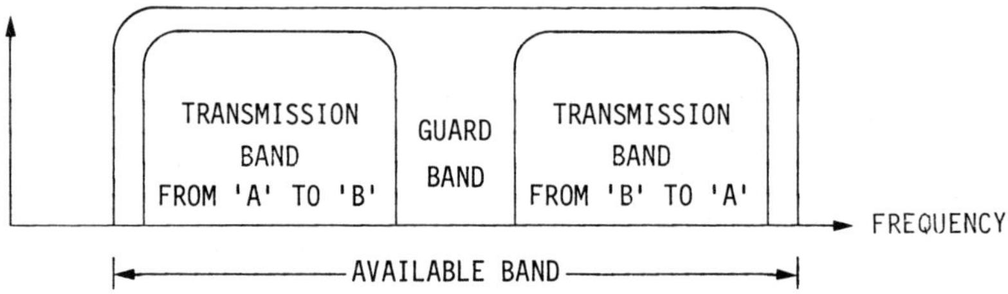

Fig. 4.10 Duplex transmission in a single communication line

Not all the physical communication lines can permit the available frequency band to be split into two separate channels. This is due to the equipment that is sometimes used by the common carrier companies to improve the voice communications. In some long distance lines, echo suppressors are installed which are applied in order to amplify the strongest signal of either direction and suppress the weak signal which flows in the opposite direction. The intention is to assist in reducing any voice cross talk created by echo signals. In voice communication this is based on the assumption that only one side is talking at any given time. If such a line is used for duplex data transmission, the result will be that the communication can take place in one direction only, while in the other direction it is suppressed.

Many lines which may seem to be equivalent in operation to a two wire half duplex transmission mode operate in parts with four wires. The reason for this is that electronic equipment can only operate in simplex transmission mode. Therefore, if repeaters and amplifiers are to be inserted in a two wire operating in full or half duplex mode, the line must be conveted in places to four wire and then back to two wire, as seen in Fig. 4.11. This is not limited to wire communication but applies to any equipment which can be inserted over the line. One such example is the satellite communication, where the satellite acts as a relay permitting the transmission in one direction only on each channel.

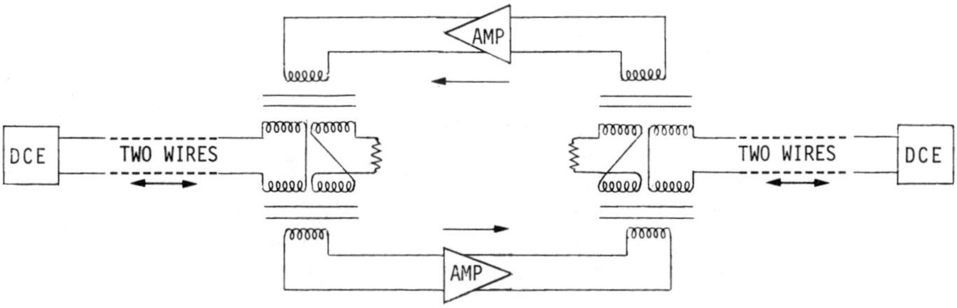

Fig. 4.11 Conversion of a half duplex two wire line into a duplex four wire line

There are some voice communication channels were the data are transmitted in a half duplex mode although there is some signalling in the opposite direction. That, is, the main portion of the frequency band is used for high speed data transmission (as seen in Fig. 4.12) while the lower portion of the spectrum is used for the transmission of supervisory signalling information. In the voice telephone network, the lower portion of the line spectrum cannot be used for fast data transmission but is concurrently used for transmitting control signals in the opposite direction at low speeds of 75 BPS. Although this communication system resembles an operation of a duplex transmission mode, it must be regarded as operating in a half duplex mode, since the actual data transmission can flow in only one direction at any given time. In this system, the supervisory backward channel is adopted for alarm notification when the receiver is out of synchronization with the transmitter, and for instructions to convert over from data to speech. This extra channel is referred to also as an outband signalling channel.

As already stated, the communication line can generally provide communication in either of the three transmission modes; electronic equipment can

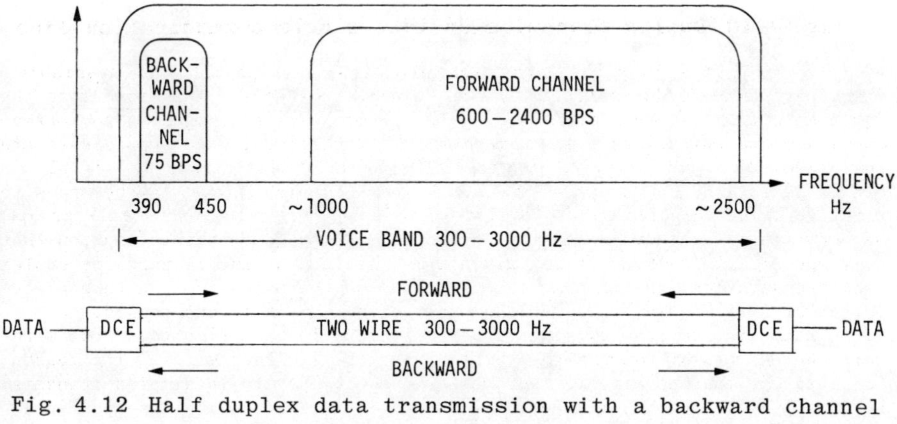

Fig. 4.12 Half duplex data transmission with a backward channel

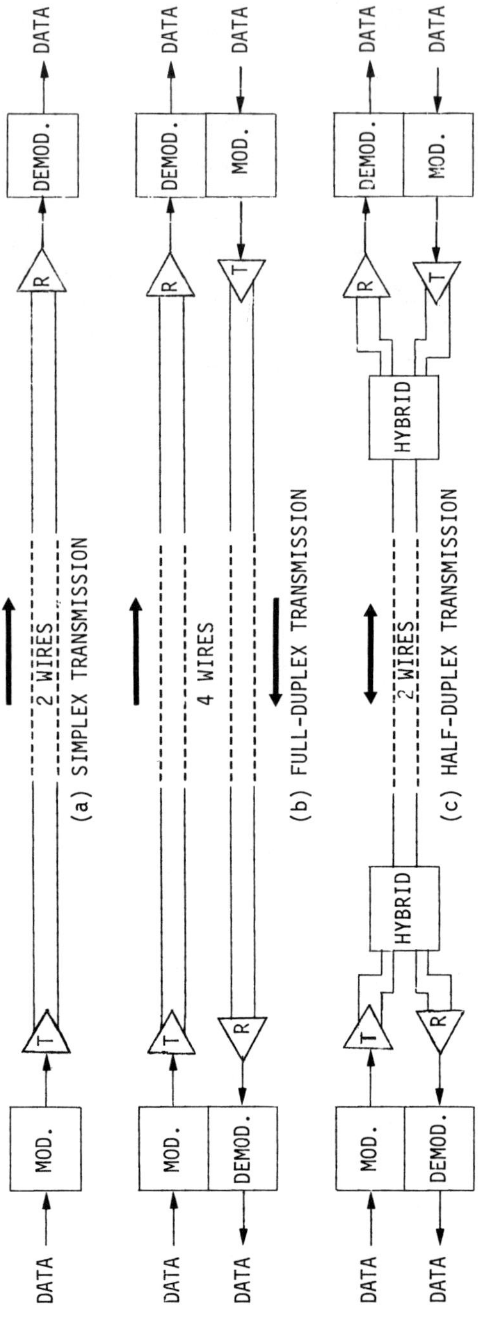

Fig. 4.13 Data communication equipment (DCE) required for the three modes of transmission

only provide communication in one single direction, or in other words, electronic equipment can only operate in a simplex transmission mode. For this reason, it is not only the communication channel which limits the direction of transmission; it is also the DCE itself which circumscribes the direction of transmission. In fact different DCE units are generally required for the three different modes of transmission, as seen in Fig. 4.13. In simplex transmission, the communication channel only requires a modulator at one side and a demodulator at the other. In full duplex transmission, each communication channel requires the same equipment as needed for the simplex transmission, that is, each side requires both a modulator and a demodulator. This means in effect that the equipment required for duplex transmission is the same as that for two separate simplex transmission channels. Although in half duplex transmission modes, the communication in both directions is over one (two wire) channel only, nevertheless the same equipment is needed as with full duplex transmission mode. In other words, the same DCE equipment is used both for half and full duplex transmission; however, a hybrid transformer must be added for half duplex transmission to enable the data to flow in both directions. This hybrid unit is the same as presented in Fig. 4.11 for converting four-wire to two-wire operation.

Despite the difference in the modem equipment that can be used for three modes of transmission, it is usual to employ a common type of modem for all the three modes. This is possible because the same DCE can generally operate as half or full duplex as well as simplex mode and this enables standardization of all the equipment.

As already stated, it is the most common practice to operate over the data communication channel in a half duplex transmission mode. This mode of transmission, however, enforces some operational problems which are due to the reversal of the direction of transmission. Each time the direction is reversed, the modulating transmitter must be switched off and the demodulator receiver switched on. Such a reversal process being time consuming is referred to as turn-around-time (t-a-t). The turn-around-time of a complete DCE unit, from its receive to its transmit mode (or vice versa) may be as much as 50-200 msec, depending on the circuits involved. In the circuits, this is affected by the hole storage effect of the transistor as well as the discharge of the capacitors. In the physical line communication, the t-a-t is also affected by the equipment along the line, such as repeaters and amplifiers, apart from the effects of stray inducatances and capacitances along the line.

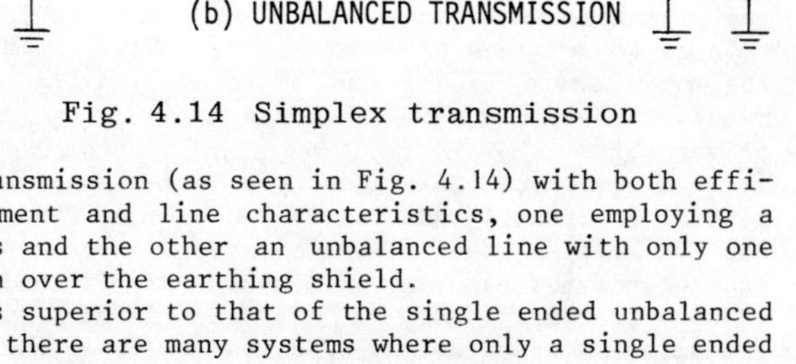

Fig. 4.14 Simplex transmission

As already explained, most if not all data communication channels, with their DCEs and lines, operate in a simplex mode. The other two modes, if required, are achieved by the joint configuration of both the equipment and lines. This characteristic of the channel is due to the electrical performance of the components. Two sets of channel configuration can be employed using the simplex mode of transmission (as seen in Fig. 4.14) with both efficiently utilizing the equipment and line characteristics, one employing a balanced line with two wires and the other an unbalanced line with only one wire having the signal return over the earthing shield.

The balanced circuit is superior to that of the single ended unbalanced configuration. Nevertheless, there are many systems where only a single ended circuit can be used, with the radio communication as a typical example.

4.5 SHARED COMMUNICATION FACILITIES

In many digital data acquisition systems there may be a large number of terminals located at a single remote site. It is a most inefficient solution to allocate separate dedicated point-to-point lines, as seen in Fig. 4.15(a), from each terminal to the central location. Instead, a shared line is proposed where the data originated from a number of terminals could participate in exploiting a common transmission channel, as seen in Fig. 4.15(b). Sharing the common channel requires special equipment at the channel junction point which is referred to as a multiplexor.

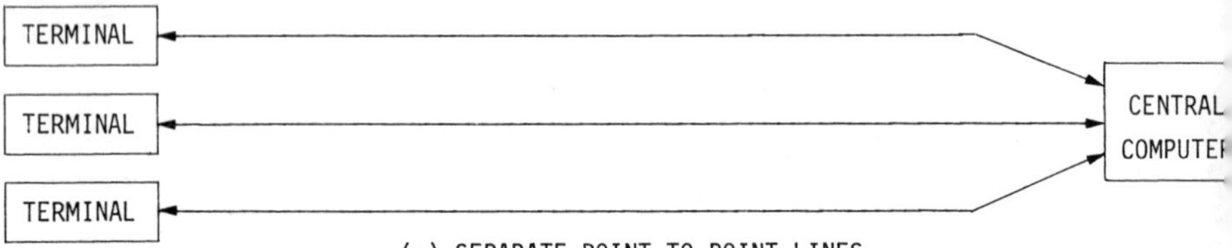

(a) SEPARATE POINT-TO-POINT LINES

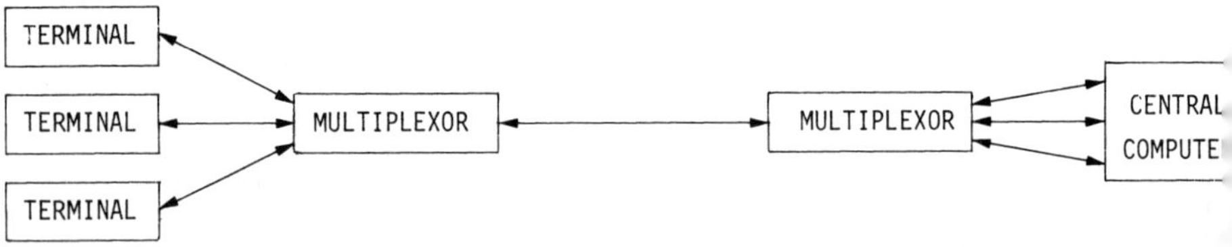

(b) SHARED COMMON LINE

Fig. 4.15 Multiplexing reduces the number of lines between the terminals and the central computer

Multiplexing enables two or more data signals to be interleaved simultaneously and transmitted over the same communication channel. The signals can then be individually recovered at the receiving end, producing the identical signals which entered the common channel. Multiplexed lines can usually transmit the data in either one or both directions; although similar to the DCE, special equipment is required for multiplexing and different equipment for the demultiplexing process.

Sharing the common channel facility may be performed either by retaining the point-to-point characteristic of each line or by employing store-and-forward characteristic, as seen in Fig. 4.16. With the point-to-point shared facility, the multiplexing process simply interleaves the signals on the common channel, while still retaining the individual direct connection characteristics between the two distant points. With store-and-forward shared facility, there is no direct connection between any of the communicating distant points, and so the combined interleaved signal can no longer be related to a given terminal source. Nevertheless, the data stream within the interleaved signal can be related to its corresponding DTE source by means of an identifying address inserted at the heading of each group of data bits.

A communication channel can be defined by the boundaries of its transmission spectrum: that is, it can be specified by its frequency bandwidth and by the time when it is available. In other words, the transmission spectrum boundaries can be regarded as a two dimension form which is a function of both frequency and time, as seen in Fig. 4.17(a), and therefore, the transmission spectrum may be divided between the individual channels either by assigning to each terminal a fixed share of the frequency band, or by a

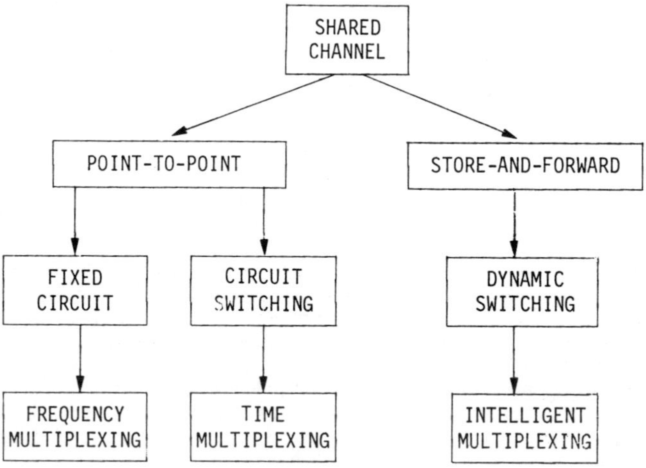

Fig. 4.16 The categories provided by the facility of a shared channel

fixed share of the available time. Frequency division multiplexing (FDM), as seen in Fig. 4.17(b), provides each terminal with a small portion of the frequency band for all the available time. Time division multiplexing (TDM), as seen in Fig. 4.17(c), provides each terminal with the whole frequency bandwidth for only a short portion of the available time. In both FDM and TDM the transmission channel is shared between the terminals by fixed and permanent allocation of the spectrum.

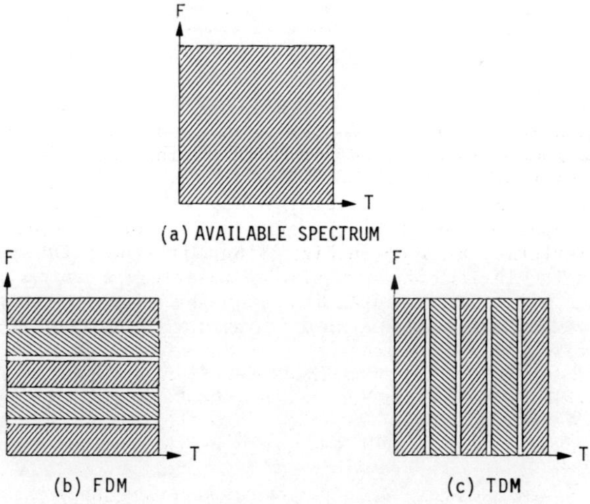

Fig. 4.17 Sharing the common channel facility by fixed allocations

Contrary to the accepted concept, fixed multiplexing data over a common channel is not intended for the efficient utilization of the data transmitted.

That is, the transmitted speed through the common multiplexed channel is not increased but travels at exactly the same speed it leaves the terminal. Both FDM and TDM only organize the data transmitted through the channel but do not alter their basic characteristic form. Furthermore, the overall summed multiplex data rate through the common channel cannot reach its maximum rate value, which could be achieved if the data were transmitted from a single source. If the maximum data rate that can flow through a channel is defined as R and there are N terminals being multiplexed, each with a data rate of r, then $R > Nr$. This results from the need to insert guard spacing between each of the shared channels so as to ensure that each of the terminal's transmission signals does not disturb the other transmission signals.

It should be pointed out that fixed multiplexing also does not efficiently utilize the common communication channel itself, since the permanent sharing of the channel may leave portions of it wasted when the terminals do not have data to be transmitted. Nevertheless, multiplexing is a most useful means of exploiting the transmission facility. Instead of employing N separate links between the terminals and the computer, only a single link is required. Although this common link requires a much wider spectrum than each of the individual links, there is the advantage that it is much easier to handle one link.

To some readers it may sound strange to regard the multiplex process as a point-to-point communication. As an illustration of this status, one may regard the common channel as a pipe which provides each terminal source with a fixed portion (or circuit) through which it can transmit its signal, as seen in Fig. 4.18. Once a portion of the channel is assigned to a terminal, it remains its exclusive property, whether it has or does not have any messages to transmit. Frequency division multiplex, in fact, operates in this form with each portion of the channel, providing a link whereby the terminal signals can be transmitted without any changes to their shape or form.

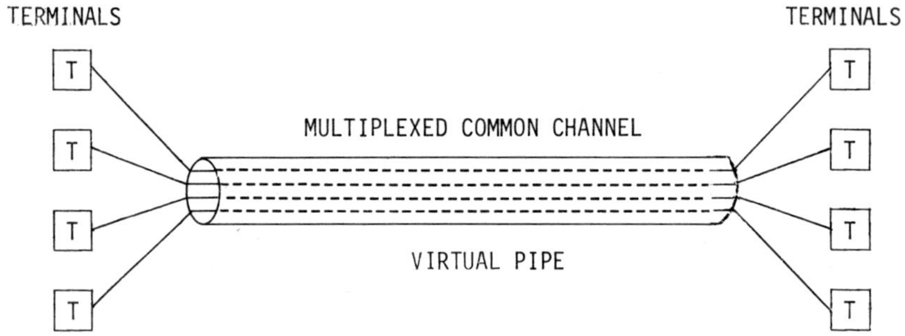

Fig. 4.18 The point-to-point concept of the multiplexed common channel

Time division multiplex can also be regarded as point-to-point communication, in the same manner as described for the FDM. That is, the data rate transmitted from each terminal travels through the common channel at exactly the same rate. However, with TDM each terminal transmits its data for only a short period, to enable the remaining time to be used by the other terminals for their transmissions. In other words, during the allocated short time period there is a direct connection between the transmitting and receiving terminals via the common channel. Accordingly the principles of the point-to-point communication are retained for TDM also, although sharing the common channel is not provided by a permanent fixed circuit but by special switching which allocates a fixed time portion to each terminal.

The operation of the TDM as a point-to-point communication is better

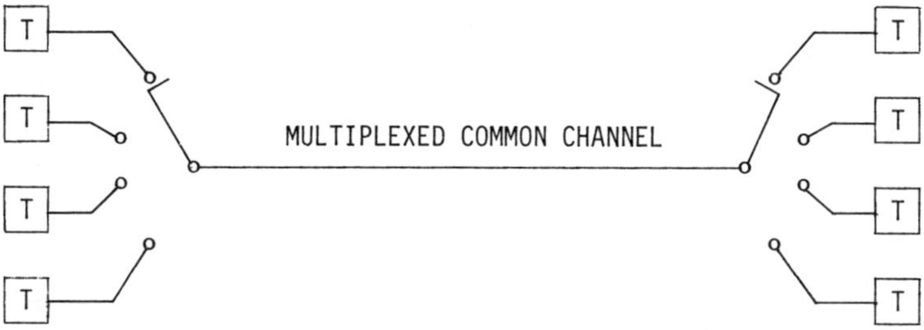

Fig. 4.19 The point-to-point concept of TDM illustrated with a selector

followed by illustrating it with a rotating electromechanical selector, as seen in Fig. 4.19. The selector scans the terminal lines in turn and transfers to the common channel whatever data is available from each terminal. For this reason it is common to refer to TDM as a form of circuit switching.

Both FDM and TDM may operate with 'dumb' terminals; that is, the terminals have no processing ability. Any data available in the terminal is transmitted in the portion of the channel which was assigned to each terminal. If the terminal has no data to be transmitted, that portion of the channel which was assigned to this terminal is left unused with no possibility of the other terminals exploiting it. In FDM the data is transmitted as it is being produced, while in TDM a small buffer must be inserted in each terminal to retain the data during the interval between the allocated transmission periods.

Although the multiplexing process in both FDM and TDM equipment requires electronic circuits, there is no processing ability required for it. Introducing processing facilities into the common multiplexing equipment enables the channel to be dynamically shared. This however, is possible only with TDM equipment.

By using store-and-forward processing techniques in the multiplexing equipment it is possible to separate the terminal transmission speed from that of the common channel transmission speed. The terminals may be scanned in the same way as TDM, although whatever data is available is first stored and then transmitted on the channel at a later stage. In other words, storing the input data in the common equipment removes any direct point-to-point communication between the terminals. In this way the common channel is not shared according to the number of terminals that are multiplexed but according to the data produced by the terminals and the maximum capacity of the channel. That is, since the terminals do not transmit continuously, it is possible to connect to a SAF multiplexor more terminals than are possible with TDM or FDM. The purpose of the SAF multiplexing equipment being to concentrate the data from various terminals and transmit them smoothly over the common channel, the SAF multiplexing equipment is referred to as intelligent multiplexors or concentrators,

4.6 MULTIPLEXING NETWORK CONFIGURATION

The addition of a processor to the multiplexor equipment, to provide the store-and-forward facility enables the common communication channel to be efficiently utilized. That is, the data flow in the common channel is made continuous even though the data originating from the terminals may be in sporadic bursts. By the addition of intelligence into the terminal equipment it is possible to further improve the whole concept of the common channel communication. That is, instead of having each terminal connected to the multiplexor by a separate line, as seen in Fig. 4.20(a), the terminals can now all be connected directly to a common line, as seen in Fig. 4.20(b).

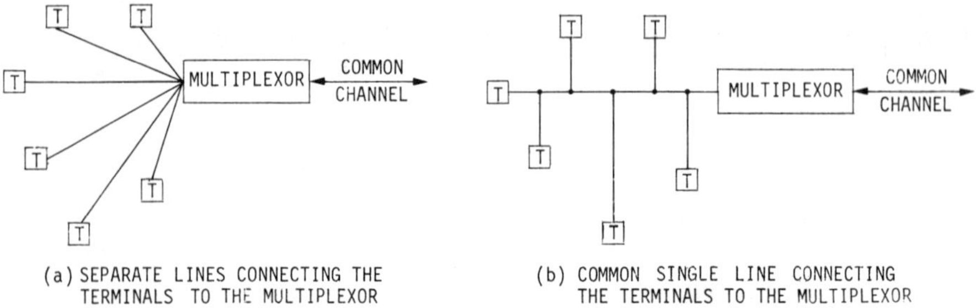

Fig. 4.20 Extending the facilities provided by the multiplexor by the addition of intelligence

In both cases, the function of the multiplexor is to control the traffic to and from the terminals for the sharing of a common transmission channel.

With FDM and TDM, where the multiplexing has no processing ability, each terminal is identified by its individual line leading to it. The multiplexor extends these lines through the common channel by sharing the wide channel resources between individual point-to-point narrow channels.

With intelligence inserted into the multiplexing equipment, it is possible to dynamically share the channel resources. That is, the common channel resources are only provided to those terminals which have data to transmit and are not wasted by idle terminals. Each terminal is connected to the intelligent multiplexor by a separate line as with TDM and FDM, although the multiplexor does not any longer provide point-to-point facilities. The data from all the terminals are first retained in the multiplexor buffer and are transmitted over it only when the channel becomes free. It is the function of the intelligent multiplexor to organize the waiting data messages in queues and transmit them one after the other with no gaps. This provides the most efficient utilization of the common channel, as there is no waste. Nevertheless, since the data transmitted on the common channel is no longer a function of point-to-point communication it is difficult to identify the source of each message. It is therefore the responsibility of the intelligent multiplexor to add an origination address to each message. This address is added automatically in the multiplexor by identifying the line from which the message arrived, as presented in Fig. 4.20(a).

The operation principles of the intelligent multiplexor are very similar to that of time multiplexing. Each terminal is asked in turn to transmit its messages within a given fixed time interval. With intelligent multiplexing, however, if the terminal is found to have no messages to transmit, the allocated transmission time is transferred to the next terminal in turn, without wasting time for the full period to elapse. In other words, the intelligent multiplexor not only efficiently utilizes the common channel but also efficiently utilizes the whole data collection operation of the multiplexor.

With intelligence inserted also into the terminals, it is possible to allow the terminals to add their own address code to each data message it transmits. This is not only a question of distributing intelligence whereby the function of the multiplexor is transferred to the terminal, but primarily of extending the efficiency of the multiplexor. This removes the need of having each terminal connected by individual lines to the multiplexor but instead having all the terminals connected to the multiplexor by one common line, as presented in Fig. 4.20(b). The intelligent multiplexor can thereby recognize the source of each message by its unique address and not by its line source. Adding processing ability to the terminals enables the intelligent multiplexor to address each terminal individually. In the same way the terminals too can now recognize the messages addressed to them, not by electrical impulses sent through the separate lines but by their unique address codes. This means that a number of terminals may be served by only one line with one computer interface. Despite the fact that a number of terminals are connected to one line they cannot transmit their messages freely. The intelligent multiplexor asks each terminal in turn if it has messages to transmit and only the terminal which is addressed by the appropriate instruction can transmit. The addressed instructions are sent to each terminal in turn as messages, a process known as 'polling'.

The intelligent multiplexors provide dynamic means of sharing the common communication channel. These multiplexors can be categorized into two classes according to the methods they employ to collect the data from the terminals, as seen in Fig. 4.21. Intelligent multiplexors which collect the data from 'dumb' terminals are referred to as ATDM or statistical multiplexors, while those which collect the data from 'intelligent' terminals are referred to as concentrators or line-controllers. (A concentrator can be made to operate as an ATDM, but an ATDM cannot operate as a concentrator.) In this respect the network associated with the dumb terminals is referred to as multipoint, while the network associated with the intelligent terminals is referred to as multidrop.

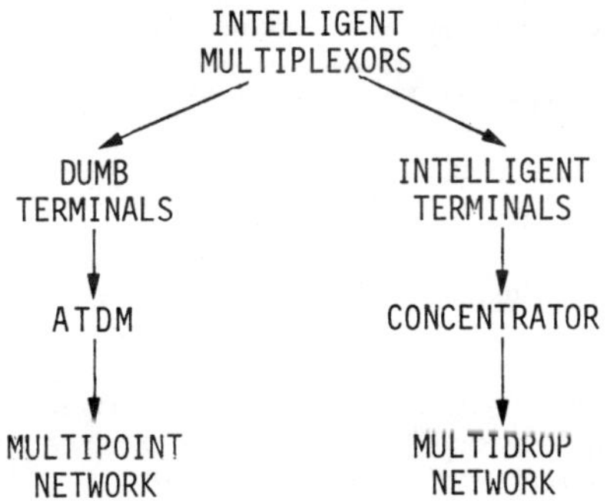

Fig. 4.21 The two categories of intelligent multiplexor operation

A multipoint network has a configuration in which a number of terminals are connected to a common channel by a single distribution point, as seen in Fig. 4.22(a). A multidrop network has a configuration in which each terminal is connected to the common channel by the shortest available path, as seen in Fig. 4.22(b). The multipoint network can also be regarded as a star network with the multiplexor in the centre of the star. The multidrop network can also be regarded as a form of multiple-access network and this will be described in the following section.

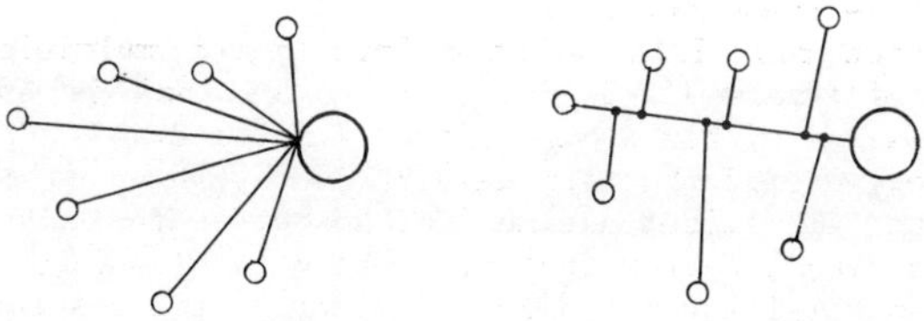

(a) MULTIPOINT NETWORK (b) MULTIDROP NETWORK

Fig. 4.22 The two network configuration offered by intelligent multiplexors

4.7 MULTIPLE-ACCESS NETWORK OPERATION

Adding intelligence to the terminal allows it to control its own operations without the need of receiving control instructions from a centre. The ability of the terminal to recognize its own address and, accordingly, with the aid of software programs, to select messages addressed to it which flow in the common channel, increases the utilization of the common channel.

Furthermore, terminals may now send messages directly to the other terminals, without the intervention of a central station. In other words, the terminals can now operate as independent stand-alone units and their transmission need not be directed any longer by a fixed concentrating centre. In fact, each terminal can extend its stand-alone capability and many contain enough intelligence to perform minor or even major processing tasks. A central station may still be used to allocate to each terminal transmission times, but within this allocated time the intelligent independent terminal does not require the central station's assistance in reaching the other terminals.

This means in effect that the whole concept of the common transmission facility can be extended. With multiplexor and concentrators, the equipment is intended to collect data from distant dependent terminals and then to direct their messages to a central computer via a common transmission channel. In this case the common transmission facility is only a branch of a large data communication network. With the introduction of independent stand-alone terminals which can communicate directly with other terminals, the common transmission facility becomes a network in itself. This network configuration is referred to as a local area network.

Multiplexing and concentration have been defined as the division of the common transmission facility into multiple sub-channels made available to each terminal by common centralized equipment. The terminals may be dumb or intelligent, although in both cases their operation is dependent on the centralized equipment, i.e., multiplexors and concentrators.

The process by which the common transmission facility is made available to multiple independent intelligent terminals, with no restriction as to their physical position relative to the communication channel, is referred to as multiple-access.

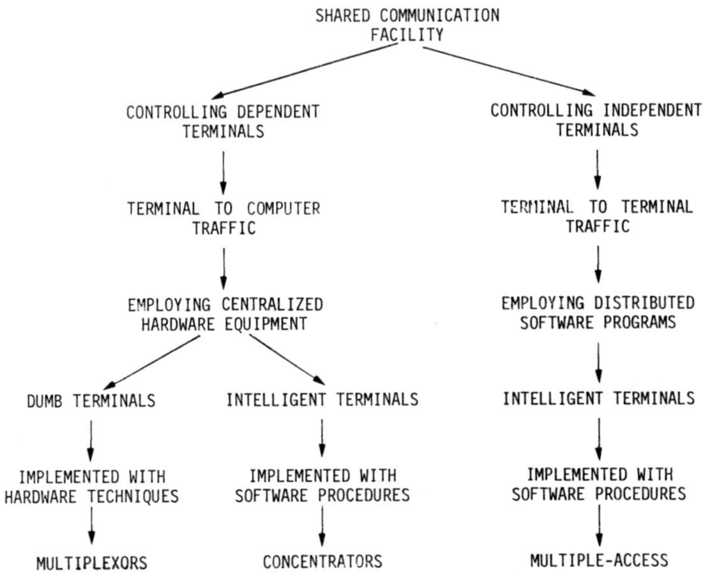

Fig. 4.23 The available choices for sharing a common communication facility

Multiplexing and concentration refer to the principles by which the centralized hardware equipment share the transmission facility, while multiple-access refers to the communication procedures which provide the means of sharing the common transmission facility. These procedures are generally controlled by software programs inserted in each independent terminal. This classification of multiplexing, concentration and multiple-access is better illustrated with the aid of Fig. 4.23. It should be pointed out that multiplexors and concentrators are generally intended to control the data traffic between the terminals and a centre location over an external common channel, while multiple-access procedures are intended to coordinate the traffic between the terminals themselves within a single common channel. Concentration may also use software programs in collecting the data from dependent terminals, but there the programs are inserted into the central concentration equipment. In a multiple-access network there is no need to employ any centralized equipment. Nevertheless, some multiple-access procedures use centralized equipment, but even then its function is to coordinate the communication and not to control the traffic.

A multiple-access configuration is a network in which a number of terminals are connected to a single communication çhannel, each terminal with direct access to the channel irrespective of their location relative to the channel. One such configuration can be regarded as the multi-drop network, shown in Fig. 4.24(a), which was discussed in the previous section. A more typical multiple-access configuration is that of a radio network shown in Fig. 4.24(b). In both examples, each terminal can communicate with any of the other terminals in the network, with the full common channel spectrum available to any of them.

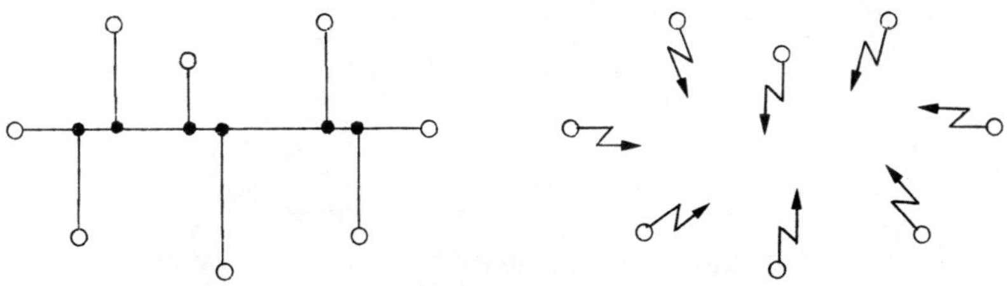

(a) MULTIDROP NETWORK (b) RADIO NETWORK

Fig. 4.24 Multiple-access typical network configuration

As already stated, sharing the common transmission facility is performed in accordance with procedures controlled by software programs inserted in each terminal. Although in a multiple-access network each of the terminals operates as a stand-alone unit, the procedures provide some transmission coordination to ensure the success of the communication process. That is, although each terminal operates independently, there is generally a degree of some synchronization which is required between them. When a number of people talk over a common facility, they can all hear each other and accept the fact that only one can talk coherently at any given time. If two or more talk simultaneously, nothing can be understood. The same applies to terminal-to-terminal communication, whereas if only one terminal transmits its messages, its contents can reach all the other terminals. Nevertheless, if two or more terminals are transmitted simultaneously, their message signals collide and consequently none of the context can be detected. In other words, any collision of messages causes the rejection of all the messages concerned, even if only parts of the messages are affected. It is the function of the procedures therefore to ensure the safe transmission of the messages with the least danger of collision. This is accomplished by using the procedures as means of assigning the common channel facility to provide the best transmission conditions.

Since the multiple-access procedures are introduced by software means it leaves the designer with a wide choice. The selection of a procedure depends on the application it is intended for. What must be then taken into account are: the real-time factor, the traffic load, the cost and the random pattern of message source. Fig. 4.25 illustrates the categorization of the procedures in respect to the way the common channel facility is assigned to each terminal. A more detailed presentation of multiple-access procedures is given in chapter 9. The channel assignment may be according to centralized control, even though then some of the terminal's independent operation is lost Centralized control is ideal for high traffic conditions, yet in most applications the traffic from the terminals is low and random. Furthermore, once a terminal starts transmitting it comes in bursts with a number of messages being transmitted. Distributed control is the best solution for low and averag traffic conditions, although it then requires high processing ability in each terminal. In the distributed control the terminals operate independently, that is, they can transmit their messages when they are ready and not only when they get permission to do so from a centre location. Nevertheless, the more independence a terminal is given, the higher the danger of collision. The collision danger can be averted by increasing the transmission restrictions, this however, will reduce the terminal's independence and require more delay before messages can be transmitted.

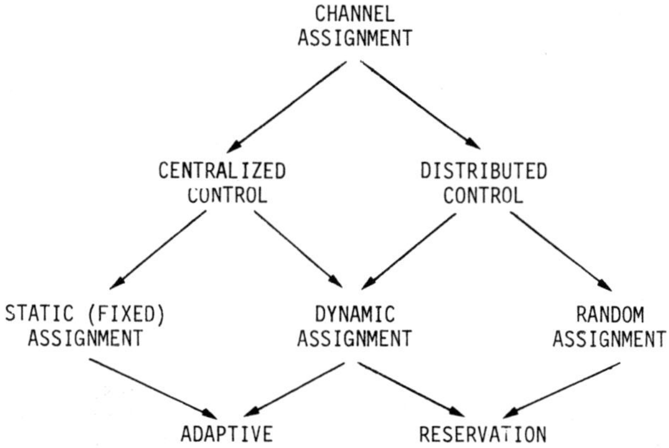

Fig. 4.25 Categorization of multiple-access procedures

The channel assignment may be provided according to static, dynamic or random assignment procedures. Static assignment ensures the best transmission conditions over the common facility. Each terminal is provided with a fixed portion of the channel spectrum for its exclusive use. As in multiplexing, this procedure is wasteful when all the terminals are not active and there are long unused delays between message transmissions. Random assignment ensures the least delay interval between the time a message is ready to be transmitted and the time the terminal can transmit it, but with an increased danger of collision. This random form of transmission is satisfactory, provided only that one terminal is allowed to transmit. Once a number of terminals transmit simultaneously there is always a danger of collision which can cause the saturation of the channel with no messages reaching their destination. Since such a situation cannot be prevented, dynamic assignment of the common facility is preferable. Nevertheless, dynamic assignement does not always guarantee against collision, nor does it prevent the delays before transmission. Furthermore, the dynamic assignments require high overhead time Despite the apparent disadvantages of dynamic assignment it is superior in most applications both to random and static assignment procedures. Dynamic assignment can provide better guarantees against wasted time intervals and at the same time better guarantees against the channel saturation. In fact, most of the procedures which have been proposed are of some form of dynamic assignment of the channel.

Message traffic transmitted by the terminals is statistically unpredicted, as it is generally dispatched in uneven bursts and at intervals. The advantages of the various assignment procedures may make them ideal for some form of traffic conditions, but unsuited for other conditions. There are, therefore, a number of procedures which adapt the channel assignment to the traffic conditions. When the traffic is generally low it is possible to use random procedures to allocate to these terminals priority reservation transmission times.

4.8 COMMUNICATION SWITCHING CENTRES

Multiple-access provides communication between the terminals over a single transmission facility. It serves to coordinate the traffic so that the maximum number of messages may flow between the terminals with the least danger of collision. Nevertheless, multiple-access enables only one single message to be transmitted and received from any terminal at any given time. If there is need for more than one message to be transmitted simultaneously, then other communication network techniques are required.

In large communication networks, where there is a considerable traffic between the terminals, there could be a vast range of terminals and computers distributed over wide geographical locations. The computers and terminals attached to these networks may either be independent processing systems or dependent processing subsystems. Nevertheless, the network provides the means whereby the computers can communicate with each other either to share the processing resources available in each installation or to exploit the data bases available in the other installations.

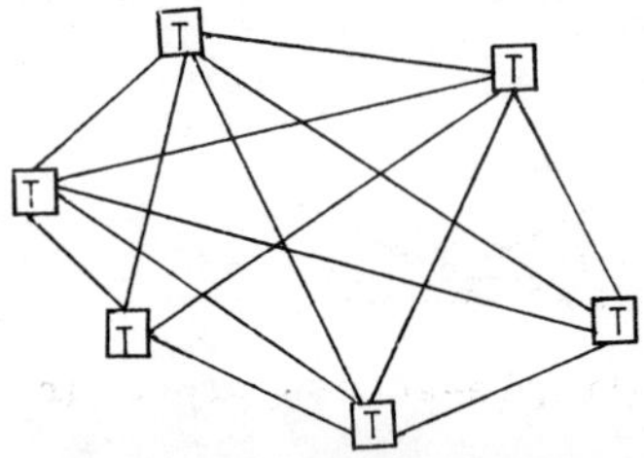

(a) POINT-TO-POINT LINKS BETWEEN TERMINALS PROVIDES COMPLETE INTER CONNECTION

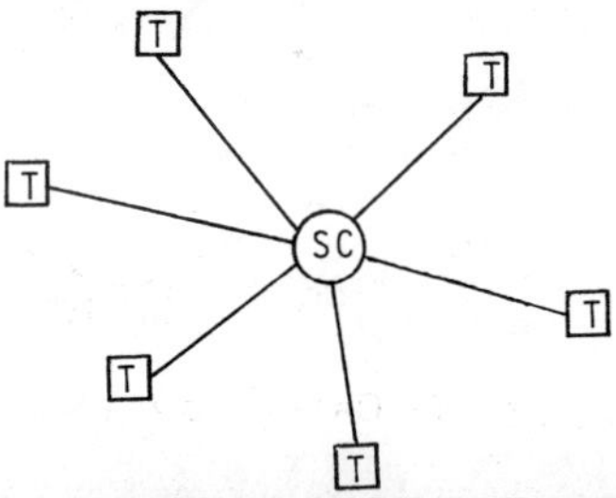

(b) A SWITCHING CENTRE IN THE NETWORK PROVIDES COMPLETE INTER CONNECTION BETWEEN TERMINALS

Fig. 4.26 Improving network communication by the addition of a switching centre

To achieve simultaneous and comprehensive computer-to-computer communication (as well as terminal-to-terminal and terminal-to-computer communication) in a single network, each computer (and terminal) must be able to be directly connected to all the others. This can be obtained only if each and every computer or terminal are linked to each other and that these links are in the form of point-to-point channels, as seen in Fig. 4.26(a). To achieve this type of network, the number of point-to-point links must be equal to $N(N-1)/2$, where N defines the number of terminals and/or computers. This network therefore enables $N/2$ simultaneous data transmissions, provided that each terminal or computer can be occupied with only one communication process.

It is obvious that as the number of points increases, the option of complete inter-connection becomes nearly impossible, since it is left to the terminal to select the point-to-point link on which it is to operate.

The only possible course in a comprehensive network is to bring all the terminal links to a central location (as seen in Fig. 4.26(b)) which can provide the full connection through a switching centre. Though the introduction of such a switching centre, adds an extra station in the centre, it effects a considerable saving in the number of communication links, i.e., reduces the number of links to N while still allowing $N/2$ simultaneous data transmissions.

The introduction of a communication switching centre in the transmission data path adds an extra stage (as seen in Fig. 4.27) which is referred to as a data switching exchange (or equipment) (DSE). The DSE centre provides the means which connect the remote terminals or computers. The centre receives all the messages transmitted from each of the remote stations and creates a communication path to the destination station, in accordance with the address as defined at the head of the message.

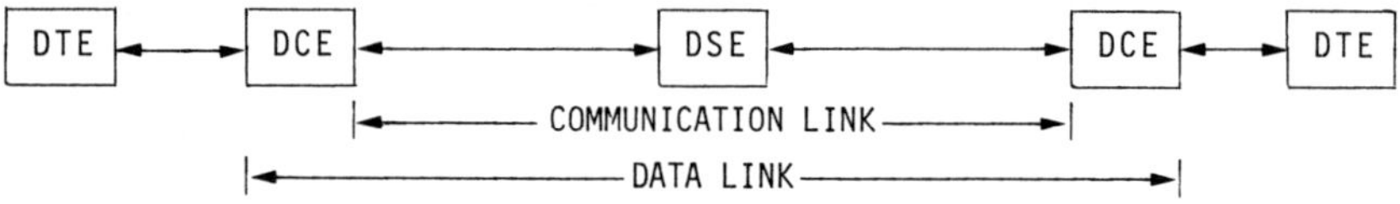

Fig. 4.27 The integration of a switching centre in the communication link

With the increase of the traffic load between the computers it is soon found that a single data switching exchange cannot cope with all the expanding traffic requirements. As the geographical spread of the computers increases, there is a need for a more sophisticated type of network. To achieve these comprehensive traffic requirements, many local switching centres may operate in the network (as seen in Fig. 4.28) while still allowing each terminal to communicate with all the remote terminals in the network. Each centre functions as a unique unit whose main purpose is to switch and re route local messages from any of the terminals and/or computers associated with it to any other associated terminal or computer of this centre. At the same time, the centre also functions as a switching agent for transferring messages between centres. That is, each centre controls the local message traffic, while acting when required, as an intermediate station for trunk messages between other centres.

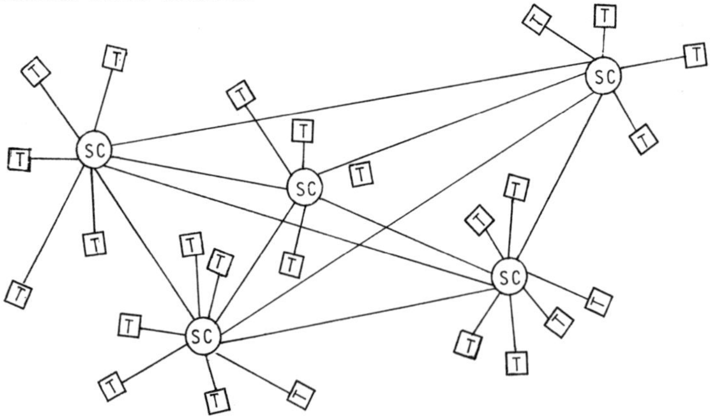

Fig. 4.28 Communication network with multi-switching centres

The introduction of the sophisticated large network promotes the efficient utilization of all the communication lines. Above all, it brings about a new communication concept, namely routing. That is, it helps to locate the shortest available transmission path when the message has to flow through a number of centres, and determines the best route when one of the links is down. By suitable network management, the messages need not always be routed by the shortest path, but may be routed over the least loaded links so as to achieve the fastest throughput over the network. Where a number of switching centres exist in the same network, the network is usually open-ended, with no single centre in control. This type of network is extremely important for many C^2 system applications since the network is inherently expandable.

Each junction point in this sophisticated network convex, where the network interferes either with the local computers or with other network junction points is known as a node. The node is a switching centre which can be regarded as the major building block of the network. Nevertheless, one must always have in mind that the node's prime function is to provide a service to the computer or terminal (i.e. the user) associated with it, and that its other network functions are secondary. For this reason the computers associated with the node are usually referred to as hosts.

In large sophisticated network configurations, a message originating from a terminal or a computer associated with one node may travel through a number of nodes before it reaches its destination, as seen in Fig. 4.29. Each node analyses the transaction heading which contains the destination address and accordingly determines the shortest available route.

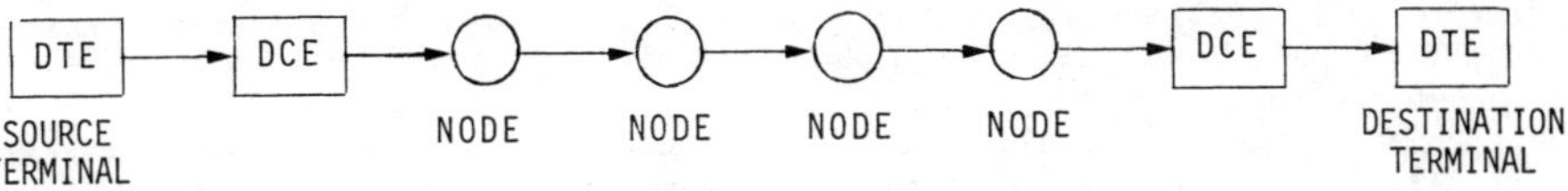

Fig. 4.29 A message route through the network

There are three basic centre switching concepts, as seen in Fig. 4.30, *viz*, circuit, message and packet switching. Circuit switching sets up a semi-permanent path within the centre between the two communicating computers for the whole duration of the transmission. This path is decided upon in accordance with the address transmitted separately ahead of the message. Message and packet switching routes the messages to their destination in accordance with the address given at the head of each message or packet. Nevertheless, message and packet switching centres do not provide a direct transmission path but may hold the message in the centre memory before routing them onwards to their destination. Here the centre function will be only briefly described, but a more detailed description of the centre switching operations will be given in chapter 8.

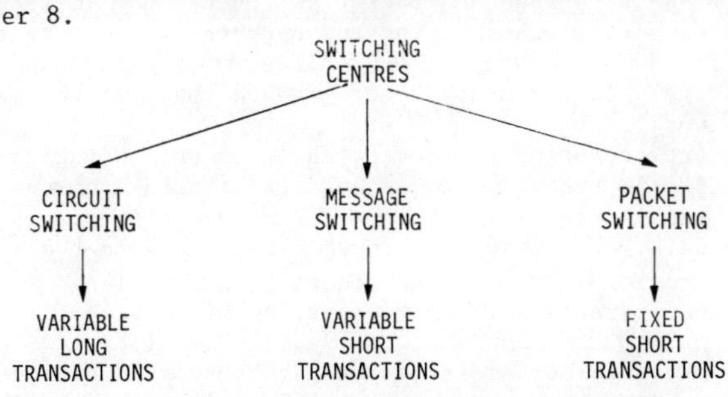

Fig. 4.30 The three basic switching centre concepts

Circuit switching is extensively used for speech communication, the best example of it being the automatic telephone exchange. It is also used for data communication for relatively long transactions. In this type of centre switching technique, a special circuit path must be set up for each communication process between any two computers. Once the path is made available between the two communicating computers, the communication proceeds directly with no further assistance provided by the centre. If, however, a path between the two computers is unavailable, the calling computer cannot off-load its messages and must retain them until a suitable path is made available. Furthermore, the computer must 'close down' its initial connection to the local node and must try again later with hopes of better success, until a complete direct free path can be located. Once the connection between the computers has been established, providing a direct semi-permanent path, the messages can be transmitted without any delays or queues along the route. The operation of making the connection in circuit switching is therefore a slow process which delays the initial transfer process. Nevertheless, once the connection is made, this centre switching technique is the fastest.

Computer communication generally proceeds in bursts, with the messages of variable length, most of them relatively very short, with occasionally some long ones. For this type of data traffic, message switching is preferable, since the node is always available to receive the messages from the sending terminal. Although the terminal can dispatch its messages to the nearest centre node, it dissociates itself from the message only after it is accepted by the final destination node. The message will then be held in the intermediate node until a suitable route becomes available. In this way the messages are passed on from node to node, where each time they are held in a queue ready for dispatching on the outgoing route path. Message switching, as distinct from circuit switching, does not provide any direct communication link between the two terminals, but only an indirect communication channel by which the messages can reach their destination. For this reason message switching does create delays between the time the message is dispatched from the terminal and the time it is received at its destination terminal. Nevertheless, the overall time of a complete message communication transaction is shorter in a message switching network than in a circuit switching network, because there is no need for setting up a direct communication link.

Packet switching is a form of message switching, except that the packets are of fixed length and therefore require the breaking up of a long message into fixed sized packets. Each packet must have its own heading which is a replica of part of the original message heading, with the addition of message assemble/disassemble information. The fact that each packet has its own heading enables each part of the same message to travel along different routes. Message and packet switching techniques hold the transmitted contents in the intermediate nodes en route till they reach the final node. There the packets are held until all the packets of the same message have been received. The advantage of packet switching over message switching is that the nodes can start transmitting the packets even before the packets of the same message have been received, since it regards each packet as an individual message. This can therefore reduce the time the message travels through the network.

An important operational diference between the switching centre techniques is the time it takes for a message transmitted from one terminal until it reaches its final destination. To illustrate this, Fig. 4.31 presents an example of the delay of a message through a three node network system.

In the circuit switching technique there is a long switching delay before the message transmission commences, required both for setting up the path and passing information that the path is available. Such a switching delay does not exist in both message and packet switching since the transmission takes place even if there is no outgoing path. Nevertheless, in circuit

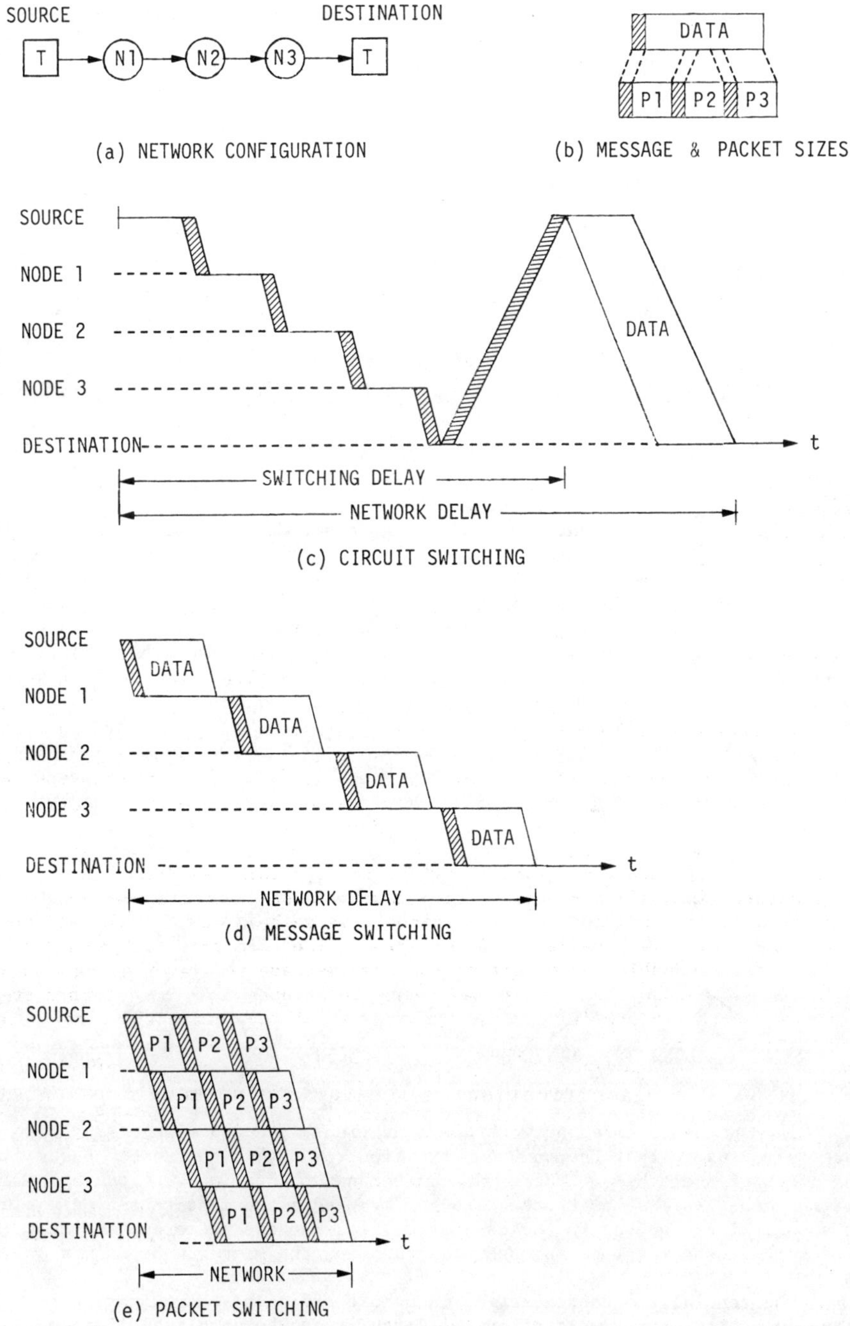

Fig. 4.31 Comparative network delay in the three network node switching techniques

switching there is no message delay in the nodes once the path is set up. Message switching is much slower than packet switching, as the whole message must be available in the node before it can be retransmitted. This is not the case in packet switching, where the packets of the same message are retransmitted before the full message has been received.

From the illustration presented in Fig. 4.31, one might get the wrong impression of the efficiency of the three switching techniques. One must always remember that in circuit switching all the paths are occupied throughout the transmission, while in message and packet switching they are only busy for the duration of the time the messages pass through them.

4.9 NETWORK CONFIGURATION COMPLEXES

The employment of communication switching centres presents new and different types of topological network configurations. The nodes of such a configuration are scattered over large geographical areas with the links between them enabling message communication to be made from any terminal associated with any of the nodes to any other terminal in the network. The network configurations can be divided into two main classes: centralized topologies and distributed topologies, as seen in Fig. 4.32. In centralized topologies the centre node controls the traffic in the whole network, either directly or indirectly. Distributed topologies enable each node in the network to operate independently and so act as a switching centre point for all the local traffic. Furthermore, each node may communicate with the other nodes via trunk communication links. In the centralized network topologies the master node is always a switching node, while the outer nodes of the network may be of a different type, such as concentrators and multiplexors. In distributed network topologies, however, all the nodes must be a switching centre of the same type, either circuit, message or packet switching technique. The employment of the distributed network configuration imposes the need for special communication protocols to ensure that no message can get lost en route, for now the responsibility of the node is for both the local and transient traffic and this responsibility ceases once the message has been successfully received by another node.

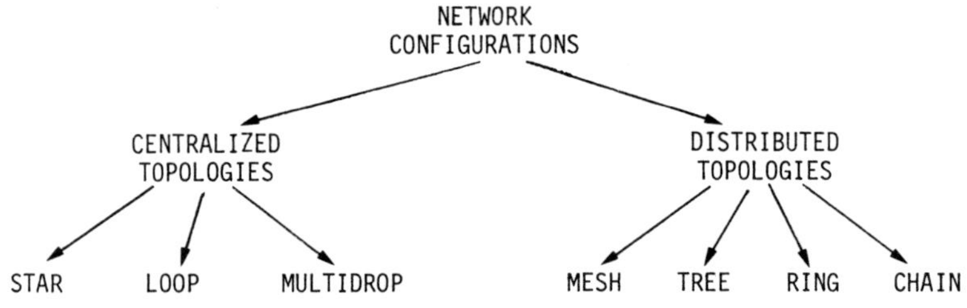

Fig. 4.32 The classification of the various network topologies

In the centralized network configurations, the function of the centre node is to ensure that the message traffic from any of the remote nodes can reach any of the others. Where the master node acts as a switching centre all the messages must be routed through it. Where the master node only acts as a control centre (as in the case of multiple-access networks) the message need not necessarily flow through it. That is, the master node only provides each of the outer nodes with adequate transmission times.

The centralized configuration is a most efficient communication technique in controlling the traffic, although it can be unreliable, since the failure of the centre node will cause complete breakdown. For this reason it is a general practice to introduce into the centre standby redundant units that can take over when the active unit is faulty.

There are three typical network topologies that can be identified in the class of centralized network configurations: star, multidrop and loop topologies, as seen in Fig. 4.33.

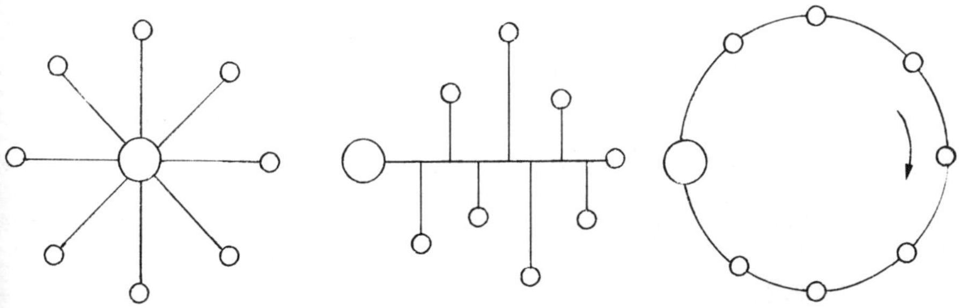

Fig. 4.33 Typical centralized network configurations

(a) *Star network*, which is a highly centralized configuration where all the messages originating from any of the outer nodes must flow first to the master centre node and only then be routed to their destination. The outer nodes may be simple terminals or even computers. They could, as well, be local centres, such as multiplexors or concentrators: nevertheless, the traffic from any of the outer nodes is always assisted by the master node. The advantage of such a network configuration is the simplicity of the operation, with all the traffic functions located in one easily reached location. Each of the outer nodes has a direct point-to-point link to the centre node and so can transmit its message as soon as it is ready.

(b) *Multidrop network*, where all the remote nodes are connected to the centre through a common shared transmission link. The outer nodes may be a terminal, computer, multiplexor or concentrator, but not a switching centre. The master node may operate either as a concentrator or a switching centre, where in both cases all the messages must first flow to it and only then be routed to their destination. The master node can also operate as a controlling node only where it provides each outer node (terminal) with transmission times, but does not necessarily perform any routing functions. In this network operation one can classify some of the time-division-multiple-access procedures which require centralized assistance. In this multidrop configuration the remote nodes always act as slave nodes since they cannot transmit their messages except when the master node allows it. This means that each of the slave nodes must include a memory unit to store the messages until transmission permission is granted. It should be added that this stipulation which is necessary for the multidrop is not required in the star network configuration.

(c) *Loop network*, where all the nodes in the network are connected in a closed annulus. That is, the transmission link threads through all the nodes in series, starting from the centre node and returning to it in a closed circle. The traffic flow is controlled by the centre node which, however, does not provide any routing or switching facilities. All the message traffic circulates within the loop in a single direction with the messages travelling from one node to the next one in the loop. Each node can thus examine or change the contents of the traffic that flows through it, enabling the nodes to receive or dispatch their messages without disturbing the traffic flow. The centre node provides the initial timing intervals for the message transmission, and subsequently the nodes can send their message if one of these timing intervals is found to be free.

In the distributed network configuration there are no master nodes and each of the nodes operates independently. Such a network configuration is intended for computer to computer communication with all the nodes operating as switching centres and the computers as hosts to the nodes attached to them Each host computer can communicate with any other computer via the node whose function it is to control both the local and the inter-node trunk traffic. To perform these functions each node must be connected to all the other nodes, either directly or with the assistance of other nodes. Connecting the nodes produces a mesh configuration of links as seen in Fig. 4.34. There are two mesh network topologies; either all the nodes are fully or partially connected. The links between any two nodes all operate in a duplex mode which enables the messages to flow in both directions.

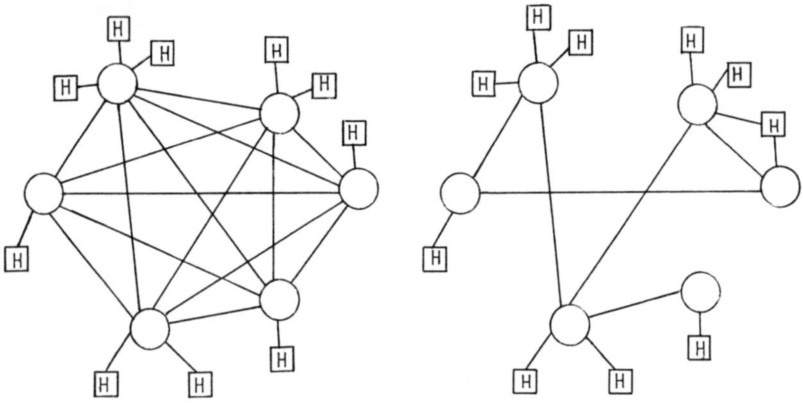

(a) FULLY CONNECTED (b) PARTIALLY CONNECTED

Fig. 4.34 Typical mesh topologies of distributed network configurations

Fully connected network configuration is a rather wasteful means of network utilization since not all the links can always be fully exploited. Nevertheless, the fully connected network does provide the best means of alternative routings and ensures that all the messages reach their destination with utmost speed. For practical application, however, the partially connected network is preferable, where links are inserted only if the traffic load justifies it. This type of network is the most popular one and is extensively used in computer-to-computer communications.

In some distributed networks which are mainly intended to serve the heavy local traffic around the remote nodes with provision in the network for the occasional inter-node traffic, there is no practical justification for fully or even partially connected mesh networks. It follows that in such a network there is no necessity to provide alternative routings. A variety of these network topologies can be seen in Fig. 4.35. The simplest topology is that referred to as a chain, which connects all the nodes in series. Inter node messages must travel from one node to the other until they reach their final distination. It is obvious that a failure of one of the links will reduce the inter-node traffic facilities. A more popular network topology is the tree configuration which is a combination of a number of chains connected together. This configuration has the advantage over the chain owing to the fact that it has shorter lines, although it has the same line failure problems as the chain. A third network topology with better failure facilities is the ring configuration. In this network, all the nodes are connected in a circle in the form of a close chain. As messages in this network can flow in either direction, with the failure of one of the links inter-node traffic can still reach all the nodes without disturbing the network operation (provided the nodes know in advance of such line failure).

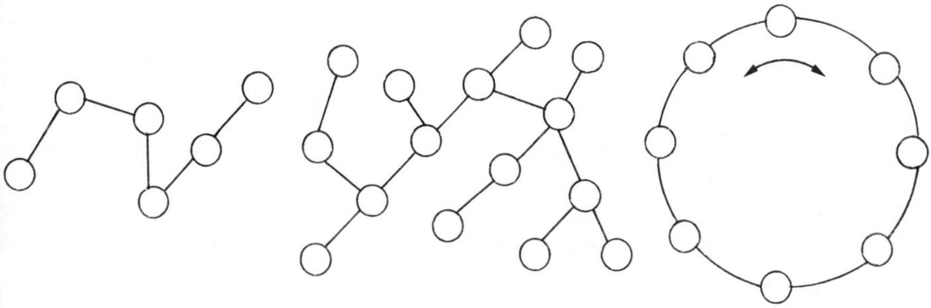

(a) CHAIN NETWORK (b) TREE NETWORK (c) RING NETWORK

Fig. 4.35 Typical topologies of distributed network configurations which do not provide alternative routings

As already stated, the chain, tree and loop topologies provide network services by allowing traffic from one node to the other but with no available alternative routes. This simplifies the node operations, since the nodes simply transfer onwards all the messages they receive which have an address that is not intended for them. In the tree network there are some routing functions required of the nodes, but again there are no alternative routing processing functions.

Another form of distributed network may be that referred to as the bus network tolology, as seen in Fig. 4.36, which is a form of a multidrop configuration. In this network, the data originating from one remote node (which could adequately also be a terminal or computer) is broadcast to all the other remote nodes. This form of network is more suited for local than long distance traffic, as it limits the number of messages that can flow simultaneously between the nodes. Such a network topology employs multiple-access techniques thus having similar characteristics to radio communication networks.

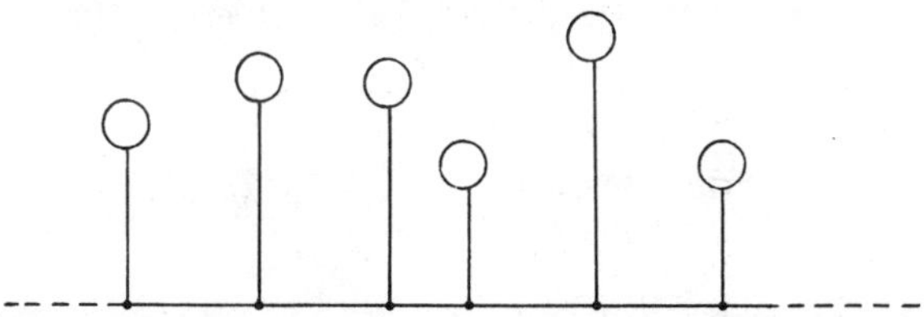

Fig. 4.36 Bus topology used in local network configuration

4.10 INTEGRATED COMMUNICATION NETWORK HIERARCHIES

The communication network elements discussed so far regarded each element as a separate entity with its advantages and disadvantages dictating its practical employment. The reader might make the wrong interpretation and infer that for each type of network application a specific element is selected exclusively. This is not so, for the full exploitation of the potential provided by the network is the employment of a number of types of elements in a single configuration, using each type to give the maximum service it is intended for. A practical network is usually constructed by an hierarchy of different types of nodes, each providing a different level of communication functions. These networks have a tree-like appearance and many sources refer to them as hierarchy-tree networks.

(a) *Level 1* is the circuit switching level which provides the physical path between the selected terminals or computers and the required lines. This level is performed by hardware means, such as multiplexors. The elements of this level are usually located near the terminals, where their function is to prevent an excess use of separate transmission lines.

(b) *Level 2* provides the logical path connection which directs the data arriving from various lines to a common shared line. This level is achieved by software means provided by microprocesses, which equipment is usually referred to by the general term of concentrators. The elements of this level normally serve as remote nodes, collecting the data from a number of widely separated multiplexors or terminals. If required, however, concentrators can also be employed instead of multiplexors, thereby increasing local line efficiency. In brief the main function of this level is to increase the utilization of the common transmission line.

(c) *Level 3* provides the actual handling of the messages, for routing them to various terminals and computers. This level calls for specially oriented computer centres to serve as message or packet switching centres whose main function is to increase the efficiency of the whole network.

A message originating from a terminal or a computer might require the assistance of all the three levels, of only one or two, or of none at all, as seen in Fig. 4.37. It should be noted that all these three levels need not necessarily operate together in each communication system or in the sequence stated, but may operate singly or in a different order of combination, as shown by the dotted lines. For example, a computer may be connected directly to the switching centre and thereby the centre will operate as a communication processor with the computer as its host.

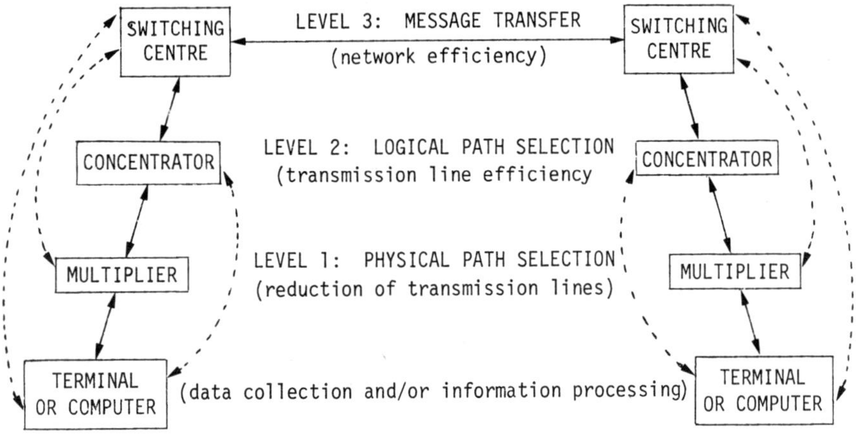

Fig. 4.37 The three distinct hierarchy levels of communication node functions

In a very large sophisticated concentrated network configuration, such as required in an airline seat reservation system, a single host computer centre is employed, with the network elements assisting in collecting the data from all the remote terminals spread throughout the world. This network configuration, as shown in Fig. 4.38, exploits the basic star topology with a switching centre in the nucleus. Each of the star lines may comprise multidrop topology with concentrators as its remote nodes. The data originating from each of the local terminals require the use of multiplexing technology to share common lines to the concentrators. Such a highly complex hierarchical configuration enables different transmission speeds to be applied on the various lines, and thereby provides a most efficient means of data collection system.

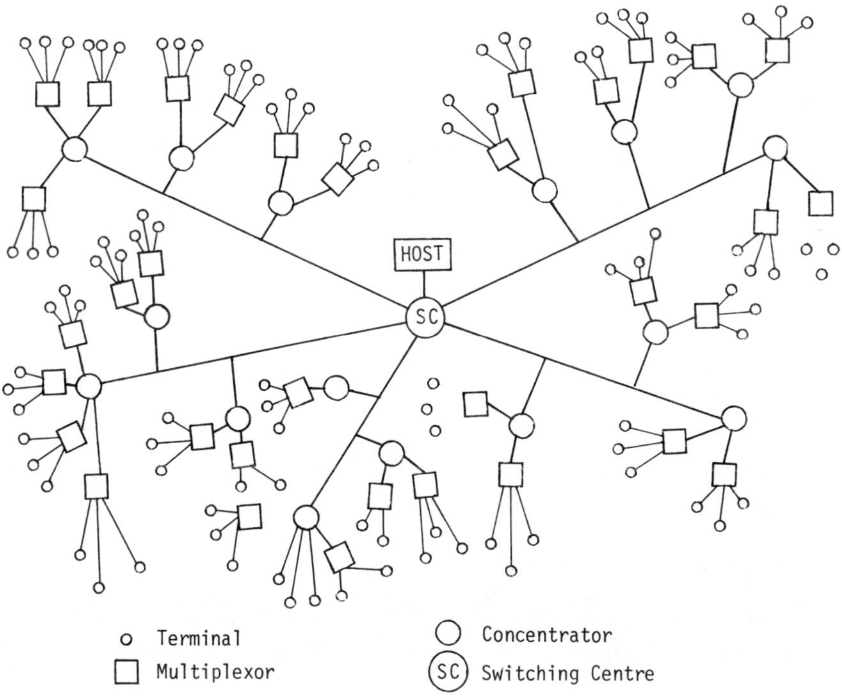

Fig. 4.38 Example of a centralized hierarchial star network

The same network hierarchical exploitation of the communication elements can also be employed in a distributed network which can offer both terminal-to-terminal and computer-to-computer communication. An example of such a network configuration is presented in Fig. 4.39, in which all the centre nodes of the network are switching centres. The computers are connected to these centre nodes via concentrators and the terminals via sophisticated intelligent multiplexors. In such a network, the switching centre nodes and their associated connecting links are referred to as the backbone of the network, while all the periphery nodes are referred to as the network access area. Both parts of this network configuration can operate independently using different and unrelated procedures.

Even in a radio communication system various hierarchy levels may be employed, as seen by the example presented in Fig. 4.40. In this example multiple access techniques are used in the radio-hop and multiplexing and/or concentrators may be used to collect the data from a number of terminals before they are broadcast in the network. One station may even act as a relay switch with the processing ability of a switching centre. The latter can therefore be associated as one of the backbones of a distributed network.

Integrated communication networks can be divided into two parts, the local network and the long distant network, Where all the traffic in the network is long distance, one can regard the local network as the access network,

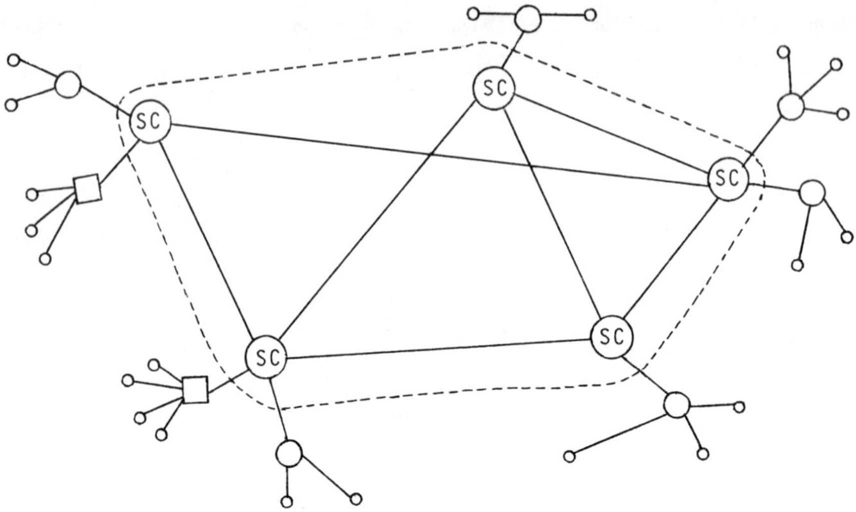

Fig. 4.39 Example of an hierarchial distributed network configuration

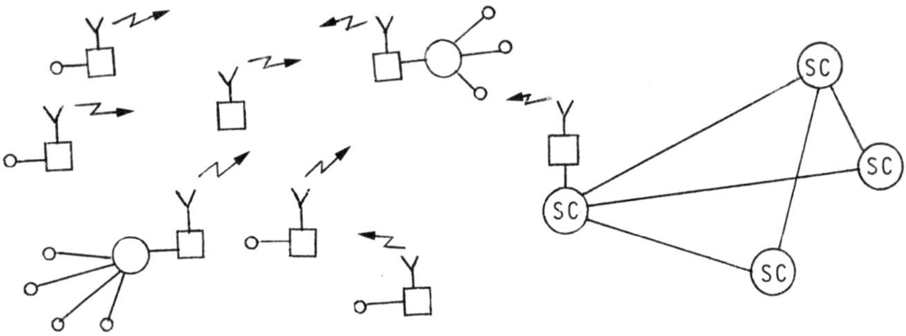

Fig. 4.40 An example of a multi-access network associated with a distributed network

but in fact in most practical integrated networks, the percentage of local traffic may be rather high and must not be neglected. That is, one must not regard the local network as an access to the long distant backbone network but essentially as the local backbone network. Although the problems involved in the design of the local network are no different than those related to the long distant network; the solutions employed in both parts of the overall network are different. In the local network lower level nodes of different types are generally used, while for the long distance network only high level nodes of the same type are employed, i.e., switching centres.

4.11 NETWORK TIME DELAY CONSTRAINTS

Circuit switching networks are based on direct communication between the sender and the receiver. This means in practice that once the call is set up, there is a direct point-to-point link between the two communicating terminals. In this case, the network configuration as a whole can be disregarded, as the only network constraints are those found in the point-to-point link; the propagation delays of the signal through the particular link. This however, is not the case in message or in packet switching, since the communication is entirely performed in an indirect operation; that is, the messages may have to travel through various number of links and stop at a number of intermediate nodes. Furthermore, in these intermediate nodes the messages are stored and may be retransmitted only when the traffic in the output link permits it and their turn in the queue comes up. All these factors add up and can cause considerable time delays between the time the message is dispatched and the time the message is received at its final destination.

The delay of the message in the network is one of the main constraints in selecting a particular communication system. In command and control systems this delay could effect the 'real-time' feature of the system; hence it is extremely important to consider it in the system design. The network delay is defined as the time required for a message to travel through the whole network until it is successfully received in the destination terminal or computer and has been properly acknowledged. In other words, it is not only one-way transmission but also includes the reply in the opposite transmission direction. If any retransmissions are required while the message is in the system, then this extra timing must also be added to the network delay.

The delay of a message in the network could be caused by a large variety of reasons, not all of them being the responsibility of the network, but of the operating system also. Since the network delay is one of the major command and control system constraints which affect the choice of the communication configuration involved, it is worth listing here some of the items which affect this delay.

(a) When a block message or packet is ready for transmission, its transmission speed cannot be calculated by the number of information bits and the links' data rate. To the information bits must be added the message or packet overhead bits needed for addressing and error recovering techniques. These overhead bits may add over 50% of redundancy and thereby increase the transmission time.

(b) The message or packet size and the communication protocol for transmitting it may also add to the delay of the message in the system. The hand shaking procedures have an inherent affect on the delay as well as the answer-repeat request procedure in which the whole message or packet may have to be retransmitted.

(c) The basic delay figure of a message in a link is defined by the links' data rate. This is a function of the bandwidth available and the type of modulation. In a sophisticated network configuration a message may have to travel through various links, each providing different data rates.

(d) In shared communication links the terminal cannot transmit when its messages are available, but only when it is provided with a suitable time slot. In these cases the slot size and the cycle duration add to the time delay of the message in the network. That is, the added delay is defined by the slot size and the gaps between the available slots.

(e) The procedure or protocol employed in the various links which defines both the delay before transmission and the probability of retransmission due to collisions. In this case the delay must also consider the estimated traffic load in the various links.

(f) The type of link operation, whether simplex, duplex or half duplex. In half-duplex communication links, one must consider the turn-around-time when the DCE units reverse the direction of transmission. This time delay, in the half-duplex links, occurs in each transmission, whether the transmission is in data, acknowledgements or control messages. The turn-around-time is not only between the sender and the destination but is repeated in each link in the network connecting any two nodes.

(g) The propagation delay along a particular link, which is a function of both the distance the message travels through the link and the type of media used. This propagation delay is pronounced in satellite links, since the distances involved are so great.

(h) The number of hops a message must make before it reaches its final destination: that is, the number of nodes it must pass through. It is obvious that a point-to-point network or a fully connected mesh network produces the least number of hops, although in practical situations such networks cannot always be provided. The system designer is left the choice of selecting the best network topology that, despite requiring the message to hop over a number of nodes, keeps the delay time below a given 'real-time' figure.

(i) In a partially connected distributed network, the designer must also consider the delays along the alternative routings.

(j) In a sophisticated network the network delay is not a constant figure throughout the whole configuration. The designer should analyze the traffic and define what is the ratio of the long distance traffic to the local traffic. It is possible to provide the long distance traffic with a priority over the local traffic so that the overall network delay is kept to a minimum. On the other hand, if the local traffic is given too low a priority it may cause extra time delays while it is required to wait for a lull in the long distance traffic.

(k) The number of communication levels the message may have to cross before it reaches the backbone of the major network configuration. It should be pointed out that it is usual to provide the backbone network with high-speed links and the access area with low-speed links. On the other hand, the lower level nodes (such as multiplexors) present negligible delay, while in the upper level nodes there is a considerable delay due to the store-and-forward nature of these nodes.

(l) The number of links assoicated with a particular node which defines the number of messages that can reach the node simultaneously and hence may affect the queue length and the time the message will be retained in the node.

(m) The processing involved in each node which defines the minimum delay time the message is to be held in the node before it is retransmitted,

The list above only brings a small part of all the items which can add to the network delay. It is up to the designer to consider as many points as possible, since the whole success of his system depends on the reaction time of the system operation.

4.12 NETWORK SYNCHRONIZATION CONSIDERATIONS

One of the crucial problems in network design is the clock timing and the mutual synchronization of all the nodes which ensures that the nodes receive and decode the incoming data streams. It is the network timing and synchronization which are the important factors in the success of message communication through a network. Due to transmission delays and DCE response times, the synchronization in a sophisticated network is extremely difficult. That is, where a number of links reach a node, the propagation delay along the alternative routes cannot be equal and the receiving node may therefore have difficulty in locking on all of them. The consequences of improperly synchronizing the nodes could result in the loss of large portions of the received messages.

The problem of network synchronization is completely different from the point-to-point link synchronization discussed in detail in the previous chapter. In point-to-point communication, using half-duplex modes, two forms of data synchronization were identified. In asynchronous data communication, the clock of the receiving DCE is free-running and adjusts its timing by means of a start bit of each character. Such a synchronization system is slow since it must be frequently updated. In synchronous data communication, the transmitting DCE's clock is in control, while the receiving DCE's clock must continuously follow suit and synchronize with that of the transmitter. In other words, the transmitting DCE's clock acts as a master and the receiving DCE's clock acts as a slave, as seen in Fig. 4.41. In nonsynchronous message transmission the transmitting and receiving DCE's clocks are fully synchronized as in synchronous transmission, although there could be gaps between the transmitted messages. During these gaps the clocks are still synchronized by the transmission of dummy data. Nevertheless, this system calls for special synchronization bits to precede the message, so as to alert the receiving DCE of the incoming message.

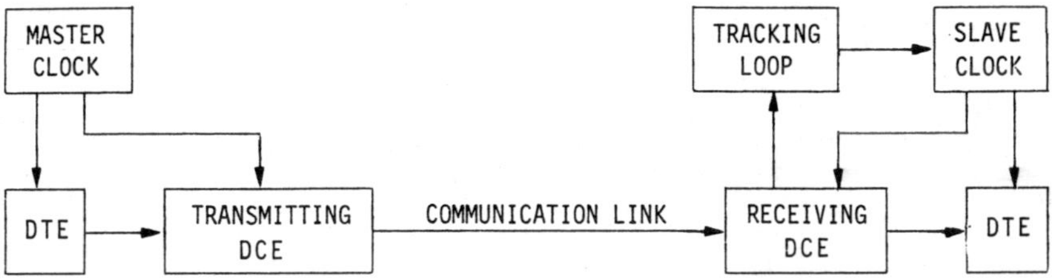

Fig. 4.41 Time synchronization in a synchronous point-to-point link

In sophisticated network configurations, it is not only the single half-duplex point-to-point link that must be considered, but also the full-duplex link. If two nodes are transmitting in both directions at the same time, which of the two is the master and which is the slave? Furthermore, it is not only the single or double node that is to be considered, but all the nodes in the whole network must be fully synchronized together. The more so, since the synchronization must include both the local nodes and the long distant nodes. A message flowing from one node must be able to travel through a number of nodes without any errors entering and changing the data contents. As a message leaves a node, its DCE (modem) converts it into analog form and then, in the next node in sequence, it is converted back to digital form. If the clocks of all the nodes are not synchronized together no message will be able to reach its destination, since it will be gradually distorted after several communication processing. Furthermore, there is no guarantee that, with the nodes scattered over wide areas and varying propagation delays in the alternative routing, the same state will be kept throughout the whole network.

For successive network operations it is necessary when designing a synchronization network, to take into account both the network topology and the network control constraints. The synchronization methods available can be classified in three groups, as seen in Fig. 4.42. In this section some of the important synchronization methods are considered and their means of implementation introduced.

Included in the technical literature there are further synchronization methods available that cannot be directly classified in the form suggested

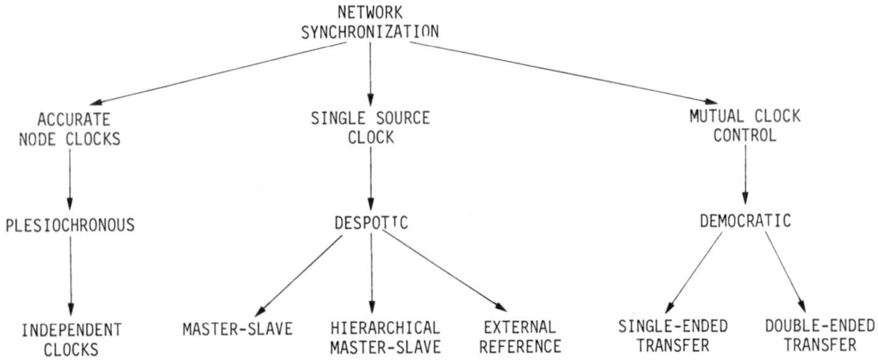

Fig. 4.42 Classification of the main synchronization techniques

above. There are also some network systems that integrate a number of techniques to solve both the local and the long distance traffic requirements as follows:

(a) PLESIOCHRONOUS METHOD

The simplest and most straightforward method is to provide each node with its own separate free-running clock, as shown in Fig. 4.43. Such a synchronization method is referred to as plesiochronous operation, or independent clocks. This technique should not be mistaken for an asynchronous operation which also operates with independent clocks, since in plesiochronous operations the data received is continuous while in an asynchronous one the data is received a character at a time, and the internal timing is adjusted for each character.

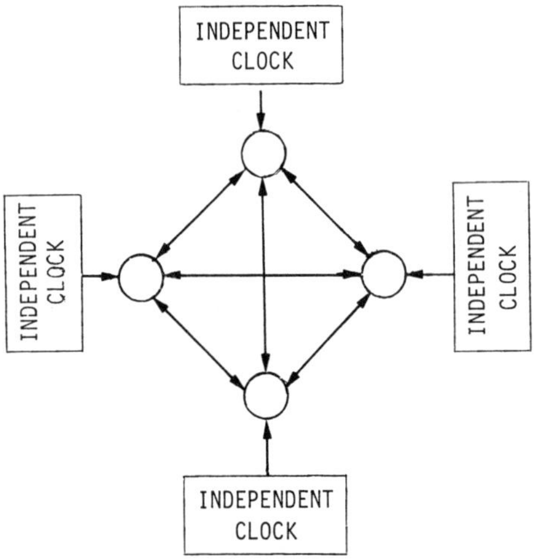

Fig. 4.43 Plesiochronous synchronization operation concept

The plesiochronous method of synchronization can only be achieved with most accurate clocks, such as atomic clocks. Any slip of the timing, even if it occurs very slowly, will cause errors in the data received. Such clocks are rather expensive and consume power and space. Since the internal clock in the node and the received signal timing differs in phase, a buffer is provided on each link to separate the link input from the node operation. This buffer compensates for the differences in timing of the clocks (as seen in Fig. 4.44) by employing different timing for shifting the data in and out. The timing for shifting in the data is controlled by the timing recovery of the received signal, while the data is shifted out of the buffer by the internal clock of the node. The data is shifted in the buffer on the basis of first-in/first-out (FIFO) principles.

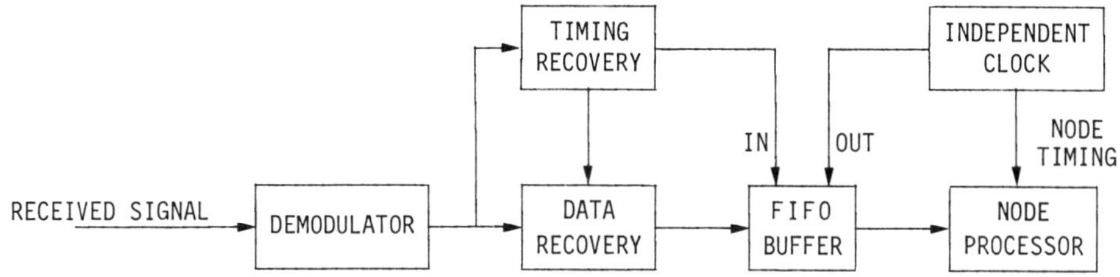

Fig. 4.44 Functional diagram of the link buffer separating the incoming links from the node's timing

With such a system there is a danger that the internal clock, relative to the external source, will differ to such an extent that it could cause the buffer to overflow. To avoid this danger it is necessary to interrupt the traffic flow so as to reset the buffers. The interval between the resetting operations depends on the accuracy of the clock and the buffer size, and it generally occurs once every twenty-four hours.

Despite the simplicity of the plesiochronous operation, it is not recommended in systems where other means can be used. This is because continuous operation can only be ensured by high reliability coupled with utmost accuracy required in each node. Such systems then are needed for applications where the nodes are unaccessible for maintenance. It is also attractive in principle for providing the interface between separate communication systems.

(b) EXTERNAL REFERENCE

The attractiveness of the plesiochronous technique derives from the independent operation of the node clocks and the separation of the node operation from the communication link operation. Its disadvantage is that the independence of the operation does not provide the required inter-node synchronization, for this reason an interruption of the traffic is essential for the buffer timing. By using an external clock reference, as shown in Fig. 4.45, all the nodes are synchronized together and it is still possible to preserve the separation of the node operation from that of the link timing. Although the clock operation calls for a centralized clock control, the network operation may be distributed with each node operating independently. As the reliability of the clock reference source and its distribution facilities must be extremely high, such a network should always have an alternative clock source, i.e. a doubled set of equipment. Furthermore, the operation may be hindered by the propagation delay of the clock

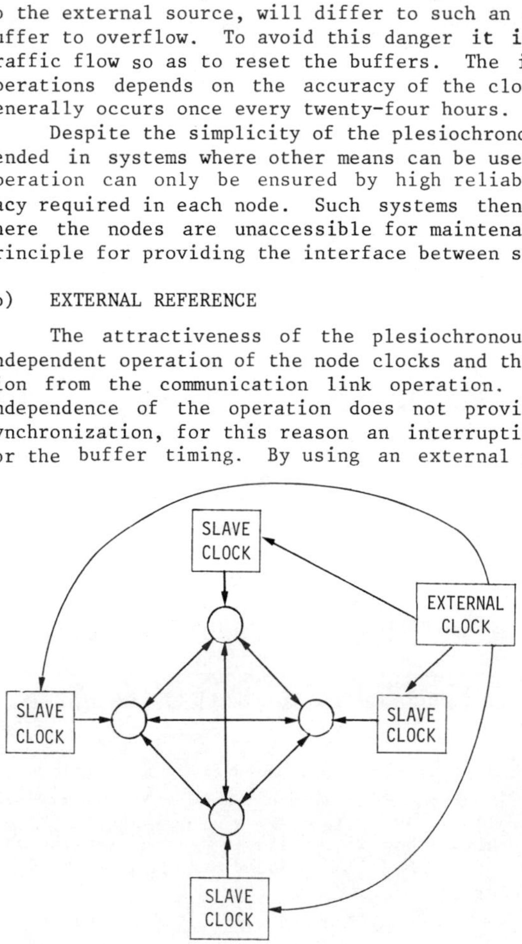

Fig. 4.45 External source reference synchronization concept

reference relative to the propagation along the transmission link. To offset this, the operation speed of the network must ensure that the relative timing difference is less than the possible error in the bit duration.

There are several external time reference sources available, not all of which have global coverage. These clock sources are primarily intended for navigation systems, with the timing calculated by the distance from the source and the time of arrival of the reference signal. Among these systems are the Loran-C system with its 100 kHz pulse signal, the Omega system operating between 10 kHz and 14 kHz, and the NAVSTAR global positioning system.

(c) MASTER-SLAVE METHOD

Instead of using an external clock reference source it is possible to exploit one of the nodes in the network as the 'master' clock source, as seen in Fig. 4.46. This is feasible only if all the nodes have direct access to this centralized node and it is suitable, therefore, for star topology with centralized network control. A clock reference is transmitted from the centre node (referred to as the master) to the closest dependent nodes, so-called slave nodes. The slave nodes obtain the timing signals by recovering them from the incoming signals, as in the point-to-point synchronous data transmission systems. The recovered timing signals is then used to adjust the internal clock of each node.

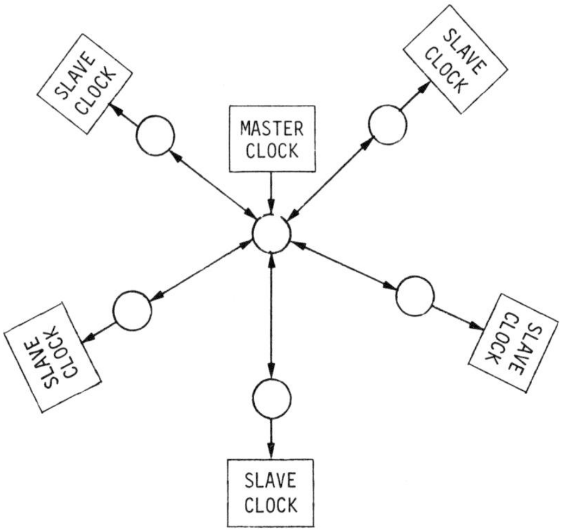

Fig. 4.46 Master-slave synchronization method concept

This synchronization method is highly recommended for centralized network operations since all the communication in such a scheme is directed to and from the centre switching node. The association of the traffic control with the timing control makes this method most reliable. In a distributed network configuration, however, the master-slave does conflict with the applications of the distributed control principles. Nevertheless, there are small distributed systems which exploit these techniques by adding special compensating circuits.

(d) HIERARCHIAL MASTER-SLAVE

There are very few pure star network configurations in practice since in most applications the outer nodes of the star serve as intermediate nodes between the terminals and the centre. That is, the centralized network configuration is generally organized in an hierarchical set-up with a number of levels in between. In this type of configuration, the centre of the star, at the nucleus of the backbone of the network topology, serves as the master clock for all the closest nodes directly attached to it. Each of the outer nodes of the main network backbone then serves as a master to all the remote nodes directly connected to it, as seen in Fig. 4.47. (In this drawing the clocks are shown as part of the node structure and the arrow indicates the direction of the clock distribution dependency.) In this way the highest node in the hierarchical chain provides the clock reference to those nodes lower in scale which are directly controlled by it. These nodes may then supply the clock reference to still other nodes of lower levels in the chain.

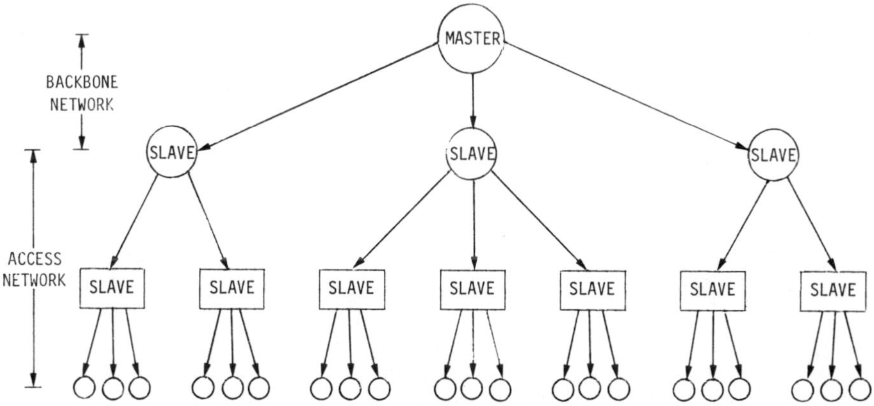

Fig. 4.47 Hierarchical master-slave synchronization method concept in a centralized topology

The problem of propagation delay does not exist in such a scheme since the traffic is always controlled by the higher level node in each particular link. Each node consists of an internal clock which is locked on the incoming regenerated clock. Nevertheless, with the lack of external timing reference, the internal node clock will continue to function with a drift determined by its stability.

(e) TIME REFERENCE DISTRIBUTION

The hierarchical master-slave synchronization method discussed before is primarily intended for a centralized network topology. In practical applications, such an organization is mainly employed for local network configurations with the intermediate nodes being of the type of multiplexors and concentrators. In order to apply this hierarchical synchronization method in a distributed network configuration to a master-slave method it must be upgraded to a system referred to as 'time reference distribution'. It should, however, be pointed out that some sources refer to this time-reference-distribution as hierarchical master-slave because of the similarities of the two synchronization techniques.

In the time-reference-distribution, as used in a distributed network topology, one of the nodes acts as a master, while all the other nodes perform

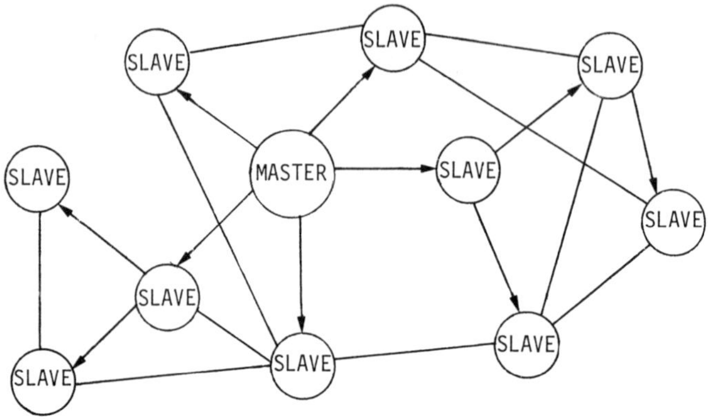

Fig. 4.48 Time-reference-distribution synchronization concept in a distributed topology

as slaves (Fig. 4.48). The arrows, on some of the links, show the timing distribution direction as in hierarchical master-slave technique. That is, the clock information is distributed in a simple tree structure organization based on the shortest links. The links without arrows are used only for data communication, but not for the initial timing distribution.

While each node may receive data from a number of input links, it receives the timing reference on only one of them. This removes most of the disturbing effects of the different propagation delays over the links. The nodes continuously measure the local clock timing differences with the selected input link timing, and correct accordingly. That is, the slave node timing is adjusted by using the techniques described for point-to-point link synchronization, with its internal clock timing locking on the recovered timing of the received signal of one particular link. This link is usually selected as being the shortest link to the nearest node next in line in the hierarchical chain. When this link (on which the clock timing is recorded) is interrupted, the nodes reverse their timing alliances to an alternative link. In such a way the node clock can always lock on one of the input signals and there is no need then for an accurate local clock. If no such link is available, after all the input links have failed, the node can continue its operation as in plesiochronous mode until the links have been repaired. This will not affect the network operation, since this particular node, with all input links disconnected, does not receive any external data which it needs to decode. If the master node fails, any one of the other nodes in the network can take over its function, but in a required assigned order.

(f) MUTUAL CLOCK SYNCHRONIZATION

In large distributed network configurations, the independence of the nodes is sometimes essential and then the time-reference-distribution cannot be exploited. In such distributed networks, the nodes are spread over the world with a variety of distances between the nodes. It is not only the differences in the propagation delays which calls for the autonomous operation of the nodes, but also the conditions where the backbone nodes may be part of the other networks controlled by different governing authorities. A country (or even a company) may not agree to have their node operation controlled by a clock in another country. The principle of having large distributed networks is not just the ability to transmit data between distant nodes, but to achieve this while allowing each node to operate independently of the other nodes in the network. A node must be able to choose any or all alternative routes on any of the output links, according to the traffic load on these links and not according to the shortest distance.

For large distributed network configurations with independent node operation, the use of the mutual synchronization method is recommended. In this method, the independence of the node operation is reserved, with each node acting as its own master, but having its timing adjusted to be in line with the timings of the other nodes; for this reason the method is also referred to as 'discreet-control-correction'. The operation principle of mutual synchronization, as seen in Fig. 4.49, is that the timing reference is carried on each line, and so can be recovered independently, as in a plesiochronous operation. The difference, however, is that in mutual synchronization, the internal clock is continuously adjusted to be as close as possible to the timing of all the incoming signals. The node measures the differences of all the incoming timing signals, and adjusts itself to the average value.

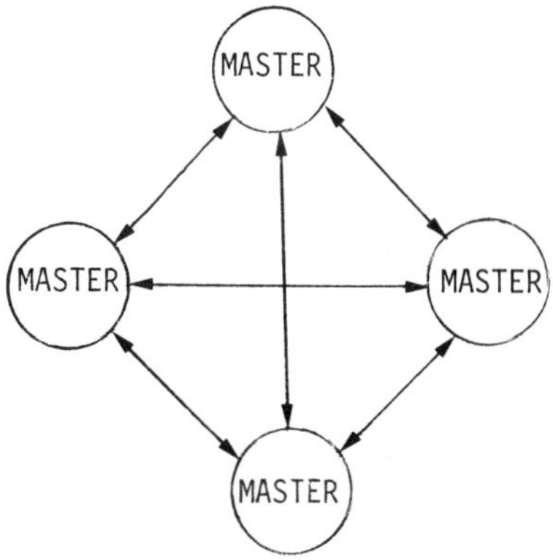

Fig. 4.49 Single ended mutual clock synchronization concept

The node employing the mutual clock synchronization receives the data stream as described for the plesiochronous synchronization. That is, the data is received and decoded according to the recovered link timing reference, and is then placed in a buffer. Once the data is in the buffer it can be transferred into the node processor at the node's own internal timing. The main difference of the mutual synchronization method is that there is no danger of the buffer overflowing, as in the plesiochronous synchronization. This is because the difference of the incoming and internal timing of the mutual clock synchronization is always kept at less than half a bit duration.

The object of the internal synchronization is to coordinate the collection of the incoming clock references so that the buffers do not overflow or deplete. This is achieved by keeping the node's internal clock with the least phase to any of the other nodes' clocks. If it was only the one link, then the internal clock could easily lock onto the received signal; since however, there could be a number of input signals, the internal clock must correct by the averaging phase of all the clock offsets. If the internal clock offset is ΔT relative to one of the input links and the number of links associated with the node is n, then the correction that needs to be performed is

$$T_{\mathbf{n}} = \sum_{\mathbf{i}=1}^{\mathbf{n}} \Delta T_{\mathbf{i}}/i .$$

The timing references of the various incoming signals are brought to an averaging comparison circuit, as seen in Fig. 4.50. This can easily be performed by an updown counter which is reset after each bit by the node timing. In practice this can be performed by advancing the counter with each incoming recovered timing pulse and decreasing the counter n times for each internal clock pulse. In that case the counter must be free running so its output must be relative to the averaging timing. The output of the counter is then converted to a smooth control signal (via a DAC circuit) so as to alter the frequency of a voltage controlled crystal oscillator (VCXO).

Although the individual nodes correct the timing according to the

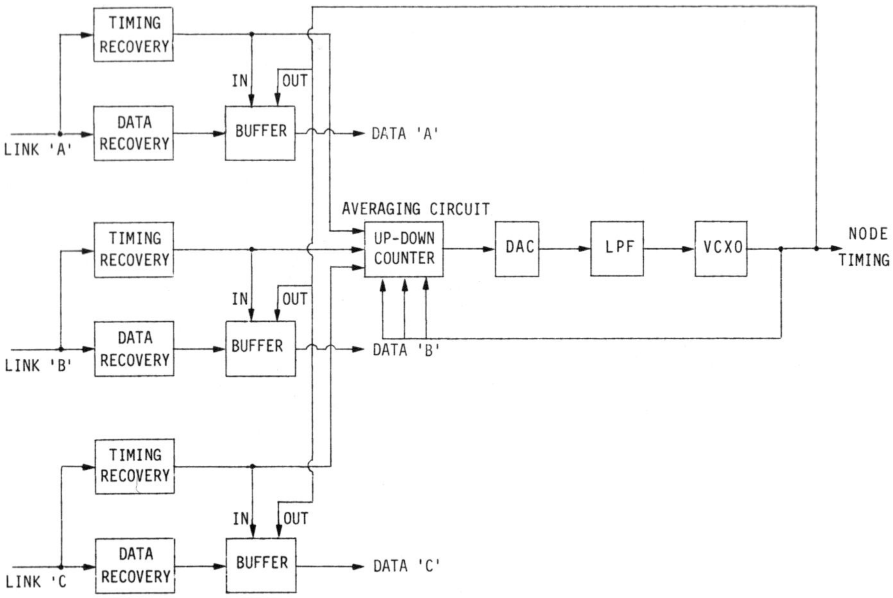

Fig. 4.50 Functional diagram of an average timing circuit in mutual clock synchronization

average phasing of the incoming signals, the difference between the various input timing can be extremely great. Lack of coordination may cause the internal timing to fail to be near the timing of some of the input signals, and may lead to errors in the system. To prevent such a situation from occurring, double ending transfer is employed, as seen in Fig. 4.51. Single ended transfer controls the errors of a particular node only from the accumulated errors of the incoming signals to this node. Double ended transfer achieves the control of the errors not only by its own measurements, but also by the accumulated errors in the nearby nodes. In this way the clocks of adjacent nodes are adjusted all the time relative to each other. A single ended transfer is usually employed with the links operating in a half duplex mode, while double ended transfer is employed with the links operating in a full duplex mode.

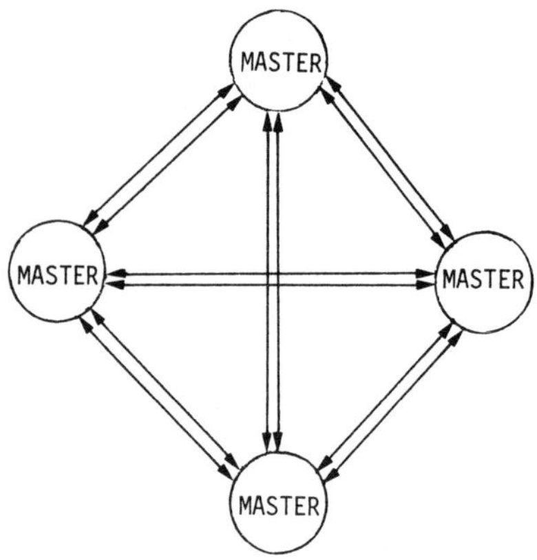

Fig. 4.51 Double ended mutual clock synchronization concept

The mutual synchronization, as described, is simple and employs readily available crystal oscillators, which are cheap and consume low power.

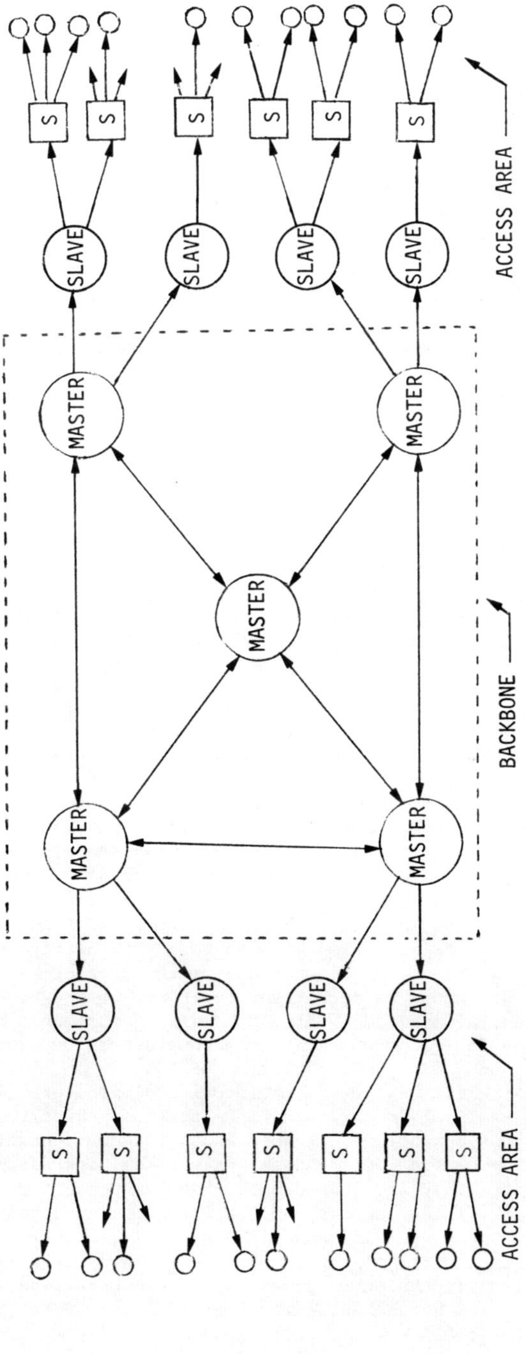

Fig. 4.52 Example of an integrated synchronization scheme employing two different concepts

(g) INTEGRATED CLOCK SYNCHRONIZATION

In large network configurations, each of the main nodes of the network backbone may also be a centre node for a local network, as seen in Fig. 4.52, or only an intermediate node to another network. In this case, it is not advisable to employ the same synchronization techniques throughout the whole network. The backbone may operate in one mode while each of the local networks may employ a different operational mode.

In the example given on page 199, the centre backbone is operating in a mutual clock synchronization technique, while the local network is operating in an hierarchical master-slave technique. In each application one must select the techniques best suited for the job, even if it calls for many different techniques in the same overall large network.

4.13 NETWORK MANAGEMENT CONSIDERATIONS

As the network configuration increases in size, the probability of a link or node failure increases. Nevertheless, even then, the rest of the network must continue its operation, irrespective of a failure in any of the other parts of the network. Furthermore, no message is allowed to be lost en route even when there is a link failure. In relatively 'small' network configurations, a message is retained in the originator's terminal memory until it receives a positive acknowledgement that it has reached and has been accepted by the destination terminal, irrespective of how many nodes the message has to pass en route. In larger network configurations, the message is kept in the sender node's memory only for the period it handles it, that is, only until it is received by the next node in the transmission chain. Once a node receives a positive acknowledgement from the next node, it clears the memory space which held the message. In other words, in these large networks no record of the message is kept once the message leaves the originator's node, and there is no means of locating the whereabouts of the message in the network in real-time. Such a sophisticated communication process is only possible if the network is properly managed, with the facilities of monitoring the system's operations in real-time.

It can be said that one of the most important elements in a large network configuration is the management ability of the system to control both the traffic operations and the communication equipment. Each node in the network is generally provided with enough monitoring and measuring equipment with which to test its local links and their associated equipment. The function of these monitoring measuring equipments is to analyze the performance of the network operation and diagnose flows in the traffic transfer procuration so as to provide rapid means of fault isolation. The monitoring operation is continuous and/or periodical, so that preventative measures can be carried out even before the traffic on a degraded link comes to a complete standstill. It is not enough to detect the fault and introduce alternative measures, it is essential also to locate the point of the fault and to identify the cause of the fault. That is, all the faults must be diagnosed so as to make possible a sound and speedy repair.

In point-to-point communication, the problem of monitoring the links is a relatively simple task, as described in the previous chapter. In networks based on point-to-point communication, such as a star configuration, the same principles apply. The centre node sends signals to all its associated terminals in turn, and there they are looped back, thereby enabling the measurements to be performed in the centre. Such a link testing technique is based on the master-slave approach, as seen in Fig. 4.53, with the centre node operating as a master and all the outer nodes as slaves operating under the command of the master. In the early days of data communication, the monitoring centre consisted of a patch board, which enabled the operator to connect the various links to the test equipment. Today it is possible to employ

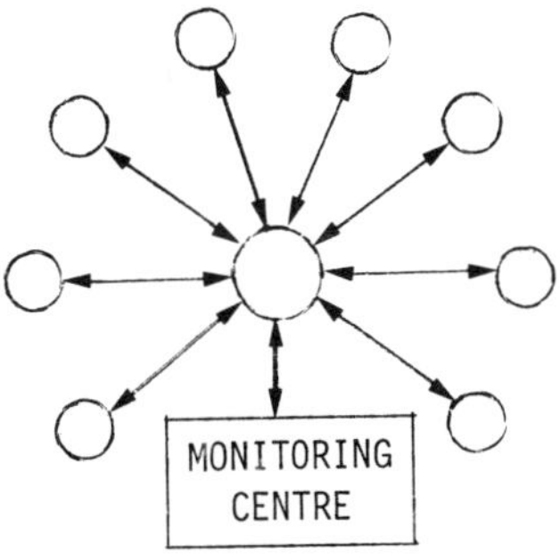

Fig. 4.53 Master-slave monitoring test technique

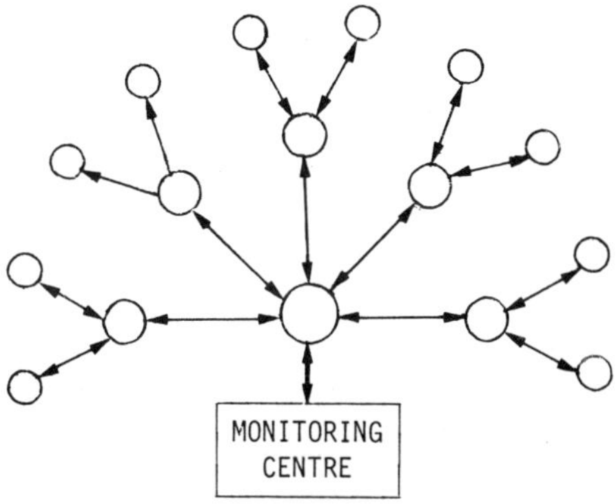

Fig. 4.54 Hierarchial master-slave monitoring test technique

computers which select the various links according to a continuous routine program.

The master-slave approach is ideal for star configurations where the test equipment is concentrated in one location and where the analytical results are readily available for evaluation. Such a technique is also suited for multi-star configurations (as seen in Fig. 4.54) where the centre node can test all the outer nodes. In this case all the intermediate nodes are transparent, that is, connecting the outer nodes (in turn) directly to the monitoring centre. It follows that all these intermediate nodes must be of lower level type, or, in other words, do not employ the store-and-forward technique.

In most communication networks, the intermediate nodes are in the form of concentrators and not multiplexors, and therefore these nodes cannot provide a direct connection between the centre and the outer nodes. In this case, the intermediate nodes which operate in a store-and-forward mode, are provided with their own test facilities. The centre node sends test instructions to its adjacent intermediate nodes and they in turn test all their local nodes, with the results transmitted back to the centre. Although the monitoring is performed with the aid of the intermediate outer nodes and concentrating the results in the centre, the operation provides identical results.

In large communication networks, most of the traffic between the terminals is not direct or through a single switching centre, but through a chain of store-and-forward nodes, with many of them operating as independently switching centres. In this case, although it employs a different communication approach, the network monitoring can still be centralized. Each node in the network has its own test facilities, with one of the nodes designed as a dedicated monitoring centre which can provide diagnostic information on the performance of all the links and the equipment throughout the whole network. This approach employs master-slave centralized monitoring techniques for testing the performance of a distributed network configuration. This can be performed in two different methods. In one approach the testing monitoring site is separate from the network, while in the more modern approach the testing site is an integral part of the network. In both approaches the nodes testing is performed by remote control techniques which operate the diagnostic equipment in each node.

In the first approach, a central site is set up which functions independently of the network communication control operations. This monitoring site is connected with each and every node in the network, as seen in Fig. 4.55, with either dialled up or dedicated lines. These lines are used only for testing purposes and no other data is transmitted over them. The central site

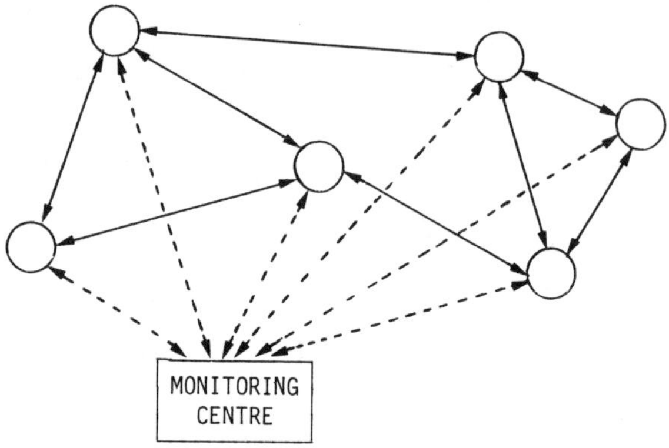

Fig. 4.55 Master-slave monitoring technique in a distributed network configuration

sends information messages to each node, which then initiates test programs for measuring the performance of their local lines.

The approach which separates the monitoring site from the network is wasteful, since it requires special dedicated links between the central site and each of the independently operating nodes. A much better approach is to exploit the existing communication network links for transmitting the test instruction messages. In this case the monitoring test site is attached to one of the nodes in the network, as seen in Fig. 4.56. That is, the nodes are tested by remote operation, by means of special test messages which are transmitted in the same way as the data messages. On receiving such periodical instruction messages, the nodes send back the results of all the local diagnostic tests of their associated links based on local loop measurements which are continuously updated. The instruction messages, in this distributed network, may have to travel through a number of nodes before they reach the node under test. The centre monitoring node can therefore build up a display of the real-time performance of all the network nodes, even if the centre is not directly connected to all the nodes.

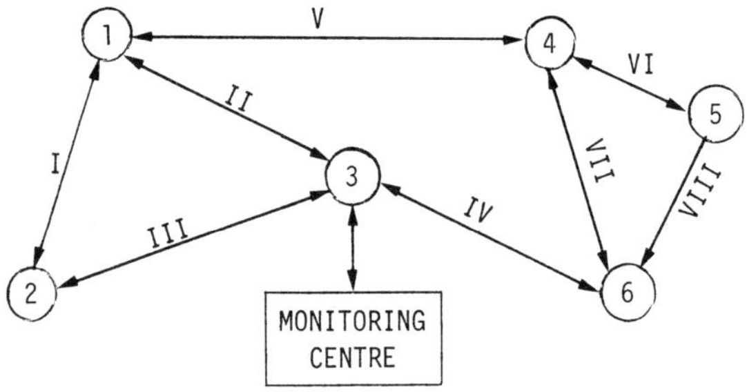

Fig. 4.56 Remote master-slave monitoring technique in a distributed network configuration

A continuous surveillance of the network performance by periodically testing all the nodes, provides the centre with all the real-time information required for managing the performance. When a malfunction is suspected, the fault may be isolated and the network topology can be reconfigurated. The centre monitoring node provides all the other nodes with the information about any faulty links or nodes so that they can redirect their traffic accordingly.

In very large communication networks, where the nodes could be in different countries, a centralized monitoring site may not be a sufficient solution. That is, the monitoring centre may be suitable for the general administrative management of the network, but it cannot provide the requirement of rapid fault isolation. In such a large network configuration with distributed communication control, each node must retain a complete real-time picture of the available routes throughout the system. That is, each node must contain a 'monitoring centre' of the whole network. The information retained in each node need not be as detailed as in the previous monitoring techniques, but need only report the availability of the various links. This is performed by special tables kept in the node's memory which are continuously updated. Besides these tables of the network link conditions, each node is fully responsible for all its local links and equipment which it continuously measures.

Each node is interested in the complete performance of the whole network, and therefore requires to know the condition of all the links. Nevertheless, it does not need to send test messages to any of the other nodes, but can build up its real-time by receiving information from its nearby nodes of the

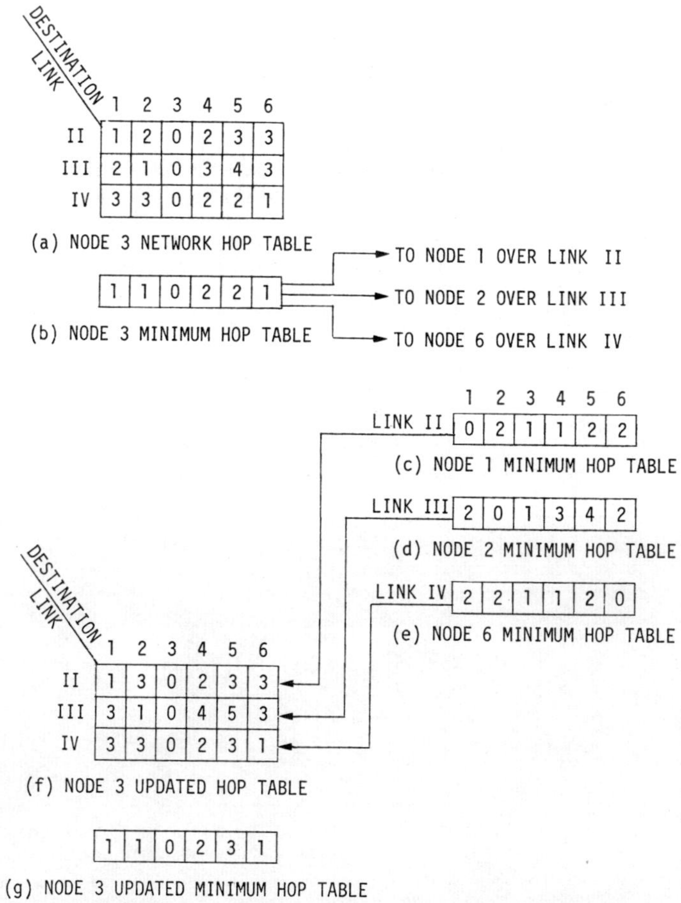

Fig. 4.57 Hop table in a distributed monitoring technique

routes available to all the other nodes. For this reason the monitoring table, as seen in Fig. 4.57, contains information of the routes to all other nodes rather than of the condition of individual links in the network. The table is built with the rows defining each of the nodes' outgoing links and the columns defining the destination nodes. The number at the intersection of the rows and columns states the number of hops the message must make in order to reach its node destination.

Each node continuously updates its own table by checking the conditions of the links which are directly under its supervision, and by the information it receives from the nearby nodes. In fact, the nodes interchange their own table contents with all the nearby adjacent nodes. There is no need to transmit to the adjacent nodes the full table contents but only a summary. This summary represents the minimum number of hops required from a message leaving a particular node until it reaches its final node destination. If any of the incoming or outgoing links of a node are faulty, as analysed by the test process, this must result in a change of the minimum hop table.

The network hop table presented in Fig. 4.57 is based on the network configuration given in Fig. 4.56. It is intended to demonstrate the updating process of the hop table stored in node 3. This node has three data links attached to it (II, III and IV), over which it can transmit messages to all the other nodes in the network, and therefore this node's table consists of only three rows. Periodically, a summary of this table is transmitted to each of the three adjacent nodes (1, 2 and 6) and it receives from them a summary of their table. These summary tables contain the minimum hop number from each of the nodes in question, to all the other nodes. The nodes receive these summarized tables and update their own table with it by adjusting the numbers in each row separately, in accordance with the links they were received from. When updating the table, the particular node adds one digit to the numbers it has received in the summarized table. To demonstrate the table updating process, the information is based on the malfunction operation of links I and VIII.

4.14 NETWORK STANDARDS CONSIDERATIONS

The networks' prime function is to assist in the transfer of data between terminals and computers of various types and sizes. Nevertheless, efficient communication can take place only if the data transmitted through the network is in a uniform arrangement. In other words, it is essential to define common standards that specify exactly the rules by which the terminals may communicate; otherwise the messages might not be accepted by the network, might get lost en route or might be delivered in an undeciphered form.

The design of the data communication networks is so flexible that it provides the designer with multiple solutions. The function of the standards is not to limit the possible solutions but to bring the solutions in line with the external requirements. Most of the communication lines are provided by the common carrier facilities; for which reason the designer must adjust his design solutions to the standard specified for the transmission of data over the communication lines. These standards specify both the physical interfaces and the procedures to be employed in making the connection. This still leaves room for the designer to specify his own network topology and the methods to be used for switching the data in the network. Further standards are necessary, however, to define the interface point between the various computers and the network they are attached to. A communication network is designed with the one and sole purpose of providing a means by which the terminals and/or the computers can communicate with each other. It is therefore not the function of the communication network designer to dictate to the computer manufacturers the rules for their input/output control. What the communication designer is called upon to do is to make the network able to operate with any type of terminal or computer. To comply with this requirement, special standards have been proposed which specify the interface point between the network equipment and the terminal equipment. Again, it is not only the physical requirements that must be specified, but also the protocols necessary for the orderly exchange of the data between the two communicating computers which use the network.

In small networks, the designer can select his own communicating and processing equipment, and can thus ensure that all the equipment is integrated for coordination operation in one system. As the data network and processing system grows in size, it is almost impossible to select all the equipment from a single supplier, which leaves the only solution that all the various equipment used operate in accordance with given standards. In this case, the network designer may instruct all the users to employ the same data communication equipment (DCE), or employ equipment which exercises identical techniques. This requirement, however, is impossible in the case of data terminal equipment (DTE), since each computer site may employ completely different types of processing equipment. Furthermore, some terminals and computers may employ asynchronous communication techniques while others apply synchronous communication techniques. It is not practical to demand from all the processing sites that they should change their equipment if they want to be connected to particular networks. Instead, it is possible to require that all the transmission formats comply with given standards. If the processing cannot comply with these requirements it might be necessary to add special hardware interface equipment which assembles and disassemble the messages in the proper network transmission format.

As already mentioned, each network may employ its own topology and its own operating system. The same freedom of choice applies for the user's computer site, where any type of processing equipment may be used. It is therefore necessary to have international agreement on the interface requirement between the data terminal equipment and the data communication equipment. That is, the communication standards must define the interface boundaries between the DCE and the DTE, as seen in Fig. 4.58. Although the DCE is situated in the user's premises, it must be regarded by the user as being part of the network.

Fig. 4.58 The communication network interface point which requires standardizations

It is generally accepted that standards are essential for the implementation of the message transfer between communication DTEs. It is, however, a paradox to find that the area of standards is full of confusions, with each organization proposing its own standards. That is, there is no single body whose recommendations are fully accepted. In fact, every country accepts different standards and even within a country various manufacturers may employ different standards. This imposes great difficulties in small countries which have to buy all their communicating equipment from foreign countries. It is not only the national official organizations that recommend standards, but many other organizations, such as professional societies, trade associations and even computer manufacturers. Among the important internationally and nationally accepted organizations are the following bodies:

(a) International Organization for Standardization (ISO).

(b) International Telecommunication Union (ITU).

(c) Consultative Committee on International Telegraph and Telephone (CCITT), which is a technical group under ITU.

(d) American National Standards Institute (ANSI).

(e) Electronic Industries Association (EIA).

(f) Federal Communication Commission (FCC), which is part of the U.S. Department of State.

(g) IEEE Standards Committees, associated with the computer and communication societies.

(h) Trade companies such as IBM and Digital.

So long as the communication networks were limited to national boundaries, the situation could be regarded as a local problem but with the expansion of the network beyond the national boundaries, the adoption of an accepted international standard for the definition of the interface between the user's equipment and the widely spread network has become a critical issue. It seems that for the first time in the communication history, there is a chance now for a world wide generally accepted standard, and indeed, the CCITT recommendation X25 for packet transmission is becoming the dominating standard in network interfacing. Even here, however, various network systems interpret the recommendation in different ways. (CCITT has also issued standards for other communication functions, but these are not as widely used.) It is hoped that one day the world will accept the need for uniformity in all aspects of data communications. The acceptance of a proposed international interface standard will solve the user's dilemma and allow him to connect his site equipment to any operating network, without the need of changing his specialized processing equipment. Each user may only have to add standardized interface equipment to enable him to connect up with all forms of networks.

As the standards must consider a large number of operational characteristics, the requirements become more complex as the network grows in size. Instead of employing a single interface standard which although simple is not independent, a hierarchical of standards is preferable. That is, various levels of standards have been proposed in defining the DTE/DCE interface. This provides for the separation of the various operational communication functions, a condition which is essential in large complex networks. It permits the segregation of responsibilities between the various equipments responsible for the transmission of the messages through the network. In special applications the segregation of levels allows the substitution of procedures or protocls as required.

The hierarchy of interface standards has been organized in clearly defined levels, as seen in Fig. 4.59. Although each level cannot exist independently of the others, it is possible to define each level separately.

The various hierarchical levels are defined independently as follows:

Level 1 — The physical interface which defines the electrical, functional and procedural characteristics of the common boundaries between DTE and the DCE. It involves the specification of characteristics required to establish, maintain and disconnect the link between the two sets of equipment,

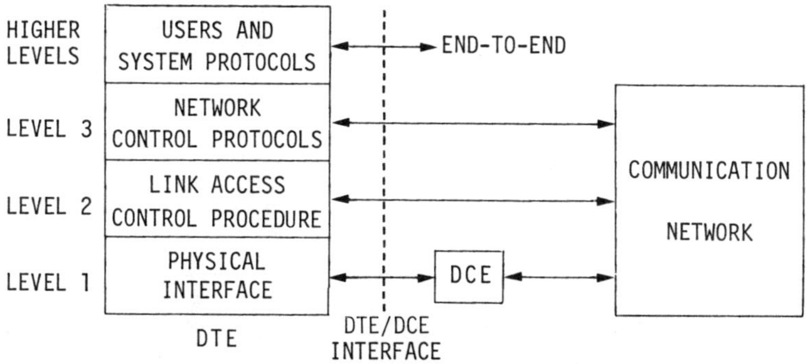

Fig. 4.59 Hierarchy of international communication standards

so as to enable them to operate together

For this level the CCITT recommends the use of standard X21 or X21-bis which are popuplar in Europe. Other CCITT recommendations in this level are X20, X24, X26, X27, X28, X29 for special transmission equipment. In the United States the EIA standards are more popular; they are RS232C, RS422, RS423 and RS-XYZ.

Level 2 – The link access control procedure which defines the handshaking and flow control functions of the communication operations which enable the interchange of data across the link between DTE and the DCE.

There are two types of data link procedures, byte control procedure (BCP) and bit oriented procedure (BOP). The BCP, also known as character oriented procedure (COP) is the oldest and most widely used procedure for many point-to-point communication systems. It defines a set of control characters which affect the orderly operation of the data link. These characters are part of the code set such as ASCII or EBCDIC. One of the most known BCP is IBM's binary synchronous communication (BSC) procedure. Today bit oriented procedures are preferable for large communication systems, due to the flexibility of the varying length fields within the controlled part of the message. The message frame used two or three specific control characters of 8 bits (see Fig. 4.60), two of them known as FLAGs at the beginning and end of the frame. There are a number of link control access procedures which employ BOP, of which the most widely used today is 'high level data link character' (HDLC), specified by ISO and also recommended by CCITT. Other procedures worth mentioning are 'advanced data communication control procedure' (ADCCP), specified by ANSI, 'synchronous data link control' (SDLC), specified by IBM and 'digital data communication message procedure' (DDCMP), specified by Digital.

Level 3 – The network control protocols which define the message and packet formating which facilitates their transmission through the communication network as a single unit.

The major standard for this level is that introduced by the CCITT recommendation X25. Although X25 deals with the three levels, it is only widely accepted for level 3, while in the lower levels other standards are still used. Many networks throughout the world are converting their communication operations to comply with this standard. Nevertherless, there are many network systems still using others for protocols. The X25 supports what is called virtual circuit service; that is, transmission facilities are only assigned when data are actually being transmitted. Other accepted protocols are the datagram, which is used in the Canadian network.

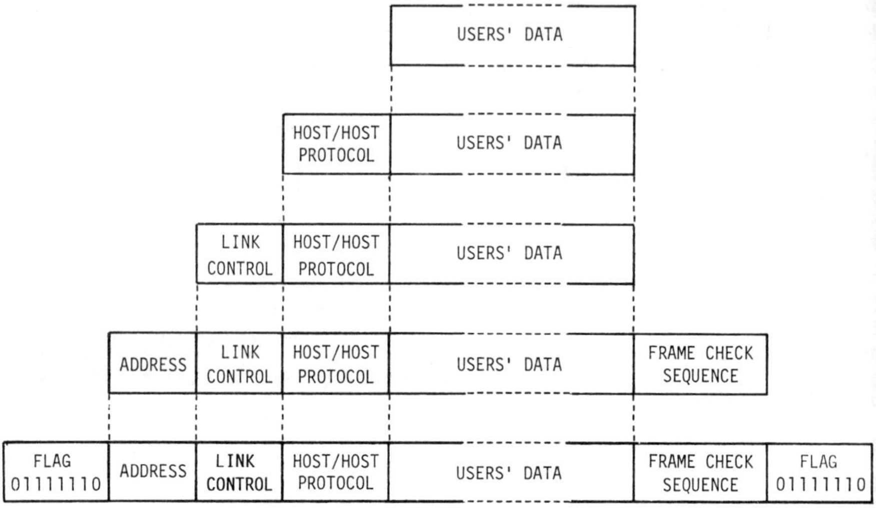

Fig. 4.60 Hierarchical formation of the transmitted message by the various links

Higher Levels — These are the user protocols and procedures which provide the means for data exchange between two operating processing systems, and are not part of the communication network. These are the host to host protocols which are defined as follows:

Level 4 — Transport protocols which allow various computer systems to interface with various data communication systems. They specify the end to end control and information exchange with network independence.

Level 5 — Session protocols which support the message dialogue between the communicating host computers.

Level 6 — Presentation protocols which translate the transmitted information of the various computer languages, data formats and codes, to allow communication between two computers.

Level 7 — Application protocols where the messages are received or generated as defined by the user's program.

The advantage of having a number of levels is the separation of functions. Each level accepts information from the higher levels and adds an extra header (as seen in Fig. 4.60), before passing the message frame, as formed so far, to the next lower level. The frame, in the example shown, starts with an 8 bit flag sequence of 01111110 followed in position by an address sequence and a link control sequence, each with one or more 8 bit characters. This is followed by the host to host control header, and then the actual data message, which may be of variable length. The message frame is terminated with two characters of frame check sequence (FCS) containing 16 bits cyclic redundancy code (CRC). The frame is then enveloped by a second flag of identical sequence.

The basic three levels may seem to be overlapping although in fact their functions are completely separated. It can be seen that level 1 defines the hardware interface, level 2 defines procedures which are generally implemented with hardware although initiated by software, and level 3 defines protocols which are purely software functions. For this reason, the author preferred the use of three separate terms for each of the three levels, as designated in the following:

(a) *Interface*, which defines the physical common boundaries between two sets of equipment. It involves the interconnection specification, enabling these two hardware equipments to operate together.

(b) *Procedure*, which defines the course of action taken for the solution of forming the communication between two terminals. It specifies the handshaking procedure and the flow transfer control using hardware and/or software techniques. Its functions are usually transparent to the higher level operations. That is, the data transfer effected in the higher levels is not restricted by the procedure operations.

(c) *Protocol*, which defines the formal set of conventions governing the format and the relative timing of the message exchange between the two communicating terminals. It specifies the formal rules which must be obeyed to ensure an orderly information transfer between two or more communicating parties.

4.15 REFERENCES

(a) DATA NETWORKS

1. McQuillan, J.M. and Cerf, B.G. *A practical View to Computer Communication Protocols*, IEEE Tutorial, IEEE Computer Society, 1978, pp.1-36.
2. Doll. D.R. *Telecommunication Turbulence and the Computer Network Evolution*, IEEE Computer Society Magazine, February 1974, pp.13-22.
3. Kahn, R.E. *Resource-Sharing Computer Communication Networks*, Proc. IEEE, vol.60, no.11, November 1972, pp.1397-1407.
4. McQuillan, J.M. *Local Network Architectures*, Electronic Design, May 1979, pp.18-26.
5. Pyke, Y.N. and Blanc, R.P. *Computer Network Technology – A State of the Art Review*, Computer Society Magazine, August 1973, pp.13-19.
6. Chou, W. *Computer Communication Networks – The Parts Make Up the Whole*, National Computer Conference, AFIPS, vol.44, 1975, pp.119-128.
7. Townsend, M.J. *Communication Control by Computer – An Introduction*, Telecommunication, May 1972, pp.33-38 and 60-62.
8. Knight, J.R. *A case Study: Airlines' Reservation Systems*, Proc. IEEE, vol.60, no.11, November 1972, pp.1423-1431.
9. Green, W. and Pooch, U.W. *A Review of Classifcation Schemes for Computer Communication Networks*, IEEE Computer Society Magazine, November 1977, pp.12-21.
10. Wecker, S.*Computer Network Architectures*, IEEE Computer Society Magazine, September 1979, pp.58-72.
11. Frisch, I.T. and Frank, H. *Computer Communication – How We Got Where We Are*, National Computer Conference, 1975, pp.109-117.
12. McGregor, P. *Effective Use of Data Communication Hardware*, National Computer Conference, 1974, pp.565-575.
13. Blanc, R.P. and Cotton, I.W. *Computer Networking*, IEEE Press, 1976, 368 pages.
14. Schwartz, M. *Computer-Communication Network Design and Analysis*, Prentice-Hall Inc., 1977, 362 pages.

(b) LOCAL AREA NETWORKS

15 Clark, D.D., Progran, K.T. and Reed, D.P. *Introduction to Local Area Networks*, Proc. IEEE, vol.66, no.11, November 1978, pp.1497-1517.

(c) DISTRIBUTED NETWORKS

16. Russo, P.M. *Interprocessor Communication for Multi-Microcomputer Systems*, IEEEE Computer Society Meeting, April 1977, pp.67-76.
17. Karp, D. and Seroussi, S. *A communications Interface for Computer Network*, IEEE Trans., vol.COM-20, no.3, June 1972, pp.550-556.
18. Frank, H. and Chou, W. *Topological Optimization of Computer Network*, Proc. IEEE, vol.60, no.11, November 1972, pp.1385-1397.
19. Kelly, P.T.F. *Public Packet Switched Data Networks, International Plans and Standards*, Proc. IEEE, vol.66, no.11, November 1978, pp.1539-1549.
20. Rybczybski, A., Wessler, B., Despres, R. and Wedlake, J. *A New Communication Protocol for Accessing Data Networks — The International Packet-Mode Interface*, National Computer Conference, vol.45, 1976, pp.68-73.
21. Durniak, A. *A New Network Tie Down Distributed Processing Concepts*, Electronics, December 7, 1978, pp.107-122.

(d) NETWORK SYNCHRONIZATION

22. Cooper, C.A. *Synchronization for Telecommunication in a Switched Digital Network*, IEEE Trans., vol.COM-27, no.7, July 1979, pp.1028-1033.
23. Harrington, E.A. *Synchronization Techniques for Various Switching Topologies*, IEEE Trans., vol.COM-26, no.6, June 1978, pp.925-932.
24. Datta, A. and Mitwally, E. *Synchronization of Multi-Exchange Local Networks by Bit-by-Bit Method*, IEEE Trans., vol.COM-27, no.7, July 1979, pp.1034-1046.
25. Perreault, D.A., Mukundan, R., Kirshnan, K.R. and Cotellessa, R.F. *Network Timing/Synchronization Evaluation Modelling*. Technical Report RADC-TR-77-10, NTIS AD A 039378, April 1977, 72 pages.
26. Alexander, P., Graham, J.W. and Miller, D.M. *System Timing and Synchronization*, Technical Report, RADC-TR-78-166, NTIS AD A 058468, July 1978.
27. de Azevedo, I. *Computer Network Analysis and Case Study System*, M.Sc. Thesis, U.S. Naval Postgraduate School. ADA 057906, June 1978, 220 pages.
28. Williard, M.W. *Analysis of a System of Mutually Synchronized Oscillators*, IEEE Trans., vol.COM-18, no.5, October 1970, pp.467-483.
29. Bittel, R.H., Elsner, W.B., Helm, H., Mukundan, R. and Perreault, D.A. *Clock Synchronization Through Discrete Control Correction*, IEEE Trans, vol.COM-22, no.6, June 1974, pp.836-839.
30. Baart, J.G., Harting, S. and Verma, P.K. *Network Synchronization and Alarm Remoting in the Dataroute*, IEEE Trans, vol.COM-22, no.11, November 1974, pp.1873-1877.
31. Harrington, E.A. *Issues in Terrestrial/Satellite Network Synchronization*, IEEE Trans, vol.COM-27, no.11, November 1979, pp.1690-1695.
32. Marlevi, A. *Directly Digitally Controlled Network Synchronization*, IEEE Trans, vol.COM-27, no.10, October 1979, pp.1561-1565.
33. Stover, H.A. *Timing and Synchronization of Switched Digital Communication Networks*, International Communication Conference, ICC 1976, pt.2, pp.25-1 to 25-6.
34. Inoue, N., Fukinuki, H., Egawa, T. and Kuroyanagi, N. *Synchronization of the NTT Digital Network*, ICC 1976, pt.2, pp.25-10 to 25-15.
35. Bradenburg, L.H. *Timekeeping in the Bell System Switched Digital Network*, ICC 1976, pt.2, pp.25-26 to 25-29.

(e) COMMUNICATION STANDARDS

36. Schutz, G.C. and Clark, G.E. *Data Communication Standards*, IEEE Computer Society Magazine, February 1974, pp.32-37.
37. Grubb, D.S. and Cotton, I.W. *Criteria for the Performance Evaluation of Data Communication Services for Computer Networks*, U.S. Department of Commerce, National Bureau of Standards, Technical Note 882, September 1975, 35 pages.
38. Cotton, I.W. and Folts, H.C. *International Standards for Data Communication*. Proc. 5th Computer Symposium, 1977, pp.4-26 to 4-36.
39. Folts, H.C. and Cotton, A.W. *Interfaces: New Standards Catch Up with Technology*, Data Communications, June 1977, pp.31-40.
40. Clark, G.E. *Computer and Data Communications Standards Organizations*, Proc. Computer Networking Symposiums, December 15, 1977, pp.1-8.
41. Folts, H.C. and Karp, H.R. *Data Communication Standards*, McGraw-Hill Publication Co., 1978, 1133 pages.

(f) NETWORK RELIABILITY

42. Morgan, D.E., Taylor, D.J. and Custeau, G. *A Survey of Methods for Improving Computer Network Reliability and Availability*, IEEE Computer Society Magazine, November 1977, pp.42-49.

CHAPTER 5

Data Communication Conveyances

5.1 INTRODUCTION

It is not enough to consider the various available communication networks, as introduced in the previous chapter, one must also analyse the actual conveyances. This chapter introduces the various forms of conveyances that may be used as communication channels and discusses the specific features which affect the communication process.

In order to fully exploit the potential of C^2 systems, it is essential for the processing centres to be able to communicate freely with each other along the communication channels. This facility is provided not only by the data communication equipment (DCE); there is also the important consideration of channel capability. In this respect, the introduction of the various forms of communication conveyances contributes greatly to the achievement of abolishing distances as a limiting factor in communication between the centres of C^2 systems.

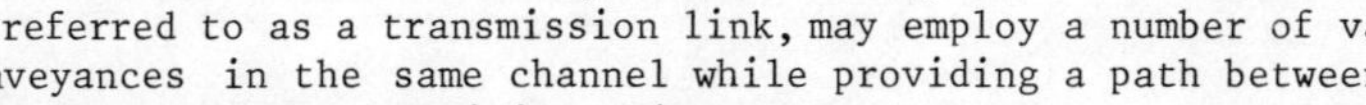

Fig. 5.1 The basic configuration of a point-to-point data communication channel

The broad function of the communication conveyance is to provide the communication channel with the facility of carrying the data signal between the two communication terminals, as seen in Fig. 5.1. A communication channel, often referred to as a transmission link, may employ a number of various forms of conveyances in the same channel while providing a path between two DCE's without the user even realizing this.

It is quite a common practice that a communication channel, provided by the common-carrier, may employ a pair of wires for part of the path then change to coaxial cable, and in many cases also employ some form of radio frequency, such as microwave or satellite in the same transmission path. It is therefore important that the user should be aware of the characteristics of the various communication conveyances media and their effect on the data transmission. A typical example worth mentioning is the relatively long delays that can be created when the channels employ satellite relay in parts of the path which can exist in some international dialled channels. The above consideration refers mainly to dialled or leased channels where the user is provided with the conveyances selected by the common carrier and it is left to him to supply preventative measures against any disturbances. When utilizing private communication networks, either local or distant, it is up to the system designer to select the best conveyance available and to introduce means of overcoming any disturbance that may be produced by the use of the conveyance (for example when using a twisted wire pair in a local installation). It is the intention of this chapter to introduce the various communication conveyance media available and discuss the physical aspect of implementing them in a communication channel.

When designing a communication network for a particular application, such as C^2 systems, the designer does not usually have to care what type of conveyance is involved provided the data can be transmitted and received between any two points without errors or emissions. To him it is the efficiency of the communication that is important rather than the means. Cost is an important consideration only in his assessment of the choices but it is the operational conveniences that have his prior attention. In other words, the selection of the conveyance media is neither a purely technical problem nor a purely economical one but a complex combination of these two factors combines with an operational need. One of the main features in specifying a communication conveyance is that of the availability of the means which is dictated by distance and geographical terrain layout.

When dealing with a data processing system, it is usual to regard the DTE's and DCE's as the hardware core of the system, while the transmission channel is considered as an auxiliary part of the system. This attitude is incorrect, since it is the way the communication conveyance is used which is the precise source of most of the troubles encountered in the communication system. While the DCE's and the DTE's are concentrated in a number of isolated geographical locations, the communication path extends for thousands of miles and is exposed to all kinds of external disturbances. A much more important fact is usually forgotten: it is the communication channel which is the most expensive part of the system and not what is regarded as the 'hardware' of the system.

The data link which serves to establish transmission over which data can be forwarded is always in a deterministic direction, as seen in Fig. 5.2. This is due to the data communication equipment being located at both ends of the communication channel. The communication conveyance, on the other hand, is undeterministic, with transmission possible in either direction. Nevertheless, despite this advantage, the available conveyance must be regarded as possessing deterministic characteristics, because it is normal practice to add all sorts of electronic equipment at various points along the communication channel to overcome the problems of attenuation and distortion.

The basic transmission feature of a communication conveyance media is the effective available channel bandwidth which imposes an upper bound on the

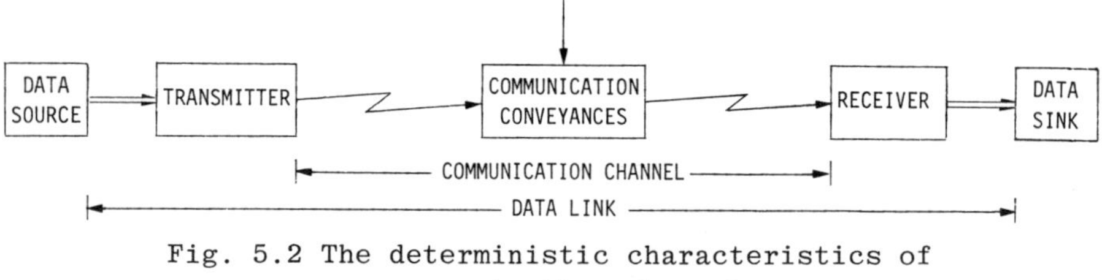

Fig. 5.2 The deterministic characteristics of a communication channel

transmission speed. Also important is the channel net loss which affects the received signal-to-noise ratio and hence defines the error margin. It is therefore essential for the system designer to understand the conveyance characteristics so that he can select the appropriate data communication equipment. For a given transmission speed and received signal-to-noise ratio, the rate at which errors occur will depend on the modulation and demodulation techniques, i.e., the DCE. Nevertheless, understanding these conveyance characteristics does not necessarily assist the designer in selecting the most suitable conveyance media for his application. This, however, is dictated mainly by the features of the system rather than those of the conveyance. That is, the essential qualification factors needed for the selection of the most suitable conveyance are the transmission distance and the data volume that is to be conveyed over the channel in a given period.

Although the main factors required in the selection of the conveyance is imposed by the system requirements, one cannot disregard the conveyance characteristics. One of the prominent features of all conveyances, whether it is made of physical or free space media, is that its transmission performance deteriorates over long distances. Nevertheless, the degree of its impairment depends on the type of conveyance employed. This feature is due to the signal propagation along the transmission path which causes losses and consequently reduces the signal strength and deforms its channel shape. The main difficulty lies in the fact that the signal deterioration is not uniform throughout the channel bandwidth. To overcome this problem, repeaters are introduced in the physical channel and relays are introduced in the radio channel. The function of both the repeater and relay is identical, that is, to perform intermediate signal amplification and signal reshaping by receiving the weak signals and retransmitting them with additive strength.

As stated above, most of the features of the various conveyances play little part in the selection of the suitable media: they are, however, important in the design of the actual communication channel. In this respect the most important feature to be considered is that noise is added to the signal while it is in the conveyance and must be eliminated in the receiver before the data signal can be used again. There are many other external disturbances that confront the signal as it propagates along the transmission path and these will be presented separately for each type of conveyance.

5.2 COMMON CARRIER CONVEYANCES

Common carrier is a general term for all associated public telecommunication companies which furnish communication services to the general public. In some countries the common carrier refers to the government postal, telegraph and telephone (PTT) services while in others it refers to the public utility authority. It is general practice for the common carriers to be the only body authorized to provide telecommunication services. In fact, in many cases they also provide the modems or specify which type of modems or other data communication equipment (DCE) that can be used to transmit data over their communication lines.

The main attractive feature of the common carrier conveyances is that they are readily available throughout the world and can provide communication between any two points situated in any fixed geographical location. A further important feature of these conveyances worth mentioning is that their maintenance is the sole responsibility of the common carrier rather than of the user. That is, if a fault is located in one path, an alternative path is soon provided, with no extra expenses to the user.

The transmission conveyance medias that can be provided by the common carriers range through all the types and forms available. The most universal available type is the dialled service provided by copper wire pairs to the local telephone exchange. Such conveyances are limited in their usefulness for data transmission by allowing successful transmission speeds over them only up to 1200 BPS. Nevertheless, the common carrier can, on request, provide better conveyances on a lease basis, which could easily be reconfigurated to the users' specification. These leased lines are usually based on the same communication channels used for dialled communication although they are wired out of the switching exchange and are specially selected to carry the data traffic over long distances at a variety of speeds. Furthermore, these leased lines are specially conditioned, both by the user and the common carrier, to accord with given characteristics. Nevertheless, in most cases these lines are limited to 9600 BPS and in some cases even less. The common carrier can also provide better transmission channels which are specially designed to carry data at higher speeds; these channels however are not readily available.

Most of the transmission channels provided by the common carrier have been developed for analog and not data transmission; that is, they are primarily intended for speech communication. As a result, the direct application of digital signals, or even analog-coded-digital (ACD) signals, to these analog channels results in excessive distortion and inefficient use of the channel bandwidth. Furthermore, these transmission channels are generally fabricated by varied numbers of conveyance medias, each of which adds disturbance effects. Above all, most if not all, the common carrier communication channels pass through the telephone or other types exchanges and use their special long distance repeater equipment. It can be said that all the electronic equipment inserted in a communicated channel to improve analog transmission could produce a negative effect on digital data transmission. Even when the common carrier leases out a channel by providing the user with a permanent communication path, this channel consists of amplifiers, modulators, active filters and other electronic equipment necessary for voice communication.

While the human ear is conditioned to tolerate most of the transmission disturbances, the common-carrier channel characteristics must be adapted for the acceptance of data transmission or to condition the data to accept possible errors. That is, noise on the transmission line usually does not affect human conversation, although it could change the intelligence of the transmitted data. The communication line possesses the characteristics of producing varying amounts of amplitude and delay distortions as a function of the frequency, and these affect the transmission of the data. These distortions must either be tolerated or rectified. A telephone voice line could accept a limited amount of data without conditioning, although only for slow speeds. As the speed increases, the required bandwidth is widened, increasing the susceptibility to noise and other distortions. Line conditioning can be provided by adding equalizers, and errors can be reduced by adding re dundant data.

The line characteristics can be classed in three groups as follows:

(a) Distortion due to the line characteristics.

1. Delay distortion.
2. Attenuation distortion.

(b) Noise induced on the line.

1. White noise.
2. Random noise.
3. Impulse noise.
4. Phase jitter noise.
5. Cross-talk noise.

(c) Equipment on the lines.

1. Filters
2. Companders

3. Echo suppressors.
4. Trunk signalling.
5. FDM carrier shift.

Delay distortion. When transmitting a spectrum of frequencies, some of them are delayed more than others a condition which is also referred to as envelope delay distortion, since it destroys the pattern envelope of the frequency transmitted and can cause differential delays between the various frequencies comprising the composite spectrum of the signal transmitted. The condition is manifested by some frequency components which are delayed much more than others and which tend to become quite appreciable at both the lower and upper edges of the voice frequency band of the transmission line. The distortion, which cannot be detected by the human ear, could create problems in data transmission, because all the frequency components do not travel through the transmission path at the same speed and these differences could cause an increase in intersymbol interference where the data signal may spill over and disturb an adjacent data signal. The envelope delay distortion is due to inductance and capacitance reactance elements of the transmission path.

Attenuation distortion. The frequency bandwidth of a transmission channel represents a varying frequency attenuation. This condition generally appears at the edges of the voice band, where there is a fast roll-off which results in larger frequency attenuation. This effect is usually caused by the filters in the transmission path which reduce the available bandwidth of the voice channel to about half the expected spectrum, i.e., about 1500 Hz, unless special treatment is provided. The effects of narrowing of the bandwidth will create transients in the data when it is transmitted at high speeds. Attenuation distortion is less a problem in voice communication than in data transmission owing to the redundant nature and low information content of speech. In data transmission any alternation or loss of the symbol shape transmitted changes the meaning of the transmitted data.

The dialled or leased transmission channel available from the common carrier does not distinguish the type of conveyance that is provided, whether radio or wire. Instead, the common carrier specifications guarantees the characteristics of these channels, i.e., the expected delay and attenuation features. In the following table, four types of channel characteristics are presented, all suitable both for voice or data transmission. All these channels have some identical characteristics:

(a) Bandwidth suitable for voice communication, i.e, 300-3000 Hz.

(b) Impedance source and load balance of 600 Ω.

(c) Maximum signal power of 0 dBm for composite data signal.

(d) Means attenuation at 1000 Hz across 600 Ω in 16 dB ± 1.

(e) Expected net loss measured at 1000 Hz for short term ± 3 dB and for long term ± 4 dB.

(f) Frequency error ± 5 Hz.

The four channel types, do vary however, in attenuation and delay characteristics, as seen in the table. The 3002 channel is used for dialled lines and the *C* type channel for leased lines. The *C* type lines allow higher speeds to be transmitted over them but they require conditioning, which the general characteristics presented in the table presume has been done.

White noise. White noise causes a disturbance which is uniformly spread throughout the frequency band and is occasionally noticeable on voice channels as a background hiss. It consists of a set of amplitudes of relatively equal magnitude with a Gaussian distribution as a function of time, and is

CHANNEL TYPE	3002		C1		C2		C4	
ATTENUATION CHARACTERISTICS	FREQUENCY RANGE	SPREAD LIMITS	FREQUENCY RANGE	SPREAD LIMITS	FREQUENCY RANGE	SPREAD LIMITS	FREQUENCY RANGE	SPREAD LIMITS
	300—3000 Hz 500—2500 Hz	-3 to +12 dB -2 to +8 dB	300—2700 Hz 1000—2400 Hz 2700—3000 Hz	-2 to +6 dB -1 to +3 dB -3 to +12 dB	300—3000 Hz 500—2800 Hz	-2 to +6 dB -1 to +3 dB	300—3200 Hz 500—300 Hz	-2 to +6 dB -2 to +3 dB
ENVELOPE DELAY DISTORTION	FREQUENCY RANGE	MAXIMUM DELAY	FREQUENCY RANGE	MAXIMUM DELAY	FREQUENCY RANGE	MAXIMUM DELAY	FREQUENCY RANGE	MAXIMUM DELAY
	800—2600 Hz	1750 µs	1000—2400 Hz 800—2600 Hz	1000 µs 1750 µs	1000—2600 Hz 600—2600 Hz 500—2800 Hz	500 µs 1500 µs 3000 µs	1000—2600 Hz 800—2800 Hz 600—3000 Hz 500—3000 Hz	300 µs 500 µs 1500 µs 3000 µs

characterized by a flat frequency spectrum at the frequency range of interest. It is composed largely of the thermal noise of the transmission path caused by small fluctuation differences of the potential of the components, which interferes much more with speech than with data. With data transmission, the signal threshold helps to prevent this noise from affecting the data reconstruction. The white noise which is always present and widely scattered is usually about 30-40 dB below the received signal level, though it may cause individual bit changes when the signal received is very weak. These random errors may then be removed by special logic circuits. When the white noise is relatively high it is feasible that a speech conversation on a particular channel cannot take place even though it is possible to transmit and receive data on this channel with no detectable errors.

Random Noise. This noise is caused by transients occurring at random and resulting in individual errors in the data stream which are usually detected and corrected in the receiver. The random noise, as distinct from white noise, is composed of varying instantaneous amplitudes embracing a wide range of frequencies. That is, while the white noise has a relatively fixed pattern, random noise occurs only seldom. It is caused mainly by thermal disturbances but could also come as a result of natural or man-made disturbances.

Impulse noise. Impulse noise causes the main troubles in data transmission. It takes the form of sharp pulse spikes which generally come in bursts of energy. These impulse pulses usually arise from the signalling equipment in the telephone network, as well as from lightning and other forms of natural and man-made disturbances. The impulse noises usually do not critically affect human conversation but can have rather destructive affects on the transmitted data. While random noise changes a bit state here or there, impulse noise changes the state of a whole sequence of data bits, i.e. the errors caused by impulse noise tend to cluster in groups. No error correction technique is known today that can guarantee the correction of all the errors caused by wide impulse noise disturbance. The only remedy then for impulse noise is to retransmit the data block once a sequence of errors is detected in it.

Phase Jitter noise. A phase jitter noise is a random jitter which may be caused by the high speed transmission of data where too few carrier cycles per bit are transmitted. It may also be caused by harmonies of the power supply frequency changing the phase of the carrier frequency. Another important cause of phase jitter is the modulation switching of the carrier when it is not in phase with the data. The latter could be quite appreciable in frequency shift keying and cause instantaneous discontinuences in the signal. The phase jitter might result in quick phase shifts in the frequency in excess of $15^\circ - 20^\circ$, which could reduce the speed of transmission when multistate binary is used. Due to the characteristics of equalizers, it has been found that phase jitter affects modems with equalizers more than those without them.

Cross-talk noise. Cross-talk is caused by coupling between adjacent lines in cables where one line picks up signals travelling on another line. Any line running parallel with another line is bound to have some cross-talk coupling between them. In theory the lines are suitably balanced and crosswired to prevent the coupling, but in practice cross-talk is unavoidable. Cross-talk is affected by the induction between lines which increases as the line length is increased.

Cross-talk could also occur between adjacent channels in a frequency division multiplex where there are a number of different channels transmitting simultaneously. Even when a guard spacing is placed between the channels some cross-talk is unavoidable. Another cause of cross-talk could be the leaking through of the supervisory control signal in a trunk channel.

Cross-talk is easily detected by the human ear, which is conditioned to weak signals. However, since the amplitude of the cross-talk noise is barely greater than that of the white noise, it causes negligible number of errors in the data transmission. It is interesting to realize that a noisy cross-talk line where conversation is barely possible could be used successfully for transmission of 4800 bits-per-second.

Filters. Most telephone lines have a number of filters inserted along them and nearly all the repeaters have some form of filtering which reduces the transmitted bandwidth. The utilization of the transmission path along a carrier system immediately calls for a number of filters, the effect of which is envelope delay, as has already been discussed.

Companders. Many trunk lines which pass through frequency division multiplex systems are designed with special equipment which compresses the loud speech signals and expands the weak ones. The special equipment is referred to briefly as companders. The companders are needed to ensure that the transmitted speech channel will effeciently utilize the multiplex line and not spill over into adjacent channels.

The companders are not required for data transmission systems, as the data signals are generally transmitted at a relatively constant output level. It is advisable to avoid the use of trunk lines which possess companders and to select lines specially suitable for data transmission, but this may be unavoidable in many cases.

Echo suppressors. On some long trunk lines, equipment is inserted to avoid the return of the echo to the sender. The equipment is known as an echo suppressor and is based on the principle of preventing transmission in both directions simultaneously. It is used in some long distance speech communication lines and will not cause interference there, since speech conversation is generally in a single direction at any given time. The echo suppressor is designed to ensure that the strong signal dominates and suppresses the weak signal. It takes about 120 m sec to reverse the direction of the voice transmission. Although echo suppressors are not needed on modern trunk transmission, they are still found in old audio circuits working on loaded cables.

Echo suppressors have two drawbacks if used for data transmission. The first is that they could prevent the simultaneous transmission of data in both directions. This means that a line which possesses this equipment could not be used either for duplex transmission or for backward supervisory control. The second is that the use with half duplex transmission will require the addition of an artificial delay to prevent loss of data after the transmission reverses direction.

When leasing dedicated lines care must be taken to ensure that no echo suppressors are inserted along the line. In dialled public lines this could by unavoidable and here, where such a line is detected, the only suggestion that can be given is to close down and re-dial.

Trunk signalling. Many carrier trunk signalling systems employ special control signalling on certain frequencies within the voice bandwidth. These discrete frequencies are filtered out at the receiver end in a way that does not affect the intelligibility of the voice transmission. Data transmission, however, could be disturbed, as the frequencies of the spectrum near the vicinity control frequencies are attenuated. In practice this effect is not critical, since most the data transmission spectrum is outside the range of the control signals even when some of the modulation signal components have some sidebands within the restricted range. There is a danger, however, that these sidebands may cause a false operation of the trunk signalling equipment in the telephone exchange.

Figure 5.3 illustrates the power level restriction for data transmission over some switched connections in the United Kingdom. The trunk

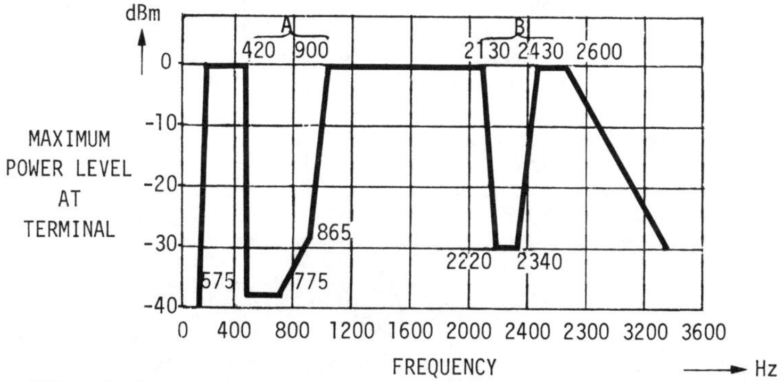

Fig. 5.3 Transmitted power level restrictions for data transmission over switched connections (Redrawn from POEEJ, London, October, 1966)

signalling equipment transmits frequencies in the ranges marked as A and B. In range A, frequency signal components up to 0 dbm may be permitted if accompanied by signals in the area of 1 to 2 Kc/s. In range B, signals are permitted if their characteristics preclude false operation of the trunk signalling equipment. In other words, the data modulated waveform must avoid the transmission of sidebands in this range for periods long enough to operate the signalling equipment. Today, this restriction is not applied on all carrier systems and it can be avoided on selected leased lines.

FDM carrier shift. In the voice transmission using the frequency division multiplex carrier system, the voice bandwidth must be modulated so that it can be transmitted over the required channel bandwidth. The carrier frequencies of the modulator and demodulator may not be precisely identical and will then cause some phase shift between them. This may not be recognized by the human ear but may still affect the correct interpretation of the recovered data.

5.3 LOCAL PHYSICAL WIRE TRANSMISSION

Local data communication usually refers to the private transmission lines within the customers' installation premises, that is, to all the data transmission lines connecting the terminals and computers operating in one centre. As such, these transmission lines are the sole responsibility of the user and are not provided by the common carrier. These transmission lines are generally characterized as carrying large volumes of data, either in series or in parallel form, over relatively short distances, usually not more than a mile.

Local data communication also refers to the part of the communication network which connects the customer with the line termination equipment in the telephone exchange. For this purpose, these local transmission lines are generally in the form of an open wire pairs and could be supplied by the common carrier.

By physical communication one refers to all metallic transmission conveyances as distinct from free-space transmission conveyances. Local physical circuits include all the transmission channels within the host computer installation and between the host computers and the network node.

Physical wire conveyance communication is inexpensive, readily available and the easiest to implement, since all that is needed is a pair of wires thrown between the two communicating points. For this reason wire lines are the most common conveyance used and are employed in nearly every communication premises. Nevertheless, one must remember that wire lines are not the only communication conveyance media, nor are they the most ideally suited for long distance digital data transmission. Narrow transmitted pulses are attenuated and spread out in time as they travel over the wire pair. This is due to the fact that the transmission channel acts as a filter and could be regarded as a delay line. For this reason, the wire channel can transfer data only at moderate speeds and for short distances. For higher speeds and longer distances, modems are required.

It is difficult to provide useful general rules for the behaviour of physical transmission lines, as the characteristics of the line depend mainly on the thickness of the copper wire. It is known that higher frequency components of the transmitted signal travel on the outer skin of the copper and suffer greater attenuation than the lower frequency components which have greater penentration features.

Physical wire conveyances provide nearly constant low impedance paths for the data signal transmission. There are many types of physical wire conveyance; the principle categories include:

(a) Single wire transmission channels.

(b) Twisted wire pair channels.

(c) Open wire pair channels.

(d) Wire cable channels.

(e) Coaxial cable channels.

Single wire channels transmit the signal over one conductor line while the return line is through the common ground, as seen in Fig. 5.4. The advantage of such a transmission channel is obvious, since it is easily constructed; nevertheless, its applicability is doubtful. Such a channel may be useful in some speech communication systems, although even there the difficulties lie in the return path. For use in any transmission application, proper earthing must be assured, a requirement that is not easily fulfilled. Further disadvantages include high vulnerability of the signal causing its fast degragation due to the large attenuation effect. Even the inclusion of modems will not increase the signal performance for data transmission and will not secure its signal against external interferences. This limits this system to very slow speeds and extremely short distances of only a few inches.

Fig. 5.4 Single line transmission channel

To allow the data to be transmitted at faster speeds although still limited to moderate short distances of a few yards; a twisted wire pair which is relatuvely inexpensive, could be used. Such channels are used for local transmission applications, between closely situated equipments. With very short distances and moderate speeds, one of the wires is used for the signal while the other is earthed (as presented in Fig. 3.4 of chapter 3); but with longer distances and higher speeds, each of the two wires transmits differential signals. That is, each wire transmits the same data signal, but as the complement of the other, as shown in Fig. 5.5, with an earthing shield around the wire pair. The signal in the receiver is detected by a differential sense amplifier which cancels out any disturbances induced in the wire pair line. Even with such an arrangement, the distance is limited to a few tens of yards, above which a modem is necessary.

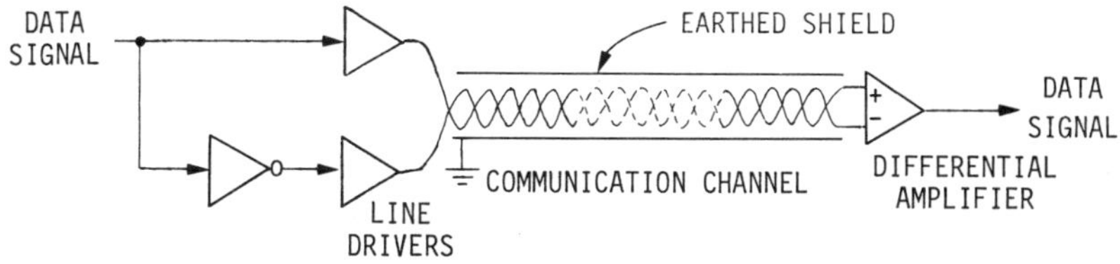

Fig. 5.5 Twisted wire transmission channel

The open wire pair channel is the most universally used communication conveyance between the local host computer installation and the local common carrier exchange. The open wire pair come in two forms, in either two or four wires. The four wire channels are usually provided for customers who require full duplex data communication, while the two wire channels are generally provided for half duplex data communication and for ordinary voice communication. Even when a customer is provided with two wires to his local premises, the transmission path is usually in parts of four wires. This is necessary for the operation of the repeaters and other electronic equipment inserted in the line. To allow the bilateral operation of the half duplex communication, i.e. transmission in both directions, different repeaters are necessary for each direction. In this case, as seen in Fig. 5.6, special hybrid circuits convert the half duplex wire pair to two simplex wire pairs. The characteristics of these open wire pairs are similar to that described in the previous section for the common carrier voice grade channels. That is, they have a frequency bandwidth of about 300–3000 Hz and can provide data transmission speeds up to 9600 BPS depending on the type of channel conditioning. Many countries still use open wire lines because of the low initial installation cost. With repeaters inserted every few miles, it is possible to transmit over the wire pairs for relatively long distances. Nevertheless, an open wire pair which carries only one data transmission channel, is ineffective when it is regarded as part of a larger network system, since it will require a huge number of arial wires all leading to the common carrier exchange. Furthermore, these lines, by nature, are susceptible to degradation of the signal and are subject to atmospheric variations. This makes this conveyance unsatisfactory for very long distances.

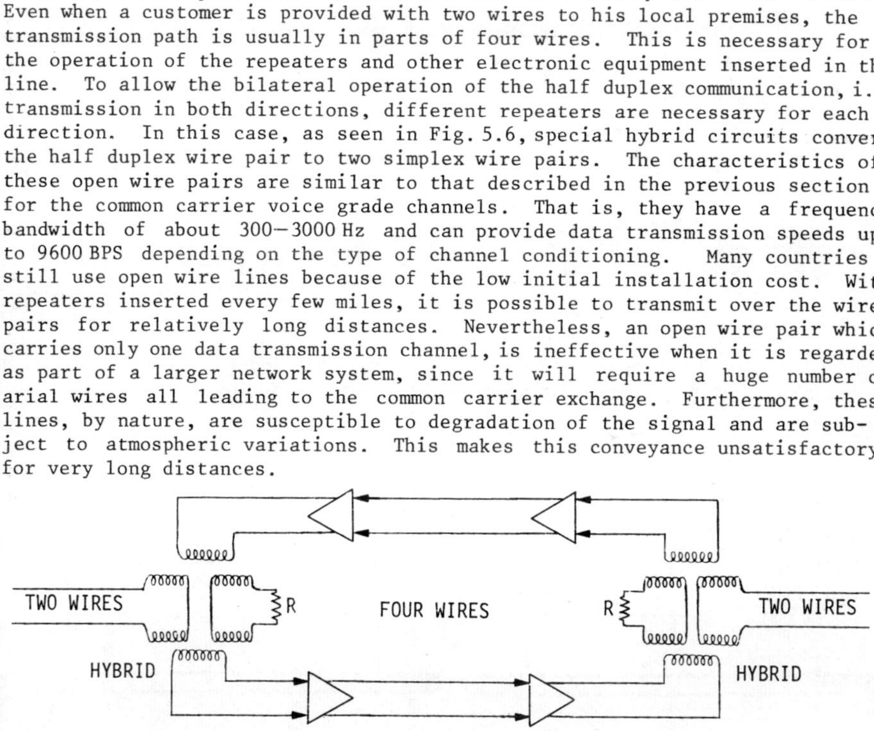

Fig. 5.6 A two wire communication channel is in fact a four wire channel

To reduce the unreliable performance of the aerial open wire pair it is recommended, where possible, to employ cable systems, which are better protected and more reliable. In these cables, multi-wire pairs are used with insulated shields to separate groups of conductors. This multi-wire cable can be regarded as an extension of the open-wire pairs and as such has similar transmission characteristics. For this reason, although it does solve the problem of the communication between the local premises of the subscriber and the common carrier exchange, it too is far from ideal. It should be added, that besides the delay and attenuation distortions, the main disturbances in a multi-wire cable is the cross-talk between adjacent wire pairs.

Another important local physical conveyance is the coaxial cable which is used in local private networks to provide wideband communication between computer installations. It can transmit data at much higher speeds than a pair of open wires and can support megabit transmission. Coaxial cables are available that can transmit successfully from 20 KHz to 5 GHz analog signals as well as data speeds up to 140 MBPS. The coaxial cable is composed of a conductor cylinder with a wire in its centre filled with a dialectric insulation material, as seen in Fig. 5.7. Its construction makes it immune to noise and is inexpensive relative to the volume of data that can be transmitted over it. Coaxial cables supersede the twisted wire by offering the flexibility and performance required in large local network systems, and not supplied by its counterpart.

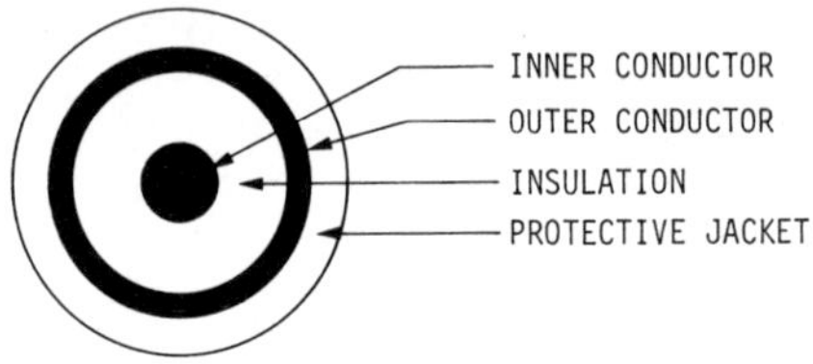

Fig. 5.7 The basic construction of a coaxial cable

Though less susceptible to external disturbances than other physical conveyances, the coaxial cable is not immune to very strong external disturbances. As low frequency signals transmitted in the cable are particularly sensitive to common mode interferences, it is advisable not to transmit them over it. This means, in effect, that when unmodulated digital signals are transmitted over the cable one should avoid using digital codes such as NRZ but special codes with a transition for every bit. Furthermore, it is important to ground the cable to the source at only one point and to ensure that all the associated electronic circuitry use as few earth connections as possible, in order to prevent ground loops. Also, the earth connection of the circuits, i.e., the negative or positive potential level should be isolated from ground but only for a single point.

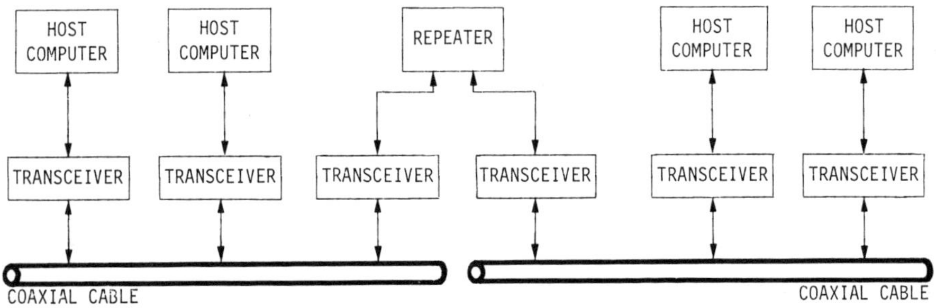

Fig. 5.8 The basic principles of the locally distributed Ethernet employing a coaxial cable data communication conveyance

One of the well-known broadband physical local communication schemes which carries digital data among locally distributed computer installations, is known as Ethernet. In the Ethernet scheme all the installations are coupled through a single coaxial cable (as seen in Fig. 5.8) which acts as a passive broadcast conveyance with no central control. Design principles of such a coaxial cable scheme can allow about 100 installations to be connected along a kilometer of cable and operate up to 10 MBPS (although it may require repeaters). Transmission may be performed to longer distances, i.e. up to 10 km, but then the speeds can only be up to 100 KBPS.

The coaxial cable channel can provide digital transmission in either direction. Nevertheless, the electronic circuitry at each node location is in a single transmission direction. For this reason, separate receivers and transmitters are necessary, as seen in Fig. 5.9. Before initiating transmission, the host computer installation, by sensing whether signals are transmitted in the Ether cable, makes sure that there are no other transmissions in the cable. The Ethernet scheme provides multiple access facilities to all the users of the system employing procedures similar to that of CSMA (which will be described in chapter 9). That is, each computer installation can transmit data messages into the cable conveyance media whenever it has messages to transmit and there are not other transmissions. These transmitted messages will reach all the other computer installations which are attached to the cable and be identified by the appropriate host computer according to the address at the heading of the message.

Each cable tap is designed to present a high impedance to the cable so as not to disturb the cable operation. Furthermore, when a transceiver is unpowered it should disconnect itself electrically from the cable.

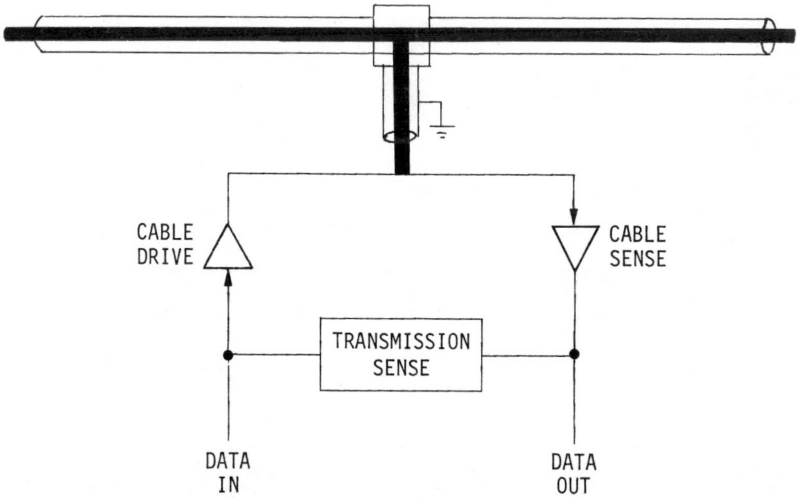

Fig. 5.9 Typical coaxial cable trans-receiver operation

5.4 CARRIER (PHYSICAL) TRANSMISSION

The use of single transmission lines for long distance communication is not always efficient since it calls for multi-lines to be laid along the path. For this reason, most long distance communications are performed by carrier equipment. By carrier communication one refers to the form of transmission in which the transmitted signal is considered as a result of frequency division multiplexing (FDM) of a number of input sources. In other words the transmitted signal in a carrier channel comes from the combination of multichannel bands translated into single wide band signal and then transmitted together, as seen in Fig. 5.10. At the distant end the carrier signal is split again and restored to the respective parallel nulti-line channels. The carrier signal can be transmitted over open-wire lines, cable, radio links or even satellite links. The latter two will be described separately in the proceeding sections.

The carrier channel is assembled in several FDM stages and is split up at the distant end by similar processes. Each stage increases the carrier signal bandwidth to contain more basic channels of 4 KHz each. This basic

Fig. 5.10 The basic principles of carrier communication

channel bandwidth also includes band spacing to guard each channel. Today a single carrier channel can provide room for thousands of voice channels. For example, twelve basic voice channels may be banked together in one fundamental carrier group (referred to also as pregroup), then five of these groups may be banked together to produce one carrier group and ten of these groups can then be banked to produce one supergroup. To increase the channel capacity, five supergroups can be modulated to produce one master group. Such hierarchical arrangement is effective only if very long distances are involved.

The carrier channel as described above was originally designed for voice communication with the FDM equipment used to combine a large number of these channels. Today many conveyance channels which were intended for carrier transmission carry instead only digital data, with the whole available bandwidth exploited for high speed digital communication. In this case the carrier channel is assembled as a result of time division multiplexing (TDM), as seen in Fig. 5.11. The transmitted carrier signal is formed by modulating the high speed data on a single carrier frequency with all the grouping performed at the digital stages.

Fig. 5.11 The basic principles of a digital carrier communication

In the local communication schemes described in the previous section, the basic data channel exploited the full channel provided by the conveyance. In carrier communication the basic data channel exploits only part of the channel provided by the conveyance. For this reason some sources refer to the transmission over a single channel as baseband transmission and the multi-channel as broadband transmission. This originates from analog communication and may be misleading when used for digital multiplexing communication. It is therefore advisable to employ these terms only for FDM carrier systems and not for TDM systems, irrespective of whether the data is digital or analog. In baseband communication, each user has the full control of the conveyance bandwidth with no danger of disturbing other transmissions. In FDM broadband communication, each user has the control of only a small portion of the conveyance bandwidth. Within this portion, the user may operate its own communication control mechanism independently of the other. In this respect TDM transmission must be regarded as baseband, since each user must use the same timing control techniques as the others. With radio communication, the terms baseband and broadband have wider meaning, since the basic baseband (either analog or digital) must be modulated to a higher frequency so that it could occupy the required transmitted spectrum.

The carrier signal, whether formed by FDM or TDM, appears in full at each and every outlet along the transmission path. That is, the signal at each outlet represents the multi-signal of all the users of the carrier system. In baseband communication, such a feature is advantageous since it provides the multi-access requirements. In broadband communication, however, this feature makes it difficult to reach the channels of individual users. Furthermore, it may call for special equipment to avoid all the other channels which are not required in a particular carrier tap. With analog signal this can simply be done by blocking with filters that portion of the spectrum band which is of no interest. In fact, it means introducing a simple single channel frequency modulator. With digital carrier signals this is more difficult, since the full spectrum must be decoded before a portion of the data stream may be blocked. The only conclusion of this discussion is that when employing broadband data transmission it is not always advisable to try to tap portions of the data stream but rather always to transfer the whole data volume as a single transmission.

In general, carrier transmission systems employ both physical wire and microwave conveyances. While microwave links are cost-effective for long distance transmissions, physical links provide cost-effective communication both for short and moderately long distances. The most popular physical carrier conveyance is that of the coaxial cable and this has become the backbone of the trunk communication in many countries, as well as of international submarine communications. There are five basic standard coaxial cable systems for FDM carrier systems, as given in the following table:

Cable Bandwidth	Number of Voice Channels
2.6 MHz	340
4 MHz	960
6 MHz	1260
12 MHz	2700
15 Mhz	3400

As for digital TDM carrier systems. the same coaxial cables, although with different electronics, could be used, with the highest speed available being 840 MBPS. In the U.S.A. there are five different hierarchical types of cables as follows:

Cable Link	BPS
T1	1544
T1C	3152
T2	6312
T3	44736
T4	274176

In Europe however there are different data rates with the highest rate that of 560 to 840 MBPS with a coaxial cable of 9.5 mm in diameter.

The bandwidth available in a coaxial cable is limited in its upper frequency range by the electronic carrier circuitry and not only by the constructional dimensions of the cable itself. The bandwidth is also limited in its lower frequency range by the actual cable characteristics. This is because the coaxial cable does not provide good shielding at low frequencies.

In the previous section the coaxial cable was described as operating in either direction. The carrier conveyance channel, however, owing to the electronic circuitry involved in producing the broadband signals, is always operated as an unidirectional channel. It is therefore usually necessary, when providing a full or half duplex data communication channel, to assign a channel pair, although each link of the pair can still be regarded as a separate simplex channel. This is important when planning a carrier system since the introduction of the repeaters is performed for each of the channel pairs separately, as seen in Fig. 5.12. The spacing between the repeaters

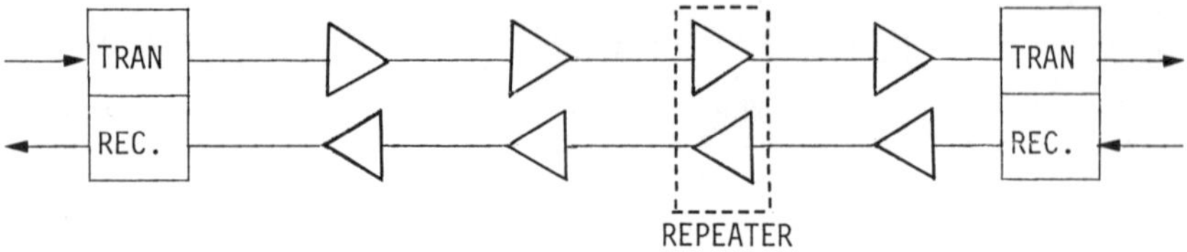

Fig. 5.12 The double channel operation in a carrier link communication

depends upon the bandwidth and the diameter of the coaxial cable. The repeater spacing distance can vary between 5 km for a 2700 multi-channel carrier link to 20 km for a 340 multiplexed link. With microwave system the spacing is greater, about 60 to 80 km, although the expense of the repeater is much higher. It is advisable, however, to employ more repeaters than are necessary in areas which are most likely to cause trouble.

The major disturbances in a carrier coaxial cable system, after dealing both with the attenuation and delay distortion, is caused by cross-talk as well as frequency and timing jitter introduced by the repeaters. A further known disturbance that affects the operation is that referred to as carrier leak. Although there are frequency guards between each modulated channel in an FDM carrier system, there are some energy leaks between channels which increase the cross-talk and also load the repeaters.

There is another important disturbance that occur in FDM carrier systems which affects the data transmission in its channels. This disturbance is a form of carrier leak but its effects are far more damaging. When transmitting data over a single wire pair, the PSK modulated data signal fills the full available bandwidth. Before this signal is remodulated in the FDM carrier channel it is passed through a band pass filter (BPF), as in any other wire channels, to confine the signals to within the allotted band. Nevertheless, the filter causes some amplitude modulation in the outer boundaries of the frequency band (as will be explained in chapter 6). These side bands are relatively very small though they do creep through the filters. When a number of channels in the carrier system carry data signals, these side bands could add up and the resultant signal could creep and disturb other channels in the system. For this reason it is customary to limit the number of allowed data channels in an FDM carrier system to less than 25%, while the other channels may be used for voice transmission. In consequence, it might be advisable to modulate the data signals using MSK techniques since this limits the signal to the assigned bandwidth. In TDM carrier systems the problem of carrier leak does not exist as the modulated data signals fills the full carrier bandwidth. It should be added, however, that many carrier systems do not assign the full spectrum available to only one common signal. In fact, for operational reasons, the full carrier spectrum is split into a small number of digitalized modulated bands.

5.5 RADIO TRANSMISSION

The communication of data over physical links provides an ideal transmission conveyance for most applications, however, its main drawback is that it requires a physical point-to-point connection between the transmitting and the receiving modems. The physical channel is limited by its available bandwidth and by the physical dimensions of the conveyance itself. The radio transmission systems provides alternative communication conveyance facilities, which in many respects are more attractive although they are also much more vulnerable to external disturbances. Their major advantage is that they require no wires, with the free open space performing the function of the conveyance and the location of the transmitting or the receiving stations may be spread over wide geographical areas. The radio transmission conveyance offers communication with less frequent repeater stages than with wire transmission, requiring repeater spacing of about 30 miles apart as compared with only a few miles apart required in cable systems. As in wire communication, it is possible to 'tap' the transmission path; in fact, in radio communication it is much easier to do so since all it requires is simply to associate a large number of digital terminal stations with one transmission system. The more so with radio communication, where the employment of multi-access techniques becomes more natural and operates at a relatively low cost.

The radio communication conveyances can transmit both digital and analog data. In fact, the first radio systems were used to transmit digital even before analog data; that is, they were used to transmit morse telegraphy signals even before they could be used to transmit voice signals. Nevertheless, the data pulses are soon distorted, and with data rates above voice frequencies (VF), no data can be both transmitted and received successfully. Therefore, as in wire communication, the radio data signal must first be modulated before it can be transmitted. The data rate that can be used in radio systems is limited by the available transmission frequency bandwidth; this, however, is defined by the transmitter rather than by the conveyance itself.

The exploitation of the full radio spectrum is performed by sharing the wide frequency band of the radio spectrum between many unrelated users. Such an allocation of the frequency bands is referred to in its broad sense as frequency-division-multiple-access (FDMA). The most attractive feature of radio communication is that its bandwidth seems unlimited and therefore always available for use. In other words, there is no need to doctor the conveyance media, as in wire systems, to achieve any required bandwidth. Nevertheless, the radio spectrum is not as available as it may seem. The actual bandwidth of the transmitted channel depends on the centre frequency selected and on the way the conveyance is used. Furthermore, those frequency bands which are so suited for communications, are over populated with users, leaving little room for any additional users. In fact, it can be said that the available transmission spectrum has reached complete saturation. Today, the available frequency band is shared out to the various users by the respective national and international bodies. To overcome the shortage of extra radio transmission bands, the spectrum is also shared with respect to signal strength and directivity of transmission, as seen in Fig. 5.13, a fact which limits the reception to given areas. This allows a number of transmission sources to use the same frequency bands, with their transmitted signal strength limited, however, to avoid disturbing each other. This precaution, nevertheless, is insufficient, and in many cases the disturbances are unadvoidable. The sole means of increasing the occupancy of the frequency spectrum is to seek for higher and higher frequencies.

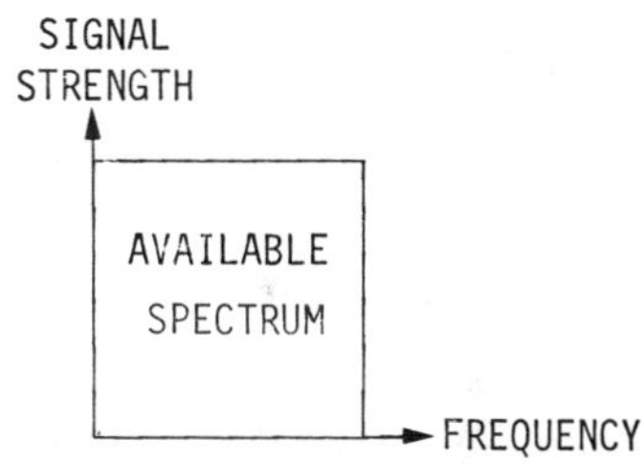

Fig. 5.13 Means of sharing the available band among many users

The use of the term radio conveyance communication refers to the transmission of radiated electromagnetic waves into the free space. The term radio communication includes various types of communication systems with extremely different characteristics. The only possible common dimension of all these radio systems is that their signal progresses through the free space from the transmitter to the receiver. This too is not accurate, since the propagation of the signal through the free space is different for each system and the air space is not 'free'; it presents varying conducting and atmospheric absorptions depending on the frequency used. Some signals may propagate along the surface of the earth while others may exploit the ionosphere for reflection purposes. Accordingly, this section and the next will discuss the general aspects of radio communication, while specific transmission techniques will be introduced in the following sections.

The actual function of a signal radiated through a space path needs to be explained. As already stated, the propagation of a signal depends on the frequencies involved. With lower frequencies, up to about 30 MHz, the propagation is along the earth surface or by successive ionosphere and ground reflections produced by the dielectric constants of the atmosphere. A signal transmitted from one point may propagate well beyond the horizon by a number of reflected hops, as seen in Fig. 5.14, reaching ranges of a few thousand miles. With higher frequencies, above 30 MHz, the propagation is basically only along direct optical line of sight paths, although longer ranges can be reached by reflections from satellite relays. Lower frequencies, however, suffer from many external disturbances which make the use of digital data communication unreliable. For wide band digital carrier communication, line of sight paths are preferable, and this includes satellite communication.

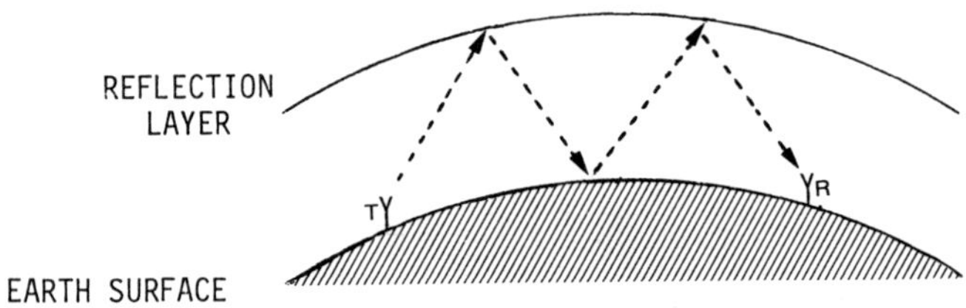

Fig. 5.14 Propagation of radio signals by a number of hops

The classification of radio channels according to the respective frequency ranges is given in the following table:

PROPAGATION BEYOND THE HORIZON		PROPAGATION OF LINE-OF-SIGHT	
Very Low Frequency (VLF)	3 kHz – 30 kHz	Very High Frequency (VHF)	30 MHz – 300 MHz
Low Frequency (LF)	30 kHz – 300 kHz	Ultra High Frequency (UHF)	300 MHz – 3 GHz
Medium Frequency (MF)	300 kHz – 3 kHz	Super High Frequency (SHF)	3 GHz – 30 GHz
High Frequency (HF)	3 MHz – 30 MHz	Extremely High Frequency (EHF)	30 GHz – 300 GHz

In the above table, the frequency band is divided into two groups according to the propagation characteristics of the radio waves. It has also been suggested that the lower frequency, which refers to waves above the audio frequencies, to be called 'ultrasonic-waves', and the upper group, which refers to those radio frequencies whose wave lengths are sufficiently short to exhibit some of the properties of light, to be termed 'microwaves'. This classification of the radio bands are shown below:

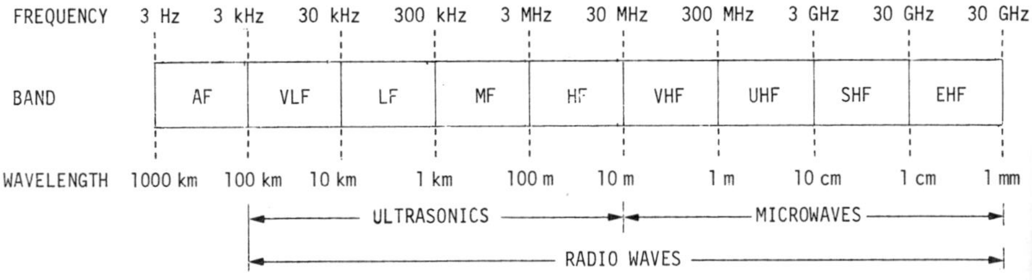

The lower frequency communication is nade possible by the existence, above the surface of the earth, of a number of ionized layers, as seen in Fif. 5.15, whose ion density varies with height and time. With higher frequencies, communication is made despite these ionized layers.

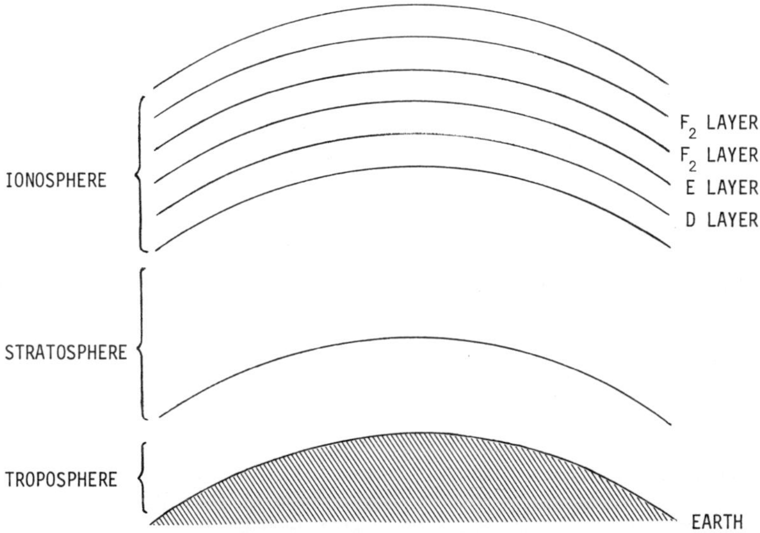

Fig. 5.15 The various atmosphere regions and layers

The atmospheric space can be divided into three basic regions, troposphere, which reaches the height of 5 km to 10 km and is used mainly for direct line-of-sight communication; stratosphere, which reaches the height of 30 km to 50 km, and ionosphere, which reaches the height of 150 km to 500 km and is used for the reflection of low frequency waves. In the ionosphere there appear four distinct layers designated D, E, F_1 and F_2, at heights of 70 km, 100 km, 200 km and 300 km respectively. The F_1 layer consists of the greatest ionization density and is therefore able to reflect the relatively higher frequencies. These layer heights are dependable on the time of the day and other atmospheric conditions. For example, the D and E layers disappear at night.

5.6 DISTURBANCES AFFECTING RADIO COMMUNICATIONS

Radio signals used for communication are extremely susceptible to all kinds of disturbances which might distort the signal well beyond its usefulness. These distortions tend to affect the lower frequency ranges rather than the higher ones, although both ranges are more vulnerable than physical communication. Some of the characteristics of the disturbances are covered below.

Path Attenuation. If the radio path is regarded as free space; the signal loss along a path is theoretically the signal strength decrease as the inverse square of the path length. The average path attenuation exceeds the theoretical value due to diffraction and multipath phenomena. Furthermore, the type of terrain over which the signal propagates adds attenuation.

Changes to the Path Attenuation. A radiated transmission signal through free space or through any uniform dielectric travels in a direct straight line with no change in its speed. Nevertheless, the atmospheric space cannot be regarded as a constant path since its density varies. When a signal wave passes from one density to another which has different dielectric constants, the wave changes direction, as seen in Fig. 5.16. Such a change of direction is referred to as 'refraction' and is directly related to the condition of the atmosphere. The change of direction is due to the change of the signal speed as it moves between two different densities. Such refraction conditions usually result from atmospheric turbulences which produce air 'blobs' whose refraction indexes are sharply different from those of the surrounding atmosphere. As these 'blobs' are unpredictable and as they are thought to occur mainly in the troposphere, this form of disturbance mainly affects line-of-sight communication.

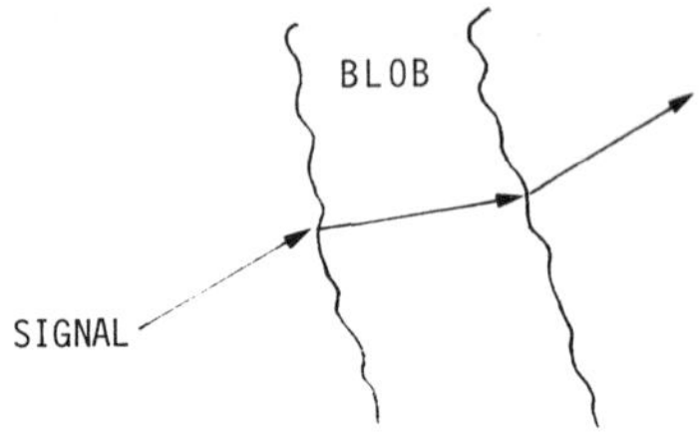

Fig. 5.16 Refraction through the atmosphere with different densities

When a signal encounters a dense layer which it cannot penetrate, it undergoes an effect referred to as 'reflection'. The amount of reflection depends on the approach angle and the reflective quality of the layer causing the reflection. When a reflection occurs the signal may result in loss of energy as a result of having some of the transmitted signal reflected back to the source. The reflection loss depends, again, both on the reflector density and the angle of incidence.

Another cause of the changing of direction by the transmitted signal is the obstacles encountered in the path. This change of direction is referred to as 'defraction', as it bends the signal around the obstacle. Of the three causes of direction change, defraction has the least disturbing effect. Even more so since it can be exploited to lengthen the UHF line-of-sight transmission path. That is, it can be used where direct line-of-sight communication is prevented by rugged terrain. Nevertheless, this technique for improving transmission beyond the horizon reduces the signal strength which consequently decreases sharply with frequencies above 1 GHz.

Absorption Losses. The loss of signal strength is due to its conversion into other forms of energy as a result of its interaction with higher densities such as atmospheric dielectric and rain. Oxygen and water vapour absorb energy from a radio signal by virtue of the permanent electric dipole moment of the water molecule. The water vapour is mainly effective at frequencies above 8 GHz. As the operating frequency increases above 10 GHz, the absorption loss increases rapidly resulting in the reduction of the transmission range.

Fading. The sudden reduction in the received signal strength caused by atmospheric conditions, reflection from obstacles, changes in path lengths, etc., is referred to as 'fading'. It is usually rendered as a result of marked changes in the refractive index through which the radio signal passes. Fading could result in short term variation of the transmission signal caused by changes in the path length or by long term variations cause by absorptions due to rain or other effects. Fading disturbance increases with distance and frequency.

From a data communication point of view, the short term fading is the most critical disturbance, since it could affect the synchronization of the receiver with the transmitter. By the time the receiver adjusts itself to the new synchronization conditions, the fading can vanish. It is very difficult to avoid fading due to sudden changes of signal strength, as well as to design the input circuits of the receiver to accommodate varying signal input levels in such a short time. The circuits involve both the automatic gain control (AGC) and the equalizer circuits.

Multipath Distortion. Radio signals travelling from the transmitter to the receiver may travel along different paths and encounter any number of hops, resulting in differing path lengths as seen in Fig. 5.17. The multipath effect of shortening and lengthening the path results in noticeable distortion of the received-signal. The longer the path, the more there are possibilities of increasing the number of hops, and consequently the signal is more vulnerable to multipath distortion.

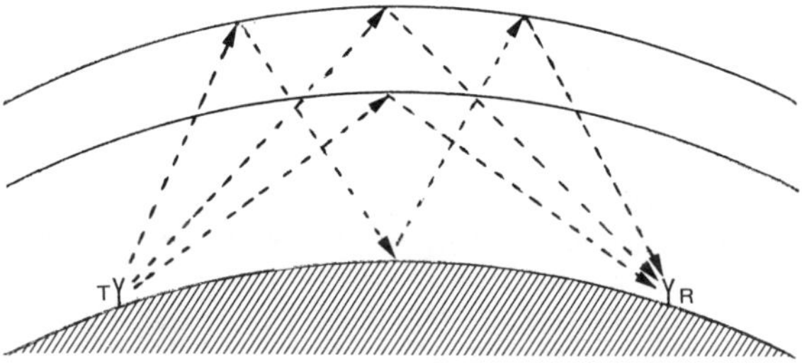

Fig. 5.17 Multipath distortion caused by varying length transmission paths

The propagation of a signal from a transmitter to the receiver over a multiplicity of paths can have serious degradation effects on a digital system performance, possibly resulting in intersymbol interference.

Multipath Fading. In many cases the signal arriving at the receiver may have been formed as a result of a number of simultaneous separate signals travelling along two or more different paths and involving different numbers of hops. Multipath fading generally increases at night time when both sky and ground waves propagation are frequently present at the same time. Since these simultaneous multipath signals all originate from the same source, the demodulated signals in the receiver tend to be superimposed, resulting in overlapping of the delay components and making its data decoding difficult and sometimes even impossible. The multipath fading effects are especially noticeable in the lower frequency band rather than the higher frequency band. With HF data communications the multipath spread could be of a few milliseconds as compared with a few microseconds that are encountered with VHF. For this reason the HF band limits the data rate that may be transmitted over this band. This effect can be reduced by using multi-tone transmission for signals in the HF band, as will be described in the following section.

Cross-Talk Disturbances. As already explained, the available frequency spectrum is polluted with a vast number of different users' transmissions. In many cases, the same frequency band may have been allocated to users who may be miles apart and/or their transmission may be restricted to a given direction. Nevertheless, it is impossible to fully guarantee that one user's transmissions will not affect transmission of others. The effect of these encroaching transmissions has a similar effect as the cross-talk in physical wire communication.

It has been known that a VHF taxi communication system, which is theoretically limited to line-of-sight, can travel sometimes, for no apparent reasons, thousands of miles well beyond the horizon and affect the communication of other systems. It has been found that the reason for this effect is that the VHF signal can be trapped in tropospheric ducts and will propagate well beyond the horizon to thousands of miles.

Atmospheric Noise. The atmosphere is full of distant noise, which is produced mostly by lightning discharges in thunderstorms and transmitted to the receiving stations via the same ionospheric layers that transmit the radio signals. Such noise sources are realized as impulse noise in the effective signal bandwidth. During the daytime, the ionosphere absorption is high and the noise level received is reduced, but at night it could have damaging effects. The received noise level decreases with the increasing frequencies and therefore has a lesser effect on very high frequency signals using line-of-sight propagation.

Galactic Noise. The type of noise is defined as disturbance in the radio frequency bands originating outside the earth or its atmosphere. The sun is the source of ultraviolet radiation causing the ionization of the layers. Sun spots are therefore believed to change the height position of the layers and consequently effect a change in the frequency that can be reflected and results in fading and other disturbances. In practice, the galactic noise is restricted by atmospheric noise to frequencies not lower than 18 MHz.

Man-Made Noise. In military operational environments, the communicating systems are susceptible to intentional interferences which are designed to jam the transmissions so as to prevent communication. This, however, is not the only source of man-made noise; there are many other sources whicb affect civilian as well as military communications. Man-made noise may be a result of unintentional interference, such as automobile ignition, electrical trains, X-ray equipment, radar, etc. This type of noise interference causes impulse noise and its effects depend on the distance of the noise source from the receiver.

5.7 RADIO TRANSMISSION PROPAGATION

The radio signal which is transmitted from one point may reach its destination along various paths. Each path may be better suited for specific frequency bands and unsuited for others. It is therefore essential to consider each radio transmission propagation media as a separate conveyance entity. The characteristics of each propagation path is vastly different from the others; both in regard to the ground distances the signal can reach and to the disturbances it encounters on its route to its destination. Most of the radio propagation techniques presented are intended for analog communication and are unsuitable for high speed data communication, they are, nevertheless introduced here for complementary material.

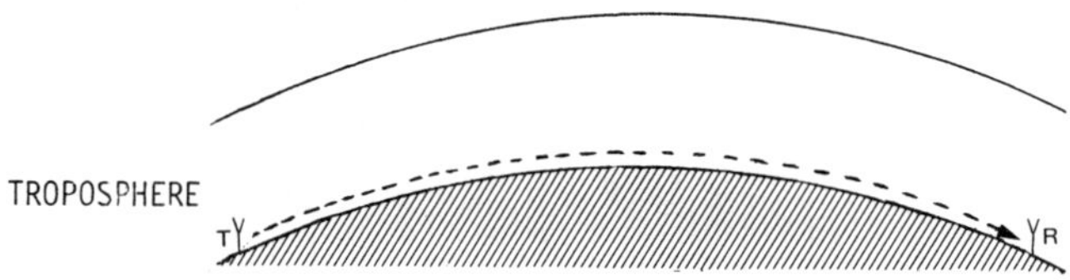

Fig. 5.18 Propagation of ground waves well beyond the horizon

Ground Wave Propagation. In the lower frequency range, in the VLF and LF bands, the radio signal has the properties that enable it to propagate over the surface of the earth. This signal propagation is ordinary effective in the lower atmosphere in the range of the ground and troposphere, as seen in Fig. 5.18, for which reason it is generally referred to as 'ground wave propagation'. There is apparently no lower frequency boundaries for ground wave, although there is an upper limit. Higher frequencies, MF and above, are attenuated more, to a point which is insufficient for satisfactory communication. Ground waves have little applications for digital communication, they can nevertheless, be used to transmit very low data rate. Their many applications are exploited for navigation purposes and this is extremely important in command and control systems. They could also be employed for special highly reliable backup systems for control purposes which require distances up to 300 km and data rates up to 75 BPS. Above 500 km there is a danger of multipath disturbances due to both ground waves and sky waves.

Ionosphere Reflection Propagation. The ionosphere is part of the earth's outer atmosphere where ions and free electrons are normally present in quantities, sufficient to affect the propagation of the normal radio signals. Communication is achieved by reflections from the E and F_2 ionized layers, as seen in Fig. 5.19, encountered in the lower frequency band from VLF to HF. For each layer there exists a maximum frequency which will be reflected at vertical incidence. With VLF waves, the reflection from the ionosphere is nearly total, although as the frequency increases this feature is reduced. The VLF band is therefore used for world wide broadcasting coverage, while the higher frequencies are used for shorter ranges up to 1600 km. Its attractiveness is due to the ease and efficiency of the transmission properties characterized by low attenuation losses.

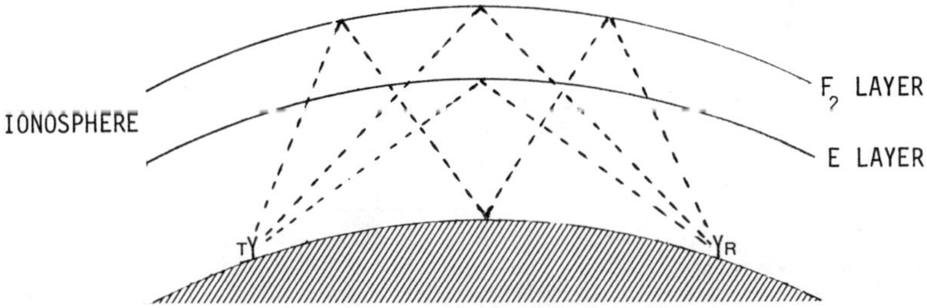

Fig. 5.19 Communication beyond the horizon by reflection from ionosphere layers

Despite the advantages of the ionosphere reflections, it is not an attractive means of conveyance for data communication because of its unpredictive performance. It can at times provide very efficient means of propagation but the condition can suddenly change. It is also vulnerable to fading, as the propagation path can vary with the time of the day. The ionosphere layers are subject to great ionization changes caused by the zenith angle of the sun resulting in height changes. The latter affects the HF band more where its signal could be absorbed instead of being reflected, thereby limiting its data communication capabilities. Ionosphere reflection propagation, due to the frequencies involved, is affected both by atmospheric and galactic noise. In long range communication it is also very susceptible to multipath fading.

The multipath disturbances affecting the use of the HF band for data communication can be overcome by a transmission scheme known as multi-tone. In this scheme up to 16 frequency multiplexed tones are transmitted in parallel, with each tone operating at 1/16 of the modulation rate, thereby increasing the duration of each transmitted data symbol without reducing the overall data rate. That is, for a data rate of 1200 BPS a modulation rate of only 75 Bauds is necessary. This advantage is achieved at the cost of very expensive modems, as each tone requires a separate modulator/demodulator. The increase in the transmission duration of each tone pulse alleviates multipath fading, for which reason it is used in marine data transmission systems which require long distance communication and have no other means of conveyance.

Ionosphere Scatter Propagation. Ionosphere scatter propagation is generally attributed to the scattering properties of the ionized lower *D* region of the ionosphere which is normally due to turbulence. The transmitted signal is scattered and this affects some of the signal waves, which need to be strenghtened. The useful bandwidth required to achieve this scattering effect is restricted to less than 10 KHz. The transmitting and receiving stations require narrow band antennas which are both directed to a common mid-point between the two stations in the *D* layer region of the ionosphere above the earth. This type of transmission conveyance permits communication both in the HF band and the lower VHF band to a distance of about 1600 km. It is used for very low data rate digital communication as accepted for telegraph communications where other forms of propagation techniques are unreliable. It is inaffective for higher data rate communication.

Troposphere Scatter Propagation. To achieve data communication of distant ranges beyond the horizon, in the frequency bands of VHF to SHF, it has been found futile to depend on the ionosphere effect to support propagation. Instead a method has been developed which enables the transmission of modulated data signals with the troposphere region over distances of 100 to 1000 km. This transmission conveyance is achieved by a technique called 'troposphere scatter'. or 'troposcatter' for short. Although this transmission technique is restricted to the troposphere, it is affected by disturbances in the ionosphere region which in turn cause disturbances in the troposphere region; this feature must be considered in the design. Fading too is one of the important problems encountered in troposcatter communication existing mainly in long range transmission systems.

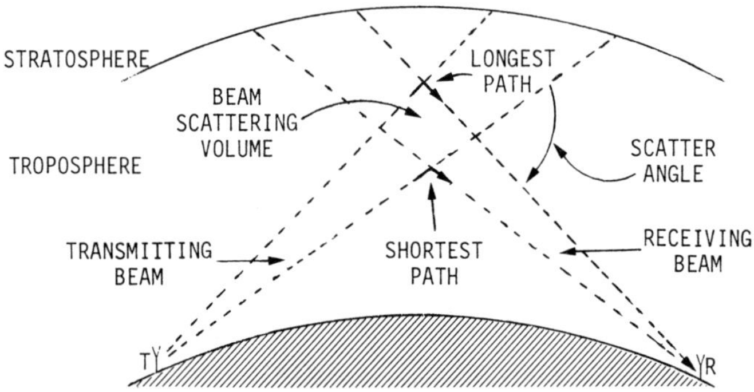

Fig. 5.20 The geometry of troposphere scatter propagation

The geometry of the troposphere scatter technique is shown in Fig. 5.20, with the scattering processes taking place below the stratosphere region. Troposcatter operation could be explained as a refraction of the signals from the horizontal layers in the troposphere where they are less dense and less humid than at the lower altitudes. These refractions cause a downward bending of the signal, effecting the signal waves to travel from the transmitter to the receiver along a number of paths. In consequence the receiver signal is composed of several components which are received in different phases and reflected from different layers situated at different heights. These differences in the path length and signal strength raise many problems. Nevertheless, troposcatter does have one outstanding feature: the signal is always there even if it is very weak and distorted. As the distance between the transmitter and receiver lengthens, and scattering volume moves to the upper troposphere thereby increasing the path loss. Achieving communication beyond the horizon calls for larger antennas, narrower frequency bands and higher transmission power.

For data transmission, the troposphere does not offer better speeds than other forms of transmission beyond the horizon. Its main drawback is the distortion caused by the various delays of the signal multipath. Some parts of the transmitted signal arrive much earlier than others, and as a result pulse stretching can occur with the tail of one pulse interfering with the next. Again the solution advised is to limit both the distance and speed. For longer distances multi-tone techniques may best be used.

Line-of-Sight Propagation. The most popular conveyance for data communication is the use of line-of-sight propagation between the transmitter and the receiver, as seen in Fig. 5.21, although as the name implies, the transmission distance is limited to about 50-70 km, depending on the land countours. This type of conveyance is ideal for high speed data transmission since it allows wide frequency band, with no definite limits, to be used. For this reason most radio carrier systems employ this technique which will be explained in the following section. The most popular frequency band to be used for data communication is that of UHF, although line-of-sight techniques are widely applied with frequencies of VHF and above.

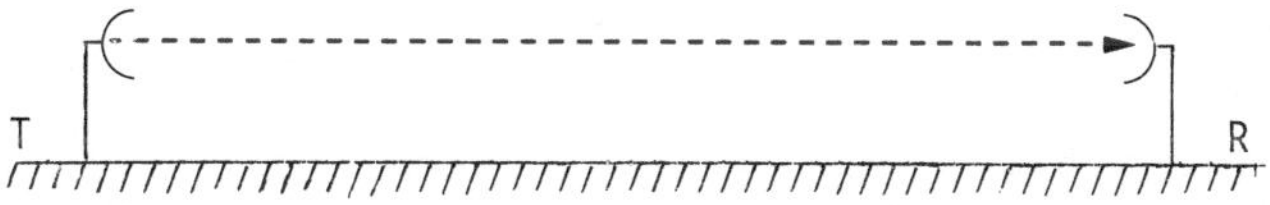

Fig. 5.21 Line-of-sight communication conveyance

Despite expectations, the line-of-sight signals do not always travel in straight lines; they are slightly bent (as seen in Fig. 5.22) due to the atmospheric refraction index which tend to degradate the desired transmission propagation. Furthermore, since the line-of-sight path is so near the surface of the earth, it also suffers from reflection from the ground surface. Line-of-sight transmission could therefore be affected by multipath fading in the same way as other radio propagation conveyances, although this could be overcome by proper design. These techniques make use of 'diversity' means of communication, as will be explained later.

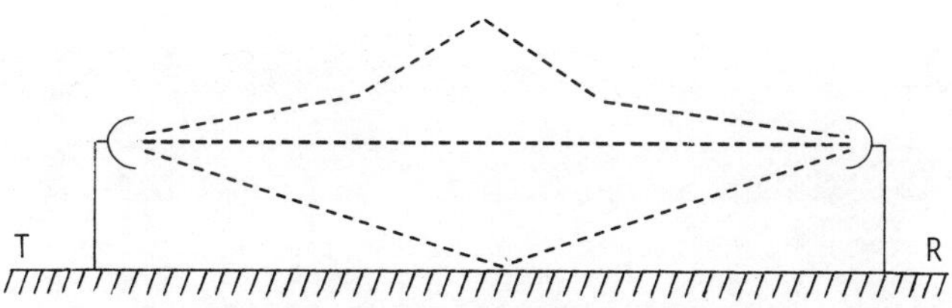

Fig. 5.22 Multipath effect in line-of-sight communication

The range of transmission distance of line-of-sight communication can be extended beyond the horizon by using diffraction techniques with the diffraction losses compensated by increasing both the transmitted power and antenna height, although no where near that required for troposcatter propagation. With normal earth surface diffraction it is possible to reach 100 km while with knife edge or rough surfaces, as seen in Fig. 5.23, the distance can reach 200 km.

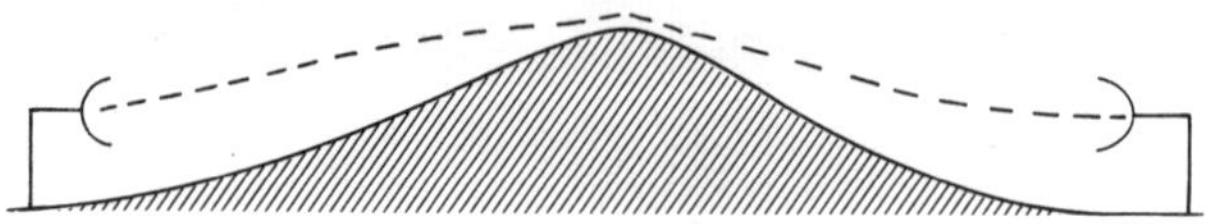

Fig. 5.23 Extension of line-of-sight to beyond the horizon by diffraction techniques

Diversity Propagation. Nearly all the propagation paths are vulnerable to fading, resulting in short or long term disturbances of the communication conveyance. It is known, however, that if two or more radio signals are sufficiently separated, either in frequency or space, the fading disturbances of the signal that occur in the various transmission paths are more or less independent. That is, in two parallel paths, one may suffer a fading while the adjacent path is unaffected. One of the best means of reducing the effects of fading and so improving the communication is selecting separate propagation paths, with each transmitting the same data signal and then combining them in the receiving station. This technique, which is known as 'diversity' propagation, increases the probability that a signal will not be lost because of fading, as the receiver automatically selects the strongest signal detected. This requires careful control of the combining of the signals in the receiving circuits so that the path differences will not introduce any phase or any other variations.

Diversity techniques are very popular with microwave communication using UHF and SHF bands although this technique is also applied sometimes for VHF and even HF communication.

There are two basic types of diversity techniques: space diversity, which employs two or more antennas at different locations, and frequency diversity which employs different frequencies. It is possible to include in one communication system both space and frequency diversities to further improve reception.

With frequency diversity, the same antennas may be used for both signals, although separate transmitters and receivers are necessary at both stations, as seen in Fig. 5.24. To give satisfactory results, the distances between the two frequencies need not be great and a distance of the order of 400 Hz in the HF band gives good diversity performance.

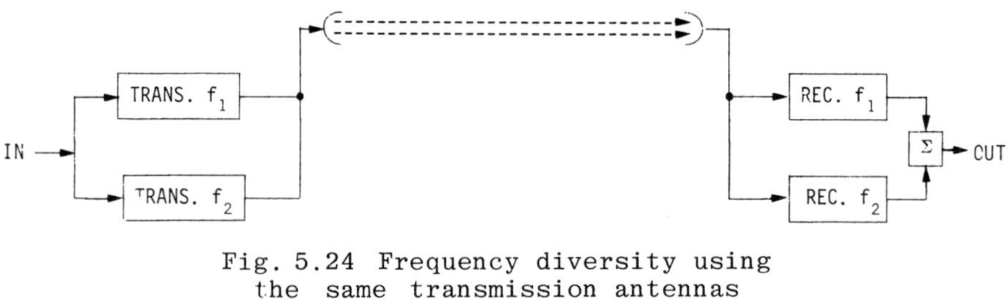

Fig. 5.24 Frequency diversity using the same transmission antennas

With space diversity only one transmitting antenna is required, as seen in Fig. 5.25, although two separate antennas are needed at the receiving station. Again, the distance spacing between the two antennas need not be great; in fact both antennas may be mounted close by on the same post, with 10 to 15 meters spacing being more than adequate.

There are two more diversity techniques that may be used, polarization diversity and angle diversity. With polarization diversity two simultaneous signals are transmitted, one with vertically polarized waves and the other

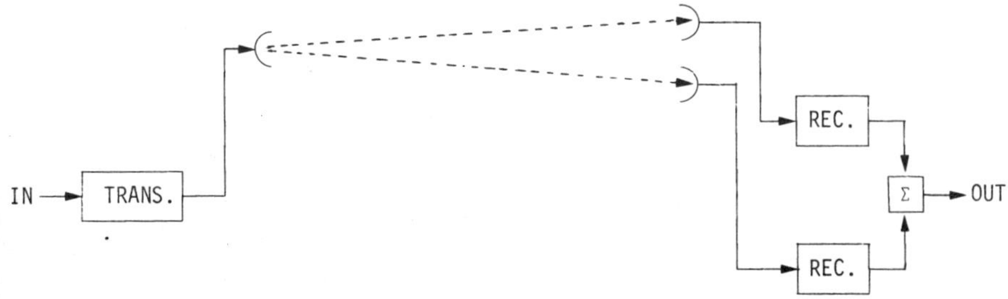

Fig. 5.25 Space diversity with separate receiving antennas

with horizontally polarized waves. The results of this type of diversity do not substantially favour this method. With angle diversity, several offset feeders produce an angular separation of the multipath signals transmitted from a single antenna. Such a technique could be used in troposcatter propagation, although it has little advantage in line-of-sight propagation.

5.8 LOCAL AREA RADIO COMMUNICATION

Local area radio communication systems have been in existence for many years. They have been developed for speech communication and have been used for fixed and mobile systems. They provide means whereby users, such as taxis, police, fire brigade, etc., could communicate with each other and with a centre. Local area communication implies that the system is intended to serve a number of stations which desire to transmit information over a fairly small area, as seen in Fig. 5.26. These local communication area networks operate by using line-of-sight techniques which employ transmission channels in the VHF and UHF bands. That is, all the stations using the same network are provided with a single frequency bandwidth which serves them for communicating with each other. In most of these local area speech system, the same frequency band is used both to transmit and receive information, although not simultaneously.

Efficiency of the local communication network is substantially increased by providing each user with direct access to the terminals of all the other users. In this respect the local area radio network differs from the local area twisted wire conveyance discussed before, as there the conveyance provides point-to-point communication between only two users. To provide full availability between all the users in a radio point-to-point network, a switching centre is essential, with separate frequency bands dedicated to each user. This arrangement is not necessary with local area network, as each user has full availability of the common communication channel and therefore may have a direct contact with all the other users, without requiring any assistance or asking for the intervention of a centre control. In other words, each of the multiple users has equal access facilities to the common transmission channel. Such a common facility of a mutual communication channel, which is characterized by providing a single frequency band available to multiple users is referred to as 'multiple-access'.

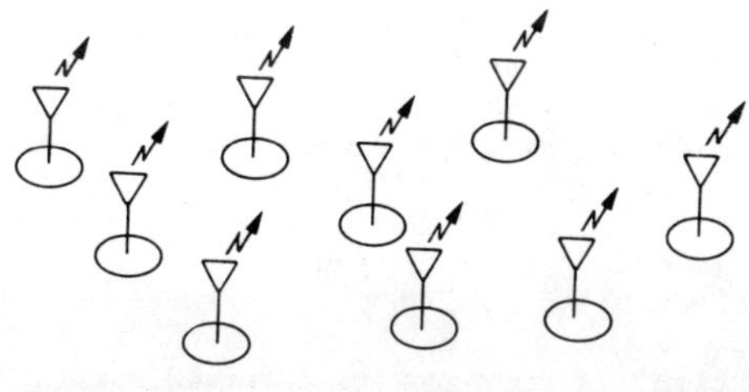

Fig. 5.26 Local area line-of-sight multi-access communication network

The multiple-access arrangement, although advantageous, has the effect that only one transmission can be performed successfully at any instance and that every user of the common channel can hear the conversation of all the other users. Furthermore, it is impossible to prevent the occurrence of communication 'collisions' between the transmissions of users who are radio 'hidden' from each other.

With multiple-access communication systems using speech transmission, the human operator can avoid the collision by using his common-sense reactions when he feels he is 'hidden' from the other users. With multiple-access communication using data transmission, however, the same techniques cannot be employed since they are 'dumb' and their intelligence is limited to what they had been programmed to perform. Nevertheless, it is possible to introduce special software programs into the terminals which define the operating procedure whereby it is feasible to achieve the required data communication with the danger of data message collision prevented or reduced. These communication procedures are presented in chapter 9 and are based on the same reactions as those of the human operator. In all these multiple-access data communication systems, the data messages are broadcast by any terminal in the network to any other terminal. An address tag label is added to each transmitted message heading, which identifies both the sender and the message destination.

The first known multiple-access data communication procedure was suggested and implemented at the University of Hawai, and was named ALOHA. Having found that the local telephone network on the island introduced to many errors into the terminal-computer communication system, they proposed transferring their communication to a radio conveyance. The common radio channel was assigned to all the users with consideration for their respective needs and according to a random assignment procedure. Since the traffic load in the common channel was low this procedure was sufficient, but with the increase in the traffic load, the collisions soon caused the channel to become saturated. Since then many other procedures have been suggested whereby the channel assignment to each user is by fixed or dynamic rather than random rules in order to reduce the danger of channel saturation.

The application of local area radio multiple-access data communication systems is becoming very popular and they are widely employed both for access by the transaction terminal to its associated local host computer and for terminal-to-terminal interaction communication. They can be used, for example, for telemetry control systems which collect data from far apart distributed data collection points, for warning systems where the terminal transmits data signals describing the stress operation conditions of the Controllee, or for police vans which transmit data regarding suspicious vehicles. They can also be used for controlling trucks or taxis, where each car can be given its instructions by means of digital radio and have the messages appear automatically on the terminal's display: this thereby removing the need for the driver to write down his instructions. The truck driver, in turn, can inform the centre of his destination and give details of his load and the information can then be transferred to the centre computer automatically. For fixed installation, as seen in Fig. 5.27, such a data communication system offers many attractive features as they can be used both for remote terminal communications and remote job entry (RJE) communications operating in a real-time environment. In the example shown, each DCE unit consists of both a transmitted and receiver operating in a half duplex mode. The DTE units may be data message entry devices (DMED), computer terminals, data concentrators or host computers.

The system for local area radio data communication employ both VHF and UHF bands. With the available frequency bandwidths, it is possible to transmit over them high speed data messages, and these days it is possible to obtain 'off the shelf' modems for mobile radio communication with data rates of 1800 BPS with FSK techniques and 2400 BPS with DPSK techniques. With fixed stations wider bandwidths are available, as for example, the ALOHA system that operates with a data rate of 24,000 BPS.

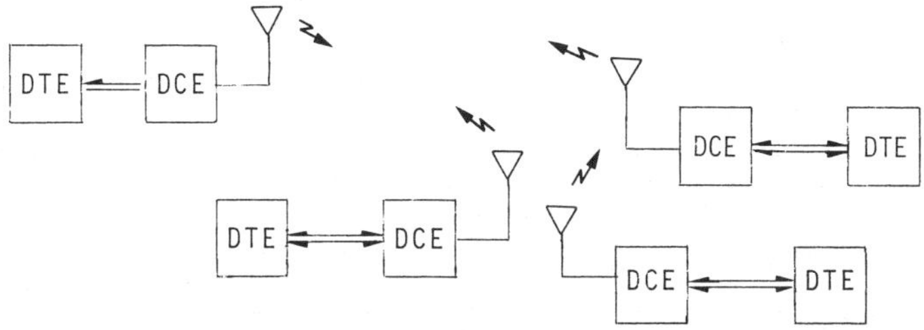

Fig. 5.27 Data terminal multiple-access local area radio communication

With all the attractive features of multiple-access schemes, their main drawback still remains that only one successful terminal transmission can take place at any given time. If there are more transmissions, collisions are then unavoidable. Central channel-assignment control is also not free of drawbacks since it makes the centre station over-active and the remote terminals too dependent. For this reason, polling techniques which are so popular for wire communication, are inadvisable for local radio communication. With polling the centre stations asks each terminal in turn whether it has a message to transmit. This means in effect that the centre will have to activate its high power transmitter at frequent intervals and thereby shortening its life term. A new solution has been introduced in the last few years employing spread spectrum techniques by the code-division multiple access procedures described in chapter 10. This solution requires high operating frequencies with much wider bandwidths, and has already been adopted by the military for their local area communication system, one of which is known by the code name JTIDS. Such solutions, however, have not yet emerged for commercial application.

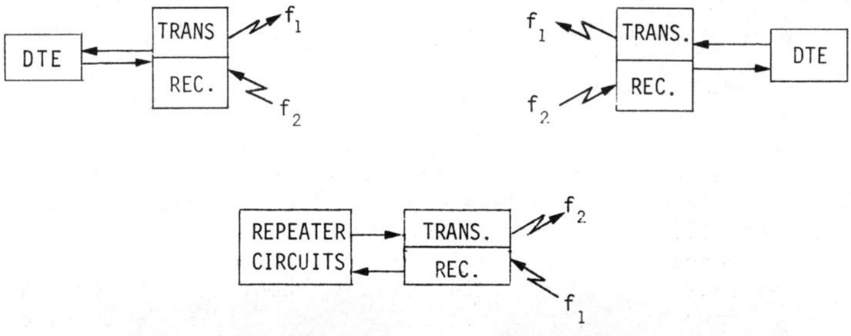

Fig. 5.28 Introduction of radio relay repeater station in the local area communication

Another drawback of local area radio multiple-access is that all the terminals must be in the line-of-sight of all the others and that the transmitted signal power is sufficiently strong to reach all the other terminals in the area. In the practical cases where the transmission ranges of the terminals are not sufficient to provide communication to the destination receivers, it is necessary to introduce a relay device which will receive the weak data signal from the transmitting terminal and retransmit it to its final destination. In a simple network operation, it is possible to extend the communication coverage range of the system by two operating frequencies with a single channel relay system, as seen in Fig. 5.28. The radio relay station only acts as a repeater in the same way as it operates in physical cable communication; that is, it repeats the signals it receives but with added power. This technique enables all the terminals to transmit at a relatively low power but of sufficient strength to reach the nearby relay station, in the frequency band of f_1. The relay then retransmits the signal in frequency band f_2 at a much stronger signal power. All the terminals are then listening to frequency band f_2 and not f_1. Such a network operation extends the multiple-access procedure but without changing its works. That is, the communications from each terminal are still in half duplex mode, (although it transmits and receives on different frequencies) since only one single transmission can be effective, at any given time in the double frequency channel.

The relay station as discussed above is passive, with no communication control functions other than the repeating duties of all the signals it receives. There are systems where active relays are employed whereby the relay performs store and forward logic functions. (The discussion of these techniques will, however, be postponed to a later, more appropriate stage.)

The operating area covered by the line-of-sight system may be extended by adding a number of point-to-point relays with a maximum spacing of about 50 km between each such relays. These relays (which will be discussed in the following section) operate at different frequency bands from the local system so as to avoid disturbing the normal operation of the terminals. Such a network, as seen in Fig. 5.29, has been referred to as 'signal leapfrogging' since it provided a conveyance media over any other terminal or relay which may happen to lie in its transmission path. The point-to-point relays transfer to the other relays all the messages which are addressed to terminals which have no local destinations. In this arrangement a message originated in one multiple-access area coverage, may proceed through a series of point-to-point relays till it reaches its final destination.

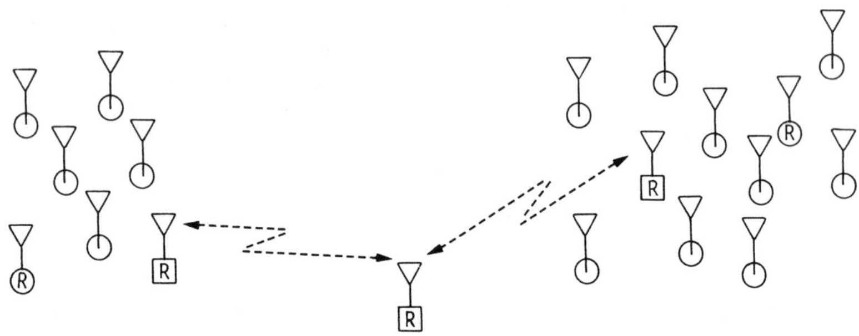

Fig. 5.29 Extension of the local line-of-sight communication to other local area networks

5.9 POINT-TO-POINT LINE OF SIGHT RADIO COMMUNICATION

Point-to-point line of sight radio communication is most widely used for carrier communication and is commonly referred to as 'microwave' radio communication because of the short wavelengths of the frequency involved. That is, it uses the UHF and SHF frequency bands which permit the transmission of very wide signal bandwidths. In these frequencies, the radio signals are easily concentrated into a beam to provide high speed data communication, provided these are line-of-sight transmission paths. Furthermore these communication channels could easily be extended for long distant links by adding relays along the path which are spaced about 50 km apart.

As stated above, microwave systems are widely used for the transmission of both digital and analog carrier signals. This form of transmission is superior to the cable carrier transmission systems already discussed since there is no need for any physical connection between the two communicating points. This means in effect that the signal can cross over the ground without the need of obtaining right-of-way permission, whereas with physical cable conveyance permission must be granted for the total ground space of the transmission path involved. All that is needed is to acquire the land for the relay stations only, with wide freedom in selecting their locations. Furthermore, the distances between the radio relay stations are much less frequent than those required for repeaters in cable communication.

Today, microwave line-of-sight is becoming the backbone of most national, as well as some international communication systems, for both public and private usage. They are employed in applications where there is a need to convey vast volumes of data, or for large numbers of independent communication channels between two long distant points. The general agreed frequency ranges for international communication is between 1 to 19 GHz but other bands are also available. Of all the communication conveyances discussed so far, there are many line-of-sight microwave links which have been designed specially for digital data communication. For example, the British Post Office has developed a microwave network using eight-phase PSK technique in the frequency band of 11 GHz and providing data communication rates of 140 MBPS. The newer digital techniques of TDM/PSK are rapidly replacing the conventional analog carrier systems of the FDM/FM and FDM/SSBSC techniques.

To preserve the signal strength and to extend the transmission path beyond the horizon, radio relay stations must be inserted within the line-of-sight transmission path (as seen in Fig. 5.30) every 40-60 km. The function of these radio relays is similar to that of the physical cable repeaters. that is, to receive the weak signal and retransmit it with additive strength. Nevertheless, the radio relays are different from the cable repeaters in the means employed to perform this task. With radio relays, different operating frequencies are required for receiving the signal and retransmitting it. This means that in the radio relay station the received signal must be demodulated to an intermediate frequency (IF) and then remodulated to a different radio frequency. Both the receiver and transmitted radio frequencies are usually in the RF band although separated by a guard bandwidth to prevent one signal from disturbing the other.

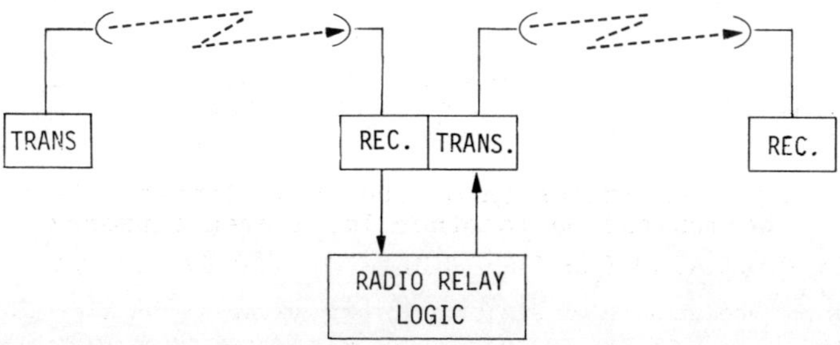

Fig. 5.30 Microwave communication beyond the horizon with the aid of line-of-sight relay stations

Radio relay stations can be classified either as passive or active. In the passive relay systems, the microwave signal is simply received in one frequency and retransmitted in another frequency. In the active relay systems, the signal is demodulated to its baseband frequency and then each individual channel is amplified separately. With digital modulated data, the active relay also adjusts the clock timing of the data pulses.

The point-to-point line-of-sight microwave communication transmits and receives the signal in one process with no means of separating parts of the signal. This is in contrast to local area radio communication where each user has equal access facilities to the common channel. In line-of-sight microwave communication, however, the common channel can provide facility to only two 'users', who are connected by a free space point-to-point path. It is possible, nevertheless, to separate parts of the signal with the aid of filters, as seen in Fig. 5.31. Such an arrangement is made possible by organizing the lower part of the frequency band to be the part that is to be separated. Consequently the separation is performed with lower and higher pass filters (LPF and HPF) and thus eliminating the need for expensive band pass filters. Such an arrangement, however, calls for the whole network to be designed initially together.

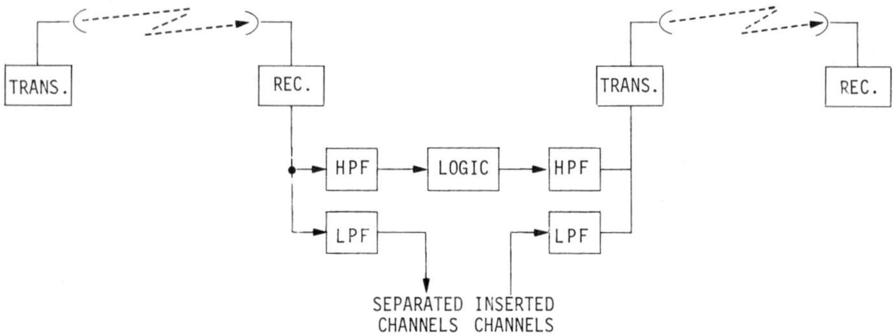

Fig. 5.31 Separating analog channels from a FDM carrier signal

With data modulated signals the separation is much easier although the signal must first be brought to its baseband and converted back to digital form. The data pulse stream can then be separated with the aid of digital timing multiplexors, as seen in Fig. 5.32, where it is possible to isolate a given channel slot by the relative timing it has been assigned.

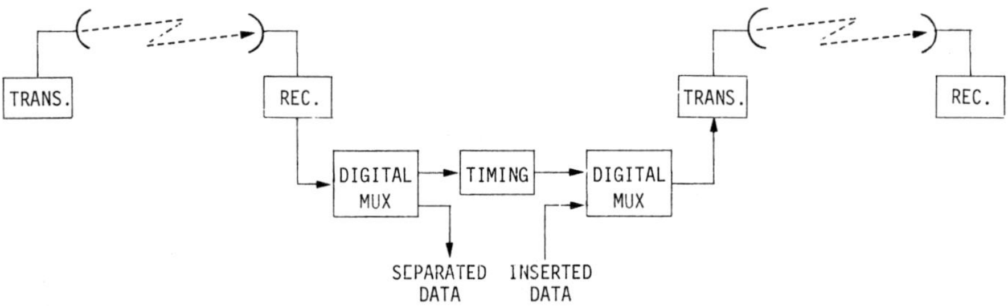

Fig. 5.32 Separating digital channels from a TDM carrier signal

The terminals and radio equipments at both ends of a path are mounted on the ground with structures supporting the antennas. Waveguides are usually used to connect the transmitter or the receiver circuits with the antennas because of the wide frequency band and the high power involved.

5.10 SATELLITE COMMUNICATION CONVEYANCE

Satellite communication is a relatively new form of data conveyance, which since its introduction, has had an impact on long distance communication and is also starting to have an affect on local wide area communication.

Initially it was intended as an extension of the microwave line-of-sight communication, with the satellite serving as a repeater between two terrestrial points. This was soon expanded to provide communication between any two points at which the satellite is mutually visible. The satellite communication systems were traditionally designed for voice traffic, with the initial satellites employing analog techniques. The newer satellites use digital techniques and are designed to carry voice and data as well as video traffic. Today, the application of satellites is further developed to include communication tasks which provide data links between computer terminals.

Satellite supplies the conveyance whereby large volumes of data can be transported at high speeds, with high reliability and at a relatively low cost. It has been found to be more cost effective, in comparison to other conveyances, for international communication links of distances over 1000 km. Its main attraction is that a single transporter can cover a wide area with its operation controlled within certain limits.

There are two types of satellites, passive and active. A passive satellite acts only as a radio reflector, while an active satellite may perform various communication tasks. The simplest active satiellite is one in which the signal is received, amplified and retransmitted (as seen in Fig. 5.33) in the same manner as microwave communication; in fact, it involves the same frequency bands of UHF and SHF, although the satellite provides better free space propagation relative to the distances of the paths. More complex active satellite provide data switching and store-to-forward tasks which enable them to function as communication network nodes.

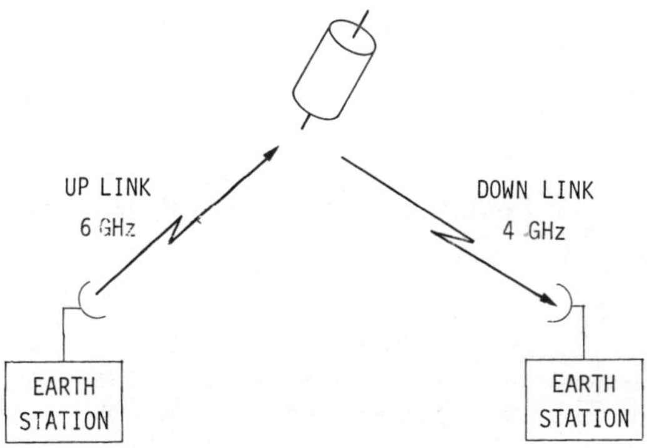

Fig. 5.33 Active satellite operating as a radio relay repeater

Most communication satellites use the geostationary orbit, where a simple satellite is orbiting over a given area and providing access to many earth stations. Such satellites have a circular equitorial orbit at an altitude of about 36 K km. The advantage of this orbit is that the satellite appears to be stationary above a single point on the earth's surface.

The main external disturbance to satellite communication is the additive white Gaussian noise (AWGN); calling for the performance to be controlled by adjustable transmitting power and by specially designed equalizers in the receivers. Another disturbance is the signal power absorption in the atmosphere which is greatly affect by rain. The most important disturbing factor, however, is the propagation delay due to the very long paths involved which could result in approximately 270 miliseconds between the time the signal is transmitted and the time it is received. This delay has an impact on many communication functions, one of which is worth mentioning is the time slot duration in a TDMA system.

Satellite long distance communications operating these days offer the services of many parallel 4 KHz voice circuits using analog frequencies with FM/FDMA techniques. The quality of these circuits is satisfactory and data can be transmitted over them at data rates up to 9600 BPS. Wider frequency bands are also available allowing data rates up to 56 KBPS, which is most useful for digitized voice communication using PCM signals. Newer and better satellites are being launched which can provide much wider channel frequency bands and can support higher data rates. These newer satellites use digital modulation techniques of phase shift keying with time division multiplexing, i.e., PSK/TDMA. The first generation satellites had a capacity of up to 6 K analog voice channels using frequency division multiple access (FDMA) while this rises to 48 K channels in the second generation. The satellites planned today are expected to be fully digital and have a capacity of 100 K to 400 K voice channels.

Most of the existing geostationary satellites use the 6 GHz frequency band for the uplink and the 4 GHz bands for the downlink, with a 500 MHz bandwidth for the transmission channel in both directions. This frequency bandwidth is already over crowded and an additional bandwidth will only be obtained by shifting to higher frequencies. Currently, the interest lies in the frequency ranges of 11/14 GHz and 19/29 GHz which will allow transmission bandwidths rising to 2.5 GHz. Although these frequency bands have the advantage of providing better channel capacity, they are known to be vulnerable to fading caused by rain. This is overcome, however, by additive transmission power which incidently increases the satellite weight. It is also possible to increase the number of channels in the existing satellites by what is known as 'frequency reuse' which uses opposite polarization to two parallel radio signals.

The efficiency of the satellite depends, to a large extent, on how it exploits the frequency bandwidth of the available spectrum. The first generation satellites simply used frequency-division-multiple-access whereby each communication channel was allocated a given slice of the available frequency bandwidth, as seen in Fig. 5.34. In this assigned band each channel could modulate its signal separately using either frequency modulation (FM) or single-side-band-suppressed-carrier (SSBSC) modulation. With frequency-division-multiplexing (FDM) all the channels are modulated together with a single carrier frequency. With frequency-division-multiple-access (FDMA) the object is to establish a number of simultaneous, but independent, communication paths between the various earth terminals. Such a multiple-access arrangement, means in effect that each channel which uses a small part of the satellite beam is modulated on a separate and different carrier frequency.

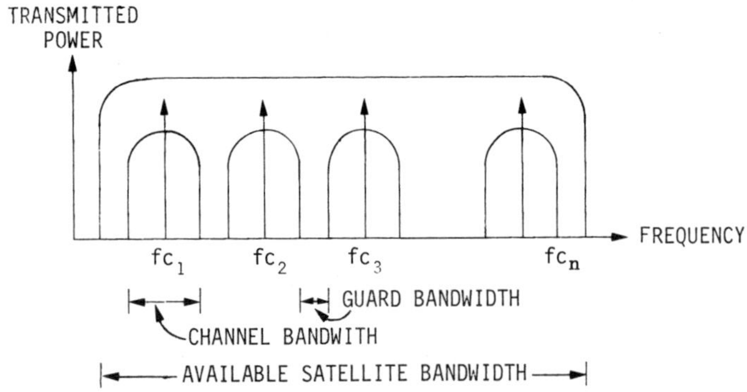

Fig. 5.34 Sharing the satellite spectrum with FDMA techniques

Consequently each transmitter can be located at various geographical locations. As this creates the danger of interference between adjacent channels in FDMA , there is a need for wide frequency bands as guards between the channels. For this reason, the number of channels over the satellite using this technique is limited.

While frequency multiplexing shares the spectrum among many channels for indefinite periods, digital multiplexing assigns the full available frequency bands to all the channels although at different sequential times. that is, each channel gets a time slot which is long enough for it to transmit all his messages. The time slots are allocated in turn to all the channels, and after a completed cycle the whole allocation sequence starts again, as seen in Fig. 5.35. Since the users of the channels are not located at one place as in TDM systems, the channel transmission time is initiated in each earth terminal in accordance with strict multiple-access procedures. This means that time division multiple access (TDMA) schemes must operate with all the earth stations fully synchronized with each other.

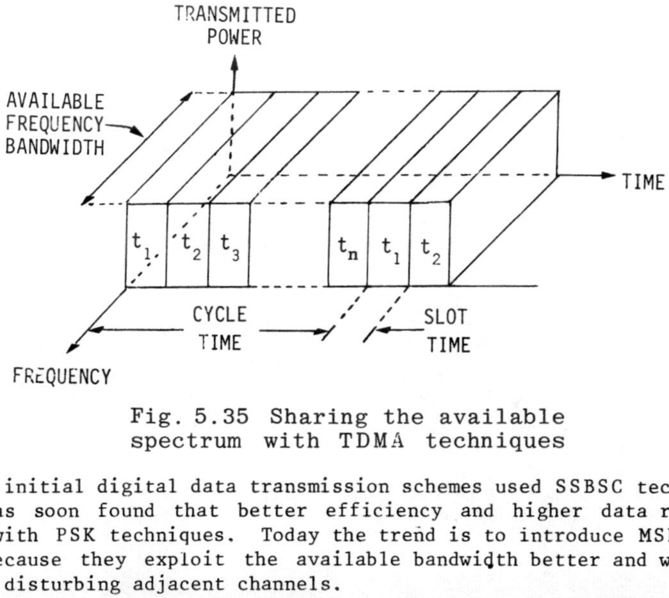

Fig. 5.35 Sharing the available spectrum with TDMA techniques

The initial digital data transmission schemes used SSBSC techniques, but it was soon found that better efficiency and higher data rates are achieved with PSK techniques. Today the trend is to introduce MSK techniques, because they exploit the available bandwidth better and with least danger of disturbing adjacent channels.

Digital multiple-access with PSK or MSK techniques are far more efficient than analog multiple-access with FM techniques. The TDMA schemes exploit the available channel bandwidth far more than FDMA and therefore provide better rates. The TDMA/MSK schemes exploit the available satellite spectrum far more than FDMA/FM and therefore can provide far more channels, nevertheless, the transmission efficiency is achieved at the expense of timing synchronization problems. This consequently means that time guarding space must be inserted so that the transmission signal in one time slot will not interfere with transmission in the adjacent time slot. It should be remembered that each time slot is assigned to different earth stations and this complicates the synchronization operation. Furthermore, provision must be given within the time slot for the propagation delay due to the long transmission paths. That is, a new time slot period should not start before the last time slot transmissions reached their destination.

Permanent assignment of a frequency band in FDMA or a time slot in TDMA to a particular earth station is rather wasteful if this station does not have continuous data to transmit. In the schemes described so far the frequency band or the time slot is assigned to a particular station, while the other stations cannot use it even if no data is transmitted in it. Instead of fixed allocation of the spectrum slices, it is possible to share it in accordance with traffic demands. That is, the spectrum slice is assigned to an earth station only for the duration of a given transmission. If an earth station has messages to transmit it demands a frequency band or a time slot in which it can transmit. The request for transmission permission can be performed on a special demand time slot, using one of the reservation procedures discussed in chapter 9. It is possible to assign one of the earth stations as a coordinator whose job it is to allocate the spectrum slices in response to the other earth station demand. Distributed control of the transmission demands provides better operation efficiency than centralized control although some of these distributed procedures are susceptible enough to cause transmission collisions. Sharing the satellite resources according to the terminal demands (but not according to fixed rules) is referred to as demand-assigned-multiple-access (DAMA). DAMA therefore offers considerable advantages by providing better utilization of both the channel assignment and the earth station interconnection facilities.

5.11 WIDE AREA SATELLITE COMMUNICATION

Satellite repeaters are best known for the long distance communication facilities they provide, nevertheless, they can however, provide communication facilities within a given wide local area and still be cost-effective. This application of the satellite repeater is consequently new and only a few such systems are available, although the results indicate that these satellite conveyances will have a great impact on computer-to-computer communication.

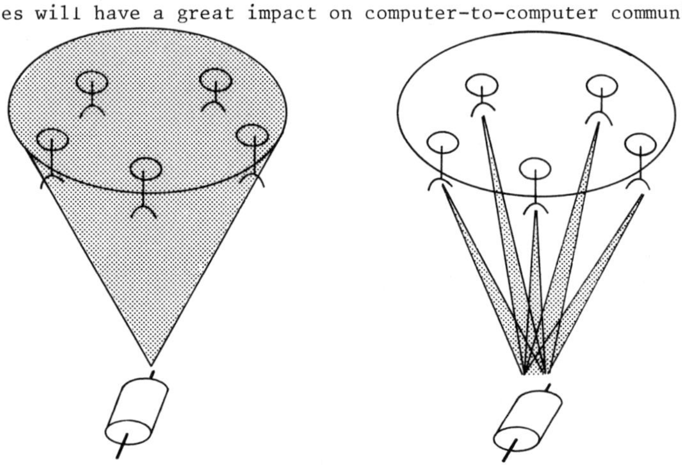

(a) UPLINK BEAMS (b) DOWNLINK BEAM

Fig. 5.36 The area covered by the satellite in a line-of-sight communication system

The area covered by a single geostationary satellite could be regarded as a single operating area in the same sense as that covered by a local area radio communication system. That is, the satellite can operate as a radio relay repeater for all the terminals communicating in a line-of-sight system, except that with satellites there are no hidden terminals and the area is extremely wide. Each of the earth stations may transmit its signal on a narrow beam directly to the satellite, as seen in Fig. 5.36(a), using the same frequency band. The satellite receives all these individual signals amplifies them and retransmits them on a different frequency as a single signal, as seen in Fig. 5.36(b). All the earth stations within the area will be able to identify those messages addressed to them by the tag at the heading of the message. In this way each terminal in this wide area will be able to communicate directly with any other terminal operating in the same area.

As all the signals in the uplink use the same frequency it is essential to use one of the multiple-access communication procedures, both to prevent collisions between the different transmissions and to exploit the available channel efficiently. In the downlink, only the satellite transmits with full control of the channel. However, if a collision occurred in the uplink the result would be repeated in the downlink. These characteristics of the satellite as a repeater allows for the dynamic allocation of the satellite spectrum by the suitable selection of the multiple access procedures. More so, the dynamic assignment of the satellite channel can be performed by the earth stations without the need for any processing ability in the satellite itself.

There is no need to allocate a specially designated satellite for wide area communication coverage since it only requires a relatively narrow bandwidth. Therefore it is possible to delegate for this application only part of the satellite spectrum, while the remainder can be left for conventional long distance communication. The available assignment bandwidth is the limiting factor in defining the transmitting data rate of this system. In one of the operating systems, a 38 K BPS is made available for multiple users, while today wider channels are being introduced.

In the area coverage communication system using multiple-access procedures, the problem of propagation delay can be quite acute. Each message received could be 270 miliseconds delayed in respect to the time it was originally transmitted. This means that multiple access procedures based on carrier sense functions (CSMA) will not be able automatically to prevent collision. In other words, a terminal may sense the radio spectrum of the channel and find that no-one is using it, while in fact a transmission has been initiated by another terminal but its signal has not yet fully propagated. Furthermore, the time allocated to each time slot in a TDMA system must take into account not only the maximum size of the message but also the maximum propagation delay.

The propagation delay also has an impact on the DCE construction itself, as compared with the DCE's used for direct terrestrial communication. Because of the long delays involved which consequently increase the slot size, it is necessary to increase the DCE's buffer size. More so, the communication procedures involving error and flow control must be modified too. With direct terrestrial communication each message is handled and checked separately and then an acknowledgement is returned. With satellite communication such a procedure would cause delays and thereby reduce the amount of data throughput. Instead, all the messages are sent one after the other with no gaps between them, using a go-back ARQ error correction procedure.

This type of network is ideal for high-speed traffic terminal-to-terminal or computer-to-computer communication using packet switching technology. Any computers or terminals in the network can reach directly any of the other computer or terminal and transfer information to them. The addressing system provides direct point-to-point contact during the time slot. That is, the addressing system enables selective dispatching of the messages through the common channel, as seen in Fig. 5.37. It also permits conference dispatching, where one terminal sends messages to a number of other terminals.

The schematic sketch shown in Fig. 5.37 gives an example of a TDMA operation. The numbers in the time slots refer to the address in the packet headings. The number on the left represents the sender's address and the number on the right the destination address. While the uplinks operate in a TDMA mode, the down link operates in a TDM mode with the satellite acting

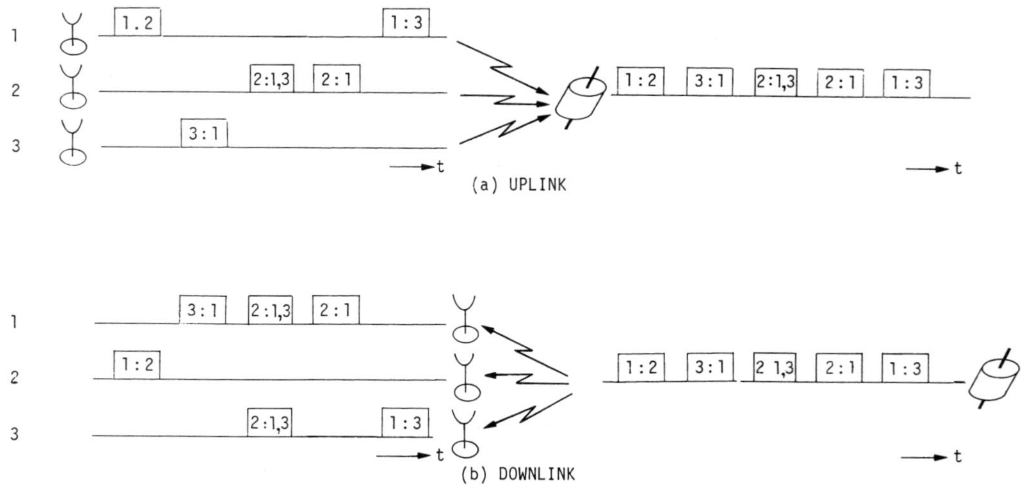

Fig. 5.37 Time slot arrangement in a satellite TDMA scheme

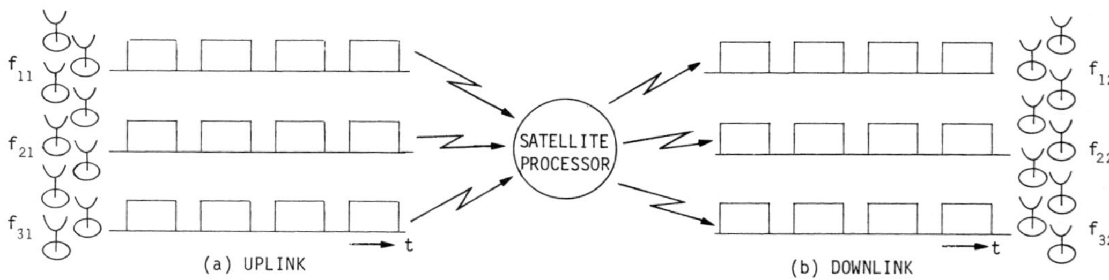

Fig. 5.38 Satellite operating as a store and forward switching communication network node

as a multiplexor.

The introduction of computer power into the satellite changes the function of the satellite from an active repeater to an active communication node. As a repeater, the satellite is part of the conveyance media and is transparent to all the communication functions, but as a communication node it can perform many communication functions and operate as a network node with store and forward facilities, That is, the introduction of a processor into the satellite enables it to perform switching tasks of accepting messages and redirecting them on alternative routes. In this application, the satellite can operate with many different paths both in the up and down links, as seen in Fig. 5.38. Both the up and down links can each serve a number of earth stations operating at the same frequency. Each up link and each down link is assigned a different frequency and therefore it is possible to regard the satellite as a centre of a star network with each link serving many terminals. The earth stations are usually part of a wider network and so the satellite operates as a communication network node connecting the various network sections. Such a system is in operation where packet switching networks, in various geographical areas, could operate together.

In the satellite communication node each of the up links operates with TDMA techniques coupled with DAMA procedures; this enables many parallel transmissions, one on each link, to be received in the satellite. The down link operates with TDM techniques, with the satellite acting as a node packet switch centre coordinating the traffic so that all the channels have a smooth and even flow. The satellite directs the down link traffic in accordance with the destination address, in the same way that it is performed in a terrestrial switching centre. To achieve this, buffers must be included in the satellite to hold the traffic queue on each down link channel. Such satellite communication schemes have been referred to as satellite switched-time division multiple access SS/TDMA.

The wide area coverage of SS/TDMA operations could be arranged to provide services for mobile communication. This would involve lower frequency bands in the UHF range, to allow smaller antennas to be used on the vehicles. This service will allow many users to communicate applying various time-division multiple-access procedures. Such a service is already being employed for marine communication, although much research is taking place to expand it for other applications.

There is a third class of multiple-access techniques besides FDMA and TDMA which further extends satellite repeater applications. This class, which is referred to as code-division-multiple-access (CDMA) uses spread spectrum techniques (to be discussed in chapter 10). CDMA seems to be ideal for satellite area communication since it does not require any satellite switching functions. The terminals need not necessary be synchronized and can transmit whenever they have a message to dispatch even if another terminal is transmitting. This shows the advantage of CDMA, permitting as it does random access to the common channel with no need for any DAMA procedures, even though this is at the cost of complex receiving circuitry. Each terminal may transmit its signal over a relatively wide frequency band and is repeated, in the downlink on another wide frequency band. In this scheme, due to the intrinsic characteristics of spread spectrum, there is little danger that a collision between two signals will disturb each other's operations despite the fact that they are both using the same frequency channel. All the terminals will receive the transmitted signal repeated by satellite; nevertheless, this signal will look to them as pure additive white Gaussian noise (AWGN), unless they know (or have succeeded in decoding) the pseudo random sequence used in the transmitters to achieve spread spectrum. At this stage, there are very few systems that employ these techniques owing to the high installment cost, so far the main interest of this technique lies in military spheres. It is, however, believed that there will eventually be many commercial systems using these techniques because of the privacy it provides and the lack of available satellite frequency channels.

The satellite area coverage could be extended by associating two or more satellites to operate together (as shown in Fig. 5.39) each covering a different area. In this application, the satellite can provide on-board processing, whereby it becomes an active part of the system. That is, it performs store and forward functions, transferring messages to the second satellite, but only those which are addressed to the terminals in the second area. In this case it is necessary to add into the satellite both processor and memory capabilities as well as demodulation circuits which bring the signal back to its baseband. It should be added that this technique could also be used for long distance communication with no need for processing ability. By special design, certain uplink signals are reflected to the second satellite instead of being reflected back on the down links.

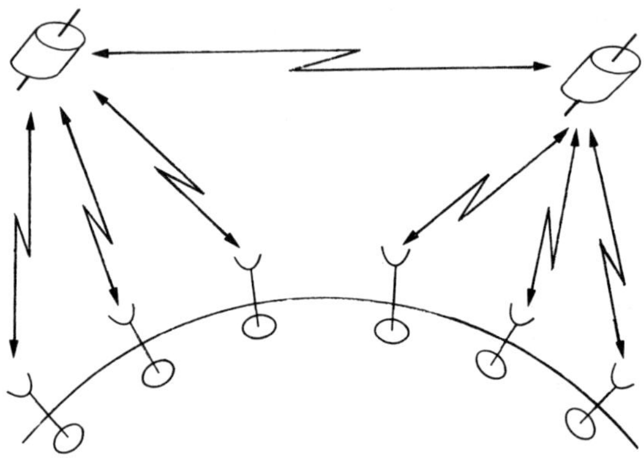

Fig. 5.39 Extension of area coverage by the association of two satellites

5.12 OPTICAL COMMUNICATION CONVEYANCE

Optical communication is a type of physical conveyance where the signal is in the form of a light wave which is transmitted through a cylindrical glass fibre core. It is distinct from all the other forms of physical conveyances by the fact that no electrical current is transmitted through the fibre and so the transmitter and receiver are electrically isolated. It is therefore more comparable with radio microwave communication conveyance, with its free-space propagation, although its signal is guided through a fibre from its radiated point to its detected point.

The fibre optical conveyance can be used to transfer both analog or digital data. It is ideal for digital data by simply switching the light source on and off according to the data 1 and 0 state respectively. The pulse shape remains in its basic rectangular form since the delay along the fibre is minute and is in the range of nanoseconds.

Although optical fibre communication, which is the most recent developed conveyance has not yet had its full impact, it has, in a very short period, progressed from a laboratory model to commercial reality. It is intended to replace the coaxial cable in its function as a carrier signal conveyance. The drive to develop fibre optical links has come from the military following the realization that these links are more secure and difficult to tap. It must be appreciated that fibre optical links have many further advantages which make them most attractive for commercial application. They are immune to electromagnetic and radio frequency interferences (EMI and RFI), free from noise and pick up, all making their error rate very small. Furthermore, they are smaller in diameter and weight as compared with coaxial cables. In the last few years fibre optics have been pushed forward with the introduction of better fibres and associated electronic components, which enable the transmission of data to be conducted with extremely low loss of about 1 dB/km.

The application of fibre optical communications is gradually increasing, but their cost is still relatively high. For many applications they are starting to prove cost effective and this is a promising factor. Nevertheless it must be noted that the fact that their main drawback is that they are fragile and can be easily damaged, and once a fibre is broken it is most difficult, although not impossible, to mend.

The fibre optical signal is in the form of light pulses which make it an ideal conveyance for digital communication. That is, it is the only conveyance where the signal can remain in its original digital pulse form and does not need to be modulated into an analog-coded-digital form. Still, data voltage pulses must be converted into optical power pulses. This is achieved by driving an electrical current pulse through a light emitted diode (LED), as seen in Fig. 5.40, which radiates light pulses into the fibre. The light signal then propagates through the fibre and is picked up at the other end by a photodiode.

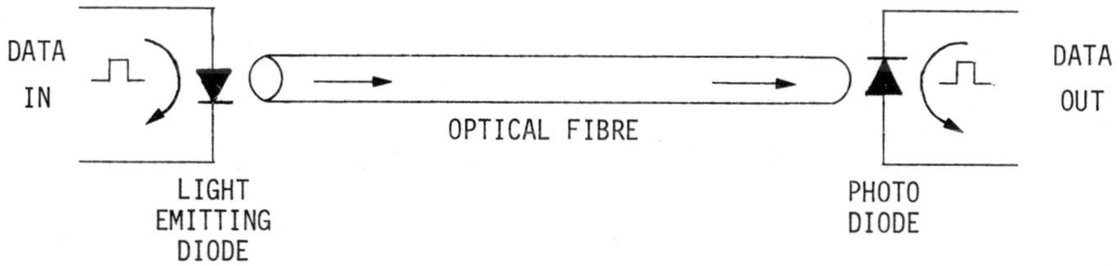

Fig. 5.40 The propagation of a light signal through an optical fibre

The light emitting diodes (LED) produce light pulses with an intensity which is a direct relation to the current magnitude that flows through them, thus causing the data link signal to be incoherent. The light beam radiated from the LED's is relatively wide and consequently the modulated rate is limited to only a few megabits per second. Because of the large angle of the beam emission, the coupling efficiency of the LED's is poor. Higher modulation rates and better performances can be achieved with injection laser diodes (ILD) which radiate narrow light bands which are highly directed and confined. The ILD is a threshold device and only emits a light signal if the electrical current through them is above a limiting level value. The threshold level is usually about 100 mA and they reach maximum emission of light at about 120 mA. This means that the ILD signals are coherent and that they are ideal for ON/OFF data transmission with speeds that can reach up to a gegabit-per-second. The fact that the production of the light signals in ILD has to be stimulated instead of instantaneous as with LED's, and the fact that their matching efficiency is better, makes them more attractive for data transmission. Nevertheless, for most applications LED's are preferable, because they are cheaper and have a longer life-time than ILD. Furthermore, high switching speeds of the ILD have the problem of cooling, due to the high power emitted from very small laser chips. In both LED and ILD the radiated light pulses are formed as a result of switching the current in the diodes ON and OFF as a function of the '1' and '0' data states.

The optical fibre is terminated with a detector in the form of either photo emitting diodes (PED) or avalance photo diodes (APD), both of which

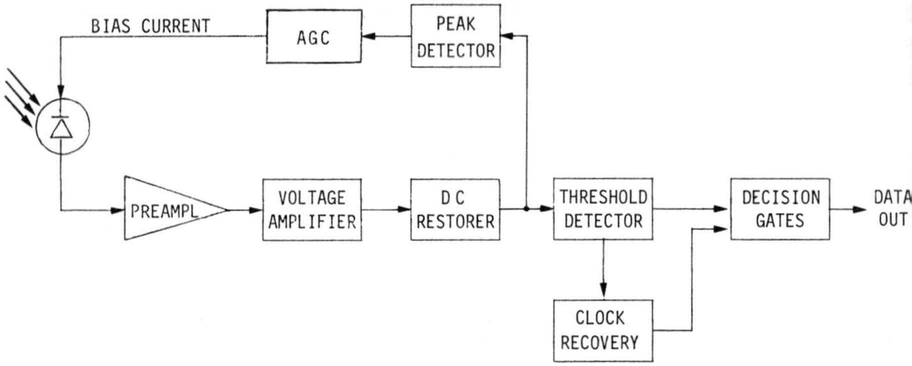

Fig. 5.41 Schematic block diagram of an optical communication detector

are sensitive to light. As a light signal falls on the surface of the photodiode, the current magnitude flowing through them changes. The diodes then convert the light input signals to variations of electrical currents. These current variations must be amplified and reshaped before they can be regenerated back to voltage data pulses. To achieve this, the photodiodes have to be permanently biased by a current that flows through them. The current variations are relatively small and so the photo-diode detector operates with a specially designed high-impedance low-noise preamplifier which is sensitive to small current changes. The regenerated waveforms are then further amplified, as seen in Fig. 5.41, and then reshaped by passing the waveform through a dc restorer and a threshold detector.

An important aspect of the optical communication is the receiver circuits which affect the format of the transmitted pulses. The regenerated waveform in the receiver is very sensitive to the signal variations and not so much to the dc level of the signal. This is because the receiver is electrically isolated from the transmitter; in consequence the receiver must be regarded as ac coupled to the fibre link. The decision whether a '1' or a '0' has been transmitted is defined by the positive or negative peaks at the clock sampling timing. That is, the state is decided by a positive or negative threshold level relative to the centre voltage level at the output of the amplifier. The fact that the preamplifier is sensitive to ac changes rather than the dc levels means that it could lose its decision level when low data rates are transmitted. This means in effect that when a series of only 1's or only 0's is transmitted, the ac coupled preamplifier will remain in its centre level and not in the positive or negative levels according to the data stream currently being transmitted. Consequently, the decision gates cannot remember the state that is being transmitted, whether it is a stream of all 1's or all 0's. Furthermore, the first 1 digit after a long sequence of 0's and the first 0 digit after a long sequence of 1's, will produce switch-on delays. Both these related effects degrade the communication performance and enforce the requirements of special pulse formating, as described in chapter 2. The most popular pulse code format used is the Manchester code although this requires a data rate which is only half the modulated rate. A much better efficiency code is that produced by the Miller pulse code format.

The optical fibre consists of a pure glass fibre core, as seen in Fig. 5.42, which is surrounded by a concentrated layer. This layer is referred to as cladding and has a different optical reflection index than the fibre so that it can reflect the light signal within the fibre core. The cladding covering therefore performs the function of a buffering coating and is usually also made of glass, although with different properties from that of the fibre core. It is fused directly onto the fibre core to produce a smooth surface. The cladding is covered by a jacket of protective material usually made of selica to strengthen the fibre. The outer jacket is made of plastic and the whole fibre cable then is not more than 2 to 4 mm.

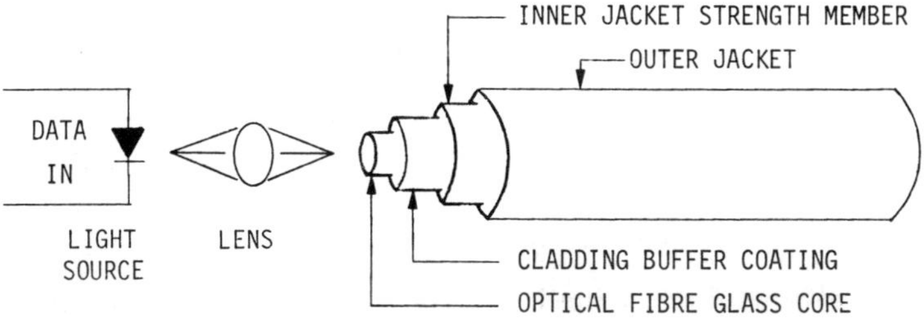

Fig. 5.42 Sketch of an optical fibre and its input coupling

A light generated at one end of the fibre will propagate along the fibre by reflection principles with only minimum loss. The angle of the light signal as it enters the fibre is important to the propagation success. Lenses are used to direct the radiated light source into the fibre to improve the input angle.

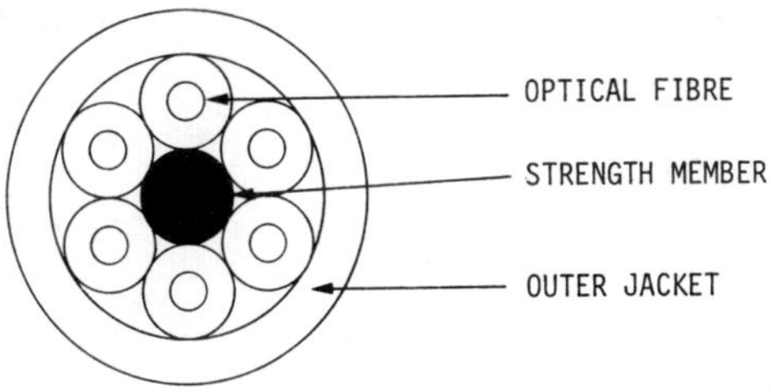

Fig. 5.43 Sketch of a multi-fibre optical cable

A number of fibres are sometimes coupled together in a single cable (as seen in Fig. 5.43) so as to increase the overall data rate. These fibres usually surround some strengthening material and then the whole cable is guarded by an outer jacket which consists of corregated steel.

The optical fibre can transmit data in either direction although the electronic circuitry on both its ends restrict the transmission to one direction only, i.e., it can be used in a simplex transmission mode. Nevertheless, by employing a special dielectric beam splitter (as seen in Fig. 5.44) the light signal can be divided into reflected and transmitted beams. Such an arrangement will then allow the transmission operation to be in a half duplex mode. Best results, however, are achieved by

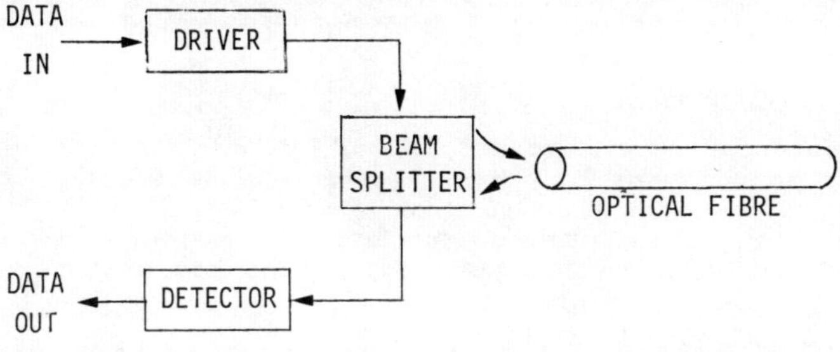

Fig. 5.44 Half duplex operation of an optical fibre using a beam splitter

twin optical fibre cable which enable the transmission operation to be in a full duplex mode. The data, in the latter, can propagate in both directions at the same time without disturbing one another.

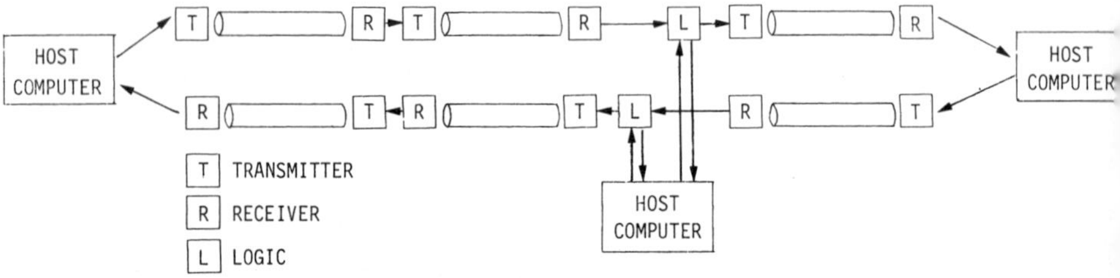

Fig. 5.45 Full duplex operation with a double optical fibre in a chain communication configuration

The optical fibre link may be used in any of the network configurations discussed in the previous chapter, i.e., star, bus, ring and loop. Their communication applications range from short data links between buildings to long distance data trunk links connecting various switching centres. Fibre optical systems are commercially available with high data rates which vary from 1 MBPS to 200 MBPS. On long distance communications, repeaters must be inserted at spaces 2-6 km apart depending on the type of fibre and the data rate. These repeaters are simply receivers and transmitters connected back to back, as seen in Fig. 5.45. When these fibres are arranged in a bus or chain configuration, the host computers are connected to the link by additive logic within the repeater to form a T-junction.

5.13 REFERENCES

(a) DATA COMMUNICATION PRINCIPLES

1. Martin, J. *Telecommunication and the Computer*, Prentice Hall Inc., 1969. pp.239.364.
2. Pierce, J.R. *Some Practical Aspects of Digital Transmission*, IEEE Spectrum, November 1968, pp.63-70.
3. Franklin, R.H. and Rhodes, J., *Data Transmission*, Journal IEE, February 1962, pp.82-85.
4. James, R.T. *Data Transmission – The Art of Moving Information*, IEEE Spectrum, January 1965, pp.65-83.
5. Hansher, D.H. (Editor) *Communication System Engineering Handbook*, McGraw Hill Book Company, 1967; Chapter 10 by Cookson, M.E. and Thompson, E.M; Chapter 11 by Wolf, J.H.; Chapter 15 by Becken, E.D.; Chapter 16 by Guenther, R., Mehlman, S.J. and Sheffield, B.; Chapter 12 by Elder, C.H.

(b) WIRE COMMUNICATION

6. Williams, M.B. *The Characteristics of Telephone Circuits in Relation to Data Transmission*, The P.O. Electrical Engineers' Journal, vol.59, pt.3, October 1966, pp.151-162.
7. Muth, R.J., Shipman, J.W. and Toffler, J.E. *Factors Affecting Choice of Modulation for Data Transmission*, Telecommunication, March 1969, pp.17-22.

(c) CABLE COMMUNICATION

8. Yeh, L.P. *Present and Future Long-Haul Communication Systems*, IEEE Mohawk Valley Communication Symposium (NATCOM), October 1965, pp.1-

9. Trompeter, E. *Noise in Cable Systems*, Tromputer Electronics Inc., July 1969.

10. Dineson, M.A. *Broadband Coaxial Area Network*, Computer Design, part 1, June 1980, pp.12-22; part 2, July 1980, pp.14-22.

11. Metcalf, R.M. and Boggs, D.R. *Ethernet: Distributed Packet Switching for Local Computer Networks*, Communication of the ACM, vol.19, no.7, July, 1976, pp.395-402.

12. Agrawala, A.K., Bryant, R.M. and Agre, J. *Analysis of an Ethernet-Like Protocol*, Proc. Computer Network Symposium, 1977, pp.104-111.

(d) RADIO COMMUNICATION

13. Gunter, F.A. *Tropospheric Scatter Communication – Past, Present and Future*, IEEE Spectrum, September 1966, pp.79-100.

14. Deutschie, C.A. *Survey of Digital Communication*, Hughes Aircraft Company, Communication Division, Task B 2061-0210, U.S. Government Report AD 277176, 1961.

15. Monsen, P. *Fading Channel Communication*, IEEE Communication Society Magazine, vol.18, no.1, January 1980, pp.16-25.

(e) PROTOCOLS FOR WIRE AND RADIO COMMUNICATION

16. McQuillan, J.M. and Cerf, V.G. *A Practical View of Computer Communication Protocols*, IEE Computer Society, Tutorial Cat. No. EHO 137-0, 1978, pp.7-10.

17. Clark, D.D., Pogran, K.T. and Reed, D.P. *An Introduction to Local Area Network*, Proc. IEEE, vol.66, no.11, November 1978, pp.1497-1517.

18. Abramson, N. *The ALOHA System – Another Alternative for Computer Communication*, 1970 Joint Computer Conference, AFIPS, vol.37, pp.281-285.

19. Binder, R., Abramson, N., Kuo, F., Okinaka, A. and Wax, D. *ALOHA Packet Broadcasting – A Retrospect*, 1975 National Computer Conference, AFIPS, vol.44, pp.203-215.

20. Gitman, I. *On the Capacity of Slotted ALOHA Networks and Some Design Problems*, IEEE Trans, vol. COM-23, no.3, March 1975, pp.305-317.

21. Kahn, R.E. *The Organization of Computer Resources into a Packet Radio Network*, National Computer Conference, AFIPS, vol.44, pp.177-186 also in IEEE Trans. vol. COM-25, no.1, January 1977, pp.169-177.

22. Frank, H., Gitman, I. and Van Slyke, R., *Packet Radio System – Network Considerations*, 1975 National Computer Conference, AFIPS, vol.44, pp.217-231.

23. Kahn, R.E., Gronemeyer, S.A., Burchfield, J. and Kunzelman, R.C., *Advances in Packet Radio Technology*, Proc. IEEE, vol.66, no.11, November 1978, pp.1468-1496.

(f) SATELLITE COMMUNICATION

24. Abramson, N. and Cacciamani, E.R. *Satellite: Not Just a Big Cable in the Sky*, IEEE Spectrum, September 1975, pp.36-40.

25. Moralee, D. *Satellites – Their Impact on Word Communication*, IEE, Electronic and Power, June 1978, pp.429-435.

26. Hindin, H.T. *Transmitting Data by Satellite Calls for Special Signal Handling*, Electronics, March 29, 1979, pp.91-98.

27. Van Trees, H.L., Hoversten, E.V. and McGarty, T.P. *Communication Satellites: Looking to the 1980's*, IEEE Spectrum, December 1977, pp.43-51,

28. Lam, S.S. *Satellite Multiaccess Schemes for Data Traffic*, International Communication Conference, 1977, vol.3, pp.37-1-19 to 37-1-24.

29. Evans, R.E. *ALOHA/TDM Systems with Multiple Downlink Capacity*, IEEE Trans. vol. COM-27, no.3, March 1979, pp.537-541.

30. Puente, J.G., Schmidt, W.G. and Werth, A.M. *Multiple Access Techniques for Commercial Satellites*, Proc. IEEE, vol.59, no.2, February 1971, pp.218-229.
31. Pritchard, W.L. *Satellite Communication – An overview of the Problems and Programs*, Proc. IEEE, vol.65, no.3, March 1977, pp.294-307.
32. Ito, Y., Urano, V., Muratani, T. and Yamaguchi, M. *Analysis of Switch Matrix for a SS/TDMA System*, Proc. IEEE, vol.65, no.3, March 1977, pp.411-419.
33. Hoversten, E.V. and Van Trees, H.L. *International Broadcast Packet Satellite Services*, 4th Int. Conf. on Digital Satellite Communication, Canada, 1978, pp.1-7.
34. Jacobs, I.M., Binder, R. and Hoversten, E.V. *General Purpose Packet Satellite Networks*, Proc. IEEE, vol.66, no.11, November 1978.
35. Retnadhas, C. *Satellite Multiple Access Protocols*, IEEE Communication Society Magazine, vol.18, no.5, September 1980, pp.16-20.
36. Long, W.G. *Technical Considerations Affecting Satellite Communication Spectrum Management*, IEEE Trans. vol.COM-27, no.10, October 1979, pp.1538-1544.

(g) FIBRE OPTICAL COMMUNICATION

37. Personick, S.D. *Optical Fibres, A New Transmission Medium*, IEEE, Communication Society Magazine, vol.13, no.1, January 1975, pp.20-24.
38. Jacobs, I. and Miller, S.E. *Optical Transmission of Voice and Data*, IEEE Spectrum, February 1977, pp.33-41.
39. Personick, S.D. *Fibre Optic Communication – A Technology Communing of Age*, IEEE Communication Society Magazine, vol.16, no.2, March 1978, pp.12-20.
40. Midwinter, J.E. *Optical Communication, Today and Tomorrow*, IEE Electronic and Power, March 1978, pp.442-447.
41. Li, T. *Optical Fibre Communication – The State of the Art*, IEEE Trans. vol.COM-26, no.7, July 1978, pp.946-955.
42. Rawson, E.G. and Metcalf, R.M. *Fibrenet: Multimode Optical Fibres for Local Computer Networks*, IEEE Trans, vol.COM-26, no.7, July 1978, pp.983-996.
43. Gambling, W.A. *Recent Advances in Optical Fibre Communication*, The Radio and Electronic Engineer, vol.49, no.4, April 1979, pp.182-186.
44. Leiceaga, P.M. *Survey of Current Technology Related to Fibre Optics*, Master Thesis, Naval Post Graduate School, 1977, U.S. Government Report No. ADA 052653.

CHAPTER 6

Digital Modulation

6.1 INTRODUCTION

After the introduction of the various forms of communication networks and conveyances, it is necessary to expand the subject of the preparation of the data to be transmitted over the communication channels. This subject, which was discussed in chapter 3, showed the need for employing modems at both ends of the channel. This chapter will analyse the various transmission modulation techniques that could be employed in the modem, taking into consideration the specific features of the conveyances.

Digital information, in the form of rectangular shaped pulses, cannot be transmitted over long distances without being distorted. After a certain distance (for example, after a few hundred metres at a rate of 1 K bit/second) the pulse shapes deteriorate to a point where they cannot be reconstructed back to meaningful information. The modem's primary function is to convert digital data into a corresponding analog signal which can be transmitted over long distance communication channels and then be converted back at the other end to the original digital data. The communication channel may be a variety of communication media such as telephone lines, wide band cables, radio or satellite. The channel itself is usually limited by its frequency band and the signal which passes through it is generally distorted due to delays and attenuation along the channel, besides noise, fading and other disturbances within the channel.

The conversion process of digital data to an analog signal is performed by means of modulation where a pure wave of constant amplitude and frequency is changed according to the data value, either by shifting its amplitude, frequency or phase. This fixed wave signal is referred to as the carrier signal $S(t) = A_c \cos \omega_c t$, where A_c refers to the constant amplitude and $f_c = \omega_c/2\pi$ refers to the carrier's fixed frequency. The modulated signal, if all three parameters shift the carrier, may take the following form

$$S(t) = (A_c + A_m) \cos (\omega_c + \omega_m + \theta_m)t ,$$

although generally only one parameter is varied in each modulation technique. The transmitted signal, although being an analog form, still corresponds to to digital information and could hence be regarded as analog coded digital (ACD).

The main task in designing digital communication systems is finding the appropriate modulation technique which can operate at the highest possible speed. However, this alone is not enough. Although the data rate could be increased by numerous modulation methods the reconstruction may still be full of errors. What we are looking for is means for transmitting maximum volume of data, within a given limited bandwidth. which can be later received reliably and decoded with minimum or no errors and at a reasonable price. The higher the speed the higher the cost of the system. Although this increase is not linear, the system to be used must be selected according to the cost one is ready to pay to transmit at a given speed, and which does not need to be the best available modulation technique. Furthermore, one must remember that the modulation technique to be selected is also a function of the communication channel employed.

In the last few years there has been much research into modulation techniques for digital communication. These research projects have not introduced completely new means of modulation but have developed better and more efficient means of exploiting the known modulation techniques of amplitude, frequency and phase. The drive for the newer modulation schemes has come from the requirement for better means of exploiting satellite communication channels, although these new developments could easily be applied to other communication media, such as wire and radio. The aim of these new modulation schemes still remains to increase the data rate by reliable means without increasing the available frequency bandwidth. As already known from the Shannon criterion, the modulation rate for a given channel is limited with no possibility of increasing it. By efficiency of the communication channel we improve our bit-rate to be as close as possible to the Shannon limitation. By efficiency of the communication system we increase our data-rate well above the modulation rate.

In this chapter the various modulation schemes are described, starting with the well known techniques of simple on-off amplitude modulation and then expanding on the most advanced modulation techniques of tamed frequency modulation. The first part of the chapter is concerned mainly with modulation schemes for wire and radio channels limited to a speech baseband of 300-3400 Hz. For these channels, the state-of-the-art has increased the speed of transmission to 9600 bits/second, although there are indications that this barrier could be overcome in the near future. In the second half of the chapter, modulation schemes for wider channels are discussed.

The development of these modulation schemes has been rather rapid and this has caused may terminology ambiguities. An effort has been made to collect together as many related names as possible so as to give the reader a key to compare similar systems with different names. The lack of international standards in this respect leaves much room for mistakes. In general all analog modulation is referred to as 'modulation' while digital modulation is referred to as 'keying' or 'shift-keying'. Nevertheless, here too there are exceptions with QAM as one example. There are other forms of terminology ambiguity all of which, it is hoped, have been properly considered.

6.2 AMPLITUDE MODULATION (AM)

In amplitude modulation, the amplitude of the carrier signal is varied according to the amount of information transmitted. The resultant wave is a direct relation to the modulation wave, producing two identical symmetrical sidebands around the carrier. As the same information is contained in the two sidebands, in effect only one sideband is required to transmit a message; nevertheless, in most systems both sidebands are transmitted, as will be explained later.

AM presents the simplest form of modulation for data transmission since all that is required is to switch the carrier on or off according to the data. If the data is in the form of 1 or 0, then the presence of a transmitted signal represents the 1 bit and its absence the 0 bit, as shown in Fig. 6.1. This form of amplitude modulation is also referred to as ON/OFF keying, or in short OOK.

As seen from the description of the system, its technical implementation is extremely simple. All that is required is an oscillator having its output switched on or off, as shown in Fig. 6.2(a). The frequency produced by this oscillator f_c is known as the carrier frequency. Although the oscillator produces either no output or the frequency f_c, the transmitted bandwidth is double that of the modulation bandwidth. If the duration of each bit is T seconds long then the data transmission rate is $R = 1/T$ bits/second. This data rate of $1/T$ occupies a bandwidth of $F_m = R$ Hz and is referred to as the data base band. Switching the carrier frequency f_c on and off causes two sidebands, at both sides of the carrier, each equal to R Hz: thus the transmitted bandwidth is $B = 2R$ Hz, as shown in Fig. 6.2(b).

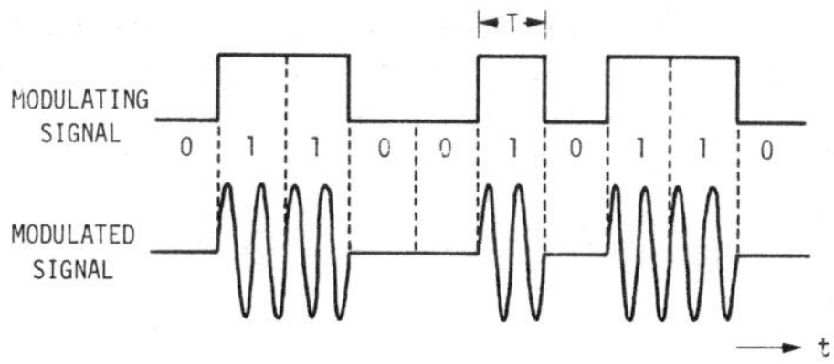

Fig 6.1 Amplitude modulation of a rectangular pulse

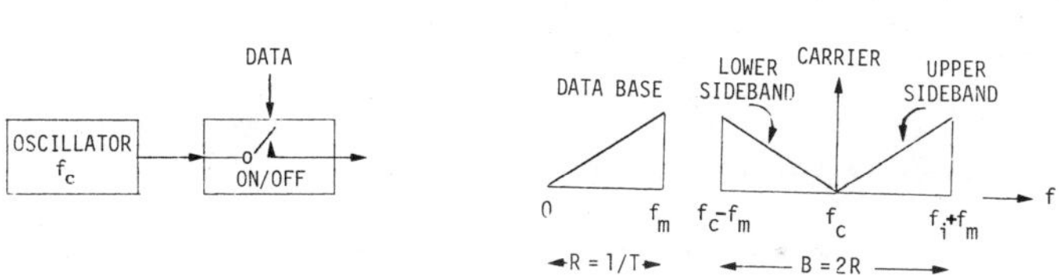

Fig. 6.2 Basic principles of ON/OFF keying

Amplitude modulation is extremely vulnerable to common type noise, attenuation distortion, delay distortion and echoes, all of which can change the shape of the signal in the transmission medium. In addition, cross-talk from adjacent channels might be superimposed and picked up in the receiver. All these drawbacks of amplitude modulation, coupled with the ineffective use of the transmission frequency band (i.e., doubling that of the data base), makes the application of ON/OFF keying rather limited. It is used only in very slow speed systems of up to 600 B/S. There are, however, other forms of amplitude modulation that are used for data transmission, and these will be discussed in later sections.

The data rate may be increased by transmitting four levels of amplitude instead of the two levels of ON/OFF information as shown in Fig. 6.3. In this case each of the four levels represents two digits simultaneously, forming what is known as a 'debit'. Theoretically the speed of transmission should be doubled, but in practice this is not the case, since the gain in speed is reduced by the decrease in the threshold margin available for detecting the pulse state. In consequence, this type of modulation is more susceptible to

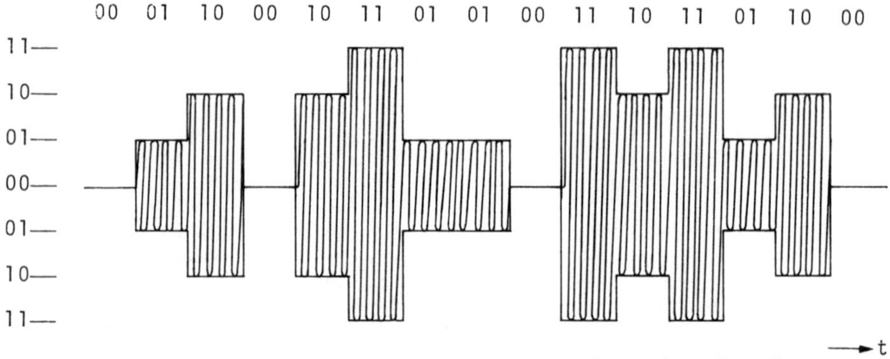

Fig. 6.3 Amplitude modulation of a four-level pulse

transmission errors, making its application somewhat limited. Nevertheless, the principle of this multilevel modulation has been applied to other types of modulation (as will be explained later).

In order to understand the operation of amplitude modulation, some theoretical background must be presented. The carrier oscillator produces a pure sine (or cosine) wave with the frequency f_c and amplitude A_c, i.e.

$$S(t) = A_c \cos 2\pi f_c t = A_c \cos \omega_c t .$$

In amplitude modulation, the amplitude of the carrier wave is changed according to the information rate $A_m(t)$ giving:

$$S(t) = [A_c + A_m(t)] \cos \omega_c t .$$

In the most general form, the data may also be regarded as a modulation source of a wave with the frequency f_m and the amplitude A_m, i.e.

$$A_m(t) = A_m \cos \omega_m t .$$

This may be better followed with the wave diagram given in Fig. 6.4. If the data transmitted is a continuous sequence of 1's and 0's, then the analog signal line represents the information rate of change. This is not accurate, since the actual wave in ON/OFF keying is square, although in amplitude modulation it could be as shown.

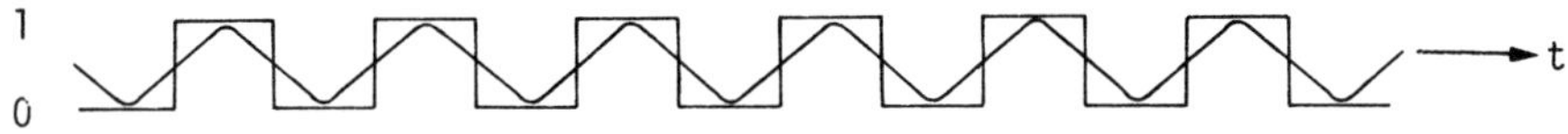

Fig. 6.4 The square wave of the data and its analog signal change

The equation of the modulating signal can be inserted into the modulation equation giving :

$$\begin{aligned} S(t) &= [A_c + A_m \cos \omega_m t] \cos \omega_c t \\ &= A_c \cos \omega_c t + A_m \cos \omega_m t \cos \omega_c t \\ &= A_c \cos \omega_c t + \tfrac{1}{2} A_m \cos(\omega_c + \omega_m)\, t + \tfrac{1}{2} A_m \cos(\omega_c - \omega_m) t \\ &= A_c \cos 2\pi f_c t + \tfrac{1}{2} A_m \cos 2\pi (f_c + f_m)\, t + \tfrac{1}{2} A_m \cos 2\pi (f_c - f_m) t \end{aligned}$$

This is the general form of an AM signal which contains upper and lower sidebands symmetrically distributed about the centre of the carrier frequency f_c. The ratio $m = A_m/A_c$ is referred to as the modulation index. Some references express the AM signal with respect to the modulation index, as shown below:

$$S(t) = [A_c + A_m \cos \omega_m t] \cos \omega_c t$$

$$= A_c [1 + A_m/A_c \cos \omega_m t] \cos \omega_c t$$

$$= A_c [1 + A_m \cos \omega_m t] \cos \omega_c t$$

From the above calculations it can be seen that the amplitude modulation signal contains three portions: the carrier at centre frequency f_c and the two sidebands at the frequency $f_c - f_m$ and $f_c + f_m$. No information components are transmitted in the centre frequency since amplitude is proportional only to the original carrier amplitude A_c, while each of the sidebands contains the same information, both being proportional to the information amplitude A_m. Furthermore, the transmitted power required at the centre frequency is much higher than that required for transmitting each of the sidebands. The power is proportional to the square of the amplitude; therefore the power at the centre frequency is proportional to $(A_c)^2$ and in each of the sidebands to $(A_m/2)^2$. In OOK, for example, where $A_m = A_c$, 50% of the power is required to transmit the centre frequency, which, as already stated, does not contain any information, while only 25% of the transmitter power is used for transmitting each of the sidebands which contain the actual information.

In view of the characteristics of amplitude modulation, various forms could be used to transmit the binary information as shown in Fig. 6.5 and described below.

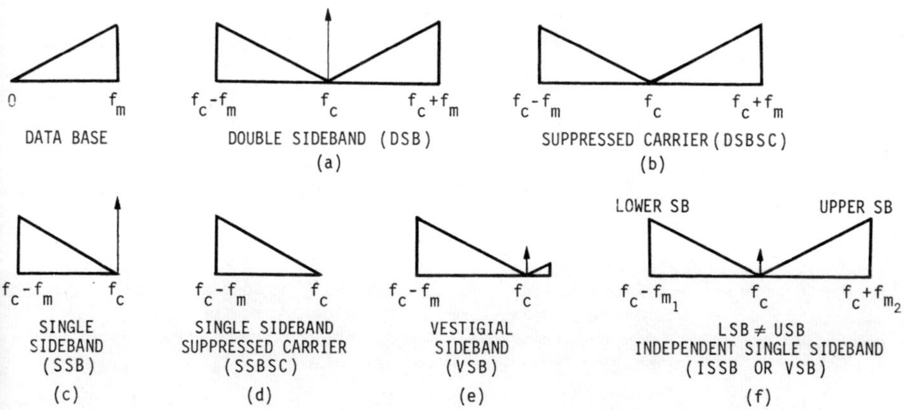

Fig. 6.5 Various amplitude modulation transmission techniques

(a) *Double sideband (DSB)*, which is used in most AM analog radio transmission systems and was also proposed in the early data transmission low-speed systems, but has since been replaced by FM methods. The four level AM data systems already described is a form of DSB. As pointed out before, the bandwidth required to transmit a data rate of R with a data base of f_m, ia $2f_m$. This technique has the further disadvantage that most of the power is wasted in the transmission of the carrier frequency which conveys no information.

(b) *Double sideband suppressed carrier (DSBSC)*, which is a form of double sideband with the carrier removed. The bandwidth required for transmission of this signal still remains at $2f_m$, but it is possible here to increase the power for transmitting the two sidebands and yet keep the transmitted power below that of DSB. Suppressing the carrier term of $(A_c \cos \omega_c t)$ is performed by a balanced modulator, as shown in Fig. 6.6. This means in effect that when no information signal is transmitted, the output should be zero; thus the OOK modulation could wrongly be regarded as a form of DSBSC.

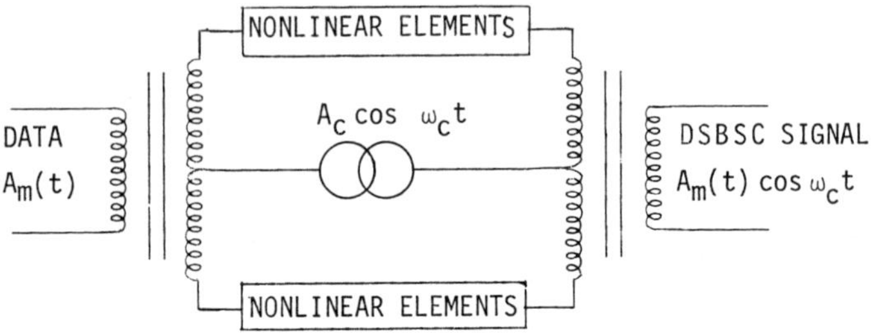

Fig. 6.6 Balanced Modulator

(c) *Single sideband (SSB)*, in which one of the sidebands is removed by suitably filtering one of them. This suppression of one sideband can be performed without destroying any of the information intelligence. Either sideband could be eased, depending on the application. This technique makes better use of the available frequency spectrum, but is less tolerable to delay distortion, amplitude variation and frequency or phase shifts. The power transmitted is divided between the carrier and the single sideband. This is an important factor in data transmission, since a large portion of the power is not useful for transmitting the actual information.

(d) *Single sideband suppressed carrier (SSBSC)*, where all the available transmission power is concentrated in the single sideband which contains the desired information. The carrier can be suppressed by the application of a balanced modulator and removing one of the sidebands with the aid of a filter. This technique, in fact, preserves the original data-base bandwidth (though shifted in frequency). As explained in the previous chapter, the spectrum of a pulse signal gradually rolls off to very low frequency including dc. This dc and low frequency cannot be suppressed with the filters. Furthermore, the removal of the carrier component makes the receiving operation rather difficult since there is no reference frequency. Although this system has better frequency utilization it is not economically practical for wire data transmission, as the circuitry required is very complex. It is used, however, for high speed wideband data transmission schemes where the advantages outweigh the disadvantages.

(e) *Vestigial sideband (VSB)*, which is a variation of SSB but also combines some DSB properties. This technique is popular for high speed modems for the transmission of two (or more) signals simultaneously. The reason for the attraction is that it permits economy both in the transmission power and bandwidth. In vestigial sideband, both the wanted sideband and a portion of the unwanted one are transmitted. The unwanted sideband is filtered out, which only partially suppresses the sideband in the vicinity of the carrier and also partially suppresses the carrier. This leaves one sideband with only a vestige of the carrier which eases the problem of frequency stability. The carrier is reduced to less than half its full DSB value and this releases more power to the sideband transmission. VSB techniques require a bandwidth approximately 1.3 times that of SSB techniques.

(f) *Independent single sideband (ISSB)*, which is based on two independent sideband amplitude modulators with suppressed carrier, as shown in Fig. 6.7. The data is fed in series at a bit rate of R, and then two consecutive bits X and Y are modulated simultaneously, thus reducing the overall modulation rate to $R/2$. In the modulator the lower sideband is eliminated, while in the other the upper sideband is eliminated. Each modulator is switched between two amplitude levels according to the data, where the two bits Y and Y are fed in parallel, one to each modulator. The two SSBs could then be combined and transmitted in a bandwidth of R Hz. In this system we have doubled the transmission rate in comparison with DSB, which is possible since there is no interference between the information transmitted in both sidebands.

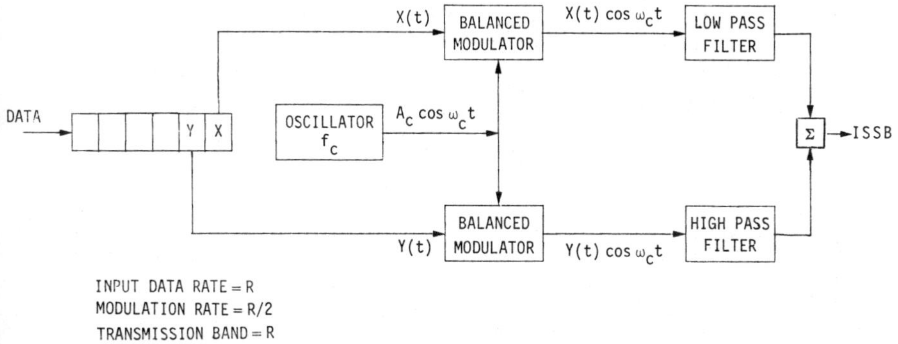

Fig. 6.7 Block diagram of independent sideband modulation

(g) *Independent vestigial sideband (IVSB)*, which is based on the principle of independent single sidebands and employs a similar circuit, as shown in Fig. 6.7. The main difference between the two circuits is that in ISSB a balanced modulator is used while with IVSB an ordinary simple modulator is used and each of the filters leaves a vestigial of the carrier, as shown in Fig. 6.8. This type of modulation is superior to DSB as it transmits double the data rate of DSB while making use of the same bandwidth. IVSB is also superior to ISSB, as the vestigial transmitted carrier assists as a reference for detecting the information from the received signal.

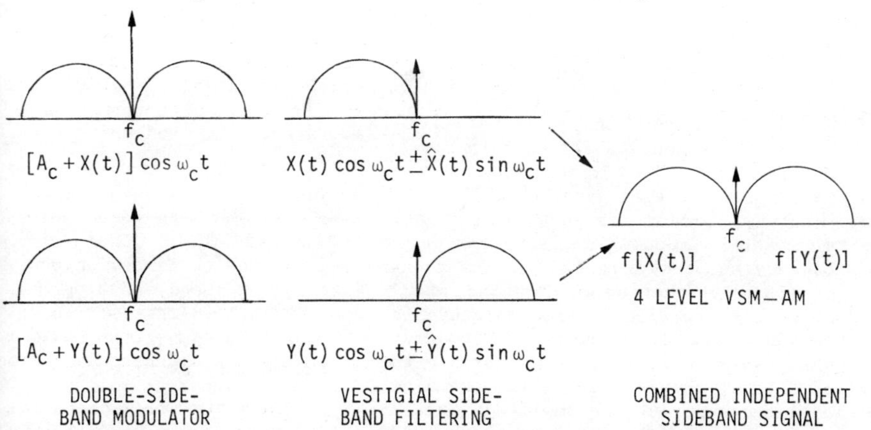

Fig. 6.8 Principle waveforms of 4-level VSB amplitude modulation

The VSB can also be used for the transmission of sixteen-level data by employing four modulators each two having a carrier with a 90° angle difference to the other two. Here the two double sidebands operate in a quadrature phase which can be separately and coherently demodulated. This is possible because the four sideband signals have no common modulated information components, as they are orthogonal or are separated by filters. The operation of the quadrature transmission will be explained later. It is important to add here that both VSB and SSB are very sensitive to phase variations where a change of 15° can cause a reduction of 20 dB in the signal-to-noise ratio.

Returning to OOK, we see that the carrier is switched on or off according to the data, which allows this modulation to be expressed as follows

$$S(t) = \cos \omega_c t \quad \text{for} \quad A_m = 1, \quad \text{and}$$

$$S(t) = 0 \qquad \text{for} \quad A_m = 0 .$$

This can be written as

$$S(t) = A_m(t) \cos \omega_c t$$

where $A_m(t)$ equals 1 or 0 respectively, switched over intervals of T seconds long, ($f_m = 1/T$). OOK modulation can be regarded as DSB with the modulation index of $m = 1$ since A_m must be equal to A_c. It can also be regarded as DSBSC, since when there is no information transmitted, i.e. transmitting the '0' bit, the output should be zero.

For fast speed amplitude modulation schemes, the 0 and 1 binary notation is replaced by +1 and −1 respectively; thus the transmitted signal can be expressed as $S(t) = \pm \cos \omega_c t$. This uses the same function as for OOK, i.e. $S(t) = A_m(t) \cos \omega_c t$, but here $A_m(t) = +1$ for the 0 bit and $A_m = -1$ for the 1 bit. This form of amplitude modulation is referred to as amplitude shift keying (ASK) and it will be described in detail in later sections. Although there are similarities of ASK and OOK they are in fact two comletely different modulation systems. In OOK no output is transmitted when the 0 bit is transmitted, while with ASK the carrier is always transmitted except that its polarity is reversed according to the data. This difference in characteristics makes the ASK superior, as it removes most of the disadvantages found in OOK.

6.3 FREQUENCY MODULATION (FM)

In frequency modulation the frequency of the carrier signal is changed according to the information transmitted. The ON/OFF data information shifts the transmitted frequency in one direction to indicate a '1' and in the other direction to indicate a '0'. The amount of frequency shift in both directions, relative to the carrier, is the same and thus we can represent the '1' by $f_c + \Delta f$ and the '0' by $f_c - \Delta f$, as shown in Fig, 6.9. Digital frequency modulation is better known as frequency shift keying (FSK).

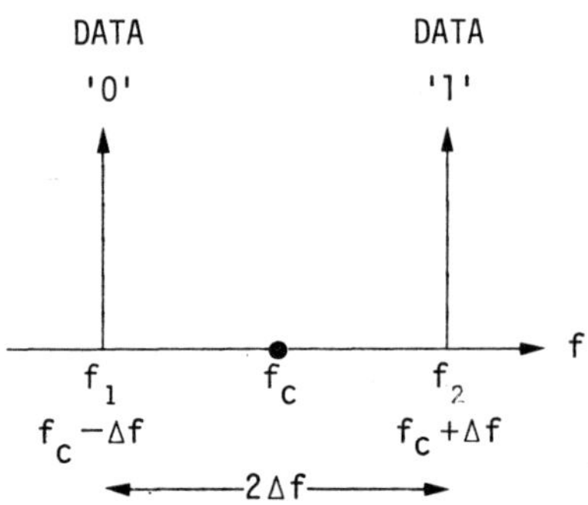

Fig. 6.9 Principle of frequency modulation

The main attraction of frequency modulation is that its transmitted amplitude is constant, and independent of the information contents. As such it is less vulnerable to amplitude level changes or to impulse noise and hence is superior to amplitude modulation, but not with respect to the data rate achieved over a given bandwidth. FSK was first introduced to replace amplitude modulation, although it generally requires a larger bandwidth

It is used today in modems operating at speeds up to 1200 – 1800 bits/second and requires approximately the same bandwidth as a double-sideband amplitude modulation.

Frequency modulation is a continuous analog process with no changes in its amplitude level, where in the absence of information the carrier is transmitted. This is one of the reasons why FM is less sensitive than the basic digital AM systems to noise, cross-talk and intermodulation. A further advantage for the continuous transmission of the carrier is that the receiver signal-to-noise ratio is constant.

Two basic types of code formats can be used for FSK, one based on the RZ code with three frequencies and the other on NRZ code using the two side side frequencies. In return-to-zero, shown in Fig. 6.10, the two discrete frequencies are used to transmit the binary states, while the carried f_c is used to transmit the intervals between the pulses. $f_c - f_m$ represents the '0' binary state and $f_c + f_m$ represents the '1' binary state.

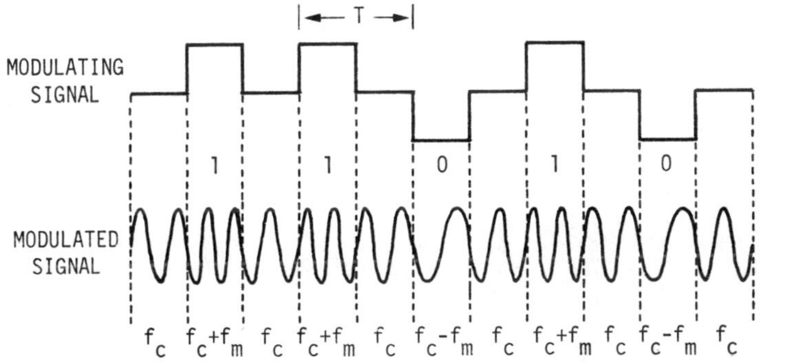

Fig. 6.10 Frequency shift keying of an RZ pulse sequence

In non-return-to-zero the frequency is shifted between two frequencies f_1 and f_2, as shown in Fig.6.11. With no information transmitted the upper frequency is usually transmitted continuously. As will be shown later, there is no theoretical difference between the three and the two frequency system since f_1 can be regarded as equal to $f_c - f_m$ and f_2 to be equal to $f_c + f_m$. When varying the frequency between '1' and '0' states, the phase must be continuous. This can be achieved by using a single oscillator and adding or deleting a capacitor to change the frequency of oscillation.

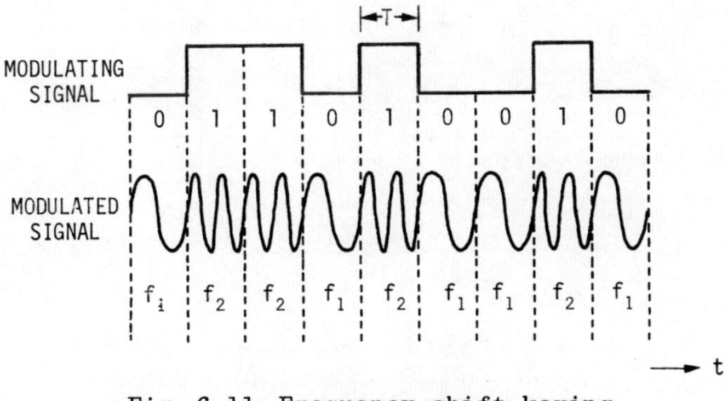

Fig. 6.11 Frequency shift keying of an NRZ sequence

The three-frequency FSK technique is used for telemetry applications and for some manually operated terminals. This type of modulation has been mainly adapted for speeds up to 150 bits/second due to the loss of time period between adjacent data pulses. The two-frequency FSK technique is much more widely used, despite the bandwidth restrictions. It is chosen for its simplicity and economy, factors which are sometimes more important than bandwidth efficiency.

For many control applications the three-level technique is very popular, because the third available state could be used to transmit additional information. The three level transmission also simplifies the synchronization off the receiver with the transmitter and the reconstruction of the clock pulses. The latter is possible because in the RZ code a separate pulse, either positive or negative, is transmitted for each clock pulse. Although the three frequency transmission is primarily used for slow speeds it could also be applied for ternary code transmission, possibly increasing the speed to 2400 bits/second. (This type of transmission uses a duobinary code and is discussed in the following section.)

FSK can hypothetically be regarded in two ways, either as switching between two frequencies according to the data or as shifting the carrier frequency higher or lower according to the data. Although both forms seem to be the same, there is a basic difference between them, and some theoretical background must be given to explain this.

Switching the signal between two frequencies, according to the data, can be viewed as the summation of two OOK signals $S_1(t)$ and $S_2(t)$ which are expressed as follows:

$$S_1(t) = A_m(t)\cos\omega_1 t \qquad S_2(t) = 0 \qquad \text{for } A_m(t) = 1$$

and

$$S_1(t) = 0 \qquad S_2(t) = A_m(t)\cos\omega_2 t \qquad \text{for } A_m(t) = 0$$

Such an approach to FSK is possible only if the distances between the two frequencies are great, the rate of change between the two frequencies is slow and the two frequencies come from two different sources. This makes the bandwidth to be $2\Delta f + 2R$, where $\Delta f = (f_2 - f_1)/2$ is the maximum frequency deviation. If, in the extreme case, the frequency deviation is much wider than the data rate, i.e., $\Delta f \gg R$, then the bandwidth approaches $2\Delta f$. Such a case is called wideband FM, and its bandwidth is independent of the data baseband. This form of FM contradicts the basic requirements of FSK since the transmitted signal must be a function of the data; also it is essential that when switching between two frequencies the phase must be continuous with no instantaneous amplitude changes.

For efficient FSK operation the frequency deviation must be smaller than the data rate, i.e., $\Delta f \ll R$ and then the bandwidth approaches $2f_m$. Such a case is called narrow-band FM and this is the form which is used for FSK. It follows that since the phase changes must be continuous (i.e., no abrupt changes) we must regard the FSK as the shifting of the carrier frequency according to the data. In other words the shift of the carrier frequency is between $f_c + \Delta f$ for the 1 bit and $f_c - \Delta f$ for 0 bit. And so the transmitted signal can be expressed as follows:

$$\begin{aligned} S(t) &= A_c \cos(\omega_c - \Delta\omega)t && \text{with } f_1 = f_c - \Delta f \\ &= A_c(\cos\omega_c t \cos\Delta\omega t - \sin\omega_c t \sin\Delta\omega t) && \text{for data equals '0'} \end{aligned}$$

$$\begin{aligned} S(t) &= A_c \cos(\omega_c + \Delta\omega)t && \text{with } f_2 = f_c + \Delta f \\ &= A_c(\cos\omega_c t \cos\Delta\omega t + \sin\Delta\omega_c t \sin\Delta\omega t) && \text{for data equals '1'} \end{aligned}$$

As already stated the maximum deviation to each direction

$$2\Delta f = (f_2 - f_1) = \Delta\omega/\pi$$

is called the frequency deviation. The ratio of the frequency deviation to the rate of change is called the modulation index and is expressed as follows:

$$m = 2\Delta f/R = 2\Delta\omega/\omega_m .$$

Assuming that the data is changing according to a sinusoidal wave, i.e. the modulating signal is sinusoidal and not square, the signal is then $S(t) = A_c \cos(\omega_c \pm \Delta\omega_i)t$ and the instantaneous frequency is:

$$\Delta\omega_i t = \omega_c t \pm \sin\omega_m t .$$

If we insert this into the signal function we get :

$$S(t) = A_c \cos(\omega_c t \pm m \sin\omega_m t)$$

$$S(t) = A_c [\cos\omega_c t \cos(m \sin\omega_m t) \pm \sin\omega_c t \sin(m \sin\omega_m t)] .$$

Solving this function can only be achieved with the aid of Bessel functions giving:

$$S(t) = A_c \left[\cos\omega_c t \left(J_0 m + 2 \sum_{K=1}^{\infty} J_k Km \cos 2K\omega_m t\right) \pm \sin\omega_c t \right.$$

$$\left. \left(2 \sum_{K=1}^{\infty} J_{2k+1} m \sin(2k+1)\omega_m t\right)\right] .$$

The conclusion that can be reached from this Bessel function is that the frequency spectrum is much more complex than amplitude modulation and has many more components. In general the spectrum is a function of the modulation index m. The modulation index was defined as the ratio of the maximum frequency deviation to the data rate. The data rate changes according to the data transmitted and hence the frequency spectrum will contain sidebands at frequencies f_c+f_m, f_c-f_m, f_c+2f_m, f_c-2f_m, f_c+3f_m, f_c-3f_m, and so on. The bulk of the transmitted spectrum is between $f_c+(\Delta f + R)$ to $f_c - (\Delta f + R)$; nevertheless, a wider spectrum is required to prevent distortion.

The relative expansion of the bandwidth required for FM, as compared with AM, is a price that has to be paid to obtain significant improvement in noise and interference rejection. For this reason only FSK is preferred to OOK. Nevertheless, by keeping the modulation index constant at 0 5 it is possible to improve the FSK, but this will be discussed in a different section.

Increasing the number of fixed frequencies, each representing a number of bits, can increase the speed but the received signal-to-noise ratio deteriorates to a point where it becomes impractical. Such a scheme is known as M-ary frequency shift keying (MFSK).

6.4 DUOBINARY FSK

Our aim in data transmission is efficiently to exploit the communication channel spectrum by striving to dispatch as much information as possible without endangering the coherent decoding of the received signal. As already pointed out, MFSK cannot provide this facility and other means have been sought for, one of which is the duobinary technique which is a form of a ternary code.

The duobinary technique code format transforms the conventional binary data into a three level waveform in a manner which allows doubling the transmission rate. While this coding technique could be applicable to AM, FM or PM, it is generally used only with FM, since it already possesses the three levels and their utilization does not increase the bandwidth. The 'duo' in the name refers to the doubling of the speed and not to the levels involved.

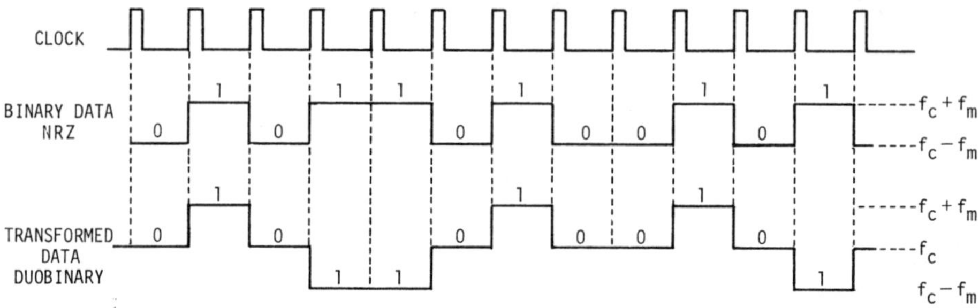

Fig. 6.12 Duobinary coding principle

The logical principle of the duobinary transmission is illustrated in Fig. 6.12. The binary data input, in the form of a NRZ train of pulses, is transformed into a specially coded bipolar RZ pulse train. The '0' data bit remains at the centre level, while the polarities of the '1's are reversed, provided that there is at least one '0' bit between them and that the number of the consecutive '0' bits is odd. The polarity of the '1's remains in the same direction if there is no '0' bit between consecutive '1's or if the number of the consecutive '0's between the '1's is even.

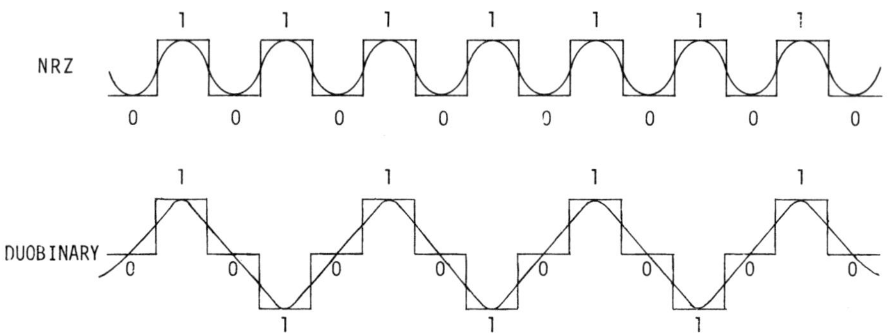

Fig. 6.13 NRZ and duobinary analog change signal

We have already seen that what limits the transmission speed in both QOK and FSK is the maximum data rate $R = 1/T$. Any increase in the data rate will consequently increase the required bandwidth which is limited by the communication channel characteristics. What really affects the bandwidth is not the maximum data rate, but the maximum modulation rate, i.e., the rate of change between the two extreme transmission states. This is better followed with the aid of Fig. 6.13, which illustrates the analog signal lines obtained from the NRZ input pulses and their equivalent transformed duobinary pulses. In both signals a stream of alternate 1's and 0's are transmitted which represent the maximum data rate. In the case of the NRZ code the data rate equals the modulation rate, but this is not the case with the duobinary code. By comparing the response time of the two analog signals it can be seen that the basic frequency rate of the original NRZ signal is twice the speed of standard transmission.

When using duobinary transmission with three level frequency modulation, the carrier frequency is allotted the '0' binary state which represents the zero centre level of the transformed duobinary pulse train. Frequencies f_c+f_m and f_c-f_m are then allotted to represent the positive and negative levels respectively. Returning to the pulse waveforms in Fig. 6.12, it can be seen that in the original NRZ waveform there could be a change from frequency f_c+f_m to frequency f_c-f_m within a time period allocated to one bit. In duobinary transmission this transition must take at least two time periods, as each transition must pass a time period with the transmission of the carrier frequency f_c. The result of transformation from the unipolar to the biopolar pulses is the redistribution of the spectral density of the original data into more highly concentrated energy density at low frequencies.

The generation of the three levels is quite simple, as shown in Fig. 6.14. The inverted data is fed into a T-type flip flop which changes its state for each '0' input pulse. (If the carrier frequency has been transmitted as the '1' state, then the flip flop must be inverted for each '1' input pulse.) The data must first pass through a clocked gate, to distinguish between consecutive '0' pulses. The two complemented output pulses obtained from the flip flop are then added separately to the original data, thus providing the upper and lower levels. The centre level is obtained either from the original data or from the zero level of the combined upper and lower levels. Also shown in Fig. 6.14 are the analog sampling points obtained from the three levels.

The transformation back to the original NRZ pulse sequence is obtained by simply reversing all the negative going pulses to positive going pulses. At the receiver this is achieved by passing the waveform obtained from the

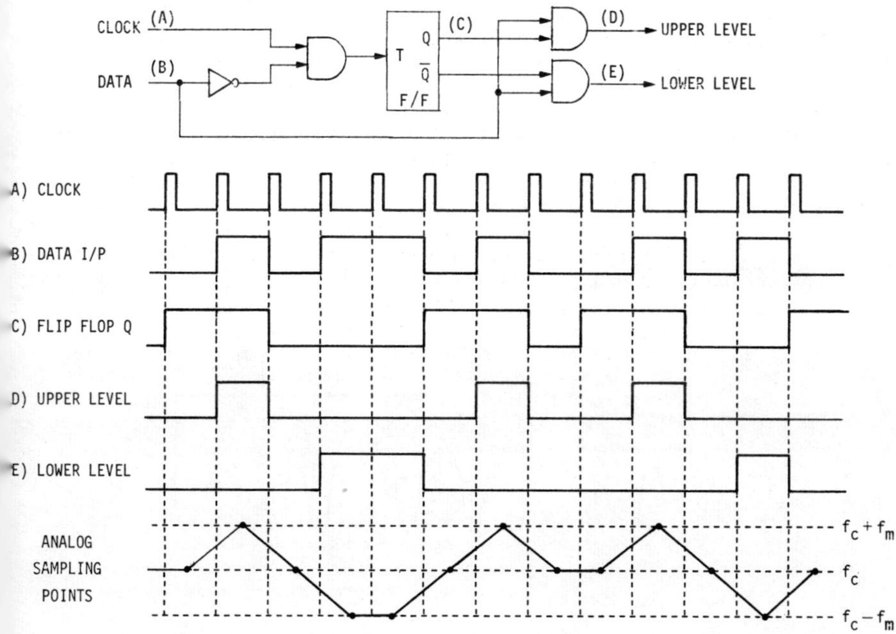

Fig. 6.14 The duobinary conversion process

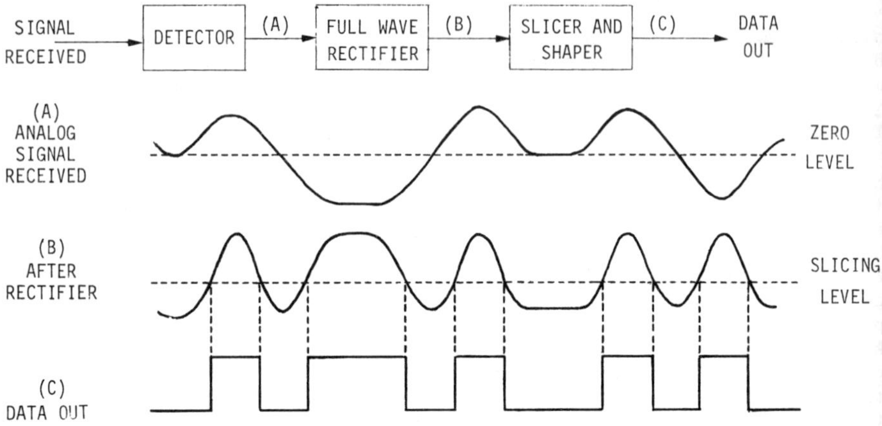

Fig. 6.15 Duobinary Receiver Process

demodulator through a full wave rectifier, then slicing and shaping the pulses (as shown in Fig. 6.15).

6.5 PHASE MODULATION (PM)

In phase modulation, the phase of the carrier signal is changed according to the data transmitted, as shown in Fig. 6.16. This modulation is better known as phase shift keying (PSK). In phase modulation both the frequency and the amplitude remain constant throughout the transmission period. In this respect PM is superior both to AM and FM techniques and so replaces the OOK and the FSK schemes for speeds above 1800 bits per second. The main disadvantage of PSK is the complexity of the equipment required, although the cost is justified wherever high data rates are required. The high speed is achieved by transmitting multistates for each phase shift. In the illustratio shown in Fig. 6.16 the carrier was shifted by 180° for each change from '1' to '0' or from '0' to '1' binary states. Nevertheless phase changes of 90° and 45° are feasible and practical and so are used for multilevel transmission.

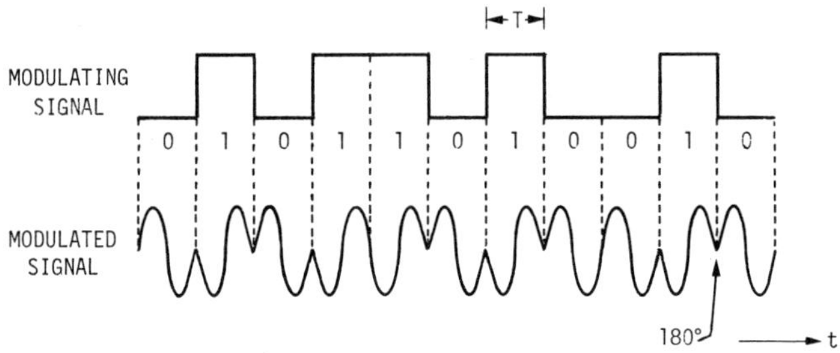

Fig. 6.16 Basic principle phase shift keying

Phase modulation is primarily used for digital transmission and is not generally intended for voice transmission or for telemetry analog informatio This is because the phase changes cannot be detected with accuracy. Althoug it is difficult to detect small phase changes required for analog transmission, it is still possible to transmit small phase changes. Phase changes of 90° and 45° or even 22.5° could be transmitted and easily be detected. With

180° phase changes, the two binary states of a bit can be represented and transmitted. With the 90° phase variation, four phase changes are obtained, allowing two bits to be represented by each phase change. In this case two bits of information can be transmitted simultaneously as in multilevel transmission. With 45° phase variation, eight phase changes are obtained and with 22.5° phase variation sixteen phase changes are obtained, allowing three and four bits respectively to be transmitted.

Phase shifts of 180° are the basic form of phase shift keying (PSK), where the two phase states are used to represent the '1' and '0' binary states, as shown in Fig. 6.17. For this reason, the basic form is generally referred to as binary-phase-shift-keying or biphase shift keying (BPSK). In practical application, phase variations smaller than 180° are obtained by a combination of a number of BPSK modulators and not by a direct phase modulator, as will be explained further on. This means that only BPSK type modulators are used, which makes the theoretical analysis of this modulator extremely important.

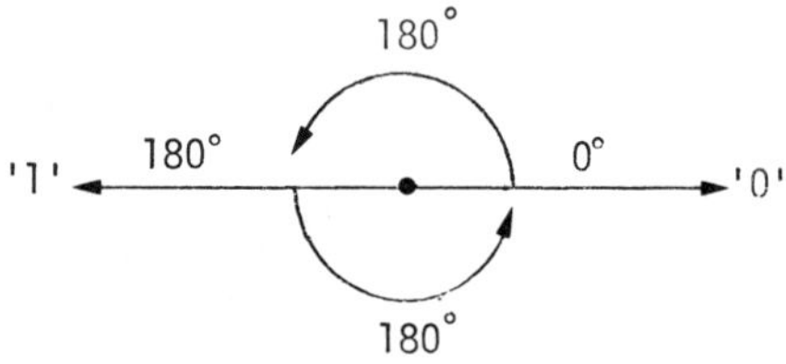

Fig. 6.17 Binary phase shift keying

In phase modulation the phase of a constant amplitude carrier signal is changed according to the information rate $\Delta\theta_m t$. Thus the transmitted signal takes the following form:

$$S(t) = \cos(\omega_c + \Delta\theta_m)t.$$

If the information is in the form of a sinusoidal wave of frequency f_m, then the phase is changed by the information signal

$$\Delta\theta_m t = \Delta\theta_m \cos\omega_m t\ ,$$

where $\Delta\theta_m$ is the maximum angle change of the phase measured in radians Inserting the information signal into the phase modulation transmission signal equation, we get

$$S(t) = \cos(\omega_c + \Delta\theta_m \cos\omega_m t)t\ .$$

The PM signal received is similar to that received for FM and it has thus an equivalent frequency spectrum. Nevertheless, for binary communication using phase modulation we only need to transmit two states, '1' and '0', which makes the signal transmitted using PSK much simpler than that of FSK. As already stated, in BPSK the carrier is shifted by 0° or 180° according to the data, which allow this modulation to be expressed in the following form:

$$\Delta\theta_m = 0° = 0 \quad \text{for data equals } 0$$

$$\Delta\theta_m = 180° = \pi \quad \text{for data equals } 1$$

The above formulae could be inserted into the original phase modulation signal, giving:

$$S(t) = \cos(\omega_c + 0)t = \cos\omega_c t \quad \text{for data equals } 0$$

$$S(t) = \cos(\omega_c + \pi)t = -\cos\omega_c t \quad \text{for data equals } 1\ .$$

The BPSK signal can thus be written as

$$S(t) = \pm\cos\omega_c t,$$

or much better, as

$$S(t) = A_m(t)\cos\omega_c t\ ,$$

where $A_m(t) = +1$ the '0' data, and $A_m(t) = -1$, for the data bit '1'. Checking back, it can be seen that the equation described for BPSK is equivalent to that of ASK. In fact, the two systems, as described so far, are identical and many references fail to discriminate between them. Furthermore, as the equations of both BPSK and ASK are a form of double-sideband-suppressed-carrier, this gives BPSK all the advantages of DSBSC (without many of its disadvantages) and gives a better signal/noise of 3 dB relative to SSB. This enables the BPSK to be produced by a simple balanced modulator, as used for suppressed carrier amplitude modulation (an example of which is given in Fig. 6.18). BSPK also has all the advantages of FSK since its signal amplitude remains constant and is as such less sensitive to noise. In all, BPSK is the best form of digital modulation technique, superior both to FSK and OOK.

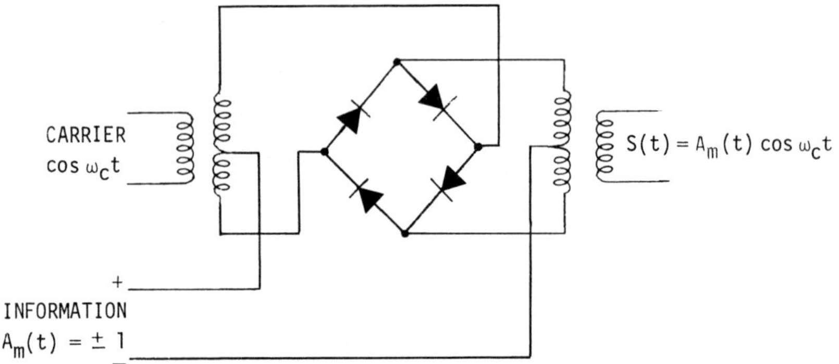

Fig. 6.18 A diode balanced modulator switching the carrier ± 180°

With BPSK, the same bandwidth is required as with DSB-AM, that is, double the bandwidth of the basic data base. In other words, no speed advantage is gained by using the basic PSK: nevertheless, with 'multiphase' PSK modulation more data can be packed into each phase state. The simplest form of multiphase PSK is with shifts of 90°, having each phase representing 2 bits, referred to as dibit. To achieve this MPSK signal we use two BPSK modulators operating in parallel, as shown in Fig. 6.19(a).

It is well known that a sine wave is orthogonal to a cosine wave, that is:

$$\int_0^{2\pi} \sin X \cos X \, dx = 0 \, ,$$

This means that the effect of phase shifts of 90° can be achieved by two orthogonal carrier waves, $\cos \omega_c t$ and $\sin \omega_c t$, each modulating a different bit. In this wave two streams of digital data, $X(t)$ and $Y(t)$ can be modulated in parallel, and thus the space summation of the results produces the desired 90°, as shown in Fig. 6.19(b) by the vector diagram of these pulses. The bit streams are modulated at the modulation rate, while they are produced by a single stream with a bit data rate of double that of the modulation rate.

The two bits of each modulation state can be in either of four phases, 45°, 135°, 225° and 315°, in reference to a constant wave according to the instant dibit state. For this reason this type of modulation is referred to as quadriphase shift keying (QPSK).

QPSK is an extension of biphase shift keying and its transmission signal can be expressed as follows :

$$S(t) = X(t) \cos \omega_c t + Y(t) \sin \omega_c t$$

where $X(t)$ and $Y(t)$ equal +1 or −1 according to the data. This equation also describes quadrature amplitude shift keying (QASK), also known as

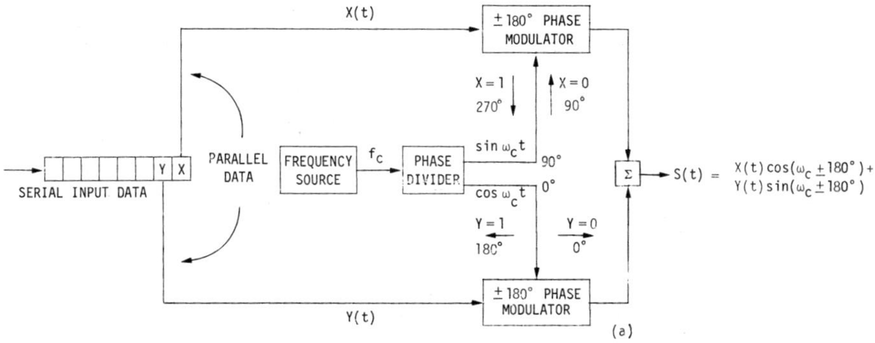

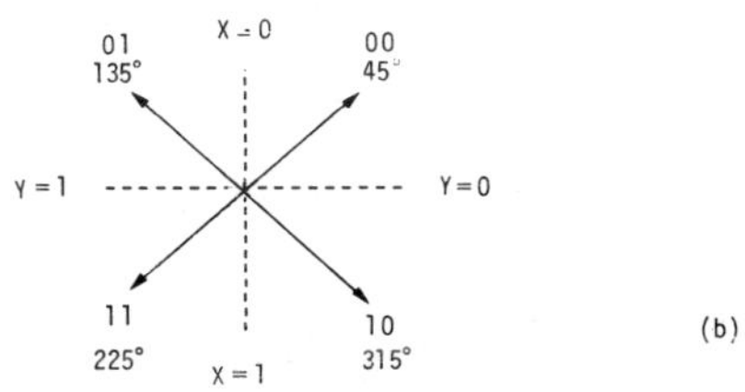

Fig. 6.19 Block diagram of a four-phase modulator transmitter

quadrature amplitude modulation (QAM). In fact, there are no differences between these two fundamental four-state modulation techniques although there are differences when the number of modulation states exceeds four: but there too, the differences are not appreciable, as will be explained later.

The operation of the QPSK is better followed with the aid of Fig. 6.20, which shows the various waveforms obtained in the transmitter circuit shown in Fig. 6.19. The two modulators are operated in parallel, each shifting the corresponding carrier phase by 180°. The outputs of the two modulators aree in phase quadrature, whereas after combining them they produce the required resultant signal shown in Fig. 6.20.

The resultant QPSK signal can be represented in either Cartesian or Polar coordination. In polar coordination the four states are shown as vectors, while with Cartesian coordination the four states are shown by the corresponding $X(t)$ and $Y(t)$ states. This means that the QPSK signal could also be expressed as a phase modulation with the signal

$$S(t) = \cos(\omega_c + \theta)t ,$$

where $\theta(t)$ takes the value of 45°, 135°, 225° and 315° according to the respective debit value of 00, 01, 11 and 10. It is worth adding that in the case of QASK (or QAM) it is accepted to express the states only in Cartesian coordination.

There are two choices for the signal encoding of phase shifts and these are known as alternatives *A* and *B*. The four phase shifts could be ± 180° and ± 90°, as used outside the United States. Alternatively, the

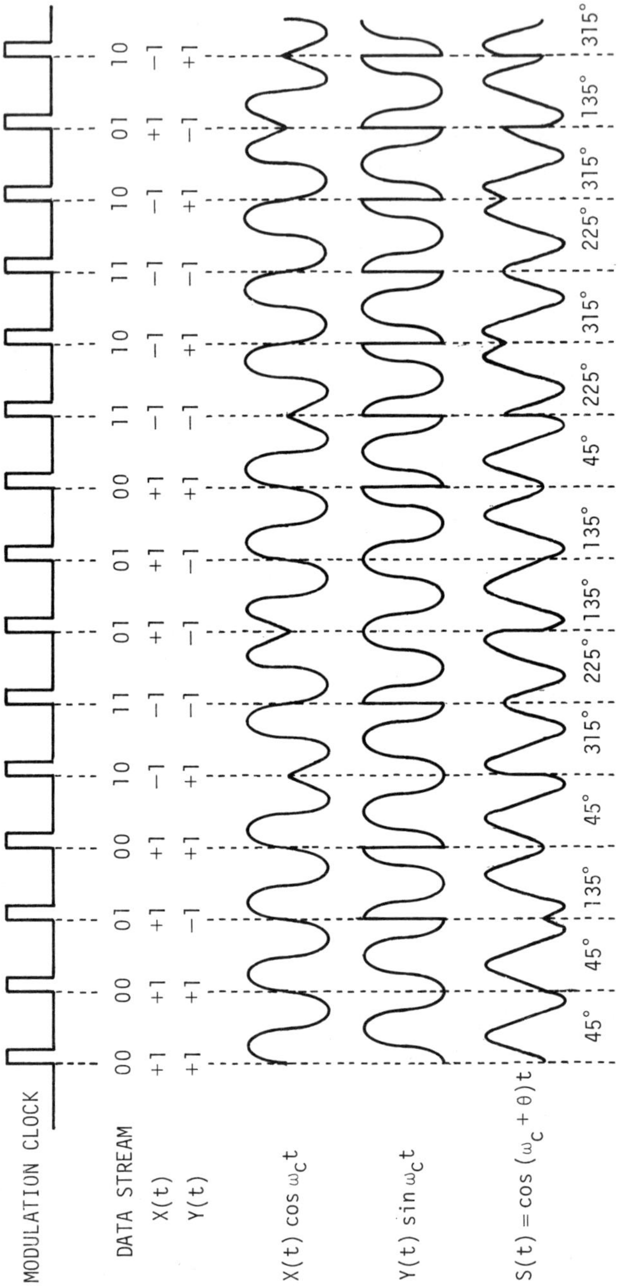

Fig. 6.20 QPSK waveforms

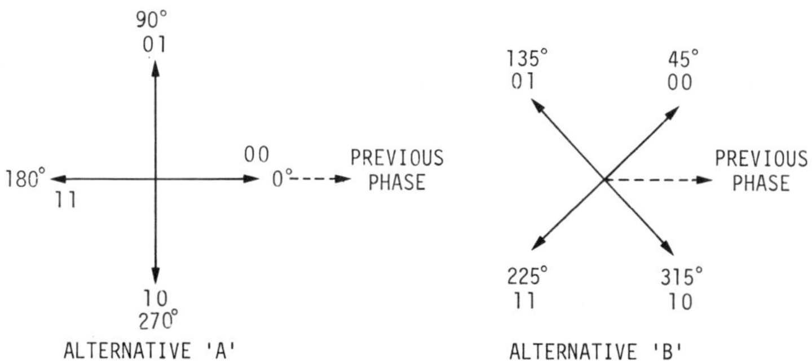

Fig. 6.21 Vector diagram of the two alternative QPSK techniques

phase shift choice could use $\pm 45^\circ$ and $\pm 135^\circ$. These two phase shift choices can be seen in Fig. 6.21. In Alternative 'B' there is a phase change for each transmitted dibit and this is the technique used in most four phase modems in the United States and presents performances superior to Alternative 'A', as will be seen later. The following table shows the codes employed for each transmitted debit information using both alternative phase shift choices. The angle stated is the phase shift occurring in reference to the steady state of the previous dibit.

TABLE 6.1

TRANSMITTED DIBIT	PHASE SHIFTS IN REFERENCE TO PREVIOUS DIBIT	
	ALTERNATIVE 'A'	ALTERNATIVE 'B'
00	0°	+45°
01	+90°	+135°
11	+180°	+225°
10	+270°	+315°

(According to CCITT V. 26)

From the above table it can be seen that when transmitting a continuous pulse train of '0's (as well as any continuous stream) there is no phase shift when Alternative 'A' is used. In this arrangement synchronization could easily be lost when a long string of continuous binary '0' is transmitted, as could occur during the interval between message transmission. This could be overcome by a special scrambler which changes the data, thus obviating a long string of '0's. Alternative 'B' has the advantage that each dibit must cause a phase shift. For a continuous transmission of '0's there is a repeated phase shift of 45° after each consecutive 00 dibit, and this arrangement is actively maintained.

An example of a four phase-shift transmitter is shown in Fig. 6.19. The example is based on the phase angle of Alternative 'B', but it could easily be applicable to Alternative 'A'. The data stream is fed into the transmitter in serial form and there it is converted into two parallel streams. Two consecutive adjacent bit pulses are always modulated simultaneously. Each modulator changes the phase by 180° according to the specific data applied to it. By combining the two modulated waves a quadrate phase change can be generated out to the line. In the combining circuit the resultant vector

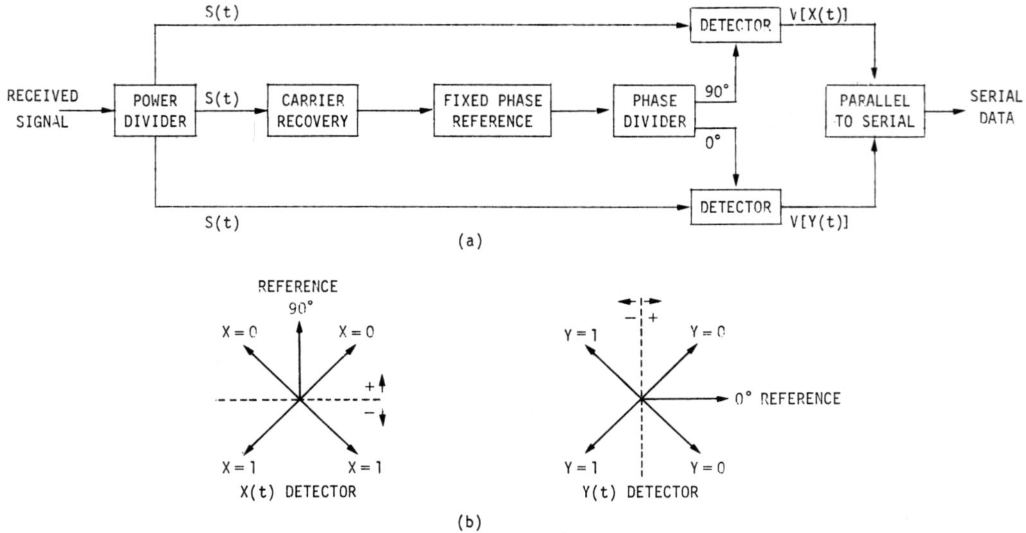

Fig. 6.22 Fixed reference detection of a four-phase signal

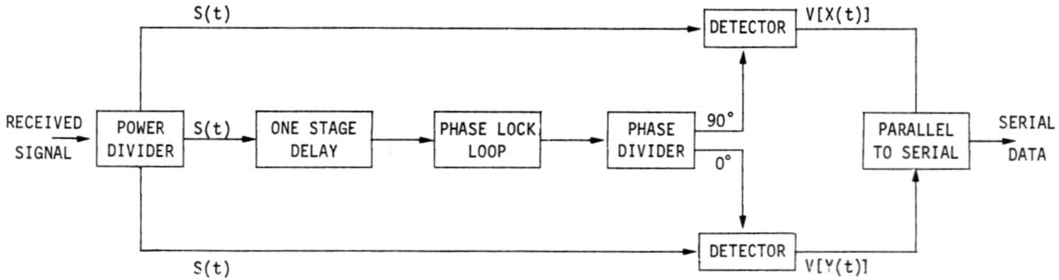

Fig. 6.23 Differential phase detection of a four-phase signal

presents the phase angle of 90° shift for each dibit, as shown by the vector diagram.

The two modulators are fed by the same frequency source except that they are perpendicular to each other. For Alternative 'A' the phase modulators is ±45°, while for Alternative 'B' it is 0° and 90°. The phase angle shift for each modulator and the corresponding resultant vector phase angle transmitted into the line is given in the following table.

TABLE 6.2

INPUT DIBIT	ALTERNATIVE 'A'			ALTERNATIVE 'B'		
	1st MOD.	2nd MOD.	SUMMED	1st MOD.	2nd MOD.	SUMMED
00	45°	315°	0°	90°	0°	45°
01	45°	135°	90°	90°	180°	135°
11	235°	135°	180°	270°	180°	225°
10	235°	315°	270°	270°	0°	315°

The receiver has no absolute sense of phase and the decoding can only be obtained by reference to another angle. For detecting the relative phase shifts there are two basic techniques:

The first method, termed fixed reference phase detection, is shown in Fig. 6.22. The fixed reference is based on the original carrier frequency which can be recovered from the transmitted waveform, irrespective of the phase shift. This can be accomplished either with a full wave rectifier, which recovers the second harmonic, or alternatively, the carrier could be recovered by a phase lock loop (PLL) which tracks the input signal frequency but not the phase. PLL cannot demodulate PSK information, as the abrupt phase shifts cause only transition changes in the voltage controlled oscillator (VCO), after which it settles back to its previous frequency. For this reason it is essential that when using fixed reference phase detection, the data transmitted system should be coherent, that is, the data clock and the carrier frequency are synchronized. For this reason such schemes are known as coherent phase shift keying, CPSK (or as the case may be CQPSK).

As in the transmitter, the recovered carrier frequency is fed to two parallel decoders, although each detector is compared with a carrier which is orthogonal to the other, that is, it has a 90° angle between them. If the incoming phases are 0°, 90°, 180°, 270°, then the fixed reference is shifted by ±45°. If the incoming phases are 45°, 135°, 222°, 315°, then the reference is based on the relation to 0° and 90° (as used in Fig. 6.22). The received waveform is compared at each modulator with the appropriate phase frequency reference, which causes the detector to put out a positive or negative dc voltage. If a positive indication is obtained from the corresponding detector, then a '0' is recorded for the respective bit of the reconstructed dibit. If a negative is obtained, then a '1' is recorded. The following table shows the relation between the input received phases, the polarity of the detector outputs and the dibit recovered data.

The fixed reference detection is susceptible to phase jitter and other channel disturbances. This creates the problem of both establishing the fixed reference and maintaining it in its fixed phase. Once the receiver loses synchronization it takes time before the synchronization can be established again. For high speeds the fixed reference is required to be accurate and stable so that it will be possible to distinguish between adjacent dibits. These problems could be avoided by using differential phase detection, as shown in Fig. 6.23, but at the risk of incurring a penalty in the signal-to-noise ratio.

TABLE 6.3

INPUT PHASE	DETECTED OUTPUT		RECORDED DIBIT	INPUT PHASE	DETECTED OUTPUT		RECORDED DIBIT
	+45°	−45°			+90°	0°	
0°	+	+	00	45°	+	+	00
90°	+	−	01	135°	+	−	01
130°	−	−	11	225°	−	−	11
270°	−	+	10	315°	−	+	10

(a) ALTERNATIVE 'A' (b) ALTERNATIVE 'B'

The second detecting method is based on the phase change relative to the last existing phase rather than to a fixed reference standard. This method is known as differential phase detection and the system is termed differential phase shift keying DPSK (or in this case DQPSK). In this modulation method each dibit is transmitted in reference to the previous dibit, and thus both transmitter and receiver must be of Alternative 'B' type. (A simplified version of a DPSK receiver is shown in Fig. 6.23). The principle of operation of the DPSK receiver is similar to that of the fixed reference method except for the carrier phase reference reconstruction. In this method each dibit is delayed and buffered for one dibit period and then compared with the following incoming dibit signal, Due to the short time available for each phase lock operation, the signal-to-noise ratio deteriorate nevertheless, the advantages gained by the positive phase correlation outweighs its drawbacks.

6.6 QUADRATURE AMPLITUDE SHIFT KEYING (QASK)

It has already been pointed out that the QPSK modulation technique described in the previous section is equivalent to the QASK modulation technique. That is, the signal produced by both modulation techniques could be represented by the following equation:

$$S(t) = X(t)\cos\omega_c t + Y(t)\sin\omega_c t$$

where $X(t)$ and $Y(t)$ take the value of +1 or −1 according to the corresponding data bit values. These four possible signals received after inserting the values of ±1 could be plotted as a function of time, as shown in Fig 6.24(a). The same transmitter circuits, as used for QPSK could be used for producing these QASK signals. For QPSK, however, a polar diagram was used to exhibit these four signals, that is, a vector was used to represent each transmitted dibit. With QASK, since it is based on amplitude modulation, a different method of representation is used to exhibit these same four signals Each of the four transmitted instant dibit states could be represented by their corresponding $X(t)$ and $Y(t)$ values and then plotted on a Cartesian diagram, as shown in Fig. 6.24(b). This Cartesian diagram is also referred to as the signal constellation of the modulation technique.

The differences between QASK and QPSK are only pronounced when the number of the possible transmitted states increases. With PSK the number of possible states may only be increased by shifting the phase by smaller angles With QASK the number of states may only be increased by adding more amplitude levels to each of the two quadriphase signals. Adding extra amplitude level to each signal produces a constellation of 16 different possible states, as shown in Fig. 6.25. Each of the two orthogonal signals has now four possible amplitude levels, as shown in Fig. 6.25(a); hence the constellation is constructed from the four possible $X(t)$ levels and four possible $Y(t)$ levels

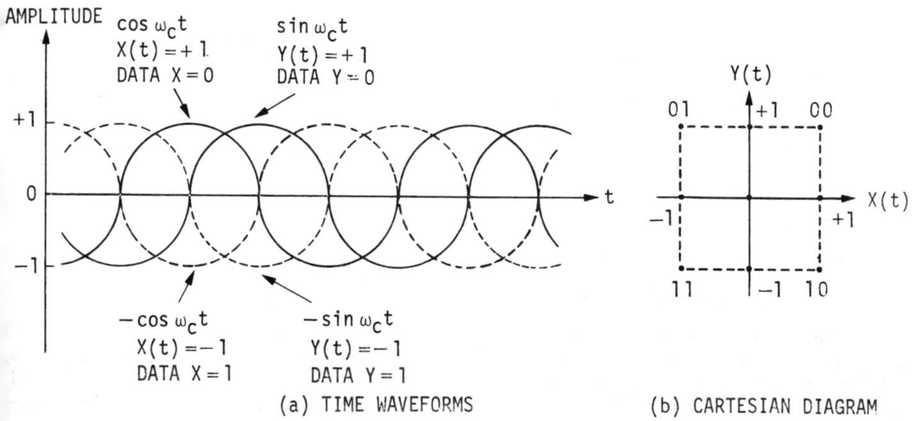

Fig. 6.24 Quadrature amplitude shift keying

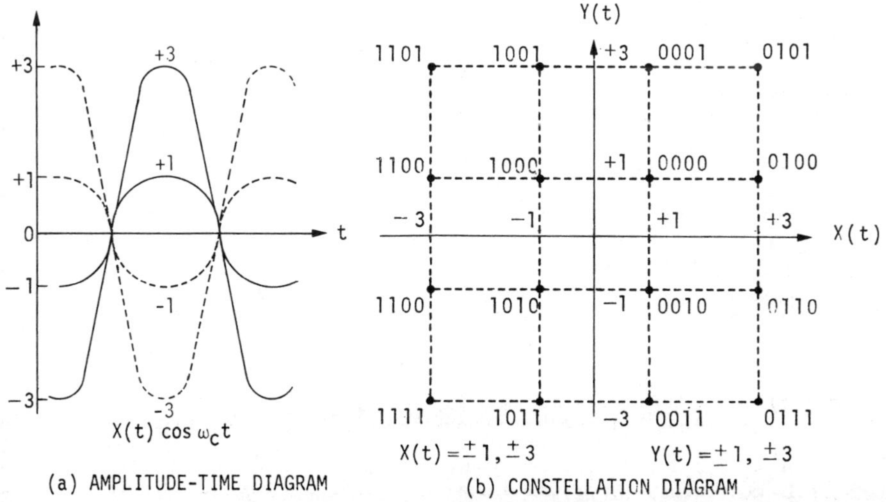

Fig. 6.25 Four level QASK

as shown in Fig. 6.25(b). This enables each constellation state to correspond to four transmitted data bits.

The values of ± 1 given to each amplitude level of the two orthogonal signals in the four state QASK/QPSK system are only representative values to describe the relative amplitude level of the transmitted signals. When adding extra amplitude levels one must represent them with values proportionally related to the original values of ± 1. It can be seen from Fig. 6.24(a) that the minimum distance between any two constellation points, i.e., any two relative amplitude levels, is $2(=|-1|+|+1|)$. This means that the minimum distance of 2 must be kept when increasing the number of levels. Hence, the next possible relative level is ± 3 and the one after it is ± 5. It can be seen from the constellation diagram presented in Fig. 6.25(b) that the constellation pattern produced by the four amplitude levels has a rectangular symmetrical shape with equal minimum distances between each two adjacent possible transmitted states.

This form of modulation is called quadrature amplitude shift keying QASK, but it is also popularly referred to as quadrature amplitude modulation QAM. Despite the similarity in names one must be careful not to confuse QAM with the four level amplitude modulation as was shown in Fig. 6.3. The latter is a form of ON/OFF amplitude level modulation.

Producing the four level QASK signal could be achieved by switching the carrier amplitude according to the level defined by two data bits while at the same time the carrier phase is shifted according to the other two data bits. This is the popular technique for modulating the signal although such an arrangement could procure distortions in the modulated signal if the two operations are not properly synchronized. There are also various other methods of achieving the required QASK constellation signal, but possibly the most novel method is the one developed in Japan using a scheme called Superposed Modulation (SPM). In this scheme two identical four-phase QPSK modulators, as described in the previous section, are employed. Each of the modulators, as shown in Fig. 6.26, shift the respective carrier signal by multiples of 90°, according to the data value of the two bits. The output signals of the two modulators are then summed together, i.e. superimposed on each other. The only difference between the output signal of the two modulators is that the amplitude of one modulator is permanently attenuated.

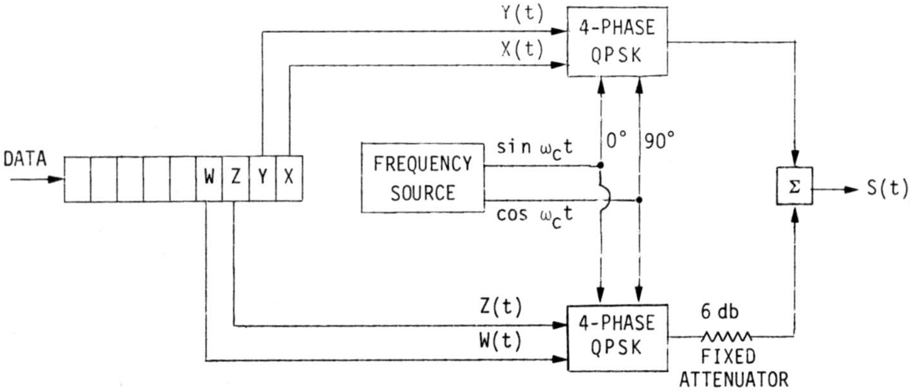

Fig. 6.26 Superposed modulation circuit

Four data bits, *XYZW*, are modulated simultaneously, with two bits in each QPSK modulator. Each modulator can shift its carrier phase, by multiples of 90°, to any one of eight possible angles, as seen by the polar diagram shown in Fig. 6.27. The output of one modulator is attenuated by a fixed resistance, thus giving a 6 dB amplitude difference between the two modulated output signals. When the two signals are superimposed, this means in effect that the larger vector is shifted according to the smaller vector. The result of superimposing the two phase modulation signals produces the same required constellation pattern required for QASK.

The Japanese have also developed a detector for this 16 state modulation scheme, which is based rather on phase discrimination than on amplitude discrimination, as seen in Fig. 6.28. The receiver consists of two 4-phase QPSK detectors. In the first detector two bits, *X* and *Y*, are ascertained as if the signal received was originally modulated by only these two bits. This means that in the first detection process we need to make sure that the received signal is in one of four possible states and not in one of sixteen possible states. This is possible since the difference between the two superimposed signals is 6 dB, which ensures that the superimposed signal remains in the *XY* square of the Cartesian diagram. The signal after detection

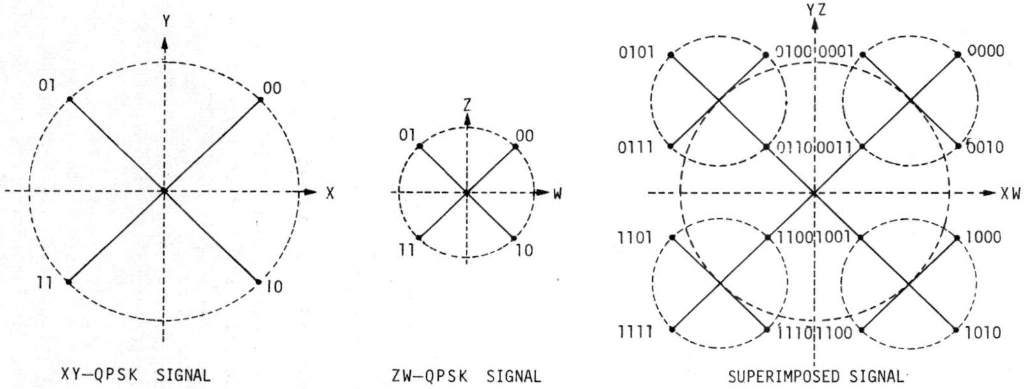

Fig. 6.27 Polar diagram of superposed modulation

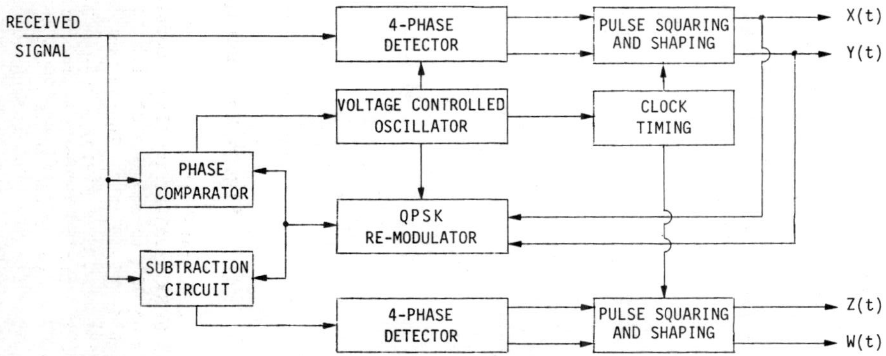

Fig. 6.28 Superimposed modulation receiver circuit

is modulated again, although this time it is only a function of X and Y bits, without the superimposed ZW components. The remodulated signal XY is then subtracted from the received signal $XYZW$ resulting in a remaining signal which is only a function of ZW. The later signal can then be discriminated, thus producing the bit value of ZW.

The whole operation described above shows that the QASK (or QAM) modulation scheme could be produced by QPSK methods. In the QASK scheme, the relative power required from each modulator to transmit the ± 3 amplitude vector is $3\sqrt{2}$. In QPSK, with superimposed techniques, one amplitude vector requires the relative power of $\sqrt{2}$ and the other the relative power of $2\sqrt{2}$. The sum of the power required in both modulators for the two schemes shows that with QASK double power is required as compared with SPM. Furthermore, with QASK, the power level depends on the amplitude value being ± 1 or ± 3, while with the superimposed modulation the same power level is used to modulate any bit combination.

The comments above still leave the question whether the sixteen point constellation is a result of phase or amplitude modulation and what is the demarcation line between the two schemes. This is a rhetorical question, since the distinction depends only on how one looks at the modulation technique, and if there is a confusion it is only a matter of terminology.

There are other forms of QASK constellations which are used to increase the speed of transmission. Because of the limitation of the channel bandwidth, the modulation rate is also limited. In PSK and ASK, the modulation rate over telephone lines is about 1200 Baud. By using quadrature modulation, where each transmitted state represents two bits, the data rate is increased to 2400 bits/seconds. By further increasing the number of possible modulation states, the transmitted data rate could be further increased. With a system with a constellation of sixteen states, it is possible to transmit at a high data rate a speed of 9600 bits/second. In such a system, each state corresponds to four binary bits, with $X(t)$ and $Y(t)$ taking the amplitude levels of ± 1 and ± 3 according to the instant value of these four bits.

There are various forms of multi-state amplitude constellation where $X(t)$ and $Y(t)$ can have different relative values. In all these constellations the relative minimum distance of 2 between any two possible states must still be kept. Another popular QASK constellation pattern uses the amplitude levels of ± 1, ± 3 and ± 5, as shown in Fig. 6.29. The combination of these levels for $X(t)$ and $Y(t)$ could give a constellation of 36 different states, each corresponding to four bits. Nevertheless, only sixteen of them are used, selected to give minimum of transmitted power and have a minimum distance of 2 between any two adjacent states.

Both the sixteen state constellation patterns described so far are referred to as QASK or QAM, except that one is based on a two level quadrature amplitude modulation and the other on a three level quadrature amplitude modulation. The constellation pattern and the minimum distance between the states define the transmitted power required and the assurance that the data can be recovered. The closer the states, the lesser the eye opening and hence the more ambiguity. There are other geometrical considerations which could also affect the received detected result, such as the angle gradient between the various states and the number of amplitude levels. It should be remembered that amplitude modulation is susceptible to noise affecting the amplitude level. To demonstrate these considerations and describe the eye opening, the two forms of QASK schemes are presented in Fig. 6.30.

Both constellation patterns shown in Fig. 6.30 were drawn using the same relative amplitude level scale. This makes the dotted boxes round each state correspond to the maximum available eye pattern tolerance available for each state. These boxes are formed by considering both the angle and amplitude tolerance. The two level amplitude QASK has square shaped boxes while

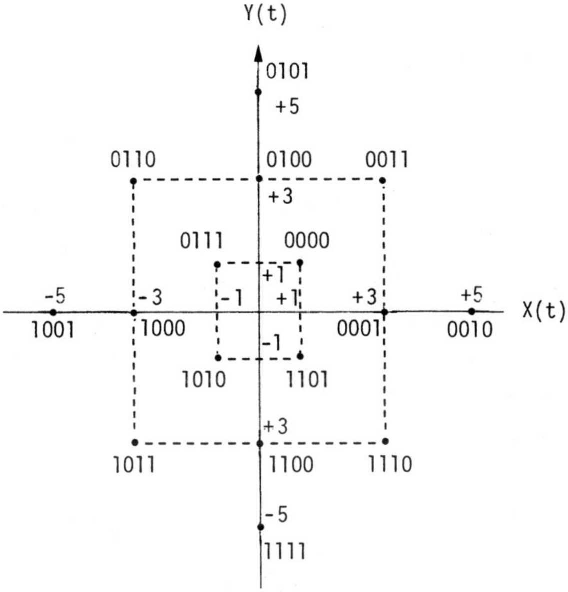

Fig. 6.29 Constellation pattern of a four amplitude level QASK

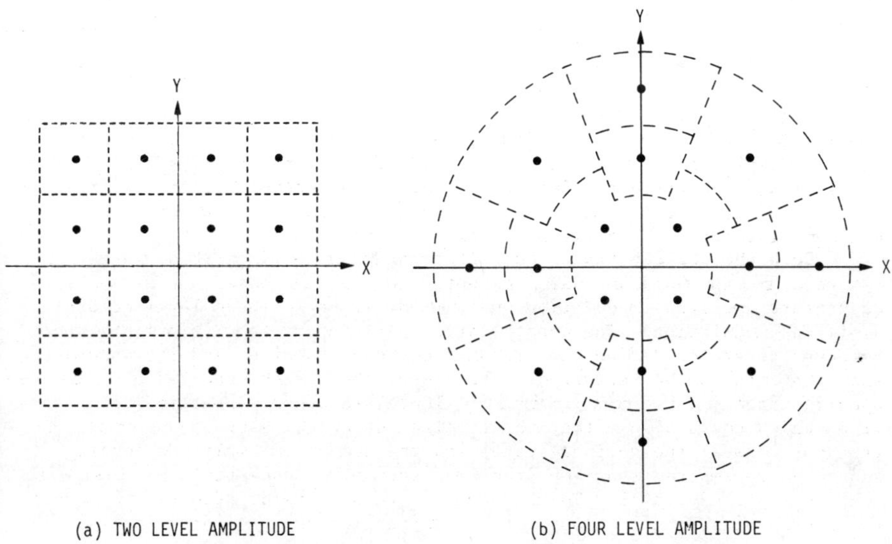

Fig. 6.30 Constellation pattern of two QASK schemes demonstrating their eye opening

the three-level amplitude QASK has wider shaped boxes. Nevertheless, the latter is achieved at the expense of extra amplitude levels. Although not theoretically conclusive (from the constellation patterns), it can be said that the two-level QASK is proof against Gaussian noise, but not against phase jitter, whereas the three-level QASK is proof against phase jitter but not against Gaussian noise. It should be added that both these QASK schemes are widely used today in high speed modems.

6.7 MULTI-PHASE SHIFT KEYING (MPSK)

In all the modulation schemes discussed so far an attempt is made to find means of increasing the data transmission speed. It has been shown that this can be achieved only by increasing the number of modulation states. In the previous section we used amplitude level changes as one of the means of achieving this task. This scheme, however, is susceptible to Gaussian noise. Hence in systems with noisy environment, phase shifts could be advantageous, even though phase schemes are susceptible to jitter. Digital modulation schemes employing multi-phase shifts are referred to as M-ary PSK.

The constellation pattern of phase shift keying is usually presented in a circle, since the signal amplitude remains constant for all transmitted modulation states. On this point, M-ary PSK differs from M-ary ASK, and the constellation of 8 and 16 states have a distinct pattern for both schemes. This circular pattern of PSK also means that for multi-phase the pattern should be represented by a polar diagram and the transmitted signal should correspond to the fuction $S(t) = \cos(\omega_c + \theta)\, t$, where the angle θ takes the values according to the data.

It is possible to transmit a tribit (3 bits) for each phase change, as shown by the polar diagram in Fig. 6.31. In this case there are 8 possible modulation states, which differ by 45° between any two adjacent states. As in QPSK there could be two possible phase shift alternative systems. In both phase modulation schemes the vectors are of equal size and the constellation is in a circle.

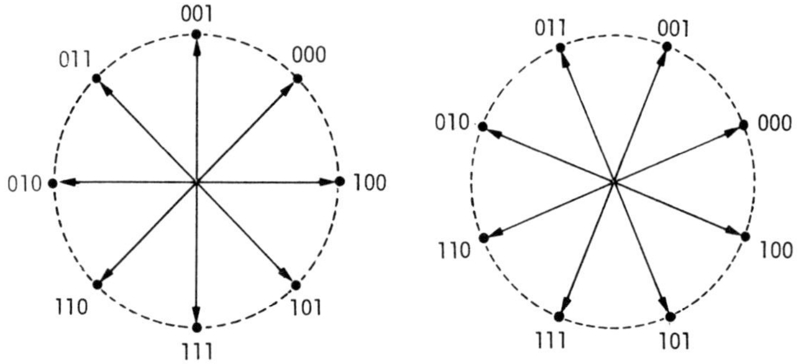

Fig. 6.31 Polar diagram of eight phase M-ary PSK

With the increase of a number of states, the eye pattern opening deteriorates, which makes the detection more difficult. This is true for all forms of digital transmission, but it is more critical in phase systems. It has been suggested that each modulation state should correspond to a grey code scale instead of the normal binary scale. In grey scale, any two adjacent tribits (or fourbits) differ in only one bit, which makes their decoding more reliable. The following table presents both code scales and their corresponding phase shifts for both M-ary PSK alternatives.

Generating phase shifts of less than 90° requires much more complex logic than that incorporated with QASK. Whereas 90° phase shifts are obtained

TABLE 6.4

BINARY TRIBIT	GRAY CODE TRIBIT	PHASE SHIFTS	
001	000	45°	22.5°
001	001	90°	67.5°
010	011	135°	112.5°
011	010	180°	157.5°
100	110	225°	202.5°
101	111	270°	247.5°
110	101	315°	292.5°
111	100	360°	337.5°

by modulating two orthogonal signals, it is difficult to achieve 45° by modulating three equal level orthogonal signals. One possible means of achieving 8—θ PSK is by using a pure phase modulator with the transmitted carrier phase angle equal to

$$\theta = (45^\circ + A\,180^\circ + B\,90^\circ + C\,45^\circ)$$

where ABC correspond to the binary scale value of the transmitted data tribit.

In the same way that it is possible to obtain amplitude-shift-keying with phase modulation techniques, it is possible to obtain phase-shift-keying with amplitude modulation techniques. This once more raises the rhetorical question of the distinction between M-ary PSK and M-ary ASK. The 8-phase PSK system using amplitude modulation techniques was again developed in Japan. In this scheme each vector is defined by the Cartesian values of $X(t)$ and $Y(t)$. Since the state pattern forms a circle, with equal distances between each state, each point can be defined by two fixed values only. These values correspond to two amplitude levels $\pm V_1$ and $\pm V_2$ and each vector angle could be defined by the function of the ratio of these two levels. As seen in Fig. 6.32(a), the first modulation state has a vector angle of $\tan 22.5^\circ = V_2/V_1$ and the second modulation state $\tan 67.5^\circ = V_1/V_2$. In the same way it is possible to define all the eight modulation states and produce them as a function of two amplitude levels.

TABLE 6.5

TRIBIT	VECTOR ANGLE	X(t)	Y(t)
000	22.5°	$+V_1$	$+V_2$
001	67.5°	$+V_2$	$+V_1$
011	112.5°	$-V_2$	$+V_1$
010	157.5°	$-V_1$	$+V_2$
110	202.5°	$-V_1$	$-V_2$
111	247.5°	$-V_2$	$-V_1$
101	292.5°	$+V_2$	$-V_1$
100	337.5°	$+V_1$	$-V_2$

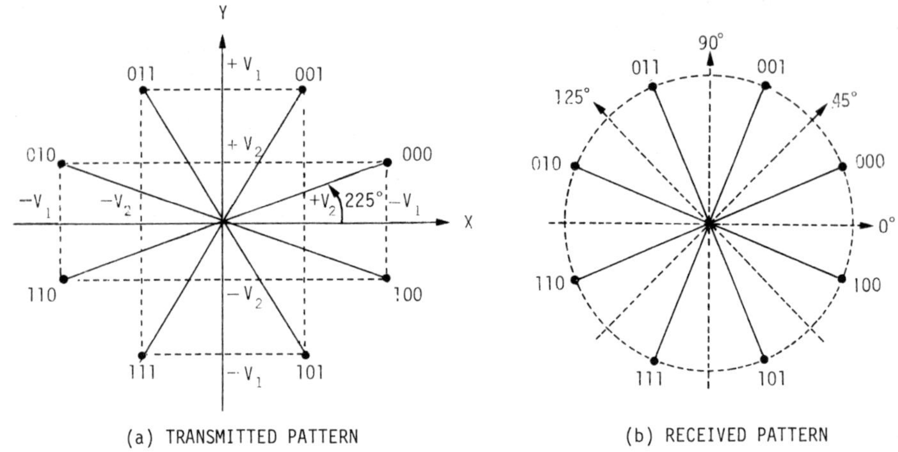

Fig. 6.32 Cartesian and polar diagram of an 8−θ PSK

Recovering the data out of the received signal is performed by pure phase detection techniques similar to that used for QPSK, as seen in Fig. 6.33. In the receiver each state is compared with four phase references 0°, 45°, 90° and 125° (see also the polar diagrams given in Fig. 6.32(b)). From the result obtained in each phase detector circuit, the three transmitted bits can be reconstructed. This recovery method has all the advantages of phase modulation where constant levels are transmitted without the disadvantages of amplitude modulation where a number of level changes are transmitted.

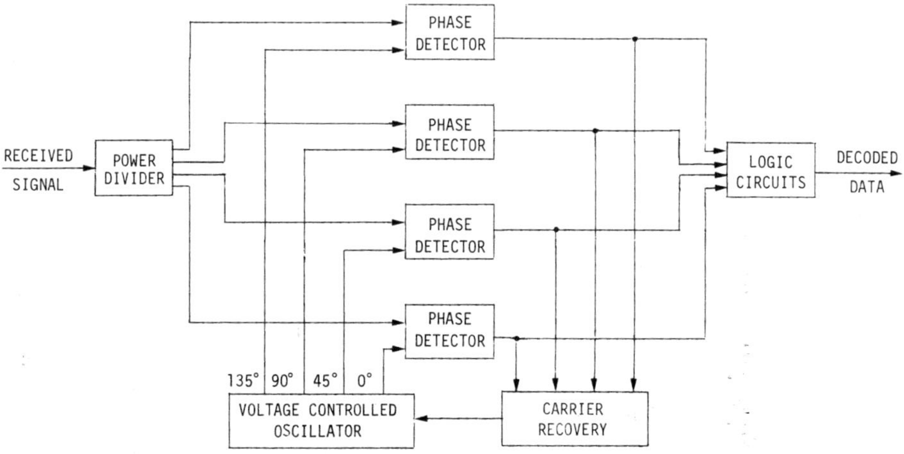

Fig. 6.33 8−θ PSK receiver circuit

A completely different approach has been developed in the United Kingdom. In this method, a phase synthesizer is used which generates all phases that may be required. In this synthesizer all the phases are continuously produced by digital means, as seen in Fig. 6.34. The synthesizer is operated by a clock source which is 8 times faster than the required signal. With the aid of a Johnson counter and corresponding decoders, 8 separate signals are generated, each delayed by T seconds relative to its predecessor, although all have the basic identical frequency.

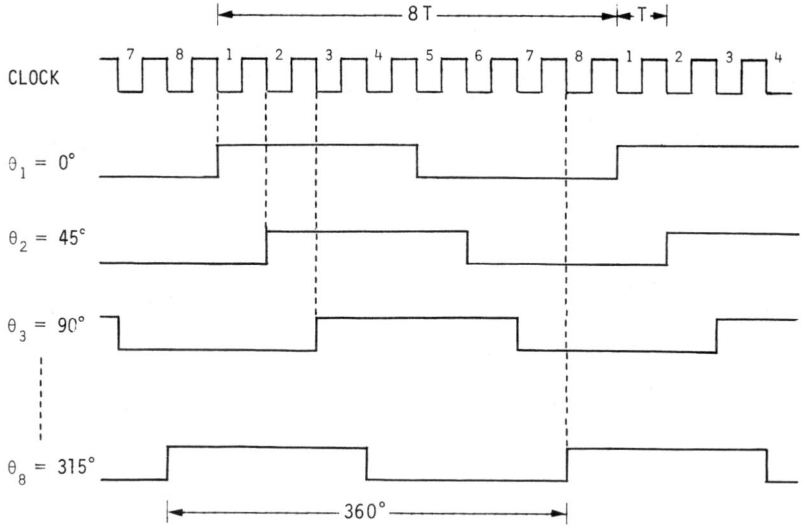

Fig. 6.34 Synthesizer producing eight delays phase signal

Each transmitted tribit selects one of eight phases, as shown in Fig. 6.35. The resulting waveform is a stream of square pulses corresponding to the transmitted binary data. Before transmission the signal must be converted to an analog signal which is simply performed by means of a low pass filter.

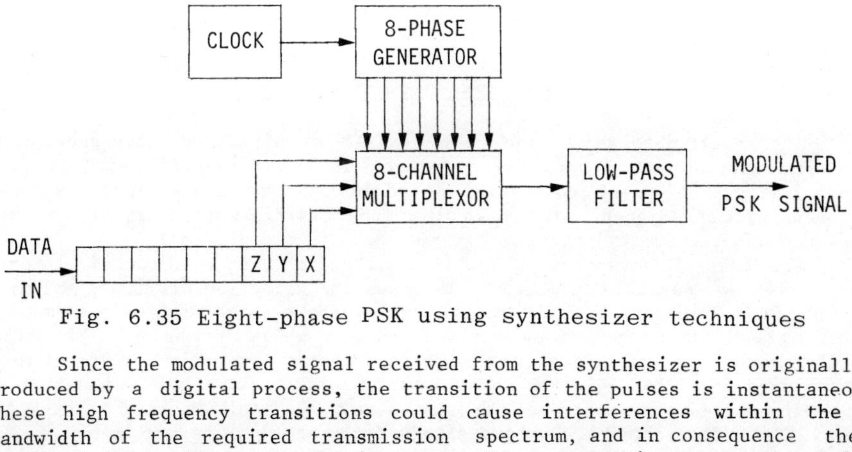

Fig. 6.35 Eight-phase PSK using synthesizer techniques

Since the modulated signal received from the synthesizer is originally produced by a digital process, the transition of the pulses is instantaneous. These high frequency transitions could cause interferences within the bandwidth of the required transmission spectrum, and in consequence the intersymbol interference could increase. To remove this danger, the whole synthesizer operation should be performed at a higher frequency and the result then transferred to the required carrier frequency by means of a mixer. The latter is not shown in Fig. 6.35.

6.8 MULTIPLE AMPLITUDE AND PHASE SHIFT KEYING

Both phase and amplitude shift keying have their limitations. Whereas ASK is susceptible to Gaussian noise, PSK is susceptible to jitter. Increasing the number of states in both schemes may increase the data rate transmitted but it also makes the detection at the receivers' end more difficult. At this stage data rate is limited to 9600 bits/second over wire channels and this could be improved only if more efficient systems are developed.

Best results could be obtained by the combination of both amplitude and phase shift keying. Such a system has been found to make more efficient use of the transmitted power, as it requires less power than PSK for achieving the same error probability. It also enables the number of amplitude levels to be optimized as compared with ASK. The received signal can be reconstructed by less signal-to-noise ratio as compared with either system.

Many theoretical hybrid amplitude and phase schemes have been proposed and they have been given various related names, some of which are presented here: Amplitude Phase (Shift) Keying (APK or APSK), M-ary (Multilevel) Amplitude Phase (shif) keying (MAPK or MAPSK), Independent Amplitude Phase Shift Keying (IAPSK), Unbalanced Quadriphase Shift Keying (UQPSK).

Independent amplitude phase shift keying is a special form of MAPSK in which the amplitude and phase modulators are separately applied by the data stream, as shown in Fig. 6.36. In this 16 state IAPSK system the carrier amplitude level is modulated by one bit, while the other three bits modulate the carrier phase by 45° shifts.

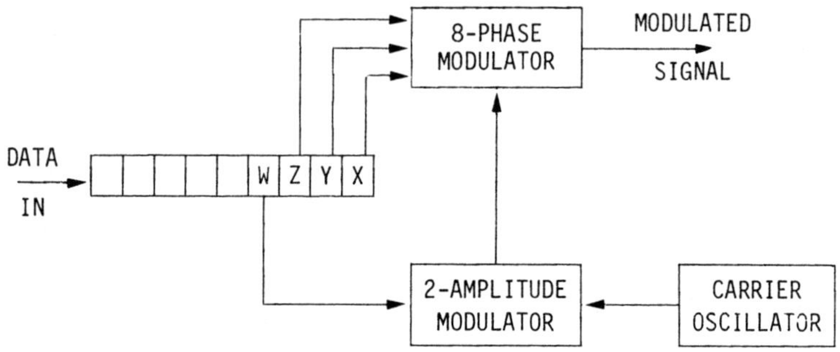

Fig. 6.36 IAPSK modulator

The result obtained by the IAPSK system as discussed above generates a constellation pattern of sixteen states in two concentric circles (as shown in Fig. 6.37) each consisting of eight states. The minimum distance between any two adjacent states in the inner circle must be equal to the distance between the two circles (not as shown in the drawing).

The main disadvantage of IAPSK is that for each transmitted vector there are two possible states. This makes the detection difficult and may result in a high error probability. To overcome these difficulties many other constellations have been proposed, two of which are shown in Fig. 6.38. In these patterns the states are so chosen that no two adjacent states may be designed in either a Cartesian or polar diagram.

The dotted lines between the states show the minimum distance between them. These distances should be designed to be as equal as possible.

Table 6.6 (page 291) gives the optimum number of amplitude levels and

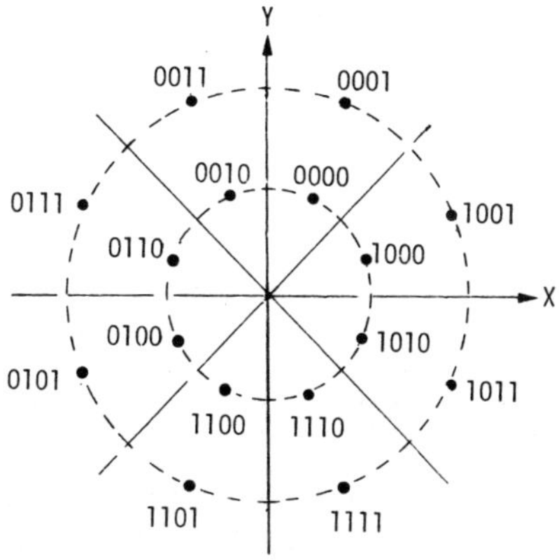

Fig. 6.37 IAPSK constellation pattern

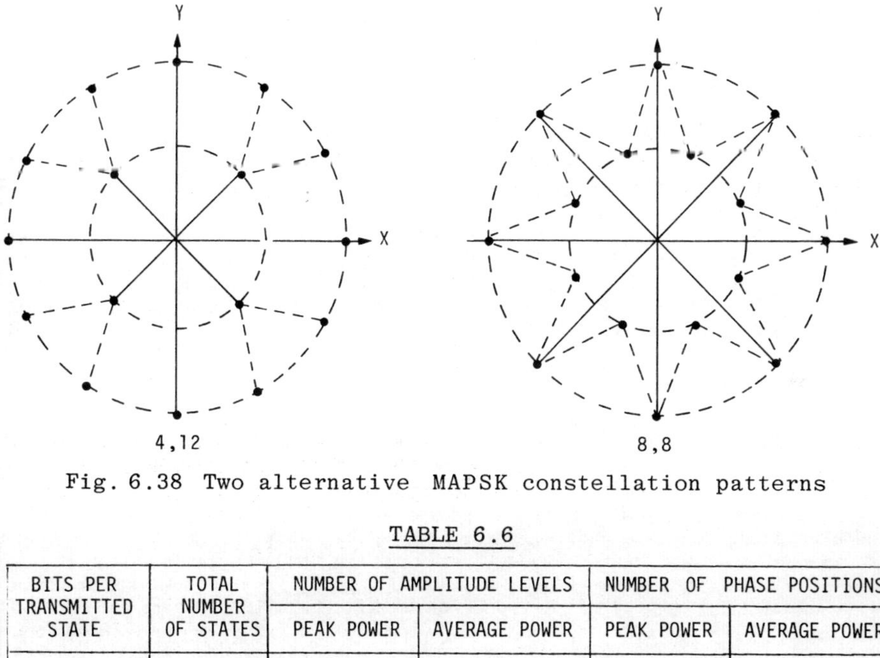

Fig. 6.38 Two alternative MAPSK constellation patterns

TABLE 6.6

BITS PER TRANSMITTED STATE	TOTAL NUMBER OF STATES	NUMBER OF AMPLITUDE LEVELS		NUMBER OF PHASE POSITIONS	
		PEAK POWER	AVERAGE POWER	PEAK POWER	AVERAGE POWER
1	2	1	1	2	2
2	4	1	1	4	4
3	8	1	2	8	4
4	16	2	2	8	8
5	32	2	4	16	8
6	64	4	4	16	16
7	128	4	8	32	16
8	256	8	8	32	32

phase position required for minimum peak and average power, as suggested by C.R. Cahn.

As already stated, the table p.291 is only theoretical, though many research projects are in progress to increase the data transmission rate. At this stage and date, however, the modems available on the market for wire transmission over the public telephone network can only transmit up to 9600 bit/second. The following table shows some of the various schemes used to achieve this speed.

TABLE 6.7

BITS PER TRANSMITTED STATE	TOTAL NUMBER OF STATES	MODULATION RATE BANDS	DATA RATE BITS/SECOND
2	4	1200	2400
2	4	2400	4800
3	8	1600	4800
3	8	2400	7200
4	16	1800	7200
4	16	2400	9600
5	32	1920	9600
6	64	1600	9600

As already explained, increasing the number of levels reduces the horizontal eye opening and consequently causes an increase in the intersymbol interference. This penalty could be minimized by specially shaping the pulses before modulation. For high speed transmission, however, more precise control of the intersymbol is required. This is achieved only by using automatically adjustable equalizers of the transversal filter type, which automatically correct both amplitude and delay distortion.

6.9 NEW MODULATION TECHNIQUES FOR RADIO COMMUNICATION

According to the stipulation of the U.S. Federal Communication Commission (FCC), 90% of the transmitted radio power must be concentrated within the allocated frequency bandwidth. The reason for this request is that as the number of available radio channels is limited so each channel is assigned a confined bandwidth having minimum spacing between any two adjacent channels. The clustering of the channels, then, is possibly only if the transmission in one channel does not disturb the transmission in other channels.

To comply with the FCC restrictions, each transmission must be first filtered so that the signal attenuation within the allotted band is zero and the attenuation outside the band is maximum with sharp transitions from minimum to maximum. Such an ideal filter however is impracticable, and so, in order to comply with the FCC stipulation, the actual transmission exploits a narrower bandwidth than the allotted one. Restricting the effective frequency band consequently limits the data rate that may be transmitted within the channel. This is the reason why the research projects in the last few years have been directed to finding new modulation techniques which will enable the efficient exploitation of the available radio channel band whereby maximum data could be transmitted without the channel sideband disturbing adjacent channels.

With wire communication there is no need to restrict the transmission, as each telephone line is spaced physically away from the others, with each line having the same frequency bandwidth, limited only by the wire characteristics. In this case the 'high' speeds are achieved by employing MPSK and MASK modulation techniques. These modulation techniques are relatively simple to implement and provide excellent ratio of errors to data rate. For wireless transmission, these modulation techniques, which were found so ideal for wire transmission, have their drawbacks, as will be explained further on. This does not mean that the MPSK and MASK are unsuitable for radio transmission (in fact, many radio systems employ these systems) but only that where the channel efficiency is the main criterion, other means may be necessary.

Most of the equipment used for radio communication employs FM techniques, and if digital data are to be transmitted with the equipment it is important to adapt the modem employed to this transmission media. In frequency modulation any amplitude changes in the analog-coded-digital signal will cause the addition of extra sidebands at the two extreme edges of the spectrum. These sidebands are usually very small and can generally be disregarded as they do not expand the main transmission spectrum. Nevertheless, when the channels are clustered together, these sidebands, generated by a number of channels, may accumulate and could then have a harmful effect. To overcome this problem, either the amplitude changes have to be restricted or the channel spacing increased. The same effect is also confronted in cable or carrier network where it is known that only one out of three channels may be used for digital data while the other channels are used for speech.

What exactly happens in ASK or PSK transmissions which can cause this effect? It has been stated that with PSK there is no amplitude level changes. This is correct. However, when passing the PSK signal through a filter (as is essential when using FM equipment) then the filtered enveloped signal has an amplitude component, as seen in Fig. 6.39. These amplitude level changes are those which could cause sidebands outside the spectrum and could consequently create errors in the data transmitted in adjacent channels.

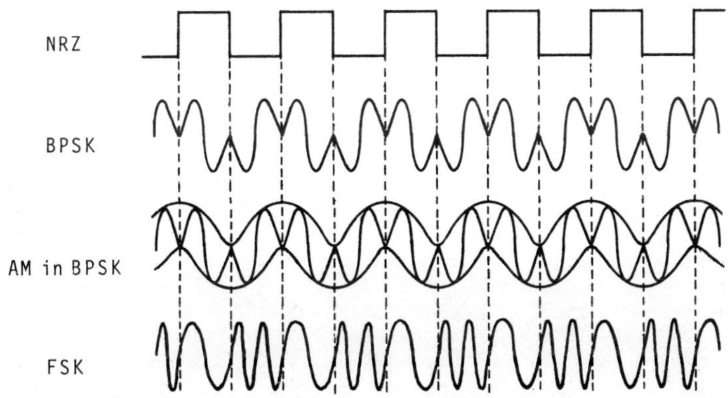

Fig. 6.39 Amplitude changes in the envelope of filtered PSK signals

In the previous sections, when discussing wire communication, we rejected the use of FSK for transmission of data at rates above 1200 B/S. The reason for this was the complex sidebands and its inferior performance as compared with PSK (where the received signal-to-noise ratio of FSK is 11 dB worse than PSK). Despite these drawbacks, it is worth considering FSK for wireless transmission as it has the essential advantage that its signal has no AM component (as seen in Fig. 6.39) and as such is ideal for FM transmission equipment.

6.10 MINIMUM (FREQUENCY) SHIFT KEYING (MSK)

It has been shown that the frequency spectrum of FSK is a direct function of the modulation index. A mathematical analysis of this spectrum requires Bessel functions, because of the nonlinear components of this signal. The analysis is more complex when transmitting NRZ digital data, since the modulation index varies constantly with the rate of change of digital data. For most FSK transmissions used in wire communication the maximum modulation index employed is $m=0.7$ to $m=1.0$. The selection of $m=0.7$ has been claimed to be optimal because it has lower intersymbol interference than that of PSK. Nevertheless, with PSK multilevel transmission could be used, and this is not possible with FSK: for this reason it has been accepted that PSK is superior to FSK. Lately, however, it has been found that with the modulation index $m=0.5$ a large portion of the energies transmitted are contained within a narrow frequency spectrum even without any band limiting. This means that this type of modulation (FSK with $m=0.5$) requires less bandwidth than BPSK and less transmitted power than any of the conventional modulation techniques already discussed. Furthermore, by differential coding of the data the modulation index can be kept constant.

The advantage of FSK modulation with $m=0.5$, whereby it occupies less bandwidth, and hence enables more data to be transmitted, has caused it to be called fast-frequency-shift-keying (FFSK). A much more appropriate title is the one based on the minimum modulation index which refers to this technique as minimum (frequency) shift keying (MSK). There are many other names that are used for this modulation scheme. Some of which will be presented later.

Before continuing the discussion of MSK (or FFSK), it is essential to present some theoretical background of this technique. The modulation has been defined as the ratio of the frequency deviation to the data base frequency:

$$m = 2\Delta\omega/\omega_m = 2\Delta f/f_m .$$

The frequency deviation is defined as the maximum shift of the carrier frequency in each direction when modulated, i.e., $f_c \pm \Delta f$. Since this shift is defined by the two frequencies allotted to the data states, the frequency deviation could be represented as

$$\Delta f = (f_2 - f_1)/2,$$

as seen in Fig. 6.40. The data base frequency f_m is equal to the data rate R which is defined as the reciprocal of the clock timing T, i.e. $f_m = R = 1/T$

$$m = (f_2 - f_1)/R = T(f_2 - f_1) = 2T\Delta f.$$

With MSK we define $m=0.5$ and thus the frequency deviation could be redefined:

$$m = 2T\Delta f = 0.5 \; ; \qquad \Delta f = 1/4T = R/4 \; ; \qquad (f_2 - f_1) = 1/2T = R/2 \; .$$

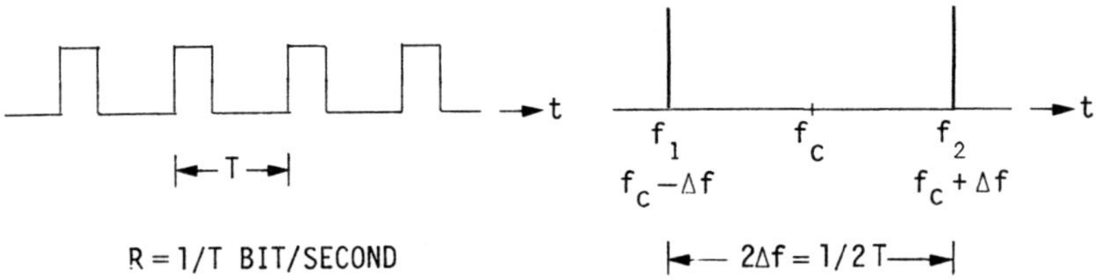

Fig. 6.40 Time and frequency spectrum in MSK

To achieve the proposed characteristics of MSK it is essential that there are no discontinuous changes in phase when shifting from one frequency to the other, i.e., from f_2 to f_1 and back to f_2. In other words the phase change must be continued without any sudden phase shifts. For this reason this form of modulation could also be regarded as a special case of what is called continuous-phase-frequency-shift-keying (CPFSK). This means that for each T duration there must be complete frequency cycles.

The frequency shift keying signal has been expressed as

$$S(t) = A_c \cos(\omega_c \pm \Delta\omega)\, t \; .$$

Although not necessary, this expression can be further expanded:

$$S(t) = A_c \cos(\omega_c - \Delta\omega) t = A_c \cos \omega_1 t \, , \quad \text{for 0 bit}$$

and

$$S(t) = A_c \cos(\omega_c + \Delta\omega) t = A_c \cos \omega_2 t \, , \quad \text{for 1 bit}$$

where $f_2 = \omega_2/2\pi$ and $f_1 = \omega_1/2\pi$ refer to the two modulation frequencies.

In MSK we define the frequency deviation as being constant and equal to $\Delta\omega = 2\pi\Delta f = \pi/2T$. The MSK signal can thus be expressed as

$$S(t) = A_c \cos(\omega_c \pm 2\pi\Delta f) t = A_c \cos(\omega_c \pm \pi/2T) t \; .$$

This transmitted signal specifies the main characteristic feature of MSK whereby there is a $\pm 90^\circ$ carrier phase shift for each bit transmitted. This means that the radian frequency being transmitted is either $(\omega_c + \pi/2T)$ for data bit 0 (corresponding to level $+1$) or $(\omega_c - \pi/2T)$ for data bit 1 (corresponding to level -1). For this reason it is possible to regard MSK also as a special case of offset PSK.

To demonstrate the MSK technique, Fig. 6.41 presents the various waveforms which compose the transmitted modulated signal. Comparing the transmitted signal with the carrier signal reveals that for each bit interval there is a $90^\circ = \pi/2$ change. In this drawing the selection of frequencies was $f_2 = 2f_1$ which is an impractical situation as, in fact, both should be whole multiples of the data rate. This may cause the necessity of adding to f_1 and/or f_2 a constant of either $X = 0^\circ$ or $X = 180^\circ$ so as to keep the phase continuous. Thus when the MSK is viewed as CPFSK the waveform can be expressed as

$$S(t) = A_c \cos(\omega_c \pm \pi/2T)\, t + X \; .$$

This extra phase constant is shown in Fig. 6.41 for f_1 and f_2 by dotted lines, although it is not necessary for that example. By properly coding the data differentially this constant phase addition can be omitted.

Summing up the essential properties of MSK (or FFSK) gives the following points:

(a) The peak frequency deviation Δf is exactly a $\pm 1/4$ of the data rate R, i.e., $\Delta f = \pm R/4$.

(b) The phase of the transmitted signal $S(t)$ changes linearly exactly $\pm 90^\circ$ with respect to the carrier frequency during each clock bit period of $T = 1/R$.

(c) There is no amplitude level change or phase discontinuities in the transmitted signal at the data bit switching intervals.

(d) The signal can be coherently detected since the frequencies f_1 and f_2 must be whole number multiples of the data rate.

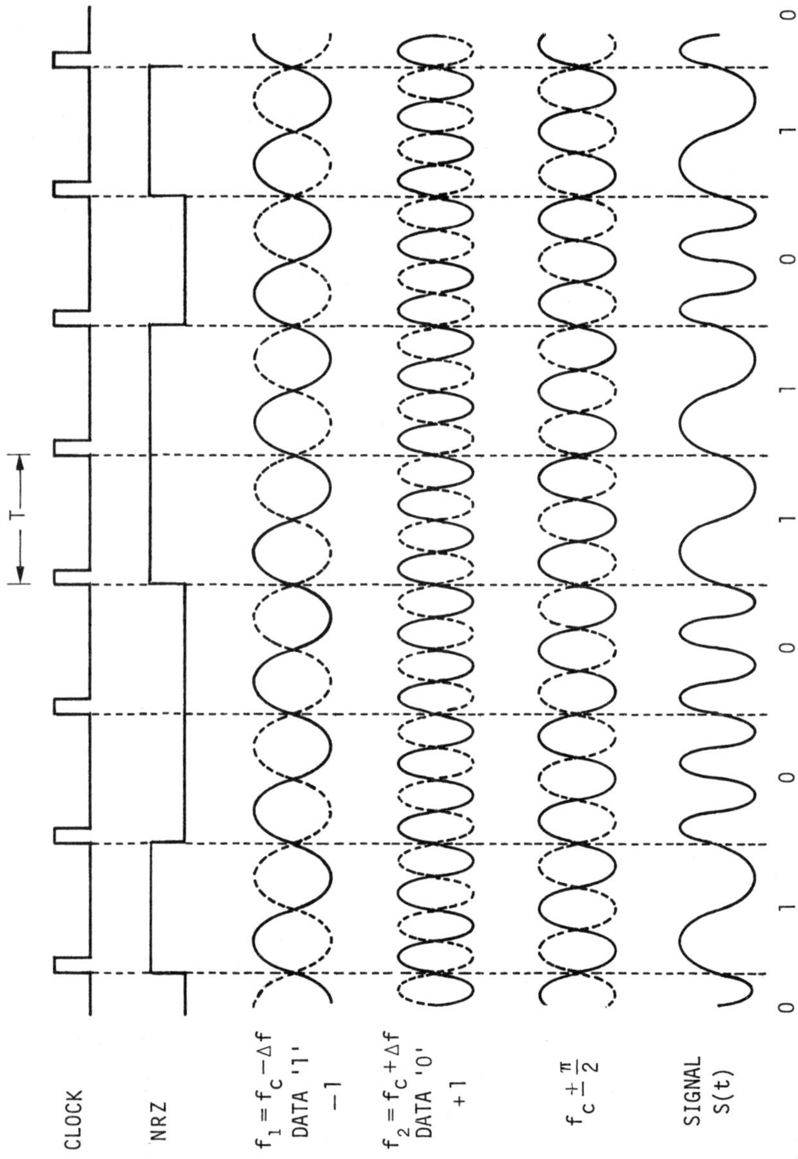

Fig. 6.41 The signals which compose MSK transmission

6.11 MSK AS A SPECIAL FORM OF CONTINUOUS PHASE FSK

It has already been pointed out that with the MSK viewed as a special form of FSK the amplitude is kept constant and also that the phase must be continuous. Although PSK can also be regarded as a special form of FSK, both the properties mentioned for MSK do not apply to PSK. This is better followed with the aid of Fig. 6.42, which shows the waveform and the polar diagram of QPSK. The four possible modulation states can be presented in a circular pattern, with each state defined by a space vector of constant amplitude and its angle to a fixed reference. Despite the fact that the states are shown in a circular pattern, the switching between the states does not follow the circumference of the circle. As seen by the waveforms, the phase shifts are rather abrupt. The polar diagram presents the vectors of the space summation of the two orthogonal X and Y signals. In certain instances the space shifts could cause transition through the centre of the circle, as for example the transition from 00 to 11 which results in a shift of 180°. These transitions result in nonlinear changes at the dibit change point. Even phase changes of 90° do not shift along the circumference of the circle, but along the shortest available path. These discontinuous transitions are magnified once the transmitted signals pass through a bandpass filter.

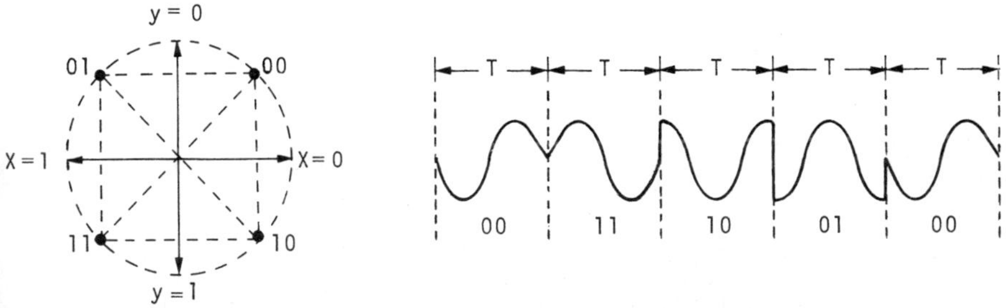

Fig. 6.42 Phase and amplitude transitions in QPSK

MSK signals can also be presented by a space vector of a constant amplitude and as such its modulation states can also be presented in a circle. MSK, however, has the property that the phase shift must be continuous, with no momentary zero amplitude signals when switching from one state to another. This property ensures that the shifting must traverse only along the circumference of the circle so as to maintain a constant vector amplitude, with no possibility of shorter paths. A further property of MSK is that the transition must be coherent, which results in a ± 90° shift relative to the carrier frequency phase for each bit transmitted. This rotation of the transmitted carrier vector can be seen in Fig. 6.43. If at time t_0 the transmitted bit was 0, the carrier vector at time t_1 must have a ±90° phase shift relative to that at time t_0. If a 0 bit is transmitted then the vector will rotate by +90° relative to the original vector and if a 1 bit is transmitted the vector will rotate by −90° relative to the original vector.

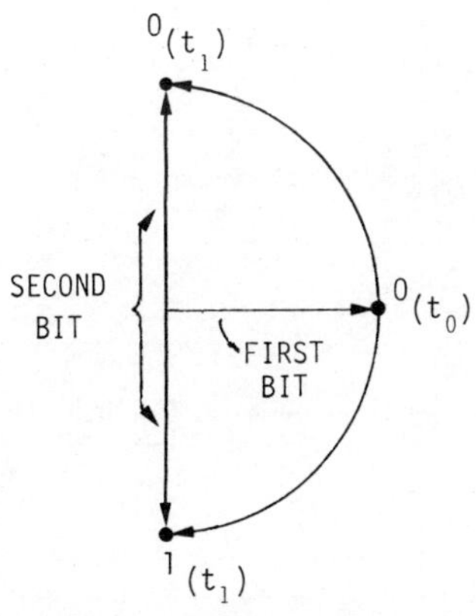

Fig. 6.43 MSK phase rotation after the second bit

The third transmitted bit rotates the carrier vector by a further ± 90°

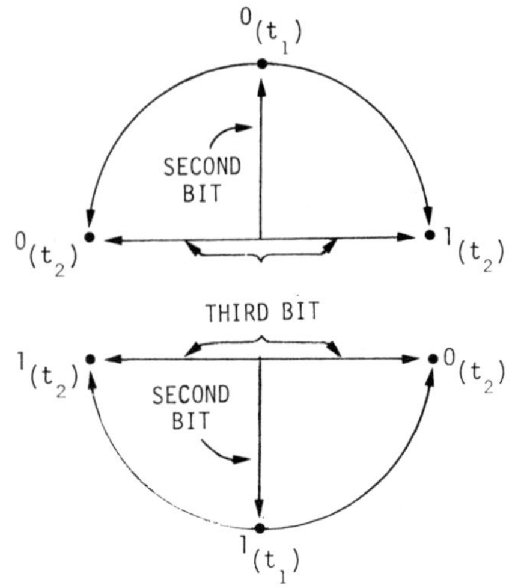

Fig. 6.44 MSK phase rotation after the third bit

phase shift, as seen in Fig. 6.44. This means that the bit transmitted at time t_2 will be in phase or anti-phase relative to the original bit transmitted at time t_0 and will be orthogonal to the vector transmitted at time t_1.

The constellation pattern of MSK is in a circle, although each point does not correspond to any given state, as in QPSK, but must always be compared with the previous transmitted bit. If the constellation is reviewed as a phase pattern, MSK appears to be differentially encoded in that the phase depends on the present bit and the ending phase of the previous bit interval. This can be expressed mathematically:

$$\theta(t_n)-\theta(t_{n-1}) = a_n \cdot \pi/2 ,$$

where $\theta(t_n)$ corresponds to the phase at bit t_n, and a_n corresponds to value +1 if the data bit 0 is transmitted or to value −1 if data bit 0 is transmitted.

The MSK transmitter is similar to the FSK transmitter having the output signal shifted between two frequencies. The main difference is that the frequencies selected are defined as a function of the data rate, i.e., $(f_2-f_1)=R/2$, and that there is a carrier phase shift of 90° for each transmitted bit. The continuous signal phase is achieved by differentially coding the data before it selects the desired frequencies. The block diagram of the essential units of the MSK transmitter is shown in Fig. 6.45. In order to smooth the frequencies change at the shifting points the actual modulation is performed at double the required frequency and then divided down to the right frequency after the analog switch. Other transmitter solutions are also possible, where a single oscillator is used and then the modulation frequency need not be doubled. In that case, the frequency of the tuned circuit is changed by adding across it an extra capacitor.

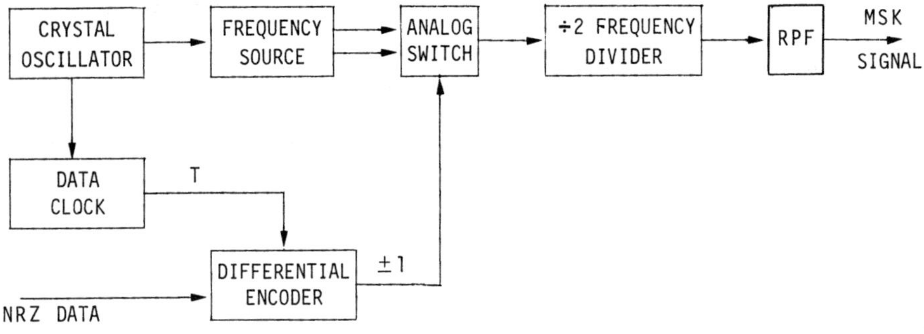

Fig. 6.45 Blocked diagram of an MSK transmitter when viewed as CPFSK

The differential encoder is intended to ensure phase continuity and is resolved by selecting the present transmitted frequency not only according to the bit to be transmitted but also according to the bit previously transmitted. This encoder, shown in Fig. 6.46, is simply the modulo-2 addition

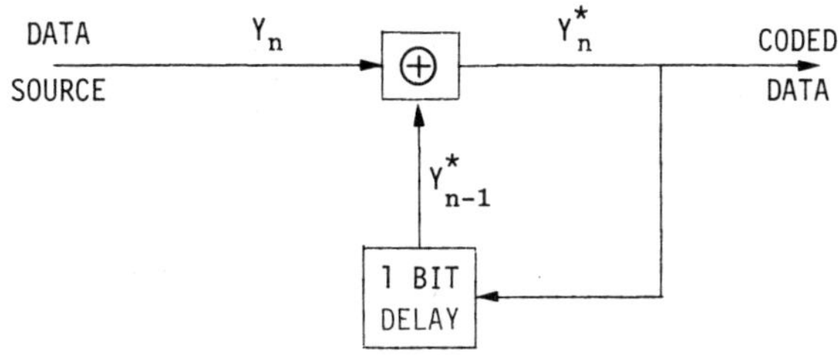

Fig. 6.46 Differential encoder

(EXOR) of the two bits, i.e., the past and present. If the present bit is defined as Y_n, then the present encoded bit is defined as $Y_n^* = Y_n \oplus Y_{n-1}^*$.

The encoder operates in accordance with the following table with the data bits of 0 and 1 represented by levels +1 and −1 respectively. In this case the modulo-2 addition can be regarded as a pure multiplication of the two levels. The table also shows the phase shifts of ± 90° in accordance with the differential coded levels.

TABLE 6.8

PRESENT INPUT BIT Y_n	PREVIOUS OUTPUT BIT Y_{n-1}	PRESENT OUTPUT BIT Y_n^*	PHASE SHIFT θ
0 +1	0 +1	0 +1	+90°
0 +1	1 −1	1 −1	−90°
1 −1	0 +1	0 −1	−90°
1 −1	1 −1	0 +1	+90°

6.12 MSK AS A SPECIAL FORM OF OFFSET QPSK

A second approach to MSK is to visualize it as a quadriphase system with offset keying of the data and sinusoidal symbol shaping. In other words the MSK can be viewed also as a special form of offset QPSK (or offset QASK). Both QPSK and offset QPSK (also referred to as off-keyed QPSK, OK-QPSK) are modulated by two streams of orthogonal carrier signals, but, with the offset transmitted signal, can only be ± 90°, while in conventional QPSK it is possible to have phase reversals of ± 90° or ± 180°. This means that both offset and conventional QPSK have abrupt phase transitions, but by sinusoidal symbol shaping of the OK-QPSK signal the continuous phase nature of MSK can be achieved.

It has already been shown that the QPSK (or QASK) signal can be expressed as follows:

$$S(t) = X(t)\cos\omega_c t + Y(t)\sin\omega_c t$$

where $X(t)$ and $Y(t)$ take the level values of ± 1 according to the data bit values of 0 and 1. Offset QPSK or Offset QASK imply that the quadrature component $Y(t)$ is delayed by half a symbol period, relative to $X(t)$. If each symbol period T_s is lengthened to $2T$ then $Y(t)$ is delayed by $T_s/2 = T$, which can be expressed as follows:

$$S(t) = X(t)\cos\omega_c t + Y(t-T)\sin\omega_c t\ .$$

This means that with OK-QPSK, the modulation time of the two orthogonal carrier signals is staggered. For this reason, this technique is also referred to as staggered QPSK (SQPSK).

To visualize the offset QPSK as MSK, the data dependent symbol is first sinusoidal shaped and thus the signal can take the following form:

$$S(t) = X(t)\cos\Delta\omega t\cos\omega_c t + Y(t)\sin\Delta\omega t\sin\omega_c t$$

To better visualize the MSK as a special form of offset QPSK, Fig. 6.47 presents the various stages in achieving this. In Fig. 6.47(a) the conventional QPSK is presented where both orthogonal signals are modulated at time T. In Fig. 6.47(b) the two orthogonal channels are staggered, with each channel modulated at $T/2$ timings delay relative to each other. In Fig. 6.47(c) the offset QPSK signal is sinusoidal shaped by an extra modulator with its frequency defined by the modulation index of $m = 0.5$, i.e., $f = 1/4T$.

The mathematical expression of the MSK signal $S(t)$, received by sinusoidally shaping the offset-QPSK, can be further expanded, giving:

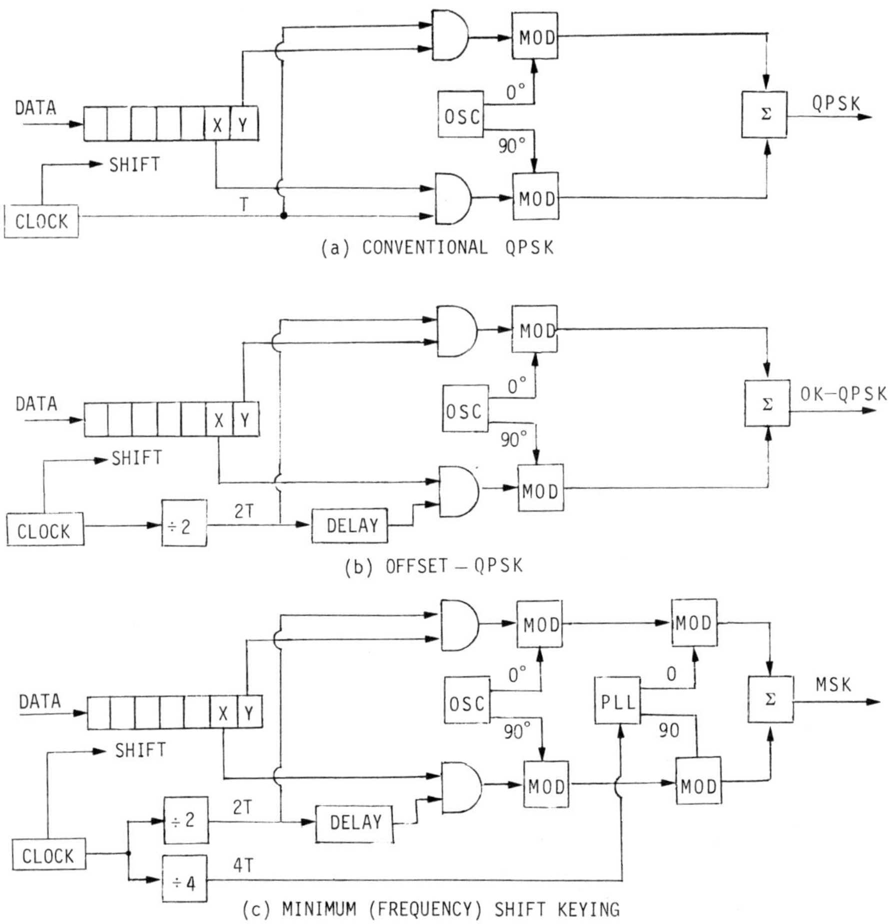

Fig. 6.47 The various stages in achieving MSK

$$S(t) = X(t)/2\,[\cos(\omega_c+\Delta\omega)t+\cos(\omega_c-\Delta\omega)t]$$
$$-\;Y(t)/2\,[\cos(\omega_c+\Delta\omega)t-\cos(\omega_c-\Delta\omega)t]$$
$$=\left\{[X(t)\mp Y(t)]/2\right\}\cos(\omega_c\pm\Delta\omega)t\,.$$

The expression received is the same expression already calculated for MSK when it was viewed as a form of CPFSK by simply replacing $[X(t)\mp Y(t)]/2$ by A_c. This is allowed, since both values can only have the values of ± 1. The sinusoidal shaping frequency has already been defined for MSK as being equal to

$$\Delta f = (f_2-f_1)/2 = 1/4T,$$

thus giving

$$\Delta\omega = 2\pi\Delta f = \pi/2T\,.$$

This can now be inserted into the MSK signal when viewed as a form of OK-QPSK:

$$S(t) = X(t)\cos\omega_c t\cos\pi t/2T + Y(t)\sin\omega_c t\sin\pi t/2T\,.$$

In each channel there is a phase change of 180° during the period of $2T$, which gives the desired carrier phase shift of 90° during the period T. The modulation timing of the two channels are staggered so that during every period T, an alternate channel is modulated. The whole process can be visualized with the aid of Fig. 6.48. The NRZ data stream which is to be modulated is presented in (a). This stream is divided into two skewed data streams, $X(t)$ and $Y(t)$, represented in (b) and (c), which are consequently used to modulate the two channels. The two sinusoidal waves presented in (d) and (e) are orthogonal to each other, both having the frequency $\pi/2T$. The BPSK modulation result of $X(t)$ and $Y(t)$ with each of the orthogonal signals is as presented in (f) and (g). If we should sum the two signals in (f) and (g), the result would be OK-QPSK, however, to obtain the MSK signal each channel must be first amplitude modulated by the carrier signal. The result of the last modulation in each channel is presented in (h) and (i). The interlace signal presented in (j) is composed by the sum of the signals in (h) and (i), This signal is the required MSK signal with constant amplitude, continuous phase when switching from f_1 to f_2, and $\pm 90°$ phase shift relative to the carrier frequency during the period of one data bit T.

By visualizing MSK as a special form of offset QPSK allows the transmitter to be redesigned, as shown in Fig. 6.49. The input digital data, shifted with clock timing T, is divided into two parallel data streams, each feeding channels X and Y respectively. The two channels are operated by clock timing $2T$, although they are aligned so that an offset of T seconds exists between them. The data in each channel represent alternative successive data samples at T seconds of the original stream. Each channel is phase modulated by a reference sinusoidal wave of frequency $f = 1/4T$ and by a carrier frequency of f_c. The reference and carrier waves of each channel are orthogonal to that of the other. The signals of both channels are then summed together so as to produce the required MSK signal.

Summing the properties of MSK when viewed as a special form of OK-QPSK:

(a) The signal produced by the MSK transmitter when viewed as an offset QPSK with sinusoidal symbol shaping and the signal produced by the MSK transmitter when viewed as CPFSK with $m = 0.5$ are identical.

(b) The transmitted signal has a constant amplitude envelope.

(c) There is a phase continuity of the modulated signal at the data bit switching instances.

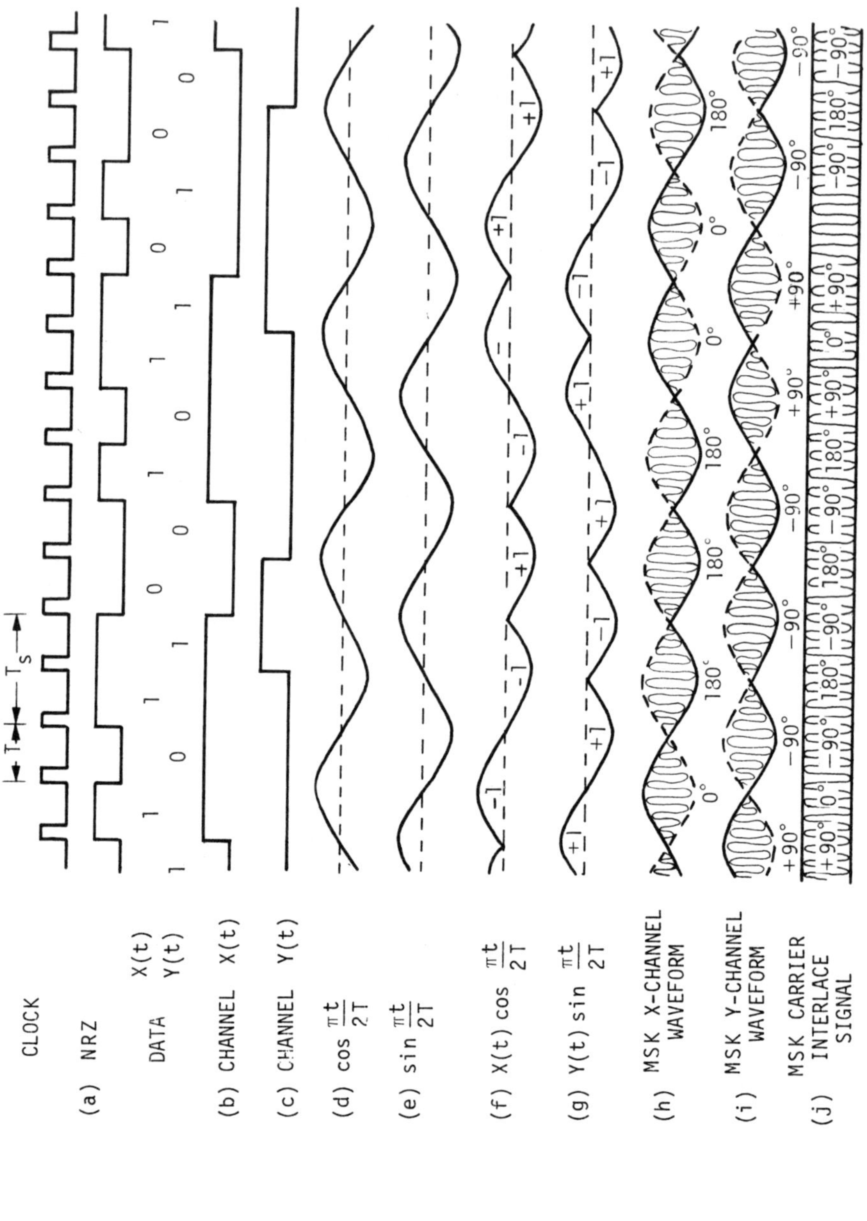

Fig. 6.48 MSK signals when viewed as a sine multiplied OK- QPSK

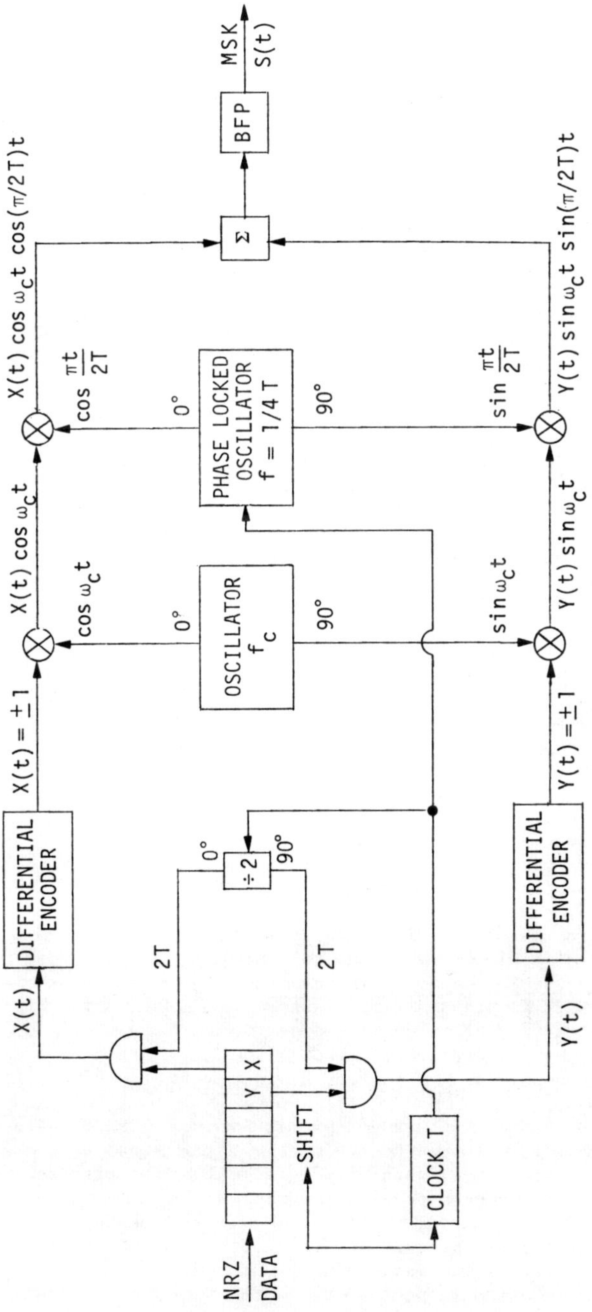

Fig. 6.49 MSK transmitted when viewed as a sin multiplied OK-QPSK

(d) The transmitted signal could be regarded as composed by switching between two frequencies, of the frequency $f_2 = f_c + \Delta f$ and $f_1 = f_c - \Delta f$, in accordance with the data.

(e) The transmitted signal could also be regarded as composed by the summation of two BPSK signals with sinusoidal symbol shaping.

(f) The two BPSK are amplitude (phase) modulated by half a sinusoidal of the frequency $f = 1/2T$.

(g) The modulation signals and the sinusoidal shaping signals applied to each channel are orthogonal to each other.

(h) The data bit instances of each channel are skewed relative to each other so that the modulation point of one signal is performed when the amplitude of the other signal is zero.

(i) The phase of the transmitted signal is always shifted by $\pm 90°$ relative to the carrier during each data bit period.

6.13 MULTI-AMPLITUDE MINIMUM (FREQUENCY) SHIFT KEYING

Since it is possible to view MSK as a special form of offset QPSK or offset QASK it is also possible to exploit the QASK properties for expanding the MSK applications. MSK has the important advantage that over 90% of its transmitted power lies in the main lobe of the spectrum. QASK is a form of phase modulation which has the advantage that many amplitude levels can be transmitted without the signal being prone to the disadvantages of ON/OFF amplitude modulation.

The system proposed here is a form of MSK with multiple amplitude levels, which as such could also be viewed as OK-M-QASK with sinusoidal symbol shaping. Although the resulting signal of this multi-amplitude MSK (referred to as MAMSK) is not of constant amplitude, its properties are superior to those of MQASK. With multilevel QASK, as with multiphase QPSK, the switching has abrupt phase changes, since the state shift sometimes passes through the centre of the constellation pattern. With MAMSK, as with MSK, there are no abrupt phase changes, as the swiching must follow the circumference of a circular path (see Fig. 6.51), which results in most of the transmitted power being still concentrated in the main lobe of the spectrum.

In ordinary MSK, when viewed as a special form of OK-QASK, two binary bits were modulated separately by half a sinusoidal, with each channel switching time staggered relative to each other. With MAMSK the same principles are kept, except that here four binary bits are modulated, having every two bits modulated by half a sinusoidal in each channel. To visualize MAMSK as a special form of OK-M-QASK, the two modulated signals must be sinusoidally shaped with the frequency defined by the modulation index $m = 0.5$. This means that the MAMSK transmitter, as shown in Fig. 6.50, is principally the same as that used for MSK when viewed as OK-QASK (see Fig. 6.49), except that the timing must be modified. Although four bits are modulated in parallel, only two channels are required, having every two bits differentially encoded.

With MSK, when viewed as a special form of CPFSK, each symbol was modulated separately so that the bit timing T was equal to the symbol time T_s, i.e., $T_s = T$. With MSK, when viewed OK-QASK, two bits were alternatively modulated which required each symbol to be lengthened to $2T$, i.e. $T_s = 2T$. With MAMSK, four bits are modulated for each complete symbol, which requires further lengthening of its timing, $T_s = 4T$. In both MAMSK and MSK (when viewed as OK-QASK) the quadrature components of the signals are delayed by $T_s/2$ with respect to the other. This means that for MAMSK there is still a $\pm 90°$ phase shift in respect to the carrier during each

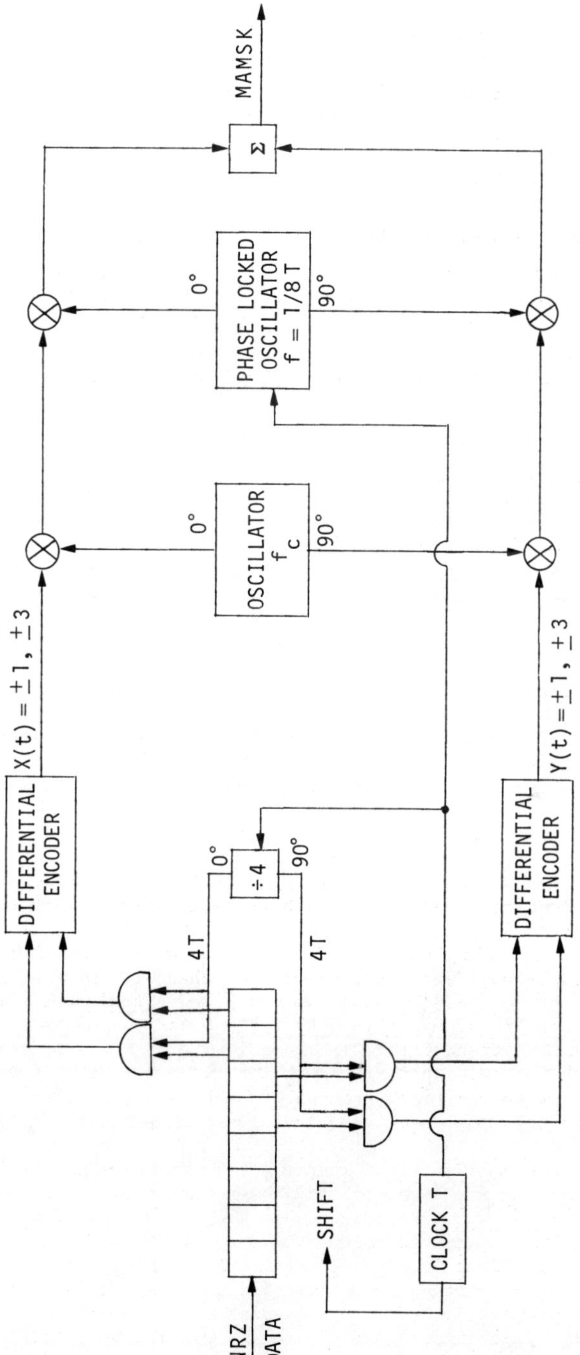

Fig. 6.50 MAMSK transmitter when viewed as a sine multiplied OK–M–QASK

dibit interval. Since two bits are differentially coded together the MAMSK signal takes the following form

$$S(t) = X(t)\cos\omega_c t\cos\pi t/4T + Y(t)\sin\omega_c t\sin\pi t/4T\,.$$

DATA	$X(t)$ or $Y(t)$
00	+1
01	+3
10	−1
11	−3

The two amplitude levels $X(t)$ and $Y(t)$ which now correspond to four digit bits, take the values of ±1, ±3 respectively. This means that every dibit is represented by a different amplitude level, as shown in the table.

Theoretically, the four values of $X(t)$ and $Y(t)$ could produce a square constellation pattern of 16 states. It is stipulated, however, that for MAMSK there is a ±90° linear phase shift with respect to the carrier during each dibit interval. It is impossible to present the MAMSK on a constellation diagram but only by a phasor-polar display. With MSK, the phase trajectory on the polar diagram is a circle, as shown in Fig. 6.51(a), since there are no amplitude level changes of the transmitted signal. With MAMSK, there are two discrete amplitude signal levels, 1 and 3; however, it is also stipulated that the change follows a sinusoidal wave with no abrupt changes. As long as there are no amplitude level changes, the switching will follow the circumference of a small (1) or large (3) circle. If there is an amplitude level change, the switching must follow the circumference of an ellipse, as shown in Fig. 6.51(b). The polar diagram points correspond to ±90° phases relative to a previous starting phase. In other words, the diagram points represent transition states rather than fixed states as described previously for the constellation pattern. For MAMSK the phasor diagram arcs correspond to data dibit inputs, as compared with MSK where the arcs correspond to data bit inputs. This pattern of MAMSK which has 16 arcs demonstrates its superior features as compared with QASK with 16 states. This means that despite the amplitude levels of MAMSK the power spectrum is still concentrated in the centre lobe.

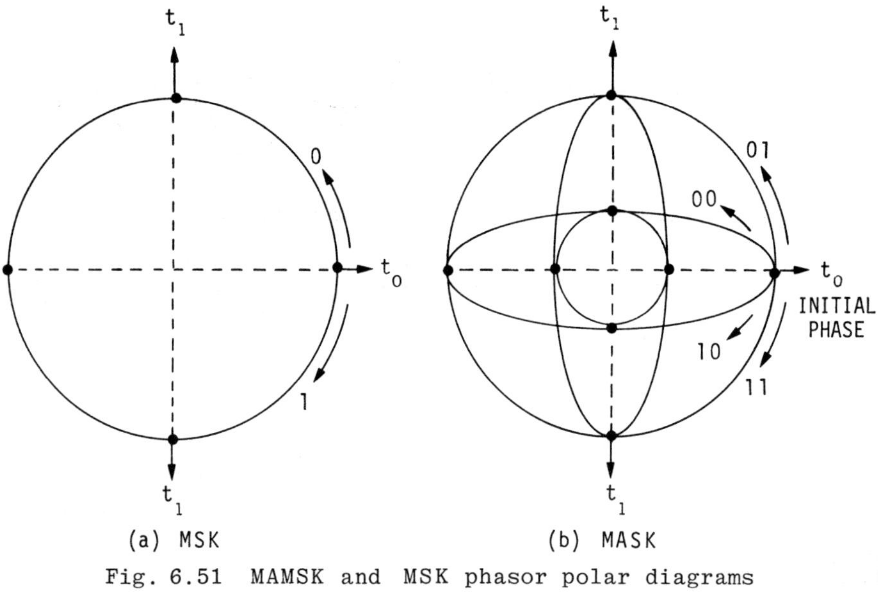

Fig. 6.51 MAMSK and MSK phasor polar diagrams

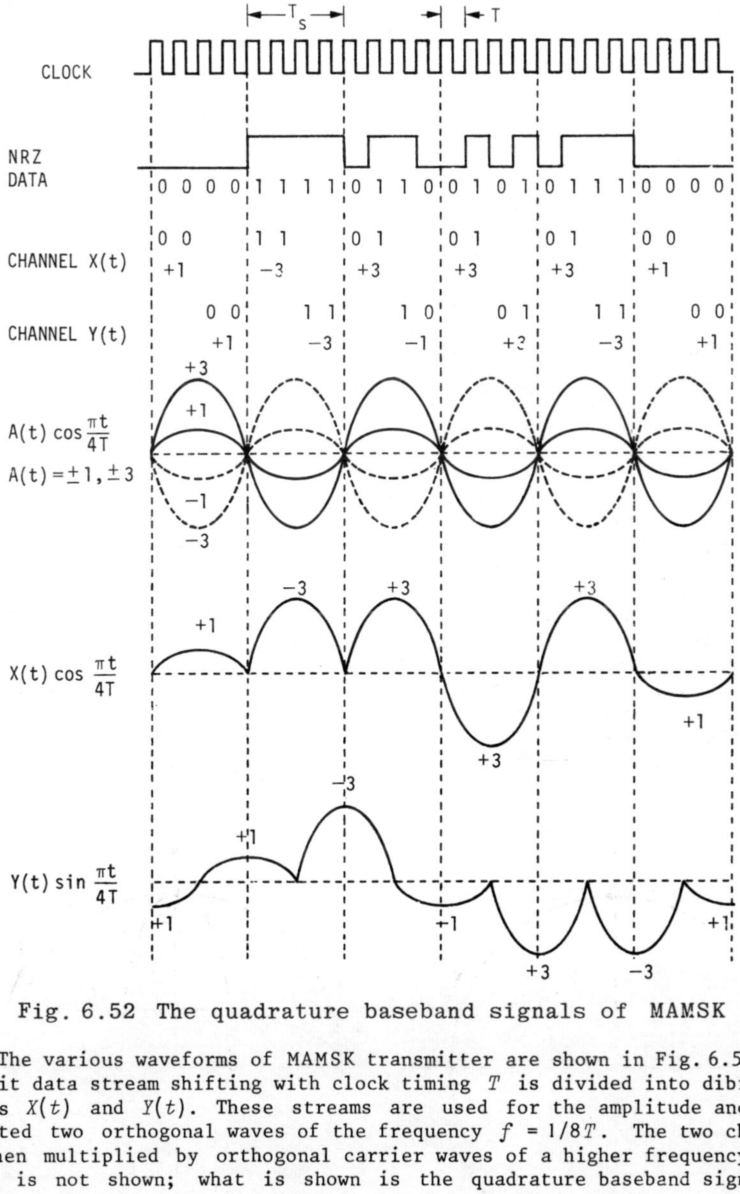

Fig. 6.52 The quadrature baseband signals of MAMSK

The various waveforms of MAMSK transmitter are shown in Fig. 6.52. The NRZ bit data stream shifting with clock timing T is divided into dibit data streams $X(t)$ and $Y(t)$. These streams are used for the amplitude and phase modulated two orthogonal waves of the frequency $f = 1/8T$. The two channels are then multiplied by orthogonal carrier waves of a higher frequency. The latter is not shown; what is shown is the quadrature baseband signals of MAMSK.

6.14 COHERENT DETECTION OF MSK SIGNALS

The fact that MSK can be viewed both as a special case of OK-QPSK and CPFSK enables the detection to exploit both FSK and PSK demodulation techniques. There have been various techniques proposed for decoding the MSK signal. Two are presented here, based on exploiting both FSK and QPSK features. First, a self synchronizer circuit which extracts the reference

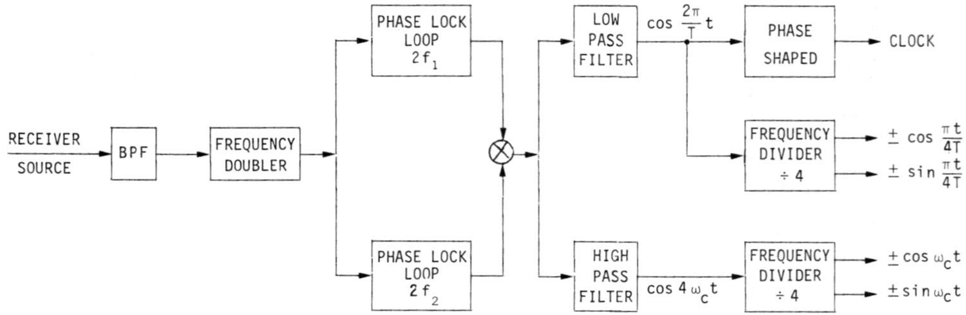

Fig. 6.53 Self synchronizer for extracting carrier and bit timing

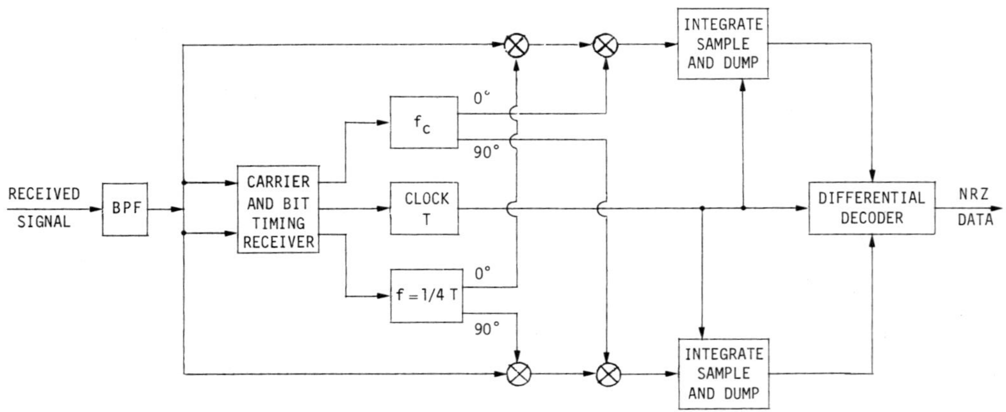

Fig. 6.54 Coherent detection of MSK signals

signals of the carrier, phase and clock, by employing coherent CPFSK techniques. Then with these references, the data can be decoded using coherent OK-QPSK techniques. Both circuits are simple and easy to construct.

The received signal when visualized as CPFSK is constructed by the switching of the carrier f_c between two frequencies f_1 and f_2. The received signal is distorted by additive white Gaussian noise (AWGN), $K(t)$.

$$Q(t) = A_c \cos(\omega_c \pm \Delta\omega)t + K(t) = A_c \cos\omega_1 t + A_c \cos\omega_2 t + K(t).$$

Since for MSK the modulation index is fixed, it is possible to regenerate the clock and carrier components :

$$f_2 + f_1 = (f_c + \Delta f) + (f_c - \Delta f) = 2f_c \qquad f_2 - f_1 = m/T = 1/2T .$$

The distance between the two frequencies is too small to be separated by a filter, but by doubling the input signal the distance could be increased :

$$[Q(t)]^2 = \cos^2(\omega_c \pm \Delta\omega) = \cos^2\omega_1 t + \cos^2\omega_2 t + \dots .$$

Part of the AWGN is filtered before the doubler and the remainder can be neglected, provided there is enough signal-to-noise ratio. The doubling process establishes spectral frequency components at $2f_1$ and $2f_2$ and these could be fed into two phase lock loops which enable the extraction of the two frequencies. The two frequencies are multiplied by means of a balanced modulator, which process can be expressed as follows:

$$\cos^2\omega_1 t \cos^2\omega_2 t = \tfrac{1}{2}[\cos(2\omega_2 + 2\omega_1)t + \cos(2\omega_2 - 2\omega_1)t]$$
$$= \tfrac{1}{2}[\cos 4\omega_c t + \cos 2\pi/T] .$$

The low frequency of the multiplier generates both the clock component, $T = 1/(2f_2 - 2f_1)$ and the phase reference which is produced by dividing the frequency into 4. The carrier component is generated from the high frequency and the result then divided by 4. The whole process is shown in Fig. 6.53.

The whole synchronizer process can employ FSK techniques while the signals obtained are then used for the data recovery employing QPSK techniques, as seen in Fig. 6.54. The phase and carrier frequencies obtained from the synchronizer can be used for coherent demodulation and detection of the data by using conventional quadriphase techniques. Two channels are employed each multiplied by the two reference frequencies, having the signals to each channel orthogonal to the other.

The process described for detection of the data still leaves room for many errors due to phase ambiguities. The phase ambibuities arise from any frequency divider. With phase-lock-loops the frequency is determined, but after the divider the frequency polarity cannot be defined and it is determined by the ± sign of the signal, which thus indicates the inherent ambiguity of the process. An alternative method has been suggested which is still easier to implement. In this new demodulation process, as shown in Fig. 6.55, the reference recovery circuit is modified. After the two double frequencies have been extracted by the phase-lock-loop they are divided by two instead of four, thus removing the fourth order effects which are the main cause of the phase ambiguities.

In this new demodulator the same doubler and phase-lock-loops are used. However, the frequencies are then divided by two to generate the two frequencies f_1 and f_2. The clock reference is obtained as before by simply multiplying the two signals

$$\cos\omega_1 t \cos\omega_2 t = \tfrac{1}{2}\cos(\omega_2 + \omega_1)t + \tfrac{1}{2}\cos(\omega_2 - \omega_1)t$$
$$= \tfrac{1}{2}\cos^2\omega_c + \tfrac{1}{2}\cos\pi t/T.$$

By filtering the above result it is possible to extract $\cos\pi t/T$ and

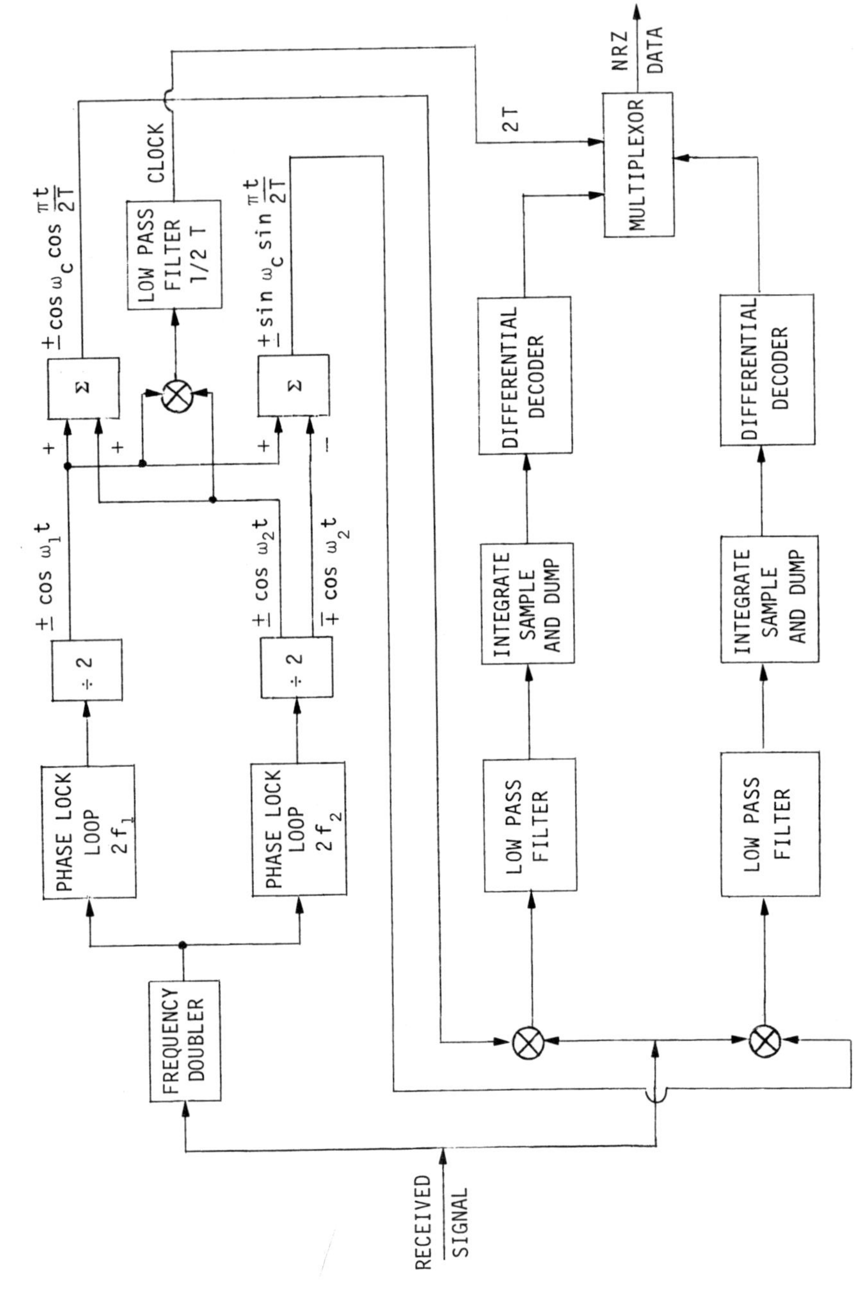

Fig. 6.55 Modified MSK demodulator circuit

from it to generate the clock reference of $2T$.

The phase ambiguities remain even after the frequency divide by the two processes, but this is of no consequence in this process. The two frequency components are added and subtracted in two separate summations, generating the following expressions :

$$\pm\cos\omega_1 t \pm \cos\omega_2 t = \pm 2\cos(\omega_1+\omega_2)t/2\cos(\omega_2-\omega_1)t/2$$

$$= \pm 2\cos\omega_c t\cos\pi t/2T .$$

$$\pm\cos\omega_1 t \mp \cos\omega_2 t = \pm 2\sin(\omega_1+\omega_2)t/2\sin(\omega_2-\omega_1)t/2$$

$$= \pm 2\sin\omega_c t\sin\pi t/2T .$$

Both the two signals obtained above have the ± sign, which means that the phase ambiguities still remain. By comparing these two signals, however, one can see that, despite the sign, they always remain orthogonal with respect to each other. This feature enables us to use the signal to multiply the received MSK signal in two parallel channels. The data can now be detected by the conventional QPSK techniques.

6.15 IMPROVING SPECTRUM OCCUPANCY

It has been shown that with radio communication the transmission must be concentrated in a narrow bandwidth if the spectrum is to be efficiently exploited. For this reason the use has been suggested of MSK techniques, which are ideal for FM modulation, and the focussing of its power spectral density in the main lobe of the bandwidth. This makes MSK an excellent radio communication technique since it efficiently utilizes the spectrum and offers good channel spacing economy. Nevertheless, the side lobes generated by MSK are still relatively 'high', and thus effort has been made to further restrict them. The MSK side lobes arose from the discontinuous change of frequencies of the transmitted signal relative to the carrier signal.

With PSK there are sharp phase changes in the transmitted signal in accordance with the data. These sharp phase changes of the signal are removed in MSK; this makes MSK spectral energy side lobe sharply below that of conventional BPSK or even QPSK signals before filtering. In MSK, however, there are still $\pm 90^\circ$ phase changes relative to the carrier in accordance with the data. It has already been shown that for each clock timing T, there is a $+\pi/2$ change relative to the carrier for data 0 and $-\pi/2$ change for data 1. These sharp phase changes can be plotted in a lattice form, known as trellis, as shown in Fig. 6.56. The trellis illustrates all the possible phase paths that the carrier signal can take after defining the initial phase as zero.

From Fig. 6.56 it can be seen that over each T second interval, the phase of the MSK carrier waveform is advanced or retarded precisely by + or - $\pi/2$, depending on whether the data input level is $+1$ or -1 respectively. It is evident that the phase at odd multiples of T will be added multiples of $\pm\pi/2$ and that at even multiples of T the phases will be real multiples of $\pm\pi$. This means the phase will be $\pm\pi/2$ at time nT with respect to the phase at time $(n-1)T$ and 0 or $\pm\pi$ at time $(n+1)$ with respect to the phase at time $(n-1)$. In general this can be defined as

$$\theta(t_n)-\theta(t_{n-1}) = a_n\pi/2$$

where a_n has the values of ± 1 in accordance with the data. This phase change can be seen in the MSK transmitted signal expression

$$S(t) = A_c\cos(\omega_c \pm \pi/2T)t$$

$$= \tfrac{1}{2}A_c\cos\omega_c t\cos\pi t/2T .$$

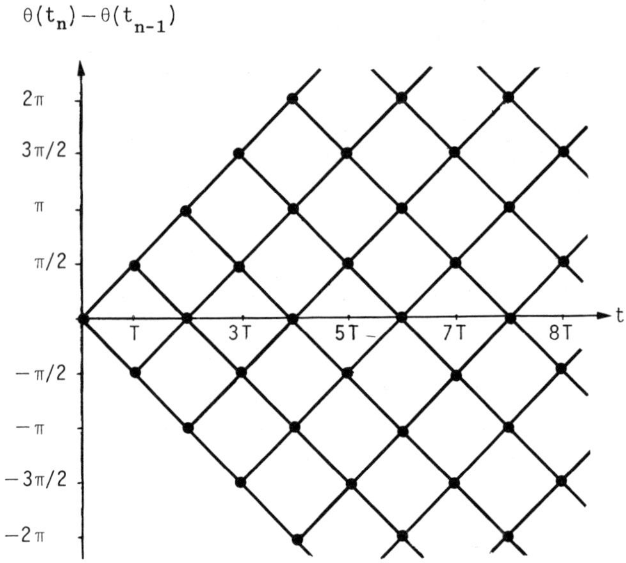

Fig. 6.56 Lattice of possible phase paths

It is possible to further improve the signal shaping, thus reducing the sidebands without sacrificing the constant envelope property of MSK. This could be performed by sinusoidally shaping the phase change by the continuous rate of data change. This means that the phase of $\pm\pi/2$ will have an extra component which will roll off the sharp edges of the carrier wave. This could be expressed as follows:

$$S(t) = \tfrac{1}{2}A_c \cos\omega_c t \cos(\pi t/2T - \tfrac{1}{4}\sin 2\pi t/T).$$

In this smoothing technique the frequency deviation, over the symbol interval, is sinusoidal, and for this reason it is referred to as sinusoidal-frequency-shift-keying (SFSK). Its power density sideband fall is substantially below that of MSK for $fT > 2$. The sharp change of the phase in MSK has completely disappeared in SFSK, as can be seen in Fig. 6.57, although the maximum slopes of the paths have considerably increased. This means that the phase changes of $\pm\pi/2$ over the period of T have remained, but the sharp edges seen in the trellis have been removed.

A better spectrum efficiency could be achieved by restricting the phase angle and thus improving both the sharp edges and the slope of the path. The phase angle of MSK is obtained by successively changing its present value in accordance with the data bit value, so as to enable the receiver to coherently detect the data. The question arises whether this change is really necessary for coherent detection. The change of phase is indispensable when a continuous stream of only ones or zeros are transmitted, so as to ensure their coherent recognition even if the same frequency was transmitted. This means that in a stream of alternate ones and zeroes, where the frequency is continuously changing, the phase reference is not essential.

When transmitting a series of data bits, the present phase of the carrier signal must be defined not only by the present input data bit value, but also by its two adjacent input data bit values. That is, if the data bit value a_n presently being transmitted at time t_n is of different values both from the bit previously transmitted a_{n-1} and the bit to be transmitted in future a_{n+1}, no phase change is necessary. Maximum phase shift is necessary

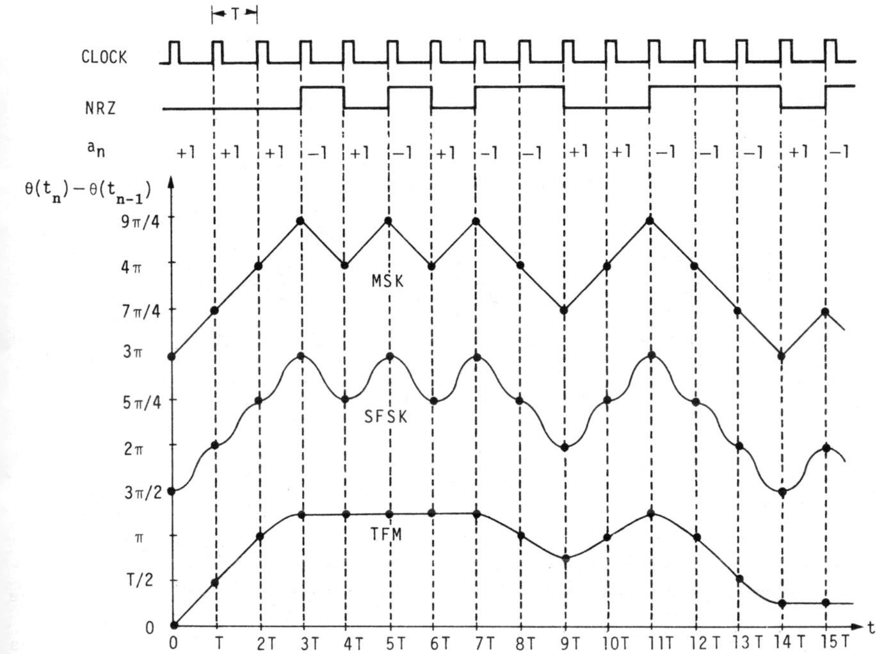

Fig. 6.57 Trellis pattern behaviour of the phase in MSK, SFSK and TFM

only if the three successive bits a_{n-1}, a_n and a_{n+1} have the same value. If the present bit differs in value from only one of the two bits, a_{n-1} or a_{n+1}, then only a phase change of $\pm\pi/4$ is sufficient. This can be concluded as follows:

$$\theta(t_n) - (t_{n-1}) = \begin{cases} 0 & \text{for } a_{n-1} \neq a_n \neq a_{n+1} \\ \pi/4 & \text{for } \begin{cases} a_{n-1} \neq a_n = a_{n+1} \\ a_{n-1} = a_n \neq a_{n+1} \end{cases} \\ \pi/2 & \text{for } a_{n-1} = a_n = a_{n+1} \end{cases}$$

A modulation technique which establishes the phase variation of the present phase change $\theta(t_n) - (t_{n-1})$ in accordance with three successive bits has been named 'Tamed Frequency Modulation' (TFM).

For MSK we have already defined the phase change as having the value of

$$\theta(t_n) - \theta(t_{n-1}) = a_n \pi/2.$$

For TFM the phase changes could be reduced and would now have the new value of

$$\theta(t_n) - \theta(t_{n-1}) = (a_{n-1}/4 + a_n/2 + a_{n+1}/4)\pi/2$$

with a_{n-1}, a_n and a_{n+1} take the level values of +1 or −1 in accordance with the data values of 0 and 1 respectively. This enables the calculation of the present phase changes for TFM for all 8 possibilities, as seen in the following table.

TABLE 6.9

INPUT BIT			LEVEL VALUE			PHASE FACTOR	PHASE CHANGE $\theta(t_n) - \theta(t_{n-1})$
a_{n-1}	a_n	a_{n+1}	a_{n-1}	a_n	a_{n+1}		
0	0	0	+1	+1	+1	+1	$+\pi/2$
0	0	1	+1	+1	−1	$+\frac{1}{2}$	$+\pi/4$
0	1	0	+1	−1	+1	0	0
0	1	1	+1	−1	−1	$-\frac{1}{2}$	$-\pi/4$
1	0	0	−1	+1	+1	$+\frac{1}{2}$	$+\pi/4$
1	0	1	−1	+1	−1	0	0
1	1	0	−1	−1	+1	$-\frac{1}{2}$	$-\pi/4$
1	1	1	−1	−1	−1	−1	$-\pi/2$

Comparing the three techniques of MSK, SFSK and TFM, as shown by the example presented in Fig. 6.57, illustrates the advantages and disadvantages of the three systems. Little improvement, however, is achieved with SFSK, since the path follows the same interval as MSK, except that by avoiding the sharp edges the slope becomes steeper. The path is considerably smoothed out with the use of TFM, since here the actual phases are reduced.

The requirements of TFM are identical with that of MSK in that continuous phase during switching is required and the signal amplitude envelope must remain constant. This means that the phasor diagram of TFM must be in a circle similar to that of MSK, as shown in Fig. 6.58.

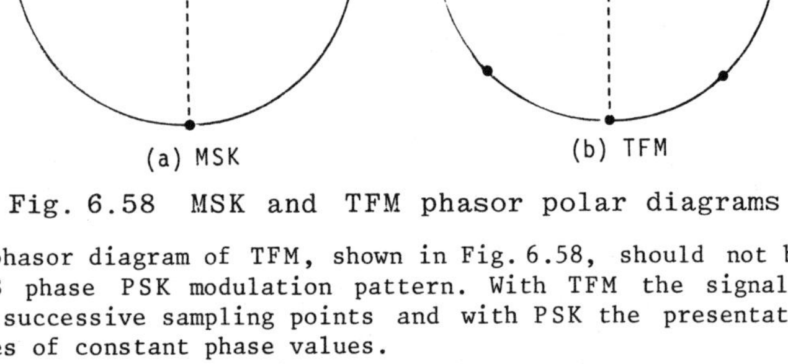

Fig. 6.58 MSK and TFM phasor polar diagrams

The phasor diagram of TFM, shown in Fig. 6.58, should not be confused with the 8 phase PSK modulation pattern. With TFM the signal pattern represents successive sampling points and with PSK the presentation represents states of constant phase values.

6.16 REFERENCES

(a) MODEMS AND MODULATION TECHNIQUES

1. Davey, J.R. *Modems*, Proc. IEEE, vol.60, no.11, November 1972, pp.1284-1294.
2. Jones, W. *Modulation Techniques for Data Transmission*, Telecommunication, June 1972, pp.63-66.
3. Lyon, D. *How to Evaluate a High-Speed Modem*, Telecommunication, October 1975, pp.35-39.
4. Dubery, B. *Data Transmission Today-AND Tomorrow*? Communication International, November 1976, pp.42-44.
5. Lowry, R.W. *On-Line Circuit Monitoring of High Speed Data Networks*, Communication International, June 1978, pp.36-43.
6. Edmonds, P. *Modems as Part of Data Communication Systems*, Communication International, March 1975, pp.27-31.
7. Edmonds, P.H. *Design Requirements for the British Post Office 2400 BPS Modem*, Communication International, November 1976, pp.21-29.

(b) AMPLITUDE AND PHASE MODULATION TECHNIQUES

8. Cahn, C.R. *Combined Digital Phase and Amplitude Modulation Communication Systems*, IRE Trans. on Comm. Systems, September 1960, pp.150.155.
9. Ottoson, R. *Group Codes for Phase and Amplitude Modulated Signals on a Gaussian Channel*, IEEE Trans. on Information Theory, vol.IT-17, no.3, May 1971, pp.315-321.
10. Foscnini, G.J., Gitlin, R.D. and Weinstein, S.B. *Optimization of Two-Dimensional Signal Constellations in the Presence of Gaussian Noise*, IEEE Trans. on Comm., vol. COM-22, no.1, January 1974, pp.28-37.
11. Simon, M.K. and Smith J.G. *Carrier Synchronization and Detection of QASK Signal Sets*, IEEE Trans. on Comm., vol. COM-22, no.2, February 1974, pp.98-106.
12. Thomas, C.M., Weidner, M.Y. and Durrani, S.H. *Digital Amplitude – Phase Keying with M-ary Alphabets*, IEEE Trans. on Comm., vol. COM-22, no.2, February 1974, pp.168-179.
13. Simon M.K. *Data-Derived Symbol Synchronization of MASK and QASK Signals*, IEEE Trans. on Comm., vol. COM-23, no.6, June 1975, pp.606-620.
14. Weber, W.J. *Differential Encoding for Multiple Amplitude and Phase Shift Keying Systems*, IEEE Trans. on Comm., vol. COM-26, no.3, March 1978, pp.385-391.
15. Braun, W.R. and Lindsey, W.C. *Carrier Synchronization Techniques for Unbalanced QPSK Signals – Part 1 and 2*, IEEE Trans. on Comm., vol. COM-26, no.9, September 1978, pp.1325-1341.
16. Sargeaunt, P.N. *A 120 Mbit/s QPSK for TDMA Applications*, ICC, 1977, pt.III, pp.184-187.
17. Cahn, C.R., May, C.L. and Robatino, V. *Bandwidth-Efficient QPSK for Multiple-Access Satellite Communication.*
18. Welti, G.R. *More Bits Per Cycle Via Satellites*, ICC, 1977, pt.1, pp.52-54.
19. Wolejsza, C., Chakraborty, D., *Performance of 120-Channel Modems for Satellite Communication*, ICC, 1977, pt.2, pp.145-148.
20. Puchkoff, S.J. *9600 BPS – A Five Year History*, Telecommunication, August 1973, pp.37-40.
21. Ishio, H., Washio, M., Inokuchi, M. and Seki, S. *A New Multilevel Modulation and Demodulation System for Carrier Digital Transmission*, ICC, 1976, vol.II, pp.29-7 to 29.12.

22. Miyauchi, K., Seki, S. and Ishio, H. *New Technique for Generating and Detecting Multilevel Signal Formats*, IEEE Trans. on Comm., vol. COM-24, no.2, February 1976, pp.263-267.

23. Washio, M., Katoh, T. and Itaya, E. *Experimental Study on a New 1.6 GPS 16-Level APSK Modem*, NTC, 1977, vol. I, pp.05:6-1 to 05:6-6.

24. Wood, W.A. *Modulation and Filtering Techniques for 3 Bits/Hertz Operation in the 6 GHz Band*, ICC, 1977, pp.5.3-97 to 5.3-101.

25. Noguchi, T. and Otani, S. *Multilevel PSK Modems for TDMA System*, EASCON, 1975, pp.139-A to 139-F.

26. Tan, V., Ryu, T., Hashimoto, S., Shinmyo, S. and Yokoyama, S. *The B-Level PSK Modem with Cosine Roll-Off Spectrum for Digital Microwave Communications*, ICC, 1976, pp.29-31 to 29-18.

(c) MINIMUM FREQUENCY SHIFT MODULATION TECHNIQUES

27 Smith, J.G. *Spectrally Efficient Modulation*, ICC, 1977, vol.I, pp.3.1-37 to 3.1-41.

28. Welti, G.R. and Kwan, R.K. *Comparison of signal Processing Techniques for Satellite Telephony*, NTT, 1977, vol. I, pp.05:1-1 to 05:1-6.

29. Van Den Elzen, H.C. and Van Den Wurf, P. *A Simple Method of Calculating the Characteristics of FSK Signals with Modulation Index 0.5*, IEEE Trans., vol. COM-20, no.2, April 1972, pp.139-147.

30. Sullivan, W.A. *High-Capacity Microwave System for Digital Data Transmission*, IEEE Trans., vol. COM-20, no.3, June 1972, pp.466-470.

31. de Buda, R. *Coherent Demodulation of Frequency Shift Keying with Low Deviation Ratio*, IEEE Trans,m vol. COM-20, no.3, June 1972, pp.429-435.

32. Rhodes, S.A. *Effect of Noisy Phase Reference on Coherent Detection of Offset- QPSK Signals*, IEEE Trans, vol. COM-22, no.8, August 1974, pp.1046-1055.

33. Osborne, W.P. and Luntz, M.B. *Coherent and Noncoherent Detection of COFSK*, IEEE Trans,m vol. COM-22, no.8, August 1974, pp.1023-1036.

34. Taylor, D.P., Ogletree, S.T. and Haykin, S.S. *A Prototype 60 MB/S Fast Frequency Shift Keying Modem*, ICC, 1976, vol.3, pp.51-26 to 51-30.

35. Mathwich, H.R., Balcewicz, J.F. and Kecht, M. *The Effect of Tandem Band and Amplitude Limiting on the E_6/N_6 Performance of Minimum (Frequency) Shift Keying (MSK)*, IEEE Trans., vol. COM-22, no.10, October 1974, pp.1525-1540.

36. Simon, M.K. and Smith, J.G. *Offset Quadrature Communication with Decision-Feedback Carrier Synchronization*, IEEE Trans., vol. COM-22 no.10, October 1974, pp.1576-1584.

37. de Buda, R. *About Optimal Properties of Fast Frequency-Shift Keying*, IEEE Trans, vol. COM-22, no.10, October 1974, pp.1726-1727.

38. Amoroso, F. *Pulse and Spectrum Manipulation in the Minimum (Frequency) Shift Keying (MSK) Format*, IEEE Trans., vol. COM-24, no.3, March 1976, pp.381-384.

39. Simon, M.K. *A Generalization of Minimum-Shift-Keying (MSK) Type Signalling Based Upon Input Data Symbol Pulse Shaping*, IEEE, Trans. vol. COM-24, no.8, August 1976, pp.845-856.

40. Gronemeyer, S.A. and McBride, A.L. *MSK and Offset QPSK Modulation*, IEEE, Trans., vol. COM-24, no.8, August 1976, pp.809-820.

41 Simon, M.K. *An MSK Approach to Offset QASK*, IEEE, Trans, vol. COM-24, no.8, August 1976, pp.921-923.

42. Keelty, J.M., Burcombe, J. and Wood, J. *Digital Realization of a Low Rate MSK Modem*, ICC, 1976, pt.1, pp.3-7 to 3-11.

43. Anderson, J.B. and Taylor, D.P. *Trellis Phase Modulation Coding: Minimum Distance and Spectral Results*, EASCON, 1977, pp.29-1A to 29-1G.
44. Morais, D.H. and Feher, K. *MSK and Offset QPSK Modulation in Line of Sight Digital Radio Systems*, ICC, 1977, pt.2, pp.24-2-140 to 24.2-144.
45. Fielding, R.M., Berger, H.L. and Lockhead, D.L. *Performance Characterization of a High Data Rate MSK and QPSK Channel*, ICC, 1977, pt.1, pp.2.4-42 to 3.4-46.
46. Regocurz, J. and Simpson, R.S. *Minimum-Shift-Keying Signal Detection with Noisy Reference Signals*, IEEE, Trans., vol. COM-26, no,6, June 1978, pp.896-902.
47. Matyas, R. *Effect of Noisy Phase Reference on Coherent Detection of FFSK Signals*, IEEE, Trans., vol. COM-26, no.6, June 1978, pp.807-815.
48. Weber, W.J., Stanton, P.H. and Sumida, J.T. *A Bandwidth Compressive Modulation System Using Multi-Amplitude Minimum Shift Keying (MAMSK)*, NTC, 1977, pt.1, pp.05:5-1 to 05:5-7.
49. Weber, W.J., Stanton, P.H. and Sumida, J.T. *A Bandwidth Compressive Modulation System Using Multi-Amplitude Minimum Shift Keying (MAMSK)*, IEEE, Trans., vol. COM-26, no.5, May 1978, pp.543-551.
50. de Jager, F. and Dekker, C.B. *Tamed Frequency Modulation, A Novel Method to Achieve Spectrum Economy in Digital Transmission*, IEEE, Trans., vol. COM-26, no.5, May 1978, pp.534-542.
51. Pasupathy, S. *Minimum Shift Keying: A* IEEE Communication Magazine, vol.17, no.4, July 1979, pp.14-22.
52. Morais, D.H. and Feher, K. *Bandwidth Efficiency and Probability of Error Performance of MSK and Offset QPSK Systems*. IEEE, Trans., vol. COM-27, no.12, December 1979, pp.1794-1801.
53. Oetting, F.D. *A Comparison of Modulation Techniques for Digital Radio*, IEEE, Trans., vo. COM-27, no.12, December 1979, pp.1752-1762.
54. Boutin, N., Dussault, L. and Morissettee, S. *Constant Amplitude PSK Signal with Minimum Out of Band Energy*, NTC, 1978, Conference Record, pt.1, pp.631 to 6.3.4.

CHAPTER 7

Communication Concentration

7.1 INTRODUCTION

Up to this chapter, communication was regarded as a transmission process with each link requiring its individual channel. This in not the most efficient means of organizing the communication networks for C^2 systems which consists of many terminals and computers spread over wide areas. In this chapter, means of sharing the channels between many sources will be presented for economizing the number of channels required.

The main factor in the cost of networks with terminals spread over wide geographical areas is the communication channels and not the data communication equipment (DCE). Accordingly, a major design consideration in adopting a network is the minimizing of the cost of the communication channels. If the volume of data originating from any point is relatively small, no design planning can reduce the cost of each bit transmitted along the transmission channel beyond a certain figure. The only way for further reducing the cost per bit transmitted is to find means of concentration of the data volume transmitted in the channel, that is, sharing the same transmission channel with other data sources. In this way, the high cost of the channel is shared, and the cost of the actual bit transmitted is reduced. Furthermore the cost of a transmission channel is based on the initial expenses of laying it along a given path rather than on the possible data volume it can carry.

Point-to-point channels are intended to establish direct contact between two communicating terminals and this calls for individual transmission channels. When a number of these channels are laid in parallel for all or only parts of the length, as seen in Fig. 7.1(a), it means in effect multiple channel costs. Instead, it is much more cost effective to install a single common channel, as seen in Fig. 7.1(b), with multiplexors or concentrators performing the data clustering operation. In this respect the equipment of the multiplexors and concentrators combines the low data volume traffic loads from each of the terminals into a single high volume traffic load. If, for example, the data volume from each terminal is labelled A, B and C, then the data volume in the common channel will be the sum of the three, i.e. $A+B+C$. It should be added that the relation between the channel and terminal data volume does not apply to the data rates. There are systems where the terminal transmitted data rates is equal to that of the common channel, while in other systems they are different. Multiplexors cannot serve more than the channel data rate capacity. Concentrators, as will be explained

CFC -K*

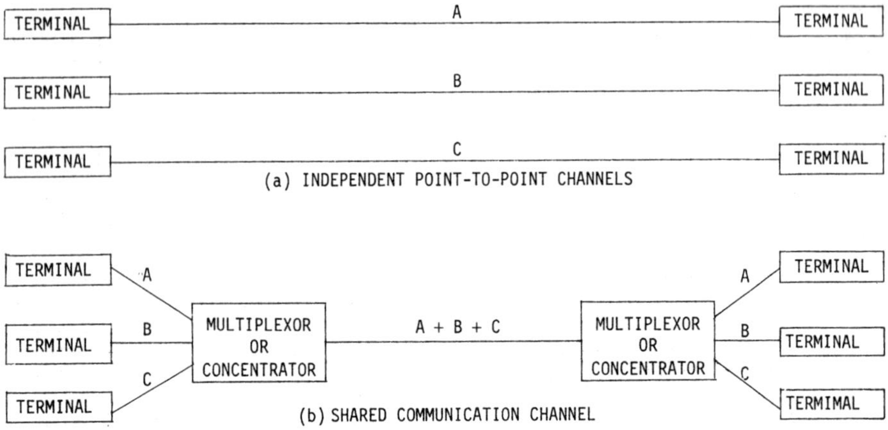

Fig. 7.1 The principle of sharing a common communication channel to reduce transmission costs

later, attempt to serve more terminal traffic than the channel can carry simultaneously.

Multiplexors and concentrators are generally employed in local area communication networks where a number of remote terminals are connected to a central host computer, as seen in Fig. 7.2. In this case, the multiplexors or the concentrators combine several different data signals from various terminals, sending them over a single transmission channel and then reconstructing them again at the receiving end. The receiving end may be a host computer or, alternatively, a communication node of a larger network which can transfer the data to other locations. The concentrator can provide further services in the local area network, but this will be discussed separately.

Fig. 7.2 Sharing a common communication channel between a host computer and its remote terminals in a local network

Multiplexors and concentrators have an identical function: providing means by which a single common communication channel can be shared between two or more terminals. For this reason, the general usage of both these terms, multiplexing and concentration, can be misleading and in many publications the terms are often interchanged. Actually they refer to two distinct types of network equipment which, although having the identical task of combining several communication channels, they differ in the means used for sharing the common communication channel. With multiplexors, the sharing is based on static allocation of the channel resources through fixed and predetermined methods. The concentrators also share out the channel resources, but in contrast to multiplexors, they utilize the channel by dynamic methods and not according to any previous arrangement.

Both multiplexors and concentrators allow a number of terminals to have access to the common channel simultaneously, although multiplexors employ transparently operated equipment, whereas concentrators provide buffering equipment for use between the common channel and the terminals. The major task of multiplexors is to assist the terminals in dispatching their data messages via the common channel with no undue delays. On the other hand, the major task of concentrators is organizing the data traffic in the common channel even if this may cause delays in the flow of messages. In consequence the data at the combining equipment arrives from each of the terminals at a staggered rate. With multiplexors, each input signal is immediately retransmitted irrespective of the traffic volume in the other inputs. On the other hand, with concentrators, the task is to smooth the data flow in the common channel, aiming at a continuous flow of traffic. In other words, the multiplexors and concentrators differ in the efficiency by which they utilize the common channel. With multiplexors, the best channel efficiency can only be achieved with all the terminal producing fixed high traffic loads, while the concentrators efficiency is best achieved with the sum of the terminal traffic equal to the maximum possible channel traffic and so enabling the concentrator to serve random traffic loads.

The available spectrum of the common channel can be illustrated as a two dimensional domain of time and frequency (as shown in Fig. 7.3) that may be apportioned in various ways for assignment to particular terminals. This means that there are two different fundamental types of channel sharing. One, known as frequency division multiplexing (FDM), assigns fixed portions of the frequency bandwidth to each terminal for all the time. The other assigns the whole available channel bandwidth to each terminal for a given period of time and is therefore referred to as time division multiplexing (TDM). Fixed time slot allocation of the common channel in accordance with a given sequence is termed synchronous time division multiplexing (STDM) while the dynamic allocation of the channel as a function of the traffic requirements of the terminals is termed asynchronous time division multiplexing (ATDM). Dynamic time slot allocation as a function of the traffic in the common channel is termed statistical multiplexing (SM) and concentrators. The term known as intelligent multiplexing (IM) is in fact a combination of both ATDM and SM. Both IM and concentrators are similar, although one operates in accordance with hard wired logic while the other in accordance with software programs. A major characteristic of all the time division multiplexors is that the bit-timing needs to be synchronized.

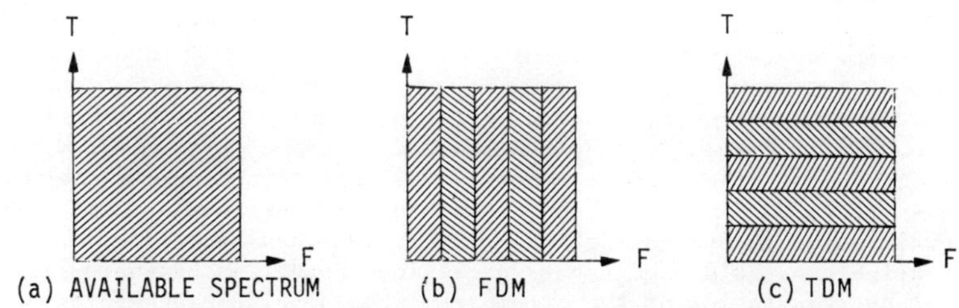

Fig. 7.3 Sharing the common channel spectrum either by frequency or time portions

A much better way of classifying the channel sharing is by taking account of the characteristics of the combining equipment instead of the way the spectrum is shared, as shown in Fig. 7.4. That is, the equipment of both ends of the channel is classified by determining whether it only performs transparent operations causing no delays or introduces buffering operations to achieve the combining function.

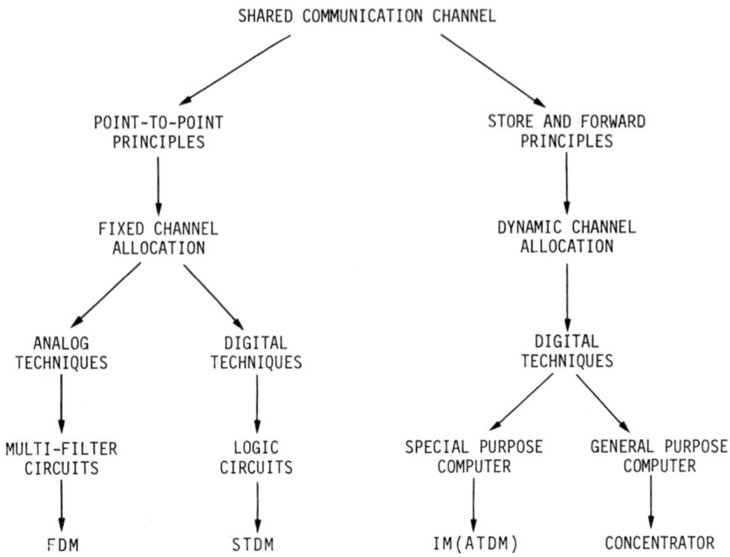

Fig. 7.4 Classification of the various means of sharing a common communication channel

There are a number of other types of multiplexors, such as code division multiplexing (CDM), phase division multiplexing (PDM), frequency-time division multiplexing (FTDM) and space division multiplexing (SDM), which could be added. Nevertheless, their unique characteristics make them difficult to be classified in the above method as will be explained later. It should also be added that the concentration has further communication functions which are not possible in any of the multiplex equipment mentioned. This too, will be discussed further on.

7.2 FREQUENCY DIVISION MULTIPLEXING (FDM)

Frequency division multiplexing (FDM) is the oldest and most common type of multiplexing in which a number of communication paths share the frequency spectrum of a single transmission channel on a continuous time basis. It was initially designed to provide a common transmission path for a number of voice grade channels between distant telephone exchanges and is still used today for most of the trunk long distance voice communication. Each signal source is permanently assigned a unique bandwidth within the common available bandwidth of the transmission channel, as seen in Fig. 7.5. The frequency translation of the original baseband voice channel (of 300 to 3000 Hz) to the allotted frequency band in the common transmission channel is performed by modulation techniques. At the receiving end a reverse process takes place and each voice channel is demodulated separately. The whole frequency spectrum of the transmission path can be handled as a single signal, despite the fact that it comprises a number of narrower signal frequency bands This wideband multiplex signal can therefore be transmitted over a conveyance media of wire-lines, cable or radio, by using a single centre carrier frequency. For this reason, a communication system using FDM techniques is often referred to as a carrier system.

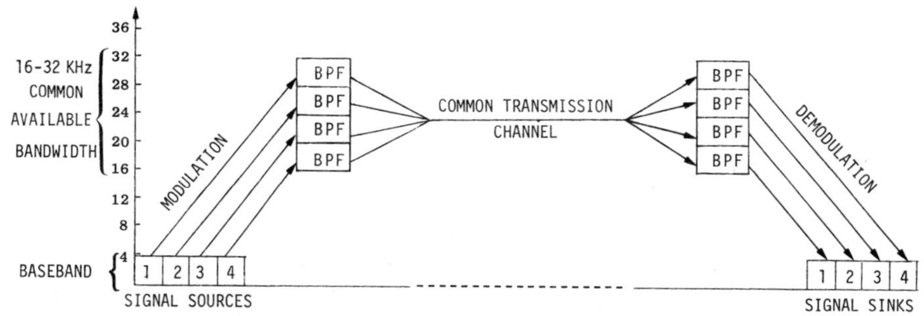

Fig. 7.5 The operational principle of frequency division multiplexing of analog signals

FDM uses pure analog techniques in performing its multiplexing operation, as seen in Fig. 7.6, with a number of parallel channels consisting of filters and modulators. The individual signal sources are fed through low pass filters (LPF) to restrict the signal spread, and are then each modulated on a separate frequency (f_c) of the band allotted to it. The modulation techniques used are generally those of amplitude or frequency, although, on most long distance, wide channel carrier systems, single-sided suppressed carrier (SSBSC) modulation techniques are preferred. For this reason, a pilot reference of the carrier signal is transmitted along with the modulated signals so as to prevent any jitter in the demodulation process at the receiving end. Before the modulated signals of each individual channel are superimposed in the common channel, they are each passed through appropriate bandpass filters (BPF) in order to prevent any sidebands of one channel disturbing the signals in adjacent channels. The output of all the filters can then be fed simultaneously into the transmission channel and be regarded as one wideband signal.

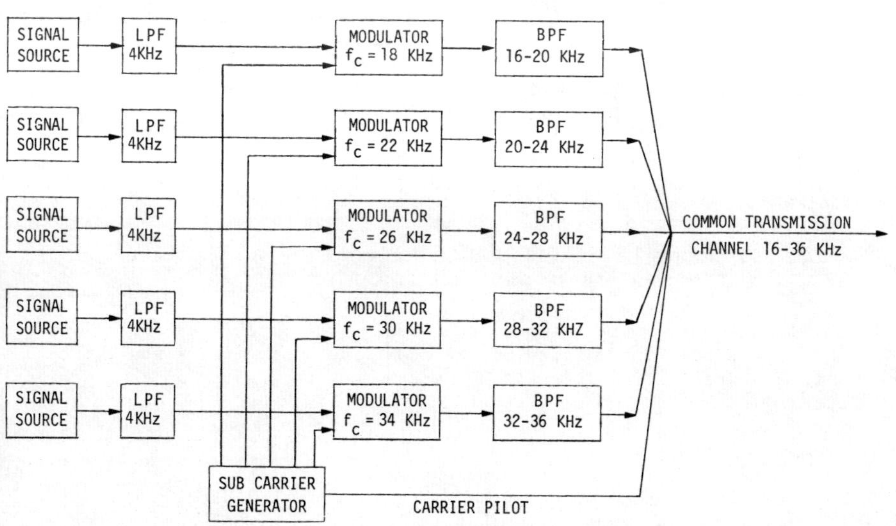

Fig. 7.6 Example of the analog circuits used in an FDM (carrier) scheme where the common channel is divided into five subchannels

Each signal source can use the allotted frequency bandwidth independently without any need for coordination with others. Even when digital data is transmitted in each of the channels there is no need for any time synchronization between their signals. Furthermore, since each channel modulator is a separate analog unit, the failure of one will not affect the operation of adjacent modulators in any way. If, however, there is a failure in the common path equipment, where the multiplexed signal is handled, this will affect all the single signal channels.

A FDM primary characteristic is the ease by which it can be cascaded to facilitate the adding of more multiplex channels along the common channel. By means of intermediate carrier frequencies, multiplex groups may be arranged. That is, the multi-channel band of a few hundred signal sources can be assembled in several modulation stages, as seen in Fig. 7.7, and then can be split up at the distant intermittent point by similar processes but in the reverse order. In the first pregroup 3 voice channels are assembled to occupy the bandwidth of 10 to 20 KHz with carrier signals at 12, 16 and 20 KHz. Then 4 of the pregroups are assembled to occupy a basic-group of 60 to 108 KHz. In some systems 12 voice channels are directly assembled into the basic-group and thus saving one intermittent stage. Mastergroups are then organized to hold 60 channels by assembling 5 basic-groups. It is possible to assemble supergroups by assembling together 16 mastergroups which therefore make up a carrier signal of 960 voice grade channels. After each assembly process the combined signal is SSBSC readjusts the frequency band.

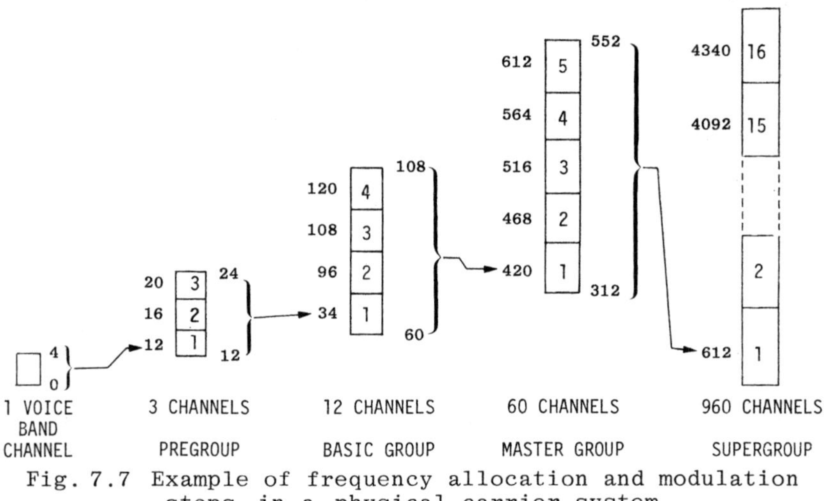

Fig. 7.7 Example of frequency allocation and modulation steps in a physical carrier system

FDM equipment can assemble both analog and modulated digital data. When a voice grade transmission channel contains a digital signal, the FDM regards the modulated data signal in the same way it regards an analog signal. In fact most of the direct distant dialling lines and many of the leased lines may progress using a number of FDM carrier signals without the user being aware of it.

Frequency division multiplexing equipment can also be used to accept the digital data signal directly with the individual data modems already tuned to the required frequency. That is, each data signal channel receives a unique frequency band which is permanently allocated to it by having the local oscillator of the modem adjusted to it. Any type of data shift-keying may be used, although frequency shift keying is the major technique employed.

In this technique the centre frequency of each channel is shifted between two frequencies in accordance with the data state of 1 or 0, as seen in Fig. 7.8. This technique is very popular in sharing a voice grade channel among a number of slow data rate terminals, such as computer teletypewriters (TTY) and telex communication.

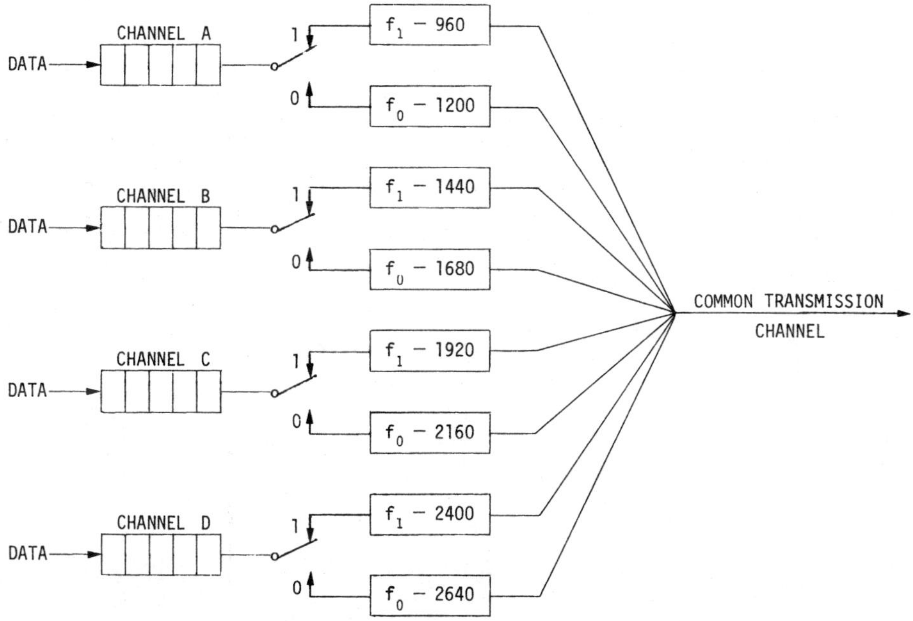

Fig. 7.8 Schematic illustration of a digital FDM with the data channels sharing a voice grade line

The full spectrum of B Hz of the common transmission channel can be divided into N data channels, with each data channel having the frequency bandwidth of B/N Hz available to it. However, each data channel can transmit at speeds far less than the frequency slot of B/N; that is, the effective frequency bandwidth of the data channel is much less than B/N Hz. The limitation is due to the need of guard bands between adjacent channels to prevent any sideband signals from overlapping and encroaching on the adjacent channels. Although bandpass filters are used at the output of each data channel to envelop the specific allocation frequency band, the slope of the filters must be within the range of the guard bands. For example, a typical TTY which transmits at 150 Baud (BPS) requires 240 Hz bandwidth channel, while the 300 Baud (BPS) terminal requires 480 Hz bandwidth. When a voice grade channel is used with frequency-division multiplexors, only a total maximum speed of about 2000 BPS can be reached.

An example where the voice grade telephone channel is shared between four data channels, is shown in Fig. 7.9. Each data channel is allocated a bandwidth of 480 Hz and can transmit speeds up to 300 BPS. The distance between the digital state '1' bit and the digital state '0' bit in each channel is 240 Hz, leaving another 240 Hz spacing between the signals of adjacent channels. The band pass filter must filter the output of each of the four individual modulators without their interfering with each other. This

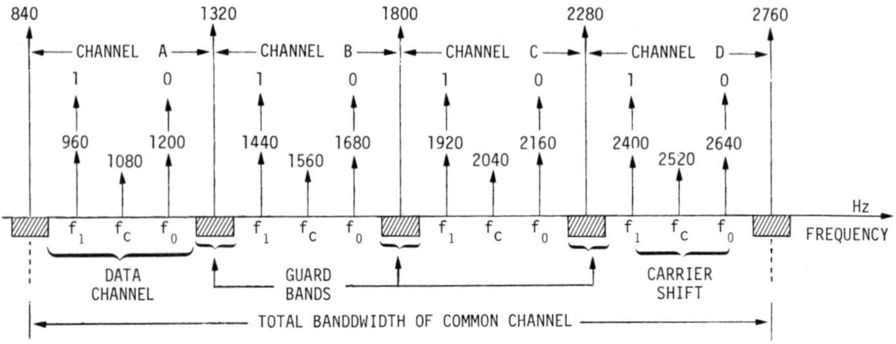

Fig. 7.9 Spectrum allocation of a voice grade channel in an example of a frequency division multiplexing scheme

necessitates restricting the effective bandwidth allocated to each channel to less than 480 Hz.

The simplicity of the operation and the construction of the FDM make it most attractive for both long and even short distances. Its primary advantage is in its relatively low cost compared with all other available means of multiplexing. It is not only the ease whereby it can be cascaded into large supergroups but also in the making of smaller sized groups of sub-voice grade channels. These channels when cascaded in a line of FDM's allow deviations at intermittent points to individual low speed terminals, as seen in Fig. 7.10. Such an arrangement allows individual terminals, spread over a wide geographical area to be added to the common channel at isolated points along the transmission channel.

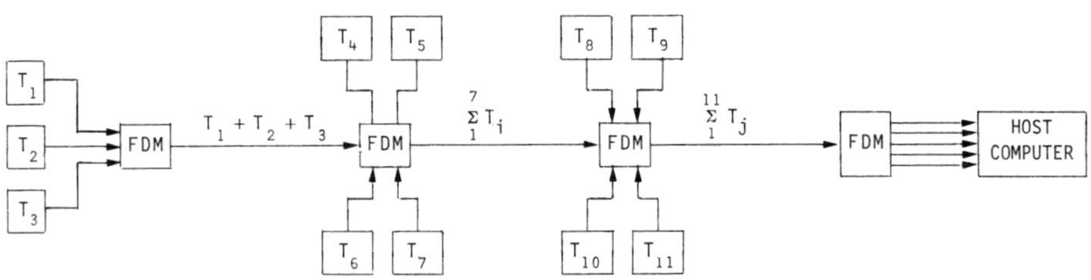

Fig. 7.10 Cascading data signals from low speed terminal with FDMs

In FDM digital transmission techniques all the channels operate in parallel form. That is, the common transmission paths can transmit different information signals over each data channel. More so, each data stream is completely isolated from all the others, and operates from different clock sources with no need for any synchronization between them.

7.3 PHASE DIVISION MULTIPLEXING (PDM)

Phase division multiplexing (PDM) is a form of multiplexing which shares the whole frequency spectrum of the common channel together with all the input terminals. Furthermore, it enables all the terminals to transmit their data streams simultaneously, each using the full available frequency channel bandwidth, without any disturbances from adjacent channels. Nevertheless, with the state of art of the techniques available today, the number of channels that can operate in parallel is limited to only four. With a conditioned transmission channel, each of the four channels of PDM can transmit up to 2400 BPS, with maximum common channel capacity rate reaching 9600 BPS.

Phase division multiplexing is based on multi-phase shift keying (MPSK) techniques discussed in chapter 6. Instead of transmitting four different levels, each representing four sequential data bits of the same data input, it is possible to transmit the same four levels but with each level representing four independent data bits of four different data sources. This is achieved by having each data source phase modulating a different carrier signal. All the four carrier signals have the same frequency, although with a phase difference of 45° in respect to each other. This means that the four carrier signals have the phase angles of 0°, 45°, 90° and 135° in respect to the internal carrier frequency source, as seen in Fig. 7.11. Each data source key shifts its corresponding carrier signal by 0° or 180° in accordance with its input digital state of 0 or 1 data bits. The outputs of all the separate modulators are then superimposed, thereby producing one combined signal of smaller phase shift. The four combined signals can be easily separated again in the demultiplexing process since there is at least 45° between each signal.

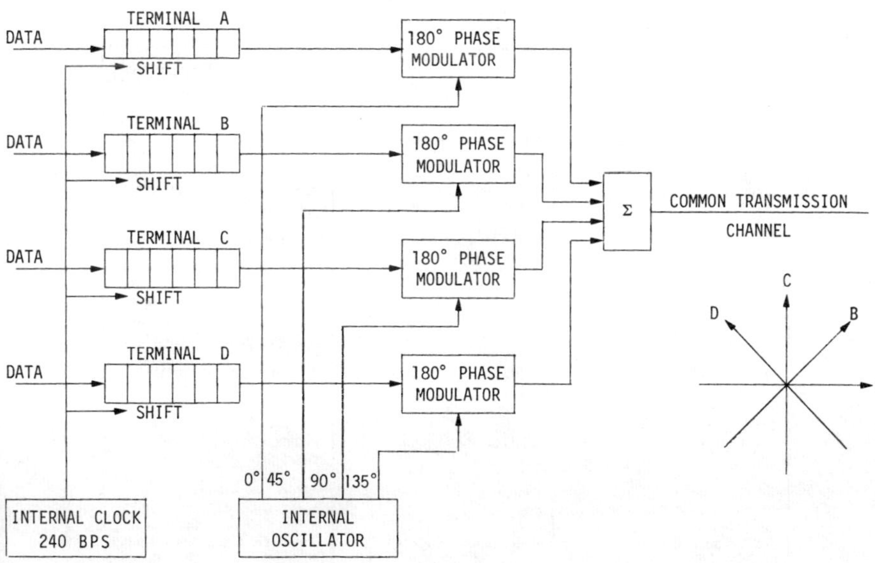

Fig. 7.11 Schematic diagram of a phase division multiplexing circuit

It could be argued that this suggested PDM scheme is the same as that using a MPSK data modem operating at 9600 BPS. This contention, however, is not correct, since instead of providing transmission services to only one terminal operating at 9600 BPS as with MPSK, the PDM provides both multiplexing and transmission services to four separate terminals each operating at 2400 BPS. It should be added that not all the terminals available on the market can operate at 9600 BPS whereas there are more terminals that operate at 2400 BPS. Usually when four terminals require the use of a single transmission channel, special multiplexing equipment must be introduced, usually in the form of STDM (which will be discussed in the next section) and then it must be followed by a high speed modem. What is suggested here is simply to convert the MPSK modem to perform both the multiplexing and transmission operation with no need for any additional equipment. Furthermore, the internal clock source of the multiplexor/modem need not be 9600 BPS but only 2400 BPS and this could slightly reduce the cost.

The data transmission rate in all the four parallel channels must be the same, i.e. coming from the same clock source (a requirement which was not needed in FDM). This means in effect that all the input terminals and the PDM equipment must be fully synchronized. The suggested maximum data rate of 2400 BPS for the terminals applies on to the upper limit; slower terminals could, of course, be more easily used. Take for example four terminals operating at 1200 BPS which are multiplexed in a 4800 BPS MPSK modem.

PDM offers many advantages over FDM, primarily in the means by which it exploits the common channel; it does, however, require all the terminals to be located at a short distance from each other FDM on the other hand has the advantage of possible spread over wide geographical areas with many cascading characteristics, two features which are impossible with PDM.

The following table shows the basic comparison points between both FDM and PDM multiplexing schemes. In both schemes it is considered that they are sharing a voice grade channel between a number of data terminals.

TABLE 7.1

FUNCTION	FDM	PDM
Type of channel modulation	FSK	PSK
Type of data terminals	Low speed terminals	High speed terminals
Exploitation of common channel	Each terminal transmits over part of the channel spectrum	Each terminal transmits over the full channel spectrum
Number of terminals	Depending on the terminal speed and the allotted frequency bands	Maximum of four only
Overall speed limitations	Variable up to 2000 BPS depending on bandwidth available	Variable up to 9600 BPS depending on the channel and modem available
Terminal speed limitations	Variable up to 600 BPS although usually 75 – 300 BPS	Variable up to 2400 BPS
Multiplexing limitation	Limited by frequency bandwidth	Limited by number of terminals
Data Sources	Completely independent sources	Fully synchronized sources
Clock sources	Independent sources	Comes from the same source
Geographical spread	Used for isolated terminals	Used for clustered terminals
Independence of terminals	Each terminal operates separately and inde-dently	All terminals must be synchronised
Guard spacing	Frequency bands between channels	No guards at all
Common channel utilization	Cannot be achieved	Maximum channel utilization
Cascading facilities	Readily possible	Impossible
Cost of equipment	Relatively cheap	Expensive

Although both FDM and PDM use analog techniques for achieving their multiplexing tasks, they are intended for completely different applications. FDM are offered for slow speed terminals located over wide geographical areas, while PDM are suggested for high speed terminals clustered at a single location.

7.4 SYNCHRONOUS TIME DIVISION MULTIPLEXING (STDM)

Time division multiplexing is a purely digital process whereby a number of separate slow data streams are interleaved together to form one fast data stream. A number of terminals are scanned one after the other in a fixed sequence, as seen in Fig. 7.12; hence the process being referred to as synchronous time division multiplexing (STDM). Each terminal is provided with an equal sized time period in which it can offload its data stream. This produces at the output of the multiplexor a continuous data stream containing a sequence of time slots, each associated with a different terminal. After a complete cycle scan of all the terminals, the whole sequence starts again. The function of the time division multiplexing is to transmit sequences of data streams deriving from many sources in an organized cycle.

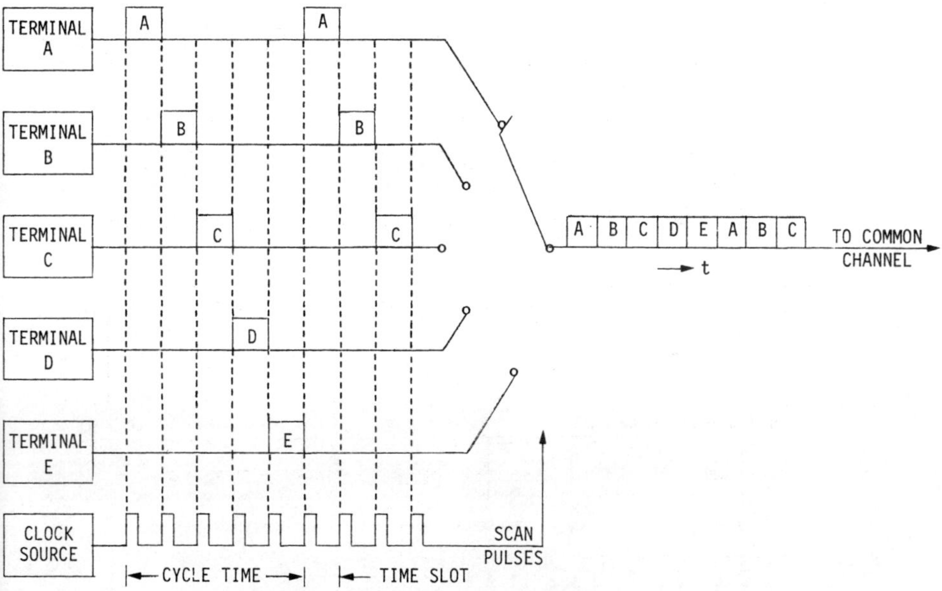

Fig. 7.12 Schematic diagram of a typical synchronous time division multiplexing scheme

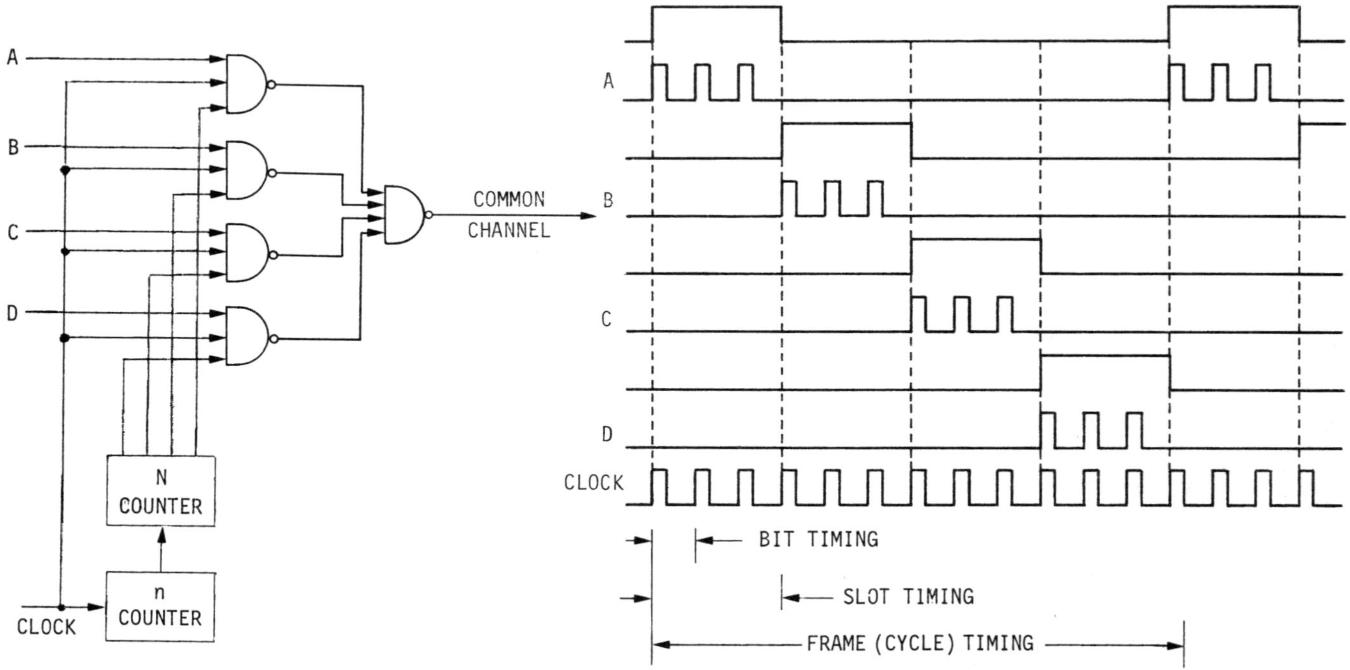

Fig. 7.13 The circuits and timing diagram of an STDM

The period of the time slot assigned to each terminal in which it can transmit its data may be short and equal to one bit or as long as required, such as the size of a character, a block or a complete message. Whatever slot size period is used, it must always be of equal duration when presented to all the terminals. Bit interleaved time multiplexing, where each terminal transmits one bit at each scan time, has the advantage of simplicity and minimum delay. (This means that a terminal may have to wait a complete cycle before it can offload its data bit, and the number of data bits in the cycle will equal the number of terminals associated with the multiplexor. In effect, each terminal cannot transmit its data bits as soon as they are ready for dispatching, but must hold them in the terminal in a buffer store until their sequence turn arrives.) Character, block and message interleaved time multiplexing are generally more expensive since they require larger buffering within the terminals and consequently present more delay.

The process of achieving synchronous time division multiplexing is comparatively very simple. The output of each terminal feeds a three input AND gate, with the other two inputs for the clock and for the counter selection pulses, as seen in Fig. 7.13. If there are N terminals associated with the multiplexor, the counter produces N wide pulses, each of the duration of a single slot size. This counter performs the scan process of all the terminals, which is a continuous process that repeats itself after a complete cycle. The slot opens the AND gate for the assigned slot period allowing n bits to proceed, where n specifies the number of bits which can be transmitted in each scan step. The period of each time slot is therefore defined by a second counter of n steps. The data stream length produced by the time division multiplexor from all the N terminals is referred to as a frame. The frame size is therefore nN bits in length, which is divided into N time slots, each of n bits.

Time division multiplexing principles are widely used in many fields. Today it is one of the most popular means of data collection especially for PCM data. Each terminal or data source is assigned a fixed permanent time slot period in which it can transmit its data. Although the data from the terminal is scattered, the speed at which it is fed from the terminal is the same as the speed at which it is transmitted on the common channel. The serial data stream within the time slot is arranged in fixed formats and each slot and/or frame is generally started by a unique bit sequence required for synchronization. It must be pointed out that time division multiplexors can operate only in a synchronous mode, as both ends of the common channel must scan the terminals using the same sequence reference, as seen in Fig. 7.14. As the time slots receive no label of the terminals they originated from, the multiplexors and demultiplexors at both ends of the channel must operate in synchronization. This means in effect that when transmitting time division multiplexing digital data, the receiver end must be synchronized with the receiver not only with its clock source but also with the time slot and cycle time period. For transmission synchronization, the receiver must therefore have means of identifying the bit timing as well as both the character (block) and frame timing.

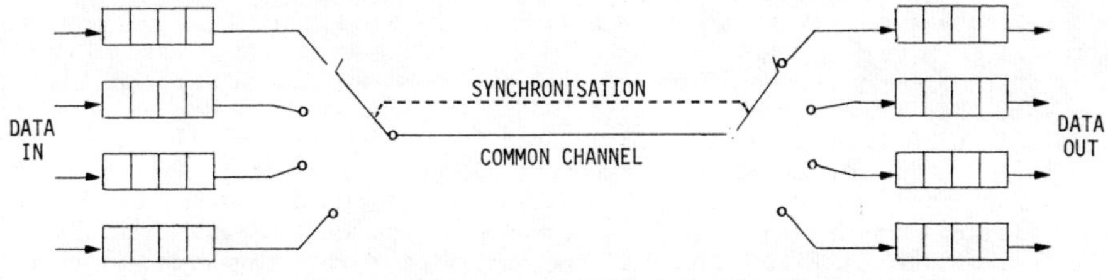

Fig. 7.14 Synchronisation need of both slot and frame periods in STDM

Although the STDM equipment at both ends of the channel must operate in synchronization, the data contents transmitted may be in the form of synchronous or non-synchronous formats. Each channel can utilize the time slot allotted to it according to the data originating from the terminal, and within the time slot period, it is independent of the multiplexor. The latter is important in PCM-TDM where each data sample originates from a different analog sampled source. It is therefore theoretically possible with TDM to transmit also asynchronous data, since each channel is said to be independent. This, however, is a most inefficient means of data transmission since it requires the receiving end to be in synchronization with the transmitter for all the slot and frame timing while additional special circuits are required to locate the bit timing of the asynchronous timing within each time slot. Efficiency of STDM can only be achieved if both the receiver and transmitter are fully synchronized for all their data formats. Bit synchronization is achieved by the data state transitions as explained in chapter 3, while character, block and message signals are recognized by special codes at the start of each period. Message synchronization in high speed transmission is obtained by the communication protocols rather than by the STDM characteristics. The synchronization codes, of a single bit or character size, used to define the boundaries of the slot and frame, are similar to what in FDM we referred to as guard spacing. Nevertheless, less overhead spectrum is used up for the guard spacing with STDM than with FDM.

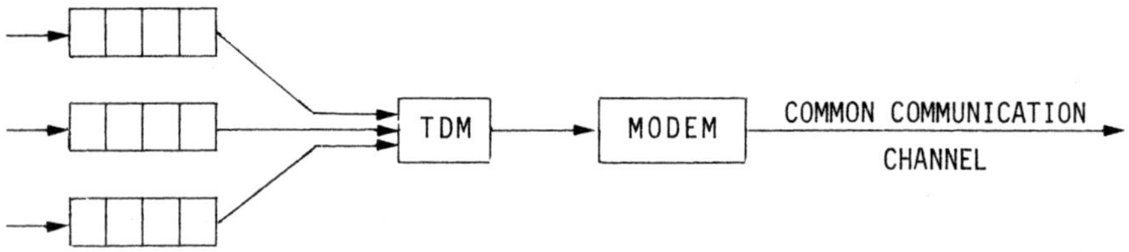

Fig. 7.15 The requirement of a modem when transmitting the STDM data over a communication channel

The time multiplexed data stream, before it can be conveyed over a long distance communication channel, must first be modulated so as to convert the digital pulses into an analog signal. That is, the digital data stream must pass through a modem before it can be conveyed over a common channel, as seen in Fig. 7.15. In this respect, the common transmission channel of the TDM signal is shared bewteen a number of terminals, in the same way as FDM. With frequency division multiplexing the common channel spectrum is shared between a number of data channels, with each terminal getting a portion of the frequency bandwidth for all the time. With time division multiplexing, the common channel is also shared between a number of data channels, although here each data terminal gets a fifferent unique transmission time slot. During the period of each time slot, each terminal gets access to the full frequency bandwidth of the common communication channel.

All the terminals associated with a time division multiplexor must be clustered together in one location, since the data that flow between the terminal and the multiplexor are in digital form. Long distance channels will only degrade the pulse waveform and disturb the multiplexor operation. Nevertheless, this does not mean that a time multiplexor cannot be connected to terminals that are some distance away from it. In this case, the signal from the long distant channel must be regenerated back to digital form before it can enter the STDM. That is, the remote terminals are connected to the multiplexor through a pair of modems, one at each end of the transmission channel, as seen in Fig. 7.16. The time multiplexor operation does not change if the data presented to it comes directly from a locally situated terminal

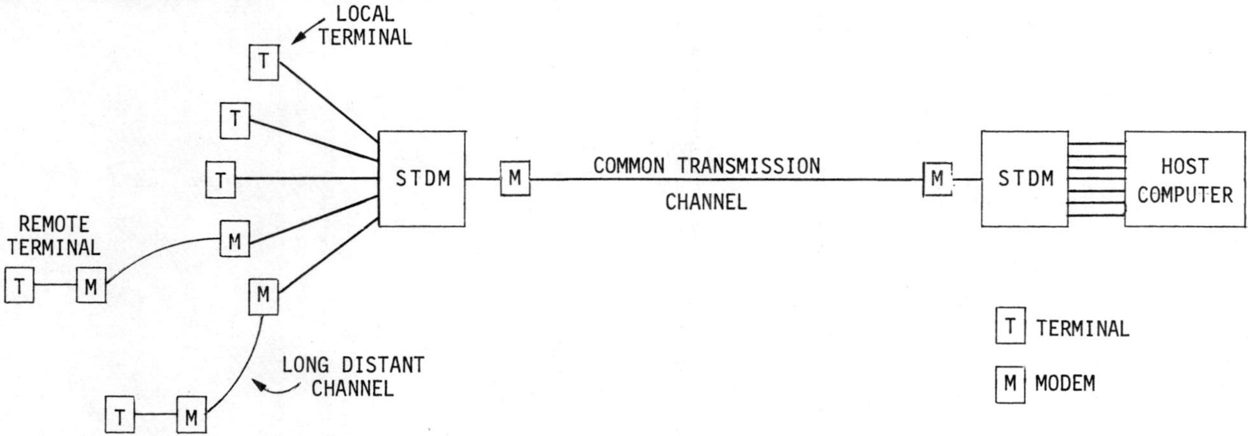

Fig. 7.16 Connecting local and remote terminals to an STDM

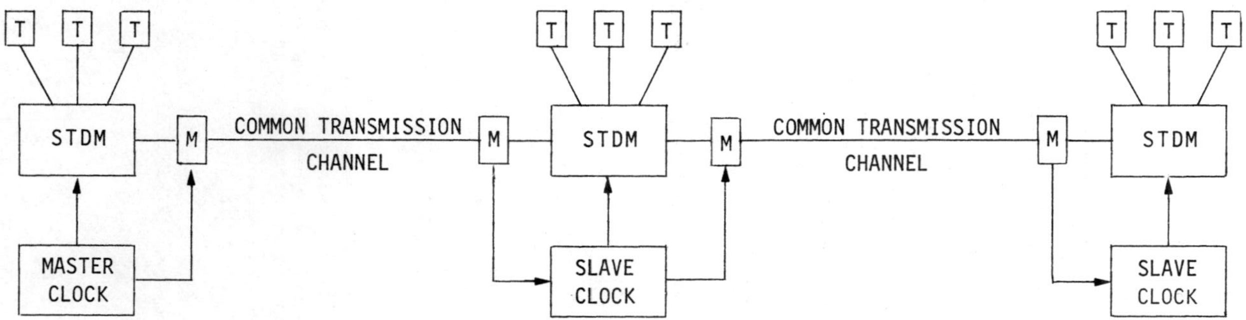

Fig. 7.17 Cascading STDM with clock recovering at intermediate stages

or from a remote long distance terminal via modems. The only two basic requirements are that all the data entering the multiplexor are in digital form and are fully synchronized.

Similarly to FDM it is possible to cascade a number of STDM, although a pair of modems must always be inserted at both ends of the channel. Cascading time division multiplexors raises the problem of synchronizing all the multiplexors in the cascade. In general this is solved by having a master clock at one end of the cascade and synchronizing the whole network with it, using a slave clock system in each consecutive STDM unit. Fig. 7.17 illustrates a typical time division multiplex cascade network with each cascaded STDM unit recovering the clock information out of the data stream.

It is possible to time-multiplex together large numbers of data channels and then transmit them over a wide communication channel. This application is widely used today for transmitting digitized voice and video sources over long distances, such as satellites. For this purpose the time multiplexors are cascaded in stages similar to those used for FDM. The basic channel starts with 56-64 KBPS, as used for PCM digitized voice with 7-8 bits/samples and reaching a few hundred mega bits per second. Digital multiplexing is superior to analog multiplexing since it is much easier to construct and separate the digital signals at the intermittent stages. Furthermore, the communication channel is much better utilized with digital transmission. There are various multiplexing standards that have been adopted such as, for example, in the USA where the multiplex sizes are shown in Table 7.2.

TABLE 7.2

MULTIPLEXOR	CHANNEL	CAPACITY
DS1	T1	2.544 MBPS
DS1C	T1C	3.152 MBPS
DS2	T2	6.312 MBPS
DS3	T3	44.736 MBPS
DS4	T4	274.176 MBPS

The main advantage of STDM relative to FDM is the overall data rate that can be reached over a communication channel. Since the rate is a direct function of the channel bandwidth, the overall rate that can be reached in FDM over a voice grade channel is limited to about 2000 BPS. In STDM the data is transmitted in series using the full available frequency spectrum of the communication channel. Thus with the modems associated with the STDM's, the data transmitted rate could reach 9600 BPS.

Another advantage of STDM is that only one modem is associated with the multiplexor unit, while with FDM each data input channel requires a separate modem. This marks at the most important feature of STDM: that it exhibits true digital characteristics and does not require any analog-to-digital conversions as in FDM. This accordingly enables the STDM data to be fed directly into the host computer, and furthermore makes the data ideally placed for handling by electronic telephone exchanges.

7.5 STDM OPERATION VERSUS FDM OPERATION

Both STDM and FDM are the most popular means of multiplexing introduced to achieve audio and video transmission economy. Reduction in the cost of the communication channels is achieved with both schemes by concentrating many individual channels into one; indeed it could easily be claimed that without multiplexing, long distance communication would be impossible. Although more FDM equipment is available today and is less expensive, STDM equipment is gradually catching up because it provides better transmission efficiency.

The superiority of STDM over FDM in long distance voice communication is marked not only in the channel utilization but also for its voice reproducing qualities. The distortion of the voice signal, is less related to distance when in its modulated digital form, than when in its original analog form. This makes STDM more immune to noise and other interference than with FDM. It has consequently been found that in a long chain of cascaded multiplexors, STDM presents better effective signal-to-noise margin than FDM. In fact it has been proved that the overall cost of equipment on long distance channels could even be lower with STDM.

In the remainder of this section, the comparison of the two multiplexing techniques is restricted to the application of STDM and FDM functioning as concentrators of data originating from digital terminals. Both types of multiplexing schemes share the common audio channel with predefined numbers of terminals. In FDM all the terminals may operate in parallel with each of its data channels assigned a portion of the common channel spectrum. In STDM all the terminals operate in series with each channel assigned a portion of the time.

Both FDM and STDM equipment use hardware equipment in performing their multiplexing operations. This means that the number of terminals associated with a given multiplexor equipment is fixed and cannot be changed after its construction; nevertheless, during the design stages some means can be introduced to increase this number. Increasing the number of the terminals associated with an FDM equipment can only be achieved by either widening the common channel bandwidth or reducing the data rate transmitted from each terminal. Increasing the number of the terminals associated with an STDM equipment can only be achieved by widening the cycle scan period of all the terminals or by decreasing the size of the time slot assigned to each terminal. It follows that increasing the number of terminals associated with a multiplexor would result in some form of degraded performance. In this respect, both FDM and STDM can be regarded as an extension of point-to-point communication, since the channel resources are shared on a permanent base. For this reason too, this form of data concentration is often referred to as circuit switching, as distinct from store-and-forward switching which uses different techniques of data concentration.

The data transmission rate of the terminals associated with FDM is much slower than the possible transmission rate over the common channel, whereas the high data rate of the terminal associated with STDM is the same as for the transmission rate in the channel. On the other hand, the data flow in the common channel associated with FDM can be continuous, while with STDM it is always staggered.

Both frequency and time multiplexing technique reduces the overall cost of the common communication network. The cost effectiveness of both systems depends on the distance between the terminals and the multiplexors. STDM is more expensive than FDM, particularly when the remote terminals are not clustered at a single location. On the other hand, if many terminals are centred in one site, as is the case in many command and control systems, the STDM presents better cost effectiveness, because the common channel can be used to its maximum capabilities, i.e. at its maximum data rate.

A most important aspect in the comparison of FDM and STDM is the position of the modems in the network. In STDM the equipment is positioned on the channel after the modem while in FDM the equipment is between the modems and the terminals. This indicates the cost effectiveness of both systems as related to the geographical spread of the terminals.

Another point of comparison is the form of a signal that is multiplexed. In FDM the signal is in the form of analog waves, which has the advantage of allowing the multiplexors to be at any distance from the terminals. In STDM the signal is in the form of digital pulses, which has the advantage that they can be easily manipulated by any logical circuitry.

The following table shows some of the points of comparison between the two multiplexing schemes:

TABLE 7.3

FUNCTION	FDM	STDM
Type of signal	Analog waveform	Digital pulses
Type of terminals used	Slow speed terminals	High speed terminals
Geographical spread	Used for isolated terminals	Used for clustered terminals
Type of transmission	Parallel transmission	Serial transmission
Relation to the original signal	The full wave, continuous	Samples only
Spectrum limitation	Limited in frequency	Limited in time
Speed overal limitation	Variable to 2000 BPS depending on the bandwidth available	Variable to 9600 BPS depending on the channel and modem available
Speed of channel relative to terminal	The channel speed is the sum of speeds from all the terminals	The channel speed is the same as the terminal speed
Position of multiplexor	On the channel after the modem	Near the terminals before the modems
Data delay	No delay at all	One cycle length
Error detection feasibility	No facilities available	Feasible and practical
Independence of links	Each parallel link operates separately and independently	The data in the common channel is interleaved
Switch facilities	Only in electromechanical or space switching schemes	Easy to manipulate on any electronic switching scheme with no extra equipment required
Connection to the computer	Requires analog to digital conversion before the computer can handle the data	Can be directly connected to the computer
Price of equipment	Relatively cheap	Expensive
Data Redundancy	None	Special data for synchronization framing and error detection
Guard spacing between channels	Frequency spacing	None beside synchronization information
Utilization of the line	The full effectiveness cannot be achieved	Efficient utilization of the full spectrum
Reliability	Susceptible to noise and distortion	Less affected by noise
Cascading	Readily possible	Requires the addition of a modem pair to each multiplex and a master-slave clock scheme

The main disadvantage of both FDM and STDM schemes, besides the limited number of terminals, is that the channel portion assigned to each terminal cannot be fully used if there is no data to transmit from this terminal. In other words, this terminal gets the channel share allotted to it regardless of whether or not it has data to be transmitted, while the other terminals have no means of taking advantage of any idle portions of the common channel.

The common principle of these two multiplexing schemes is that each terminal always has a transmission channel available to it, since a shared part of the line is fully confined to each terminal. For this reason the terminal can never find the line busy, as may occur in other schemes. Although this may be regarded as an advantage, it could conversely be regarded as a disadvantage since the line is not always efficiently utilized.

In conclusion, it can be said that FDM is intended for slow data rate terminals up to 300 BPS, spread over wide geographical areas, while STDM is intended for high data rate terminals of 1200 BPS and above, clustered in one location. Both FDM and STDM are only cost effective if the terminals produce continuous data traffic. In other words, both forms of multiplexing, FDM and STDM, are not cost effective for terminal to computer type of data communication where the traffic source comes in random bursts.

7.6 STDM VERSUS SIMULTANEOUS MULTIPLEXING OPERATION

STDM, as explained before scans the various data sources and thereby produces the data in the common channel in a staggered form. This results in delays between the time the data is ready for dispatching in the terminal and the time it is transmitted in the common channel. Such a situation calls for a buffer store in the terminal of the size at least that of a time slot, This problem does not exist in FDM, where all the terminals may dispatch their data messages simultaneously as soon as they are ready; here however, the speed of data transmission from each terminal is limited.

Simultaneous transmission has the advantage of no delays in data dispatching, and so can proceed in transmitting, immediately data is made available in the terminals. Furthermore, the transmission in the common channel is parallel rather than in series. The data volume from each terminal, however, is limited, since the common channel capacity is circumscripted according to Shannon. As a result, the common channel efficiency when using simultaneous transmission cannot always be attained.

There are a number of simultaneous transmission schemes besides FDM which may offer better characteristics, one of them being the PDM technique. Another technique worth mentioning is the code division multiplex (CDM) scheme which is discussed in chapter 10. CDM, similar to FDM, utilizes the common channel in a most inefficient way, although it does offer means of privacy that are not found in any other form of multiplexing.

Phase division multiplex (PDM) offers many features which are similar both to FDM and STDM. Although both PDM and FDM provide simultaneous transmission, the characteristics of PDM are more like STDM since they both provide the concentration of high data rate terminal traffic. Furthermore, they are both intended to supply multiplexing services to terminals clustered in one location. The number of terminals that can be serviced by one PDM unit is, however, limited with todays technology to four only.

A comparison of STDM with PDM, when they both provide multiplexing to four high speed terminals, as seen in Fig. 7.18, indicates that the channel utilization is identical in both examples. Nevertheless, in this example only, the PDM is proved to be the more cost effective since it requires less equipment in the network and no buffering stores in the terminal. At the same time, with PDM the data flowing in the common channel is simultaneous but with STDM it is staggered. Furthermore, to ensure the full channel utilization, the data rate leaving each terminal must be four times faster in the

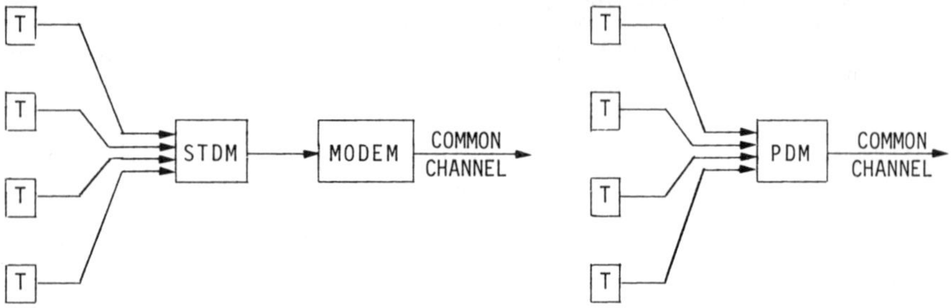

Fig. 7.18 The equipment used in both STDM are PDM schemes

STDM scheme than with PDM scheme, while the data rate in the common channel is then identical for both schemes.

Comparing STDM with PDM, when they both provide multiplexing services to a different number of terminals is in most cases in favour of STDM rather than of PDM. Nevertheless, for command and control applications, PDM is sometimes preferred because of its simultaneous characteristics. That is, PDM may offer better real-time features than STDM.

The following table presents some of the main comparison points between both STDM and PDM characteristics.

TABLE 7.4

FUNCTION	PDM	STDM
Type of transmission	Parallel transmission	Serial transmission
Type of input signal	Digital pulses	Digital pulses
Type of common signal	Analog waveform	Digital pulses
Delay in transmission	Instantaneous	Up to one cycle delay
Geographical spread	Clustered terminals	Clustered terminals
Exploitation of the common channel	Each terminal may transmit over the whole channel spectrum all the time	Each terminal may transmit over the whole channel spectrum for only part of the time
Relationship to the original signal	Continuous data stream	Samples of the data stream
Adjustments to the terminal	None required	Added buffer and increase of transmission rate
Multiplexor/modem relationship	The modem is part of the multiplexor	An extra modem is required before the multiplex signal may be transmitted over the common channel
Number of terminals	Limited to four only	Unlimited in design although number affects cycle time

Similar to all the other multiplexing schemes, PDM as well as STDM suffers from the disadvantage that the portion of the common channel assigned to each terminal is permanent. The other terminals, associated with the same multiplexing unit, cannot exploit the other portions even when they are not being exploited by their assigned 'owners'.

7.7 SPACE DIVISION MULTIPLEXING (SDM)

The multiplexing schemes discussed so far share a single common communication channel between a large number of data terminals. It is sometimes necessary to share more than one communication channel between a number of data terminals, thereby allowing many simultaneous parallel transmissions. Furthermore, the previously discussed multiplexors enable the signal originating from any terminal to reach only one destination, and if required to be directed to a number of destinations, a switching unit must then be added. This requirement can be met directly by a single multiplexor which provides simultaneous access to a number of communication channels. Such a form of multiplexing is generally referred to as space division multiplexing (SDM) since the sharing process is separated in space rather than time or frequency. It is possible to regard the space division multiplexors as a multi-stock of frequency or time multiplexors, as seen in Fig. 7.19. That is, in SDM any one of n input channels can have access to any one of the m output channels.

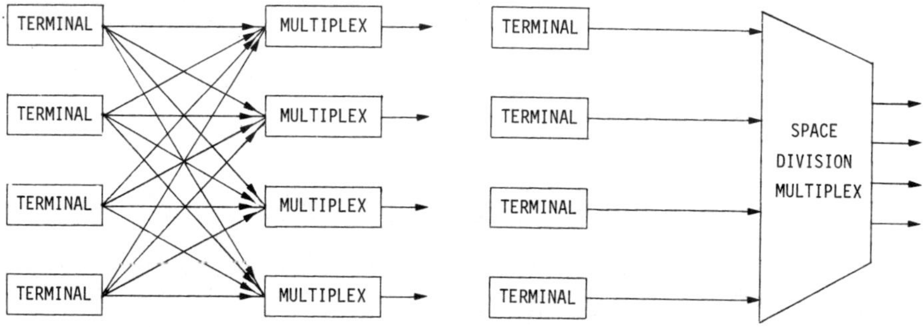

Fig. 7.19 Space division multiplexing as a multi-stock single multiplexing

The assumption that space division multiplexing can be regarded as a multi-stock multiplexor, is not completely accurate. SDM differs from the multiplexing techniques so far discussed in the way the common output channels are shared between the input terminals. In FDM or STDM, the common channel resources are assigned to each terminal on a permanent basis. In SDM the common channels are assigned by a semi-permanent arrangement. In other words, any output channel can be assigned to any of the input terminals and then remain permanently connected until the assignment is cancelled. Once an output channel is assigned to any of the terminals, the terminal has the full control of the channel spectrum for the full assigned time. The characteristics of SDM make it more like a multi- STDM although the assigned time duration in SDM is extremely long and much beyond that of STDM.

There are various forms of SDM schemes which can concentrate both analog or digital signals.

Analog space division multiplexing (ASDM) equipment is most popular in telephone circuit switching as well as in telemetry systems for concentrating and routing measurement readings. The simplest form of ASDM is the crossbar switch, as shown in Fig. 7.20, which consists of isolated horizontal and vertical bars. At each cross-point there are relay contacts which can connect the two perpendicular bars, thereby providing the means, when the relay is energized, for a signal to flow from any input channel to any of the output channels. By suitably selecting the cross points, it is possible to provide a number of parallel paths between any of the input rows with any one of the output columns. These relay contacts at each cross-point can be regarded as 'space' switches, and they are usually activated by an address formed by the coincidence of two sources in accordance with the coordinance of the space switch positions. In this configuration, the SDM acts as a matrix switch, providing a permanent path from any selected terminal to any one of the output channels, for the full duration of the space switch operation.

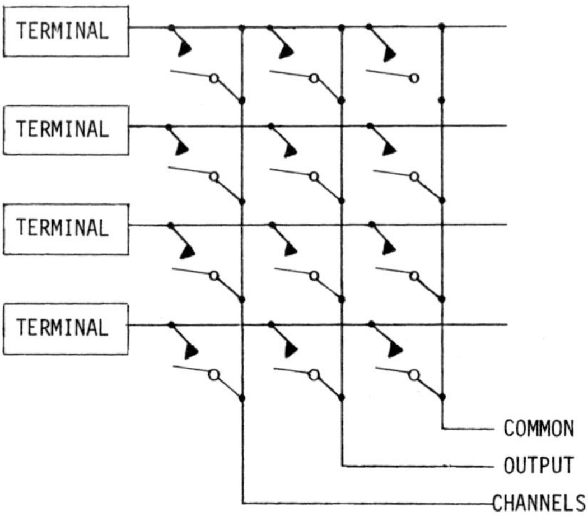

Fig. 7.20 The basic principles of a cross-bar switch

The accepted representation of SDM, as shown in Fig. 7.21, is based on the cross-bar matrix. The number of input channels is represented by n, the number of output channels by m and the number of levels in each cross-point b g. The number of cross–points in th cross-bar matrix is therefore $n \times m$ which allows m simultaneous transmission paths within the matrix, provided $n > m$. These m physical connections between the output channels and input terminal are hence separated in space.

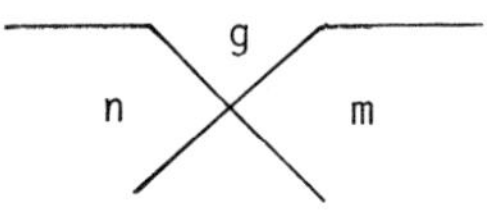

Fig. 7.21 Accepted representation of space division multiplexing

Each cross-point in the cross-bar matrix consists of more than a single relay contact; in fact each cross-point is formed by a stand alone relay with a number of contacts. Most semi-electronic cross-bars in the last two decades employ a dry reed relay for this purpose because of their small physical size and relatively fast operation. When a cross-point is selected, the relay at the junction of the row with column of the matrix is energized and thereby sets simultaneously all its association contacts. Selecting a cross-point must enable the required signal contact to be opened or closed and then remain in this state even after the selection operation is removed. This memory operation is achieved with the aid of the relay at each cross-point. The relay is selected by appropriately choosing a row and column of operating lines which energize the relay at the junction of these two perpendicular lines, as seen in Fig. 7.22. Two extra contacts on each relay hold the relay energized through two separate rows and column hold lines, until the path is no longer needed. After the required relay is energized, the current in the column and row operating lines can be removed without releasing the relay. This will allow other cross-point relays to be selected without any danger of them being affected or their affecting the relay that is already operating.

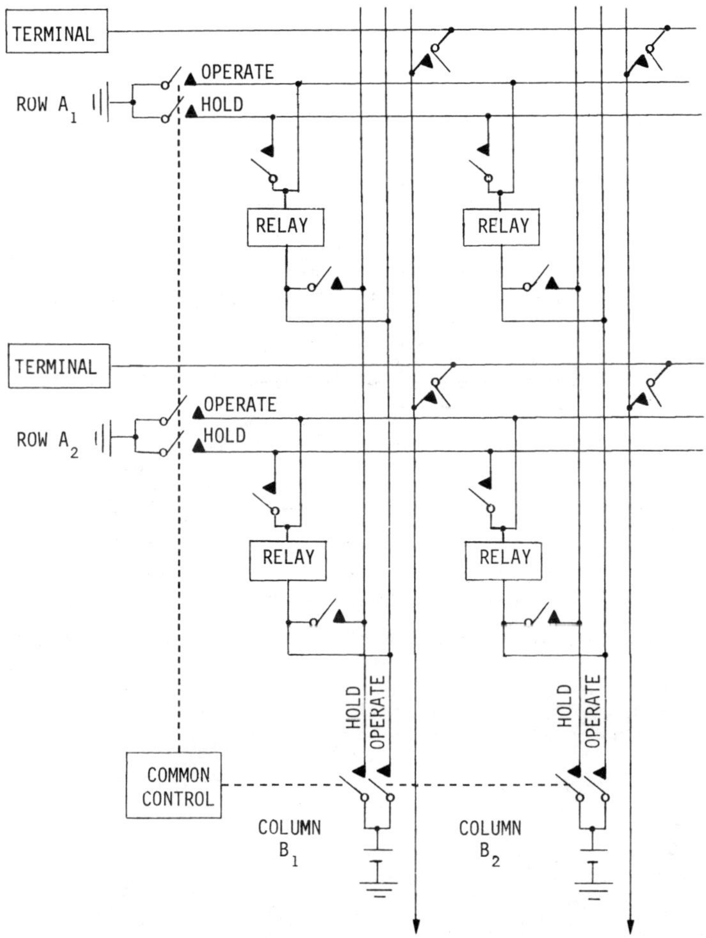

Fig. 7.22 The operational principle of the cross-points in a relay cross-bar

One of the essential parts of the SDM is the common control unit, whose main purpose is selection of the space path. This control unit generally gets its instructions from any of the input channels by requiring a free output channel. In this arrangement, the common control can select only one cross-point in each operation stage; nevertheless, since the cross-point remains held, the matrix allows many relays to continue to be operated simultaneously. Consequently the common control unit, often referred to as a marker, sets up a number of parallel paths, with only one, however, set up at a time.

Today, many of the analog cross-point systems are replaced by solid state integrated switching circuits which have many similar characteristics to those of the relay contact. While the selection of these solid state switches is made by digital means, the signals which pass through them are analog. Each cross-point also contains memory facilities, so that they can remain in the closed or open state even after the slection currents have been removed. The integrated space switches are therefore analog gates which are opened and closed according to a program map contained in the memory of the common control unit.

There are many different circuit configurations that can be fabricated as electronic solid state cross-points, as seen in Fig. 7.23. Most of them are based on four layer diode devices such as thyristers and SCR devices. The device, by injecting a high current into it, is selected and is then held by a small dc current flowing through it. Removing this current will switch off the device. The use of integrated circuits reduces cross talk problems which are unavoidable in electromechanical switches. It has also the other advantages of simplifying the matrix network, ease of expansion, small in size, economical and above all reliable.

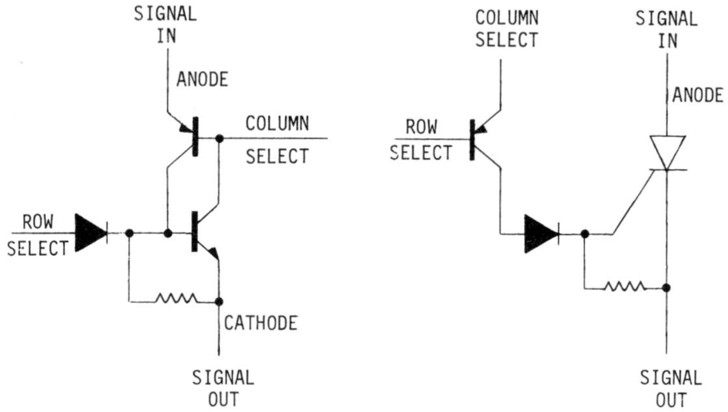

Fig. 7.23 Examples of solid state analog cross-point switches

Digital space division multiplexing (DSDM) has the same matrix properties as the relay cross-bar. That is, each relay is replaced by a flip-flop and the contacts by AND gates, as seen in Fig. 7.24. The basic difference between DSDM and ASDM is that the signals that can be transferred through the matrix must be in digital pulse form. Nevertheless, DSDM is still used to switch voice signals, although this time it is in the form of digital voice. The DSDM is ideal for PCM switching, since it can transfer the data in its pulse time division form. Furthermore, space division enables the transfer of PCM signals to be made in parallel, a form which is impossible in the simple time division. Since the cross-point switches in DSDM use flip-flops which are semi-permanently connected, the pulse formats that can use this matrix need not necessarily be time synchronized. This means in effect that DSDM can be used to transmit not only PCM type of digital voice but also PPM, PWM and PDM (see chapter 2). The advantage of DSDM over ASDM is that DSDM can be designed as a single integrated circuit with a simple address register and decoder as the common control unit.

In today's electronic telephone exchanges not only are the data switching control pulses that pass through the exchange but also all the data voice pulses fully synchronized in the time domain. In such schemes, the STDM equipment is used for multiplexing the various data sources into serial time slots, while the DSDM is used to provide the semi-permanent connection with one of the selected output channels. Such a time-space multiplexing scheme is shown in Fig. 7.25. Although primarily developed for telephone exchanges, it could be most useful in some C^2 system applications for the data acquisition subsystems. In these applications, each controllee provides a number of measurement results, all of which may be directed on request to various

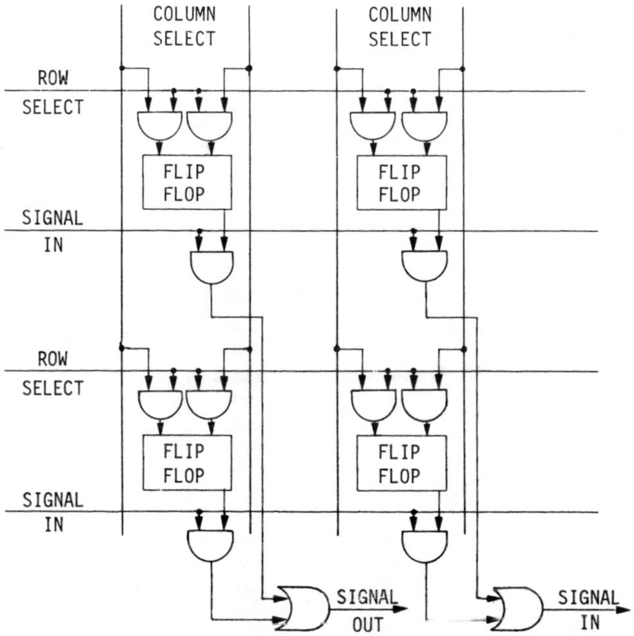

Fig. 7.24 Example of a digital space division multiplex network

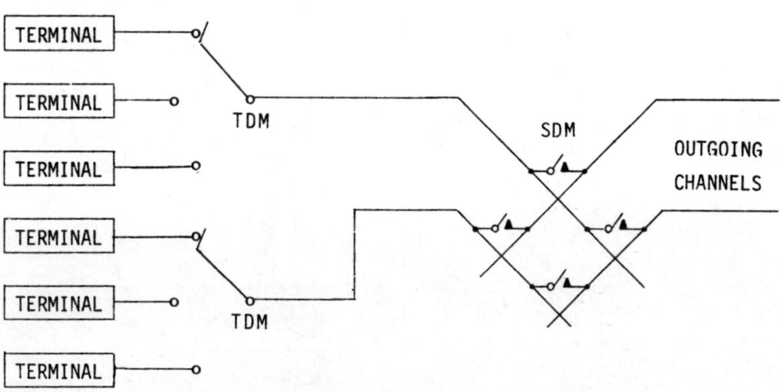

Fig. 7.25 Schematic diagram of a time-space division multiplexing scheme

destinations. The time switch scans all the controllee terminals, thereby providing a stream of pulses. The space switch then provides a longer connection which enables all the results of a single controllee to be directed to a given destination. For the full use of the DSDM in telephone exchange, an extra switch is necessary, namely, time slot interchange (TSI) which will be discussed in the next chapter. Such a TSI is necessary to nullify the semi-permanent arrangement of the SDM.

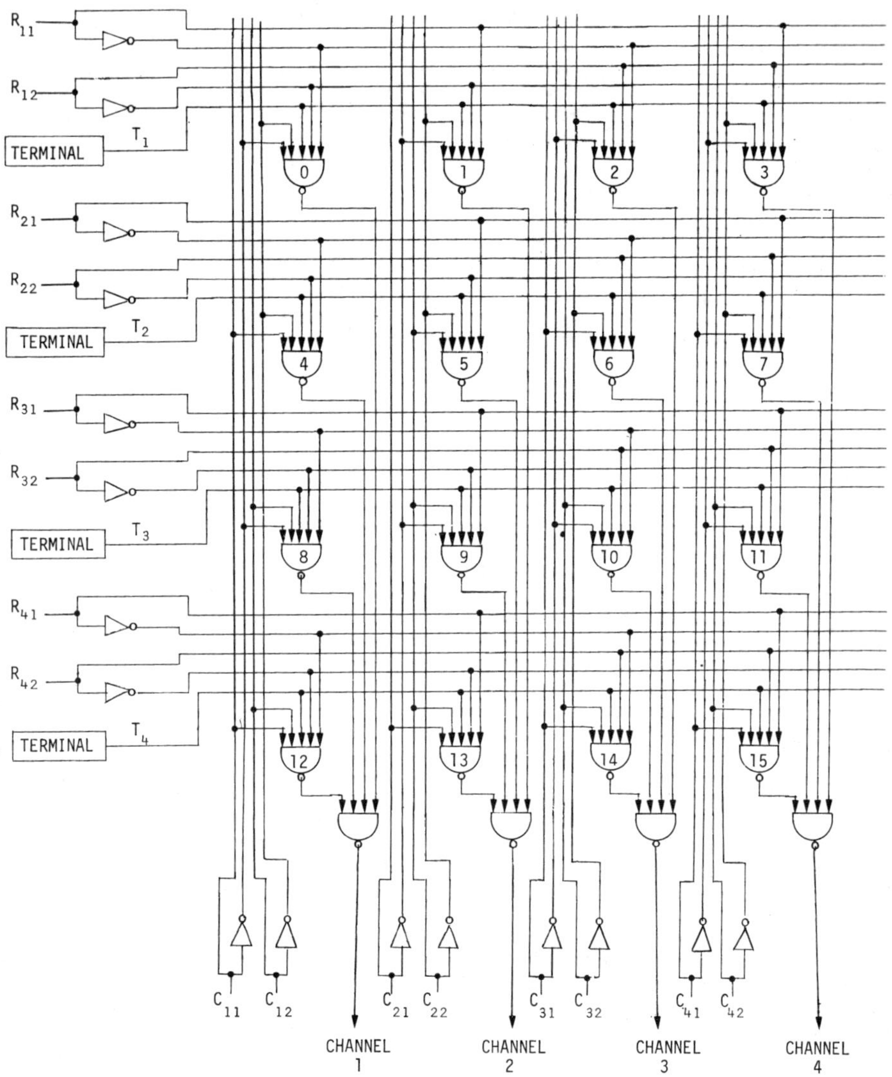

Fig. 7.26 Blocking space division multiplexing circuit

All the analog and digital space division multiplexing schemes discussed so far can be classified as a form of non-blocking techniques. That is, the currents holding the cross-points in the operation state do not block or affect any of the other cross-points in the matrix. Nevertheless, the feature is achieved at the expense of requiring some form of memory in each cross-point, and also, for their selection to be performed one at a time. The advantage of such a configuration is the provision of many parallel paths on a semi-permanent basis.

There are, however, digital space division multiplexors in which the connection is constantly changed to accommodate different paths in accordance with a variable common control program. In this case, the space cross-point switches open and close for short periods only and therefore there is no longer any need for a memory at each cross-point. This means that the cross-points remain operated as long as current flows in the two perpendicular selection lines. To enable the matrix to allow parallel transmission while the selection currents are flowing, each of the coordinate lines must not only choose the junction cross-point but must also block all the unselected cross-points attached to these two lines. In this case the selection lines will consist of a number of wires which provide the supply of a control current signal to only one cross-point in a row or column, while inhibiting all the others in the same row or columns. This selection is based on a binary code with an address of k bits for the selection of k^2 cross-points. In other words, each selection line will consist of $2k$ wires which can then be used to identify one of k^2 cross-points on each row or column. To achieve a successful operation, all the data signals as well as the control signals must be fully synchronized.

In this blocking space division multiplex matrix, each cross-point consists of an AND gate, as seen in Fig. 7.26, with the whole matrix operating as a multi-TDM stack. That is, each column can be regarded as a single STDM unit, indentical with that which was shown in Fig. 7.13. Nevertheless, by constructing a number of STDM units in parallel, and thus forming a matrix, it is possible to associate each input channel with any output channel. Although only one cross-point is selected with each row/column selection operation, it is possible to select in the same manner a number of simultaneous paths through the matrix without one affecting the operation of the other.

To follow a selection operation, take for example that terminal 2 is connected to output channel 3. In this case, AND gate 6 at the junction of row 2 and column 3 is selected. In the row 2 control lines the voltage state $\overline{R_{21}}$ and R_{22} are selected while in the column 3 control lines the voltage state C_{31} and $\overline{C_{32}}$ are selected. By R_{22} and C_{31} one refers to high voltage levels and by $\overline{R_{21}}$ and $\overline{C_{32}}$ one refers to low voltage levels, i.e. 01 and 10 binary states.

Non-blocking SDM, when operated by itself, is inefficient from a traffic standpoint. However, by using a combination of the semi-permanent space connection with interchangeable time slot, a most efficient network switch is provided. Blocking SDM is an efficient form of multiplexing but it requires complex selection methods. Non-blocking SDM is therefore used for long duration switching applications, such as exist in voice communication. Blocking SDM are therefore used for short duration switching such as needed in computer communication controllers.

7.8 STORE AND FORWARD CONCENTRATORS

As the transmission channels are the major components of command and control systems and other distributed computer systems, a basic objective in the communication network design is minimizing the channel cost. Much progress in this direction has been made with the adoption of multiplexors and having common communication channels instead of a large number of separate channels. This, does not, however, make for improved efficiency of the communication process; because the multiplex common channel is formed by combining numerous single point-to-point channels, with each data terminal receiving a permanent share of the common channel jealously regarded as a privately owned property. Consequently, if a terminal has no messages to transmit, then the 'privately owned' share of the channel remains idle and untouchable, with no other terminal having any facilities for exploiting the idle share.

One is tempted to say that multiplexing is possibly the most inefficient means of sharing a common communication channel. This treatment should not imply that multiplexing is always operated in an inefficient manner. In fact there are applications where the multiplexing process is the ideal solution. In applications where the terminal sources produce a continuous data stream, the multiplexors operate most efficiently, for the simple fact that in these applications there are no idle portions created by the transmission of these terminals and so the common channel is exploited to its full capacity. For this reason, multiplexing is primarily used for PCM type data source.

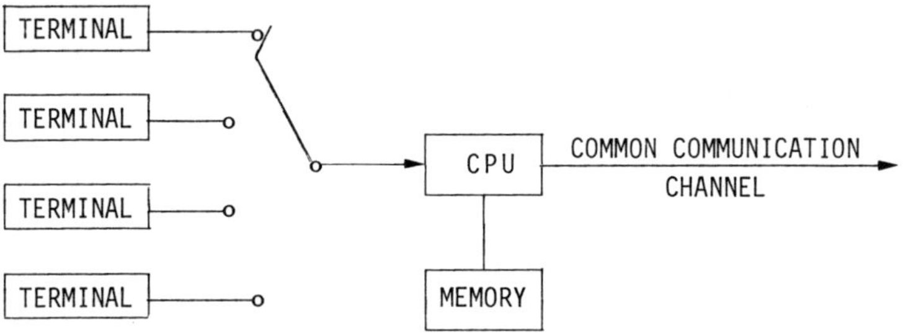

Fig. 7.27 General configuration of a store and forward concentrator

For terminal-to-computer data communication networks, however, the multiplexing solution proves to be most inadequate, since in this application, the average data volume produced from each terminal is extremely small. Terminal-to-computer traffic comes in bursts of sudden volumes which are produced at random with unpredictable intervals. In other words, the channels leading from each terminal remain idle for most of the time. Statistics indicate that the terminal-to-computer traffic demands represent only 3% of the overall channel time, with peaks requiring up to 20% of the full load, leaving the remainder of the channel time wasted. It is impossible to increase the channel efficiency by adding a number of terminals to each multiplexor since the channel resources are shared on a permanent basis. Instead, it is suggested that other means of sharing the common channel be employed, namely, the use of store and forward (S & F) techniques. That is, to introduce, instead of a multiplexor, a concentrator in the form of a small (mini or micro) computer which provides its memory as part of the network facilities, as seen in Fig. 7.27. In this case, each terminal can off-load its messages into the memory where they can be held for short periods. The CPU can then transmit the messages on the common communication channel, one after the other with no idle times or any unnecessary spaces between the messages. In this way, the messages can be sequenced into a smooth data stream that can be transmitted at a uniform data rate with high channel utilization.

The common communication channel is shared now in accordance with a dynamic assignment rather than a fixed assignment basis. This makes the channel operation most efficient, as it can be made to function at its maximum utilization even when some of the terminals associated with it are idle. Furthermore, since the traffic in the common channel can no longer be directly associated with any terminal source, the number of terminals that can be connected to the store and forward concentrator is unlimited. The only restraining factor in the number of terminals is that of delay; that is, it is essential to ensure that the data messages will be retransmitted in a given time period. If the common channel has a maximum capacity enabling it to transmit C messages in a given period, say an hour, and each terminal transmits only c random messages in the same period, then the number of terminals that can be associated with the concentrator for efficient operation will be $m = C/c$. More terminals may be further added to the concentrators, if longer delays are allowed.

Data messages, using store and forward concentrators are therefore confronted with delays between the time the messages leave the terminal and the time the messages are transmitted on the common channel. This delay is a function of the number of terminals and the number of messages each terminal has to dispatch. The minimum delay is one time slot while the longest is N time slots, where N is the number of terminals. The calculation of this delay is based on the statistical distribution of the message dispatch. Command and control systems are designed so that the delay will have no appreciable influence on the real-time operation of the system. At all events, the disadvantage of the delay (which is usually relatively short) is overshadowed by the considerable saving in the network cost by introducing efficiency into the communication.

Multiplexing operates by using point-to-point techniques with the process completely transparent to each terminal. Introducing S & F concentrators into the network results in the equipment being no longer physically transparent. That is, the messages in the common communication channel can no longer be associated with any particular terminal; thus requiring extra means to be inserted into the common data stream so that the messages can be identified. This is achieved by means of an address that is added by the concentrator at the heáding of each message, with a unique code dedicated to each terminal. The address could adequately be added by the terminal, but this would require special circuits, a provision which does not exist in every terminal.

Removing the point-to-point facilities that characterize multiplexors, introduces further advantages for the concentrator. In multiplexors, the data transmission speed from each terminal coincides with that of the data travelling along the common channel. This means in effect that each multiplexor must be tailored to deal with specific types of terminals, defined by their data rate. This requirement is unnecessary with store and forward concentrators. The concentrator acts as a storage buffer between the terminal and the common channel, thereby isolating the terminal operation from the channel operation and enabling it also to perform speed conversions. Adequately it can collect data from slow and/or fast data rate terminals and transmit them at the fixed constant data rate which is used in the channel.

Scanning the terminals to see whether they have data to transmit can be sequential, as in STDM, or in any other order required. It is possible, however, to provide further facilities by taking advantage of the buffer characteristics. Since the common channel is no longer transparent to the terminals, it is possible to design the concentrator in such a way that the terminals may always transmit their messages with no delays, disregarding the situation whether the common channel is free or busy. That is, the concentrator must always be free to accept all the terminals' data regardless of the common channel traffic. This is achieved by providing each terminal with individual buffer spacing within the concentration memory thereby having it always available for the terminal. The terminal dispatches its messages directly into the buffer, as seen in Fig. 7.28, and so the concentrator can scan the buffers instead of the terminals. The messages are thence arranged in a quene and transmitted over the common channel in the sequence of first-come-first-out. All the concentration facilities are provided by software programs, taking advantage of their processing abilities.

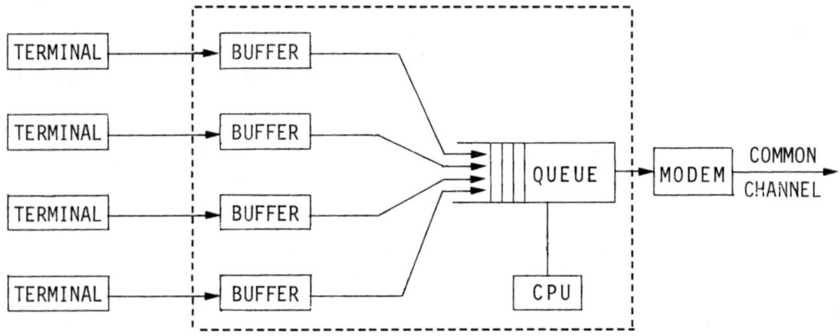

Fig. 7.28 Basic principles of a store and forward concentrator with a buffer store

Although obvious, it should be pointed out that the data handled by the S & F concentrator is purely digital and therefore has similar characteristics to those discussed for STDM. That is, all the terminals must be clustered near the concentrator and a modem must be employed before the data can be transmitted over the common channel. Remote terminals may be associated with the concentrator, but then, only through a pair of modems.

7.9 CONCENTRATORS VERSUS MULTIPLEXORS

Most of the comparison points between multiplexors and concentrators have already been discussed in the previous section; however, since these two techniques are the most popular local network equipment and have their terms commonly interchanged, it is advisable to sum up their comparison points in a single table. Both techniques are used for the purpose of providing means of sharing a common communication channel; nevertheless, they are distinguishable by the way they share the resources of this channel, as classified in Fig. 7.29. Broadly, concentrators assign the channel dynamically on a task basis, while multiplexors do so rigidly on a fixed basis. Multiplexors share the channel using permanent hardware techniques whereas concentrators perform the same functions but use software techniques.

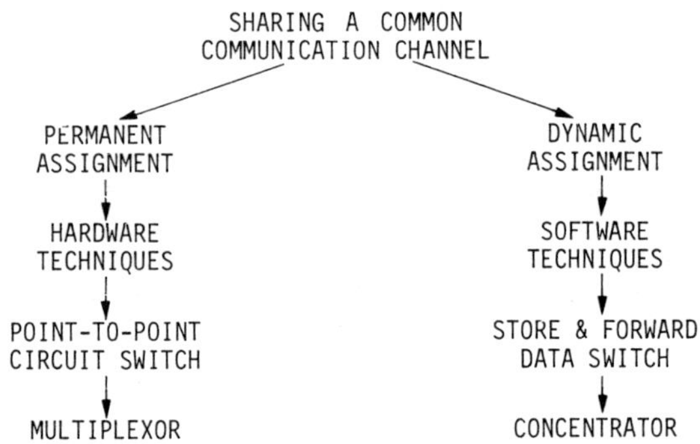

Fig. 7.29 Classification of both multiplexors and concentrators

Concentrators and multiplexors are both intended for different applications and they both use different techniques in catering with them. Multiplexors are intended to provide a service to the terminal by presenting them with means of sharing a common communication channel, although they do not perform any communication tasks. Concentrators are intended to increase the efficiency of the communication channel and therefore can be regarded as a form of traffic cop directing the communication at a common junction.

Concentrations could be visualized as intelligent multiplexors, as presented in chapter 4, because of the common task of sharing a common channel. Nevertheless, since concentrators, by exploiting their programming ability, can provide many more facilities, it is advisable to refer to them by a separate term. Concentrators can operate as multiplexors, since they are more flexible whereas multiplexors cannot act as concentrators, because their operational functions are static.

The following table sums up the major points of comparison of both multiplexors and concentrators.

TABLE 7.5

FUNCTION	MULTIPLEXORS	CONCENTRATORS
Network performance	Point-to-point	Store and forward
Physical performance	Transparent	Via memory
Network delay	No delay at all	There is always a delay
Channel data rate	The same as the terminal	Independent of terminal data rate
Channel assignment	Permanent	Dynamic
Data source	Continuous data	Random data
Identification of data	In accordance with timing	In accordance with an address tag
Communication tasks	None at all	Translation of formats and speeds
Number of terminals	Fixed	Variable
Network efficiency	Poor	Good
Efficiency with high even traffic	Good	Poor
Efficiency with random low traffic	Extremely poor	Excellent
Type of terminal	Simple	Intelligent
Operating technique	Hardware	Software
Type of data	PCM type data	Terminal-to-computer

7.10 ASYNCHRONOUS TIME DIVISION MULTIPLEXING (ATDM)

As already pointed out, the widely used synchronous time division multiplexing operation is the most inefficient means of exploiting the common communication channel and is therefore unsuited for terminal-to-computer applications. On the other hand, concentrators, which are ideal for these purposes, are rather expensive. A different approach to the common channel utilization is offered by the synchronous time division multiplexing (ATDM) which is a combination of a time multiplexor and a concentrator. The purpose of the ATDM is to apply TDM principles more efficiently while transmitting the data as a concentrator.

The fundamentally inefficient operation of STDM can be seen in

Fig. 7.30, in which the unused time slots remain idle and wasted. The terminals are scanned one after the other at a fixed rate, with each terminal receiving a fixed equal sized time slot. In other words, the common channel is connected to each terminal for the duration of the time slot and is then transferred in turn to the next terminal at the end of the time slot. (Since the scanning of the terminal is performed in a fixed sequence and each time slot can be identified by its relative timing position, there is no need for an address label for each message transmitted.) With STDM the scanning of the terminal is independent of whether or not there are messages to be transmitted. This means that the multiplexors will remain connected to each terminal for the full period of each time slot, irrespective of whether it is required or not. It can clearly be seen in Fig. 7.30 that if there are no messages available for transmission, then the specific time slots remain empty. This creates economical problems when using STDM for terminal-to-computer type communication, although it is still useful for continuous type data streams as produced by PCM terminals.

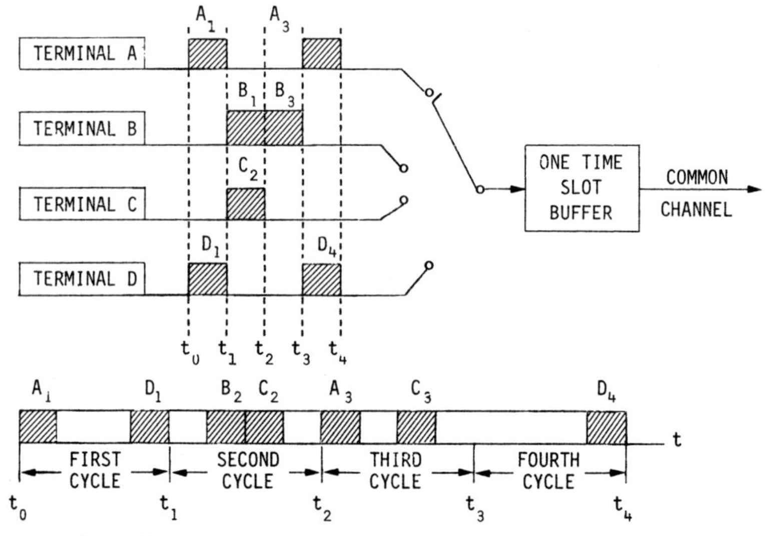

Fig. 7.30 The operational principle of STDM

To solve the random arrival of the data messages in terminal-to-computer communication, a more flexible system is required. The scheme, which is known as asynchronous time division multiplexing (ATDM), transmits the data messages in the common channel in a continuous stream, one after the other, with no empty slots, whereas it scans the terminals in the same fixed sequence as STDM. The difference, however, is that if there are no messages to be transmitted from a specific terminal, the multiplexor will not halt but will immediately continue to the following adjacent terminal in the sequence. In this manner this multiplexor switches from terminal-to-terminal at a fast speed and will only halt at a specific terminal if there is a message to be transmitted. The time allocated to each terminal for transmission is fixed and is long enough for the terminal to unload the message, which could be of the duration size of a bit, a character or a block, depending on the multiplexor design. With this arrangement there is a great saving of time.

The principle of the ATDM scheme is demonstrated in Fig. 7.31, where the time saving can be seen from the data sequence. The time saved by this

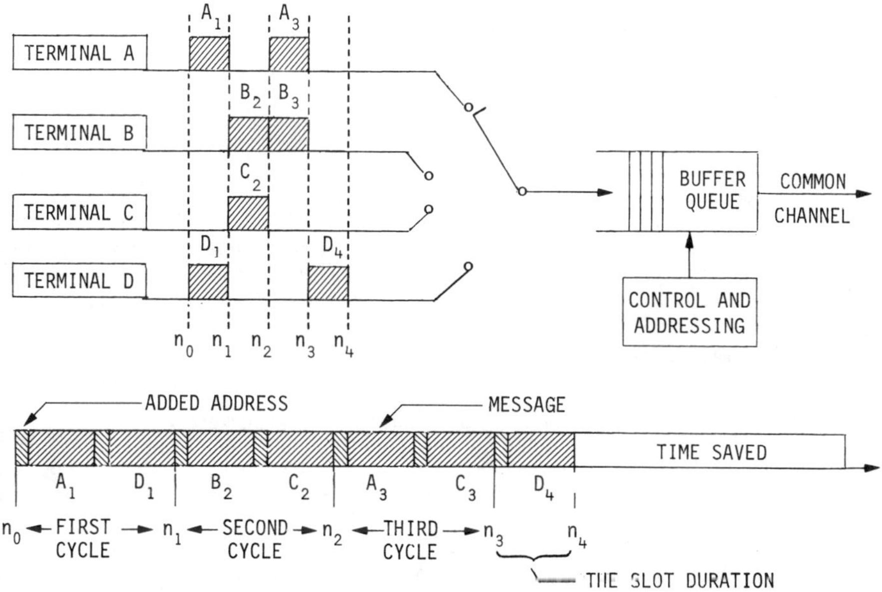

Fig. 7.31 The operational principle of ATDM

scheme could be used for serving more terminals and is similar to that achieved by the concentrator discussed in the previous section. In both schemes, each terminal is granted access to the common channel whenever it has a message to dispatch.

As in concentrators, a labelling address must be added to each message as it is accepted from the terminals so that it can be identified by the computer. The message with its associated label tag is located in the memory and placed in a buffer queue ready to be transmitted on the common channel. As soon as the channel becomes free the message is transmitted with no gaps between adjacent messages, at a constant continuous data stream.

The control unit which supervises the sequence operation of the scanning sequence also controls the transmission sequence and adds the address label. The design of the ATDM is based on the asynchronous arrival of the message (i.e., at random timing) and their synchronous transmission dispatching (i.e., at a constant speed) over the common channel. The buffer must be of reasonable size so as, on the one hand, to prevent long queue delays and, on the other hand to prevent an overflow when the traffic volume is large. The design of the buffer size is therefore based on the statistical behaviour of the random message arrivals, the number of terminals scanned, the volume of data from each terminal and the maximum allowable delay in the system. The design trade-off is then calculated on the buffer size, the number of terminals and the relative data rates of both the terminals and common channel.

Since both STDM and ATDM have many similar characteristics, with both scanning the terminal at a fixed sequence and producing fixed duration time slots, it is important to compare the two techniques. The following table presents some of the main points of comparison of the two schemes.

TABLE 7.6

FUNCTION	STDM	ATDM
Type of connection	Point-to-point	Store and Forward
Scanning procedure	Sequential	Sequential
Scanning rate	Fixed	Dynamic
Common channel allocation	Permanent time slots	Random time slots
Time slot duration	Message size	Address and message size
Identification of time slots	In accordance with timing	In accordance with an address tag
Data delay after dispatch	No delay at all	Delay depends on traffic
Terminal dispatch delay	One cycle delay	Delay depends on traffic
Physical performance	Transparent	Indirect via a buffer
Common channel efficiency	Inefficient and wasteful	Most efficient
Application	PCM type data	Terminal-to-computer data

The major features which characterize both ATDM and concentration, are as follows:

(a) The terminals are completely isolated from the common communication channel.

(b) Delays occur between the time the terminals dispatch their messages and the time they are transmitted over the common channel.

(c) An address tag must be added to each message to identify the terminal the message originated from.

(d) A buffer store is required to hold the queue of messages which arrive randomly.

7.11 STATISTICAL AND INTELLIGENT MULTIPLEXING

ATDM equipment is intended to serve not only terminal-to-computer but also computer-to-terminal communication, as seen in Fig. 7.32. The common communication channel may operate in a half or full duplex mode, whereas in both cases the same channel transmits messages in either direction. The traffic contents, however, in both directions, is not the same. Terminal-to-computer communication is generally based on short enquiry messages with the time slot size easily satisfied by character interleaving time multiplexing. On the other hand, the computers' reply is usually long.

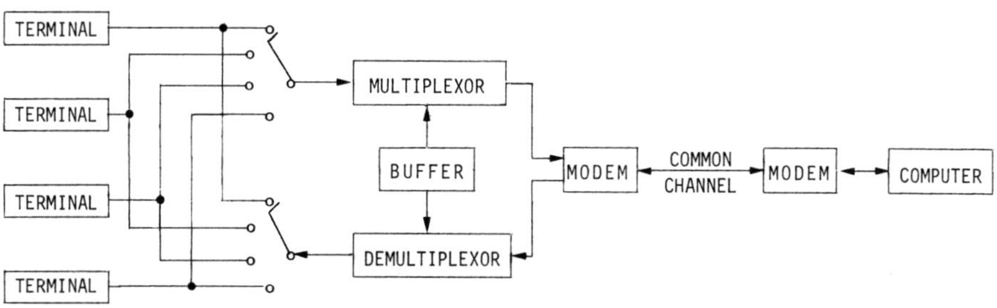

Fig. 7.32 Multiplexing and demultiplexing with a half duplex common communication channel

The same multiplexing equipment must serve both the transmission and the receiving equipment. With STDM the problem seldom arises, since the PCM data is usually operated in a simplex mode. Even with duplex transmission, the problem is not critical, since the terminals are slow and transmit or receive a character one at a time with the same sized slot duration used in both directions. This is not the case in terminal/computer communication with different sized messages in both directions. The data produced by the terminals is usually constant and the result of a keyboard operation by humans; hence the time slot duration may be wide enough for only a single character. Even with intelligent terminals with displays which allow the operator to prepare his messages beforehand the time slot size will be relatively short, possibly only long enough to transmit one line on the display. The computer's reply is generally variably long, with its contents easily filling up the whole display. These two types of slot size which can be located into the common channel are shown in Fig. 7.33, and are characterized as follows:

(a) Constant short duration time slots in accordance with the terminal originated messages.

(b) Variable length duration time slots in accordance with the computer originated messages.

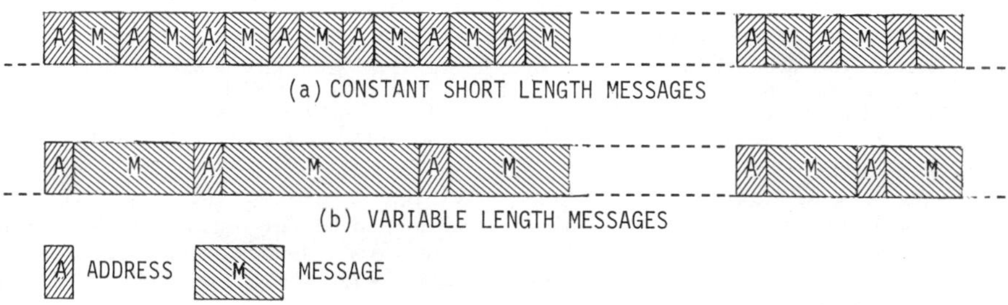

Fig. 7.33 Time slot sizes in terminal/computer type communication

The constant shot messages, which correspond to terminal-to-computer traffic, are wasteful owing to the high percentage of space taken up by the address. The variable long messages, which correspond to computer-to-terminal traffic, complicate the storage allocation. In this case, special characters will be required to identify the start and end of each message and the queue will take the form of a complex model each time it handles a different sized length message.

Such controversal requirements in the time slot sizes bring forward a number of operating problems when designing the ATDM equipment. These are magnified when the same memory unit is used to store the traffic contents in both directions. As ATDM uses wired-up programming techniques in performing its multiplexing operation, the control unit must be designed to deal with both the constant short messages and the long variable messages. The multiplexing efficiency could be increased by segmenting the messages into fixed sized blocks, which, while yielding the minimum operating cost, could satisfy the basic operating requirements. The last could be achieved by means of what is known as a statistical multiplexor (SM), which is broadly an ATDM with the fixed time slot duration defined by the statistical analysis of the traffic in both directions.

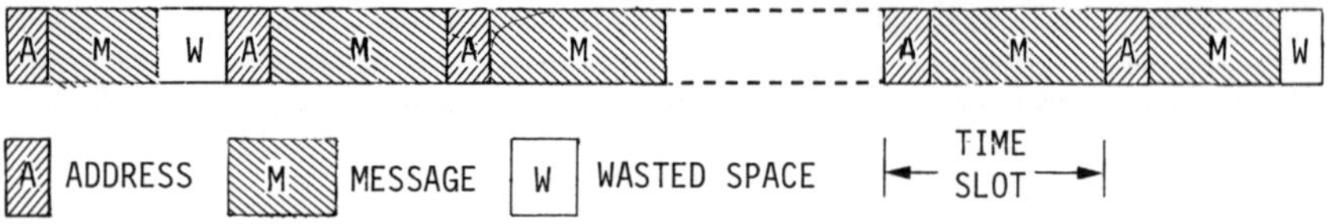

Fig. 7.34 Random length messages segmented into fixed length blocks

The fixed length block sizes simplify the buffer allocation and thereby also the multiplexor design for both terminal-to-computer and computer-to-terminal traffic. The segmenting, however, may cause some wasteful space, as seen in Fig. 7.34 with messages smaller than the fixed block size. Nevertheless, the segmenting of messages presents the most generally economical means of multiplexing.

The segmenting of random length messages into fixed sized blocks could be performed in two schemes. The first builds the block immediately after arrivals of the messages and transmits them regardless of whether the time slots are full. This type of transmission could be effective for computer-to-terminal traffic, where the messages emerge in long bursts, but is uneconomical in the reverse direction, since the terminals produce their messages at a slow rate. In this scheme, the blocks are transmitted only if the time slots are full or only after a fixed time has elapsed, thus reducing the possibility of transmitting half filled blocks. The second scheme will inevitally produce some delays in transmitting the terminal originated traffic, but will produce minimum delays in the computer originated traffic, since the blocks there are filled as soon as the messages emerge. This waiting period, despite delays, is the best compromise between performance and storage costs.

The memory in the statistical multiplexor, as distinct from that described for the ATDM, is not only intended as a buffer before transmission but also as a storage where the short messages are assembled into fixed sized blocks. Only when a full block is assembled and when the common communication channel becomes available are the messages transmitted.

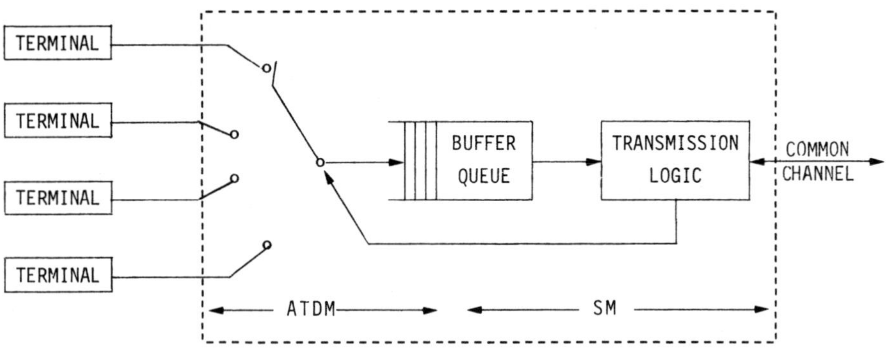

Fig. 7.35 Principle operation of intelligent multiplexing

Multiplexors which combine both ATDM and SM, as shown in Fig. 7.35, are known as intelligent multiplexors (IM). That is, the scanning of the terminals is performed as described for the ATDM, while the transmission of the messages on the common communication channel is performed as described for the SM. Intelligent multiplexors can therefore be regarded as a hybrid between multiplexors and concentrators, with similar operating characteristics.

In the literature, intelligent multiplexors are many times referred to indiscriminately by either of its two components, namely, ATDM and SM. Nevertheless, IM seems the most appropriate since ATDM only signifies the technique by which the terminals are scanned and SM signifies the technique by which the messages are transmitted.

Intelligent multiplexors, (or statistical multiplexors, as they are commonly referred to) are highly recommended for terminal/computer communication since they are cheaper than concentrators and eliminate the wasteful channel utilization that exists in multiplexing.

7.12 PROGRAMMABLE CONCENTRATORS

Intelligent multiplexors are most effective forms of concentrators, however, all the communication facilities are generally controlled by wired-in programs, which does not leave room for any operational flexibility. It is therefore not always possible with intelligent multiplexing to provide further options at a reasonable price. If the intelligent multiplexor is to be designed for all the possible market needs, it would require a wide range of different types of equipment, a solution which is uneconomical. Instead, it is suggested that a general purpose computer be used which could operate as a concentrator with all its tasks controlled by software programs. The added flexibility of the programmable concentrator, as opposed to the fixed performance of the intelligent multiplexor, enhances its usefulness by its ability to add processing. Wherever the traffic justifies the use of a concentrator, the designer should ascertain first whether an intelligent multiplexor could solve all the requirements. Only if the available multiplexors cannot do so should he turn to a programmable concentrator.

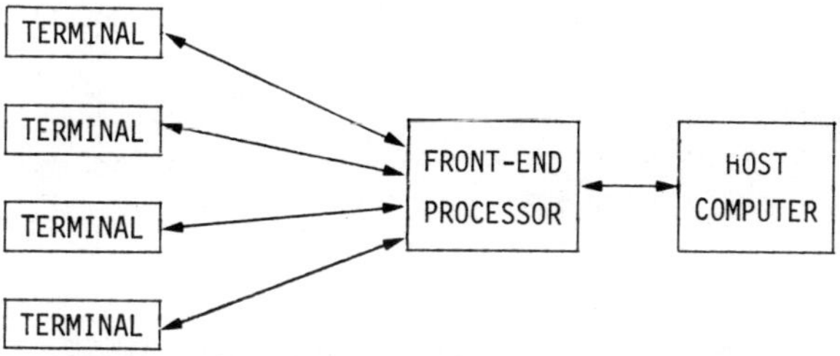

Fig. 7.36 A programmable concentrator operating as a front-end processor

For most terminal-to-computer applications, the services provided by the IM are sufficient and there is no need to search for other, better forms of data concentration. That is, wherever the communication equipment is needed to collect data from a known number of terminals and transmit over a common communication channel, the IM is the best solution. The function of the IM here is to smooth the data flow in the high speed communication channel by eliminating any transmission peaks which may occur when collecting random data from the terminals. There are, however, other communication tasks which the intelligent multimplexors do not provide. One of these applications is the remote concentrator which accepts messages from many terminals which operate at different speeds. In this application the programmable concentrator not only provides a common channel but also regulates the message flow at a constant high speed. Another example of the programmable concentrator is the front-end processor, as seen in Fig. 7.36, which collects the data from a number of remote terminals and transfers them directly to the host computer. In this specific application, the concentrator also provides means by which the host computer can communicate with the terminals. The independent operation of the front-end processor releases the host computer from all the communication tasks and thereby releases significant amounts of memory and valuable processing time. The concentrator performs all the terminal transactions while leaving the host for processing operations only.

As explained above, the programmable concentrator is usually a minicomputer which can present large scale service functions, not provided by the schemes previously discussed. A concentrator of this kind can perform many functions, such as sorting, formating, error detection, error correction and simple processing.

It is important to distinguish between the concentrator and the intelligent multiplexor. The intelligent multiplexor is a special hardware system containing wired-in programming logic. Concentrators use a small general purpose computer which enables both hardware and software to be used for providing more facilities. Furthermore, the facilities in the concentrator can be changed by simply altering the program, which is not possible in intelligent multiplexors. In multiplexors the terminal lines are scanned sequentially, while in concentrators, the lines may be scanned either sequentially or operated by request.

The main characteristic operational distinction between the multiplexors and programmable concentrators is in fact the technique that is used in scanning the terminals. With multiplexors, each terminal is connected to the common equipment by means of separate point-to-point physical communication lines. The scanning is then performed by special electrical pulses flowing in these separate lines. This is unnecessary with concentrators, since all the functions are performed in software, with the terminals too scanned with software instruction, which therefore abolishes the need for separate physical lines. That is, all the terminals could be associated on a single collecting common channel, as seen in Fig. 7.37, in a network configuration known as multidrop. In this configuration each terminal is connected by the shortest path to this collecting channel thus providing a considerable saving in channel costs. The terminals may be scanned in sequence as in the multiplexor multipoint configuration but, instead of using selecting electrical pulses, special instruction messages check whether the terminals have messages to dispatch.

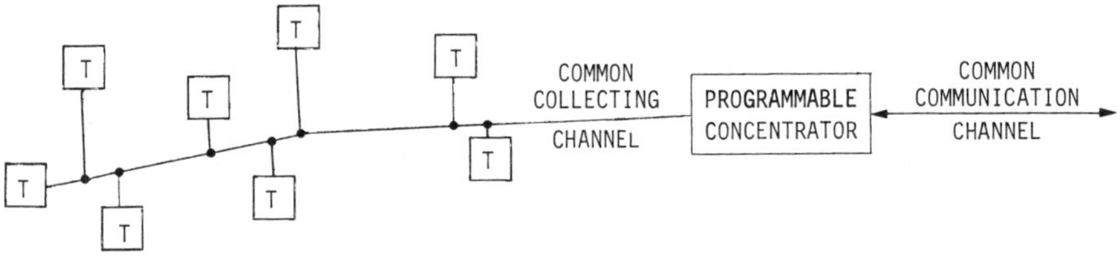

Fig. 7.37 Programmable concentrator used in collecting data in a multidrop configuration, employing software techniques

The form of the software technique used in scanning the terminals is known as polling and selecting; its use is possible only if the terminals have enough intelligence to identify their unique code address and follow the instructions sent to it by the concentrator. Furthermore, it is essential now that the terminals (and not the concentrators) add their identification address. This is because the programmable concentrator has no means of identifying the message source on the common collecting channel. In other words, this form of concentration is only possible with intelligent terminals.

Multidrop is the most advanced local network configuration and today is widely used for communication between terminals and computers. It is considerably more economical than the multipoint separate line configuration and is therefore used in remote intelligent terminal operations; there are many programmable control procedures that may be used for scanning the terminals, such as IBM's B.S.C. for synchronous communication, or Bell's B.A. 1 for asynchronous communication. In the following section the polling scheme based on B.S.C. is described.

Polling may be regarded as another method of multiplexing where the concentrator scans the remote terminals sequentially, requesting data. If the selected remote terminal has no messages to dispatch, the succeeding terminal will be asked whether it has one to transmit. If any of the checked remote terminals has a message to dispatch, the complete channel control will be transferred to this node for a fixed period of time. The scanning of the terminals need not be according to a fixed sequence as in time-division multiplexing, but in any sequence required. Furthermore, this scheme allows certain remote terminals to be scanned in each cycle more frequently than others and in any appointed sequence, with provision for easily means of changing the sequence. The sequence is performed according to a polling list which is controlled by a software program stored in the memory. The polling sequence can take into consideration the relative priority of the terminal and the amount of traffic which originates from each terminal.

The major disadvantage of the programmable concentrator is that it is more complex than that of intelligent multiplexors. The programming tasks in IM are wired-in while in the concentrators the tasks must be specially written to suit the specific applications. This calls for specially designed equipment for each concentrator application whereas with multiplexors they could be bought programmed ready for operation. Furthermore, the concentrators are much more expensive. Nevertheless, the concentrators' unique advantage is its flexible operation, as it can be designed to suit any communication situation and provide further features which are not possible with intelligent multiplexing.

The operations of the programmable concentrator and the intelligent multiplexor are very similar and it is therefore important to compare these two schemes. Both schemes use store and forward techniques and both employ programming techniques in selecting the terminals. Nevertheless, both schemes are completely different, as can be seen by the following table.

TABLE 7.7

FUNCTION	INTELLIGENT MULTIPLEXOR	PROGRAMMABLE CONCENTRATOR
Type of terminals	Simple terminal	Intelligent terminals
Addition of the terminal address	By the multiplexor	By the terminal
Physical connection between terminal and common equipment	Separate channels from each terminal	Single common channel connecting all the terminals
Terminal to equipment network configuration	Multipoint (star network with the multiplexor in the centre)	Multidrop
Means of scanning the terminals	ATDM	Polling
Equipment facilities	Fixed	Flexible
Programming techniques	Wired-in	Software
Control facilities	Special purpose computer	General purpose computer

7.13 ROLL-CALL POLLING

Polling is the most commonly used terminal control procedure where the front-end processor dominates the operation of the locally operated terminals. Polling is used also for remote terminals with the concentrator acting as the master station of the network, controlling the traffic flowing on a common channel by sending out suitable operation commands. The terminals act as slave stations, as they cannot initiate any channel operations but must only respond to all the concentrator's commands.

There are various forms of pollings, some of which are described in chapter 9. The polling scheme described here is known as roll-call polling with the terminals scanned in a fixed sequence as in ATDM. In all these procedures there are two types of commands. The first applies where the concentrator asks the remote terminals whether they have a message to transmit. (This command—request prompts the selected terminal to send a message and is referred to as 'polling'.) The second applies when the concentrator itself has a message to transmit to a particular remote terminal. (This command—instruction prompts the particular terminal to receive a message and is referred to as 'selecting'.) With these two types of commands, the programmable concentrator can control the traffic and the direction of the transmission on the common channel.

Each terminal is allocated a unique code, referred to as the terminal's address, which contains sufficient information to make possible automatic response when it appears on the common channel. The address information, when transmitted on the common channel, will reach all the terminals connected to the common channel, but only a single terminal may react to its unique code address. All the terminals will tend to decode the address, but only the terminal addressed will be able to react and connect itself to the common channel. Accordingly, this operation is essentially polling and not scanning.

A special synchronous code is sent at the head of each command message which will alert the terminals to check the proceeding address characters. If the message is not intended for a particular terminal, it will recognize this by the address, and will relax its attention till an end-of-message code is transmitted on the channel which tells it to wait for the next synchronous code.

Each remote terminal, on the receipt of a concentrator's inquiry intended for it, must first acknowledge, and only then, if it has a message to dispatch, can it transmit it. If it has no message to transmit, it must also send an acknowledgement, with an indication that it has no message to transmit. When the concentrator receives a message from a terminal, it must acknowledge that the message has been received correctly or, if an error was recorded, to send a request for retransmission of the message. The operation indicates the main constraints of the polling and selecting, which are the following:

(a) a large portion of the transmission time is wasted on the questions and answers; and

(b) delays are encountered between each question and answer for the turn around time period.

The following procedure schemes are introduced here to demonstrate, with the aid of an example, how concentrator control is performed. The scheme should be regarded as an initiation version and not as an authoritative definition of the actual line control procedures. The main control characters used for this example are:

SYN — Synchronous character, which precedes the transmission for the purpose of obtaining or maintaining synchronization.

EOT — End of transmission character, which marks the final end of a transmission sequence which may contain any number of messages, in multidrop lines, this character is also used as a signal to all the remote terminals to inform them that the previous communication to a particular terminal has been completed and that a new polling command may follow.

ENQ — Enquiry character, used to solicit a response from a remote terminal.

ACK — Acknowledgement character, used to affirm the positive reception of a message with no errors detected.

NAK — Negative acknowledgement character, used to request retransmission of the last block when a transmission error was recorded.

STX — Start of text character.

ETX — End of text character.

SOH — Start of heading character.

BCC — The total number of characters included in the actual message, i.e., the message heading and text, but excluding any synchronization characters.

The operation sequence of polling and selecting can be explained with the aid of the example shown in Fig. 7.38. Before each transmission message a SYN code is sent from the originating node to the receiving node for the purpose of obtaining synchronization. The EOT code is used to call the attention if all the nodes to be ready for an addressed command. With the transmission of an EOT followed by a particular address, a virtual point-to-point connection is established between the concentrator and one of the remote terminals. All the other terminals will be inhibited for the duration of the transmission connection, which is maintained until the next EOT code is transmitted on the line. The polling sequence of the example is as follows:

(a) The concentrator sends a message by a SYN code which synchronizes all the terminals. This is followed by an EOT code which sets all the terminals on the channel in preparation to decode the proceeding address. The address characters which follow select only one terminal while inhibiting all the other terminals on the channel. Following the address, a command code is transmitted, which in the case of polling is an ENQ checking whether the terminal has a message to transmit.

(b) The terminal which was addressed must reply to the centre's command. This reply is again headed by a SYN code, but it will only synchronize the concentrator, since all the other terminals are inhibited until the next EOT code. If the addressed terminal has no message to transmit, it will return a negative reply. There is no need to send a NAK code, as the message can be conveyed by an EOT code which indicates that the transmission has been completed with no messages to transmit.

(c) The concentrator sends another poll equiry message to a second terminal at another address.

(d) The new terminal has a message to dispatch, which it transmits immediately after a SYN code. To save time there is no need to send a positive reply (ACK). The message transmitted consists of the heading and then the text. The message ends with a BCC code which specifies the number of characters in the message.

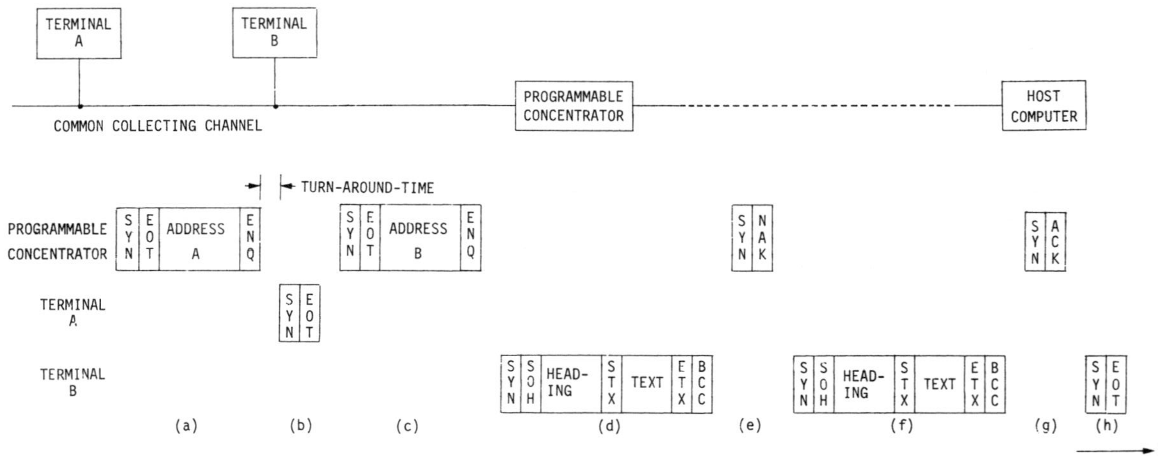

Fig. 7.38 Multidrop network configuration and the polling sequence of events when the concentrator scans the terminals

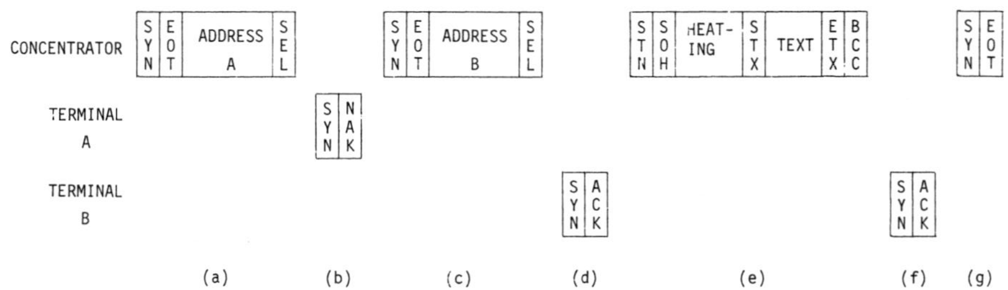

Fig. 7.39 Sequence of events of a selection procedure when the concentrator has a message for a particular terminal

(e) The concentrator checks whether there are any errors in the received message. If any errors are detected, the concentrator will then transmit a negative acknowledgement, NAK code, which would be recognized by the terminal as a command for retransmission of the message.

(f) The terminal retransmits the original message.

(g) The concentrator rechecks the retransmitted message and if no errors are recorded, it transmits a positive acknowledgement,ACK code.

(h) The terminal on the reception of the ACK code can clear its buffer, since the message has been successfully transferred to the concentrator and so there is no need to hold it for error enquiries. The terminal having completed its tasks will transmit an EOT signal on the channel which will tell all the other terminals that the transmission has been completed and that they should be on the alert for a new poll on select command.

Polling suffers from a major disadvantage of transmission delays due to turn-around time. The remote terminals, in many of the multidrop networks, are connected to the concentrator by means of a half-duplex operational mode. That is, the same common collecting channel is used for both the concentrators' queries and the terminals replies. The terminals, during most of the time, must remain in their receive state, and then, only after one terminal has been properly addressed, can it convert to its transmission state. At the same time, the concentrator too, has to convert from its transmission state to its receive state in anticipation of the terminal's answer. The conversion period of the modem in the terminal and the concentrator is referred to as turn-around-time. As already discussed in chapter 3, the time delay due to the turn-around-time could be quite considerable.

The polling scheme, by virtue of its software characteristics suffers from further delays due to the wasted overhead of question and answer. Even if no terminals have any messages to dispatch, the terminals must still be queried to check if they have any outgoing traffic. The delay timing consists of two turn-around-time periods and the question and answer procedure for each terminal in the system. Such delays do not exist in ATDM, although this delay is the trade-off needed when using a concentration which provides the saving in the number of separate channels that are required in ATDM.

Both ATDM and polling are employed for the same type of application, e.g., collecting messages from remote terminals and transmitting them over a common communication channel. Both techniques scan the terminals sequentially and provide transmission facilities only to the terminals with messages to dispatch; that is, they both collect the messages in asynchroneous steps. Furthermore, both employ programming techniques coupled with store and forward schemes. Nevertheless, in operation both of these data collecing techniques are completely different, as summed up in the table on the next page.

Polling is only used when the programmable concentration scans the terminals to check whether they have a message to dispatch. When the concentrator has a message to send to a particular terminal, it uses a different procedure, as seen in Fig. 7.39. This procedure is referred to as 'selection', since the concentrator addresses the selected terminal directly without any need to consult the other terminals.

The selection procedure is identical with the polling procedure except that the command is now SEL, which informs a particular remote terminal to expect a message. The terminal must acknowledge such a command, either positively or negatively. A negative acknowledgement means that the addressed remote terminal is busy with other tasks. A positive acknowledgement means that the node is ready to receive the message.

TABLE 7.8

FUNCTION	ATDM	POLLING
Network configuration	Multipoint	Multidrop
Transmission on the common channel	Statistical multiplex	Programmable concentrator
Connection to the terminals	Separate channels	Single common channel
Interface with terminals	Electrical pulses	Software instructions
Scanning technique	Wired-up programs	Software programs
Type of terminals	Simple	Intelligent
Number of terminals	Limited by design	Limited by programme
Address insertion	By the multiplexor	By the terminal
Program overhead	None at all	Questions and answers
Delay before dispatching	Only as a result of traffic	Delays caused by traffic, questions and turn-aroond time
Advantages	Saving in transmission time	Saving in communication channels

7.14 RANDOM TIME DIVISION MULTIPLEXING (RTDM)

A completely different approach to local network communication is offered by the loop network which is intended for both terminal/computer and computer-to-computer communication. The scheme is ideal for system interaction communication in a distributed command and control system, where a number of computers can operate independently and at the same time be part of a comprehensive system, sharing data with other computers.

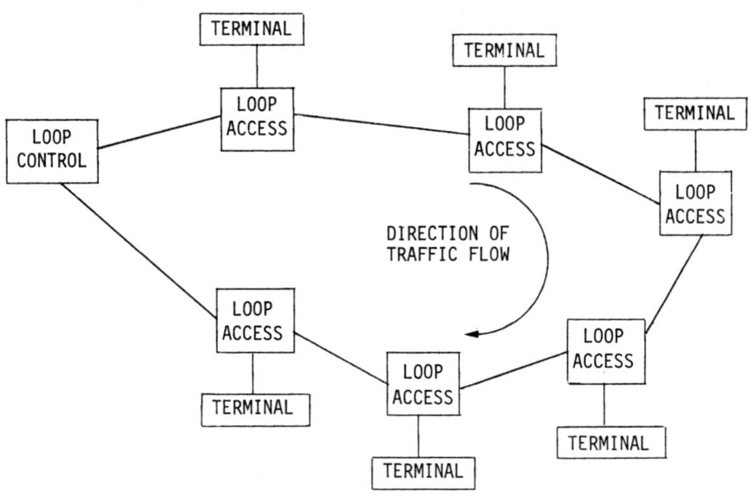

Fig. 7.40 The closed loop network configuration

The loop network is constructed by means of a closed common communication channel configuration that starts from the loop control station, runs through a number of access stations and then returns to the loop control station, as seen in Fig. 7.40. The data flowing in the loop is in one direction only, and theoretically, any message originating from any point in the loop could return to the same point after a complete circulation of the loop. The loop channel can, if so, be regarded as a common communication channel, which provides the same concentration features as the multiplexor. Nevertheless, the multiplexor operates in a centralized network configuration, while the loop offer the operation of a distributed network configuration.

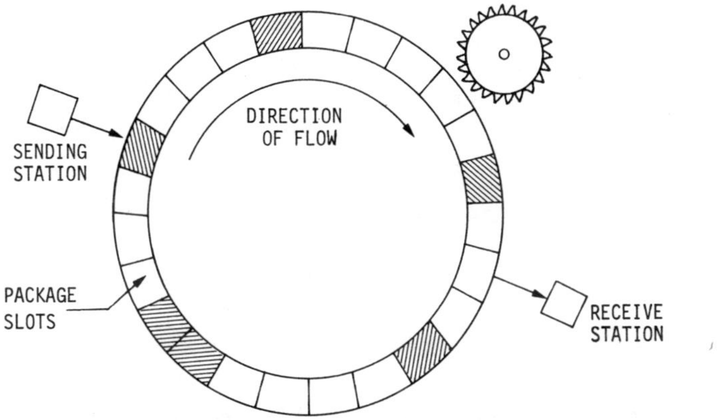

Fig. 7.41 The operational principle of the loop communication demonstrated by means of a conveyor belt

The loop system operation can be clearly understood if expounded with the aid of an example. A conveyor belt, which has framed slots inserted in it, is driven continuously in a circle, as shown in Fig. 7.41. At one point of the circle, there is a control unit which pushes the conveyor belt along with a smooth continuous movement. The conveyor frame slots might contain packages for conveyance between stations. Whenever a frame passes a station, the label on the package is checked to see if it is directed to that particular station. Any packages intended for that station are then taken off the belt and dispatched to their local destination. If a station has a package to send to another station, it just waits for a vacant frame slot, deposits the package in it and the conveyor belt then transfers it to the other station.

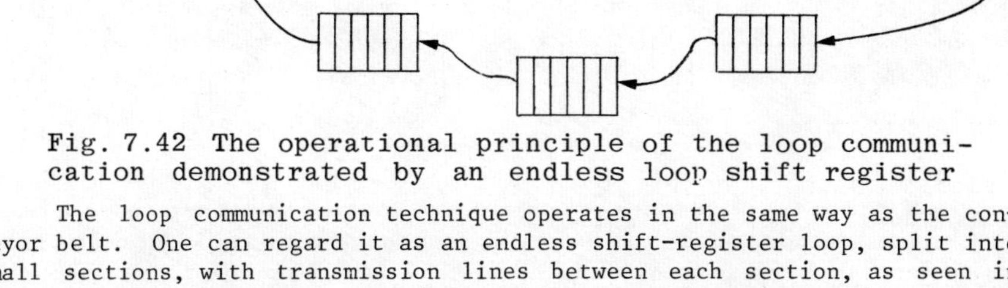

Fig. 7.42 The operational principle of the loop communication demonstrated by an endless loop shift register

The loop communication technique operates in the same way as the conveyor belt. One can regard it as an endless shift-register loop, split into small sections, with transmission lines between each section, as seen in Fig. 7.42. Each terminal is associated with one section of the shift register, which is of the size of one message length. The operation of the loop is similar to that of time division multiplexing, with time slots (of the size of the shift register sections) made available for transmission. A control unit ensures that the traffic in the loop is continuously on the move, ensuring that at every period, a different time slot is placed in each shift register section. That is, the fixed size message slots are transmitted continuously around the loop in a single direction. If the sending terminal has a message to transmit, it looks for a free vacant time slot to put the message in. To ensure that the message placed in the loop reaches its destination, at the receiving terminal, every occupied time slot is inspected, to check whether its message address is intended for that particular terminal. If the destination address of the message tallies with that of the terminal, the message is lifted off the loop, thereby freeing the time slot.

The operation of the loop, being based on fixed size time slots, is the same as that suggested for STDM and ATDM. Nevertheless, although there is a centralized loop driver there is no centralized equipment whose function it is to assign the time slots, and each terminal has an equal probability of reaching a free available time slot. In STDM the time slots are assigned according to a fixed routine, with each slot dedicated to a given terminal. in ATDM the time slots are assigned dynamically according to the traffic, with each terminal scanned in a fixed sequence requiring the terminals to wait their turn before being provided with a time slot. In the loop operation, the time slots are moving continuously and the terminals can select any time slot on a random basis. For this reason, the loop operation can be regarded as random time division multiplexing (RTDM).

As already stated, the major distinction of the RTDM in comparison with both STDM and ATDM, is that the logic operation of the common channel is not based on any intelligent centralized equipment, but only on the terminals themselves. STDM and ATDM are intended for local centralized network configuration, where the centralized equipment collects the data from all the terminals and directs them to a common destination. That is, the STDM and the ATDM can both be regarded as a form of master-slave operation. RTDM, on the other hand, is intended for a distributed network operation, with all the terminals operation independent of the others. That is, in RTDM network all the terminals are their own masters, with no equipment needed to control or to coordinate their operations. This operation, however, means that each terminal must be intelligent, with the addition of a loop access unit.

The operations of all three schemes, STDM, ATDM and RTDM, are based on fixed sized time slots assigned for message transmission: mevertheless, they all three employ different techniques. The following table shows the basic differences between these three schemes.

TABLE 7.9

FUNCTION	STDM	ATDM	RTDM
Network operation	Point-to-point	Store and forward	loop
Network control	Centralized	Centralized	Distributed
Common channel assignment	Fixed	Dynamic	Random
Time of dispatching	When sequential time slot arrives	When its turn to be scanned	When there is free time slot
Type of terminal	Dumb	Simple intelligent	Highly intelligent
Application	PCM type data	Terminal-to-computer data	Terminal-to-terminal data

It is somewhat misleading to compare the RTDM operation with that of STDM and ATDM since they are intended for different types of applications. It is far more important to compare the loop operation with that of message switching (which will be discussed in the following chapter) since they are both intended to solve the problem of terminal-to-terminal applications.

7.15 LOOP TRANSMISSION OPERATION

The time slots in the loop operation are in a continuous movement around the loop. The time slot format, of the 1's and 0's state, is renewed each time it passes through the loop control unit, irrespective of whether it is free or holding a message. This ensures the cancellation of any delays or distortions created by the common channel characteristics. The loop control unit, shown in Fig. 7.43, has two main functions: maintaining synchronization by renewing the slot timing, and supervising the loop performance by preventing errors in the traffic from blocking the loop.

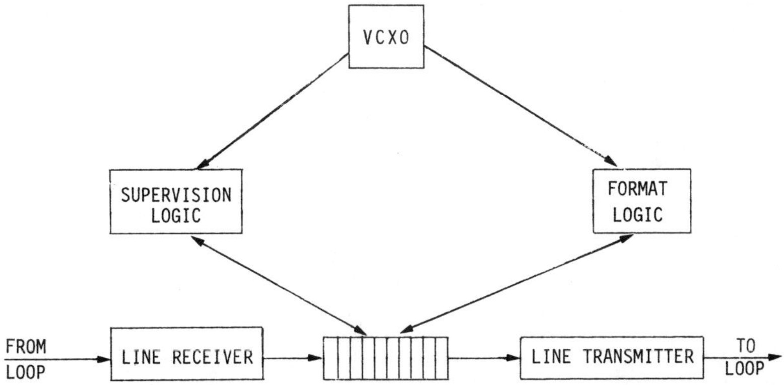

Fig. 7.43 Basic components of the loop control unit

The synchronization operation of the loop configuration is based on the shift register buffer, the size of a slot in the loop control. It accepts the data from the loop at its distorted timing, but retransmits them by the right timing. This timing is controlled by a master voltage control crystal oscillator (VCXO) which enables the slot to be reproduced accurately and consequently to synchronize the whole loop. All the terminals in the loop are synchronized with this timing by reconstructing it from the state transitions of the transmitted bits.

The loop controls' function is intended also to guard that the loop is not saturated with messages that have not reached their destination; in other words, to ensure that the messages do not circulate in the loop indefinitely. The first time a message passes through the loop control, a marker is placed in the slot to indicate this fact. If the message passes through the control a second time this would imply that the message has not reached its destination and in fact is 'lost' in the loop. Since a terminal must always be free to accept a message, it can only be assumed that this particular terminal is out of commission, or else there is an error in the destination address. The loop control, on finding that the circulating message has not reached its destination, may destroy it by erasing the message, thereby liberating the slot.

The loop control must inform the supervisor of all the control functions it performs. That is, when it frees a time slot, it must report the fact to the sender and/or perform other measures required. The loop control unit also keeps track of all the faulty terminals and alerts the supervisor to this fact. It could also send special messages to the terminal's access unit which tests all its communication tasks.

The format of the data in the time slot contains not only space for the message but also room for control characters, as seen in Fig. 7.44. At the heading of the time slot a special synchronization code is transmitted to enable the terminals to synchronize their clocks and to define the start of the time slot. This is followed by a data bit informing the terminals whether the slot is free or busy and they by a few control characters which are for the sole use of the loop control, to enable it to control the loop operation. Each slot must contain addresses of both the destination and the sender. Only after all the peripheral data is transmitted can the message itself be dispatched. The slot size is of no major importance and could be adjusted to each application. Nevertheless, once the slot size has been selected for a particular communication system, it must remain fixed.

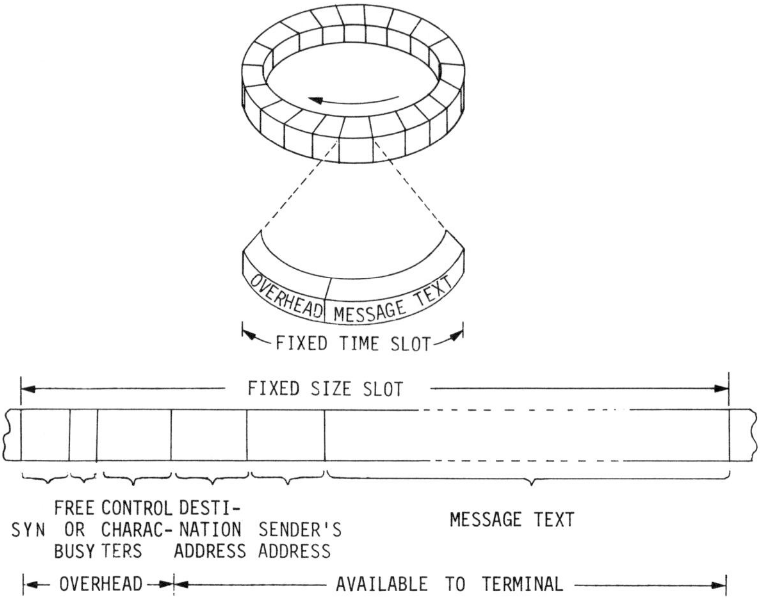

Fig. 7.44 The format of the loop's time slot

All the terminals have equal importance, with no special duties attached to any particular slot in the loop, hence, there are no priority scales both to the terminals or the slots. Thus all the terminals have an equal chance of reaching a free slot to transmit a message. This random seizure of the slot provides high operational efficiency of the system.

Theoretically, each time a slot passes a terminal it must wait for it to check the address and to lift the message from the loop. This may cause delays, which will increase as more terminals are added to the loop. This problem, however, is easily overcome by arranging the logical operation of the terminal to be much faster than the data transmission speed of the loop. In this way the whole operation may be concluded before the slot leaves the terminal.

The terminal's access unit is very simple (as seen in Fig. 7.45) consisting of a pair of modems and a buffer. The message is lifted off the buffer (after the address has been identified), in parallel. Then, before the slot leaves the terminal, the whole section of the slot available to the user, will

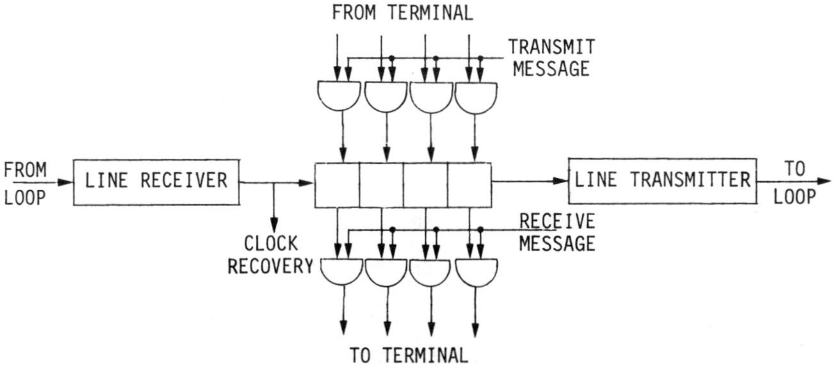

Fig. 7.45 Basic components of the terminals' loop access unit

have to be erased. When a message is to be transmitted, it is inserted into the buffer also in parallel.

As the buffer in each terminal can allow a large number of messages to be in the loop simultaneously, there may be added delays in the system's operation. This could be avoided by reducing the buffer size to that of only the overhead and address characters. Once a terminal recognizes its own address, it could lift off the messages in series rather than in parallel, and at the loop transmission speed.

The major disadvantage of the loop network operation is that of reliability. Any single failure in the system will cause the entire system to fail. To overcome such a situation, it is advisable to employ two or even three loops operating in parallel.

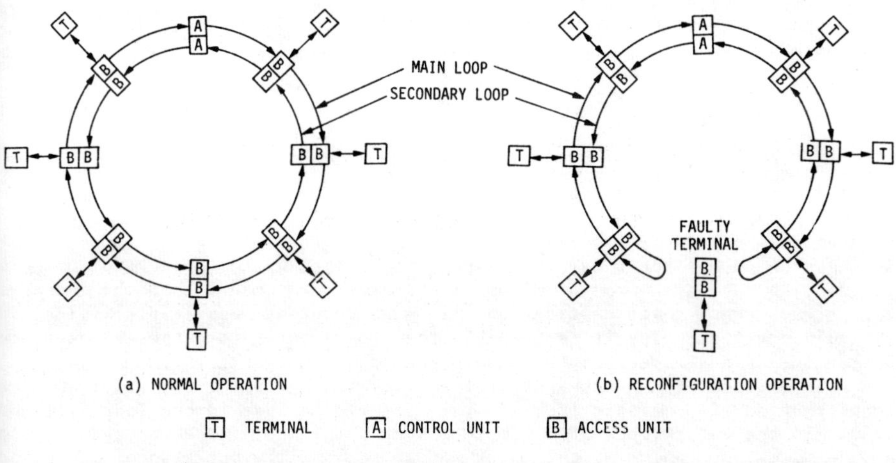

Fig. 7.46 Dual loop network redundancy to increase system reliability

In the double loop redundancy scheme, shown in Fig. 7.46, the whole loop network is duplicated, including the control and access units. These two loops could thus be regarded as operating completely independently of each other, with the data flowing in opposite directions in each loop. Normally, both loops are active, with data flowing in both, but with only the main loop carrying actual messages while the standby for dummy data. Since the channel path in both loops is along the same route, a failure in one loop or terminal will normally result in both loops failing. The control units are soon alerted to the existence of the fault and can locate the failure point by sending test messages to each terminal in turn. The terminals receive these test messages on the main loop and answers on the standby loop. Once the failure point is located, the network operation can be reconfigurated by appropriately closing the two loops at both sides of the faulty terminal The operations of the remaining terminals in the reconfigurated network, will proceed as before, as they will still receive and transmit messages on the main loop. The standby loop is then used as means of closing the faulty loop leaving it with no further transmission facilities.

An even better approach for solving the problem of reliability is to employ three loops operating in parallel, as seen in Fig. 7.47. While such a network requires triple loop control, it only needs to double the terminals' access units. In other words, the equipment of the triple loop network is very similar to that of the double loop network. The main loop runs through all the terminals in series, as in the double loop network, while the two standby loops run only through alternate terminals. The advantage of this network configuration is that the paths of the three loops need not necessarily travel along the same routes nor do the standby loops string the same terminals. Each terminal is interfaced both with the two terminals closest to it and with the two of their nearest neighbours. In the event of a single failure, the operation can revert to the standby loop. In the event of a multiple failure at various points of the triplicate network, the terminals may transmit their messages on any of the remaining loops. It is then up to the control unit to redirect the messages from one loop to the other, to ensure that the messages will always reach their destination. In the event of a multiple failure in two adjacent terminals, it is possible then to reconfigurate the loop network operation, as described for the double redundancy network,

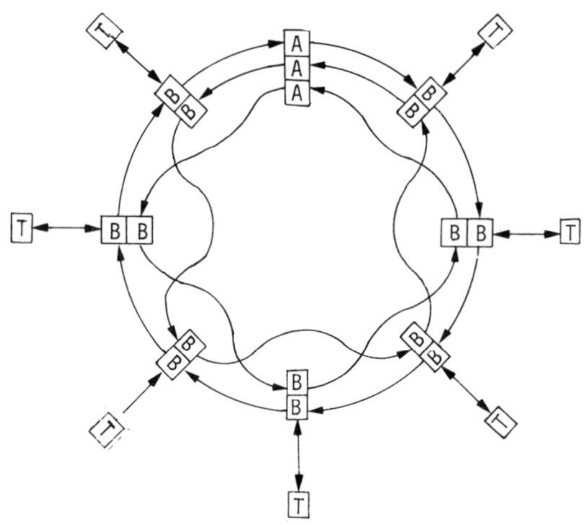

Fig. 7.47 Triple loop network redundancy

An alternative approach to reliability is to form a number of smaller loops instead of a single one. In this case, a failure in one loop will not result in the catastrophy of a complete system failure. The advantage of such a scheme is not only in the increase of reliability but also of increased efficiency. Each loop can operate at a different speed, without it affecting the operation of the other loops and thus each loop can serve different types of terminals. A message originating in one loop, could easily reach any terminal in any of the other loops, by means of intermediate loop transfer units.

The major attraction of the loop network is the simplicity of the system in comparison with other similar systems, and the relatively low cost required to install such a system. The numbers of terminals attached to each loop is variable and could be as high as a hundred. The major offset of this scheme is its low reliability, although this could be improved considerably by using redundancy techniques. Loop netwroks are intended for both terminal-to-terminal and computer-to-computer communication. They are ideal for local network installations, where a large number of terminals are situated in a single building or compound which has less risk of channel failure.

7.16 REFERENCES

(a) MULTIPLEXING AND CONCENTRATION TECHNIQUES

1. Martin, J. *Telecommunications and the Computer*, Prentice-Hall Inc., 1969, pp.276-292.
2. Martin, J. *Teleprocessing Network Organization*, Prentice-Hall Inc., 1970, pp.184-210.
3. Doll, D., *Multiplexing and Concentrating*, Proc. IEEE, vol.60, no.11, November 1972, pp.1313-1321.
4. Rudin, H. *Data Transmission: A Direction for Future Development*, IEEE Spectrum, February 1970, pp.79-85.
5. Louchez, S. *Basic Communications to Help to Understand Data Transmission*,
6. Hersch, P. *Data Communication*, IEEE Spectrum, February 1971,
7. Rocha, I.A. *Computer Network: Analysis and Case Study Design*, Master Thesis, Naval Post Graduate School, June 1978, U.S. Government Report ADA 057406.
8. Townsend, M.J. *Communication Control By — An Introduction*, Telecommunications, May 1972, pp.33-38 amd 60.62.
9. Rudin, H. *Data Transmission: A Direction for Future Development*, IEEE Spectrum, February 1970, pp.79-85.
10. Hardwick, K. and Federbusch, W. *Local Networking: The Missing Link Energes*, Data Communication, July 1980, pp.81-89.

(b) INTELLIGENT MULTIPLEXING

11. Chu, W.W. *A Study of Asynchronous Time Division Multiplexing for Time Shared Computer System*, 1969, Spring Joint Computer Conference, AFIPS Conf. Proc., vol.35, pp.669-678.
12. Chu, W.W. *Demultiplexing Consideration for Statistical Multiplexors*, IEEE Trans. on Communication, vol. COM-20, no.3, June 1972, pp.603-609.
13. Cowsley, D. and Wolf, J.K. *On Statistical Analysis of Queue Length and Waiting Times for Statistical Multiplexors with ARQ Retransmission Schemes*, IEEE Trans., vol. COM-27, no.4, April 1979, pp.693-702.
14. Towsley, D. *The analysis of a Statistical Multiplexor with Nonindependent Arrivals and Errors*, IEEE Trans., vol. COM-28, no.1, January 1980, pp.65-72.

(c) PROGRAMMABLE CONCENTRATORS

15. Rudin, H. *Performance of Simple Multiplexor-Concentrator for Data Communication*, IEEE Trans. on Communication, vol. COM-19, no.2, April 1971, pp.178-187.
16. Newport, C.B. and Ryzlak, J. *Communication Processore*, Proc. IEEE, vol.60, no.11, November 1972, pp.1321-1332.
17. Ball, C.J. *Communication and the Minicomputer*, Computer, vol.5, no.5, September/October 1971, pp.13-21.
18. Matteson, R.G. *Computer Processing for Data Communication*, Computer, vol.6, no.2, February 1973, pp.15-19.

(d) SPACE DIVISION MULTIPLEXING

19. Andresen, S. and Hareson, S.R. *Toward a General Class of Time-Division Multiplexed Connecting Network*, IEEE Trans., vol. COM-20, no.5, October 1972, pp.836-846.
20. Pippenger, N. *On Crossbar Switching Networks*, IEEE Trans., vol. COM-23, no.6, June 1975, pp.646-659
21. Leakey, D.M. *Switching in Space and Time*, Proc. IEE, vol. 124, no.1, January 1977, pp.17-24.
22. Marcus, M.J. *The Theory of Connecting Networks and their Complexity: A Review*, Proc. IEEE, vol.65, no.9, September 1977, pp.1263-1270.
23. Masson, G.M., Gingher, G.C. and Nakamura, S. *A sample of Circuit Switching Networks*, IEEE Computer Magazine, vol.12, no.6, June 1979, pp.
24. Skaperda, N.J. *Some Architectural Alternatives in the Design of a Digital Switch*, IEEE Trans., vol. COM-27, no.7, July 1979, pp.961-972,
25. Charransol, P., Hauri, J., Atheres, C. and Hardy, D. *Development of a Time Division Switching Network Usable in a Very Large Range of Capacities*, IEEE Trans., vol. COM-27, no.7, July 1979, pp.982-98

(e) LOOP TRANSMISSION

26. Doll, D.R. *Telecommunication Turbulence and the Computer Network Evolution*, Computer Magazine, February 1974, pp.13-22.
27. Penncy, B.K. and Baghadi, A.A. *Survey of Computer Communication Loop Network: Part 2*, Computer Communication, vol.2, no.5, October 1979, pp.224-241.
28. Hays, J.F. and Sherman, D.N. *Traffic Analysis of a Ring Switched Data Transmission System*, Bell System Technical Journal, vol.50, no.9, November 1971, pp.2947-2978.
29. Avi-Itzhak, B. *Heavy Traffic Characteristics of a Circular Data Network*, B.S.T.J., vol.50, no.8, October 1971, pp.2521-2549.
30. Piercem J.R. *Haw Far Can Data Loops Go*, IEEE Trans. on Communications, vol. COM-20, no.3, June 1972, pp.572-536,
31. West, L.P. *Loop-Transmission Control Structures*, IEEE Trans. on Communications, Vol. COM-20, no.3, June 1972, pp.531-539.
32. Anderson, R.R., Hayes, J.F. and Sherman, D.N. *Simulated Performance of a Ring-Switched Data Network*, IEEE Trans. on Communication, vol. COM-20, no.3, pp.576-591.
33. Spragins, J.D., *Loop Transmission Systems – Mean Value Analysis*, IEEE Trans. on Communications, vol. COM-20, no.3, pp.576-591.
34. Pierce, J.R. *Network for Block Switching of Data*, B.S.T.J., vol.51, no.6, July-August 1972, pp.1133-1145.
35. Kripfl, W.J. *An Experimental Data Block Switching System*, B.S.T.J., vol. , no.6, July-August, pp.1147-1165.
36. White, H.E. and Maxemchuk, N.F. *An Experimental TDM Data Loop Exchang* International Conference on Communication, 1974, pp.7a-1 to 7A-4.
37. Hayes, J.F. *Performance Models of an Experimental Computer Communication Network*, B.S.T.J., vol.53, no.2, February 1974, pp.225-259.
38. Majithia, J.C. and Dube, J.D. *Simulation Results for a Loop Network with a Hybrud Message Handling Protocol*, Computer Networking Symposium, 1976, IEEE 76 CH 1143-7C, pp.19-23.
39. Majithia, J.C. and Dube, J.D. *Analysis and Simulation of Message-Switched Loop Data Networks*, Proc. IEE, vol. 124, no.3, March 1977, pp.193-197.
40. Rawson, E.G. and Metcalf, R.M. *Fibernet Multimode Optical Fibres for Local Computer Networks*, IEEE Trans., vol. COM-26, July 1978, pp.983-990.

41. Oh, Y. and Liu, M.T. *Interface Design for Distributed Control Loop Networks*, National Telecommunication Conference, 1977, pp.31:4-1 to 31:4-6.

42. Farber, D.J. *Networks: An Introduction*, Datamation, April 1972, pp.

43. Martin, James, *Teleprocessing Network Organization*, Englewood Cliffs, N.J., Prentic Hall Inc., 1970, pp.122-125.

44. Pierce, J.R., Coker, C.H. and Kropfl, W.J., *An Experiment in Addressed Block Data Transmission Around a Loop*, IEEE International Convention Record, March 1971, pp.222-223.

CHAPTER 8

Communication Switching Centres

8.1 INTRODUCTION

In the previous chapter, means of sharing a single communication channel between many sources were introduced. It was shown that these means increased the network efficiency for traffic between the terminals and a central computer. Such a communication network operation is regarded as 'local', since the traffic can flow only in the direction of the centre. That is, the local area network is centralized, controlled by a single centre. Large C^2 systems, however, may consist of many such centres spread over wide areas, each operating independently with their own local networks. In such a widely distributed network, each centre must be able to communicate with any of the other centres. The means of providing this service will be presented in this chapter.

The communication network serves to provide means whereby messages transmitted from widely situated terminals can reach their destination. The application of the communication facility is not only between two specific terminals attached to the network but also between any of the terminals with any other terminal. The design goal of these networks extends beyond the provision of a transmission path towards achieving maximum data efficiency. One contributing means of providing this efficiency is multiplexors and concentrators, as suggested in the previous chapter. Multiplexors and concentrators, however, are intended for the simplest network configurations; namely, remote terminal to host computer communication. That is, they provide the common means by which the terminals can communicate with centralized equipment but not with any other terminal. Over this common facility, information can be transmitted either from terminal to host or host to terminal. Furthermore, the multiplexors and concentrators allow only one transmission transaction to be performed in any given time. This type of network configuration is ideal for master-slave type communication such as required in data acquisition systems and interactive time-sharing systems, both of which are essential in C^2 operating systems. Nevertheless, in large C^2 systems, when the demand for comprehensive communication network is considerably increased, the network efficiency with only multiplexors is correspondingly reduced.

Command and control systems are not only planned for centralized operations where all the data flows to-and-from a centralized processing command post. Most C^2 systems are designed with a number of hierarchical command posts at various operating levels, all operating in harmony, and with a number of adjacent system command posts operating in conjunction with each other. The data that flows in the sophisticated communication network, serving both the interaction and interoperability messages between the various C^2 systems and subsystems, is not only from the terminals to a centre, but also between centres. Such a comprehensive requirement calls for additional communication facilities to ensure overall transmission efficiency of the message flow through the common network.

In large, sophisticated command and control systems there may be many processing installations which operate either as independent systems or dependent subsystems. Each such computer installation, serving either as a data collecting post, a data query post, a supervisory post or a command post, may be situated adjacent to the other computers or may be miles apart. In all cases it is the communication network which connects these widely distributed installations into one complex, and it is the communication network which must provide the means for all the computer installations to cooperate by sharing the resources stored in their respective data bases. The means of communication for sharing these data files, as indicated for computers, should be equally provided for the terminals to enable them to communicate both with the computers and other terminals. The transfer of messages between computers and/or terminals must be performed in real-time so that the information may be made available where they are required and when they are needed. Such a communication network requirement cannot be commissioned by a centralized configuration, as provided by multiplexors and concentrators. This call for a distributed network configuration allowing each computer installation to operate independently while being able at the same time to communicate with each other. The so-called independent computer installations may be concurrently centres of local networks with many terminals associated with them, possibly via multiplexors and concentrators.

The simplest way of providing computer-to-computer communication is to design a comprehensive network configuration, in which each computer is connected directly with all the others. This network, as shown in Fig. 8.1(a), known as a non-switched configuration, entails the use of point-to-point communication channels between each and every computer installation and thus allows many simultaneous transactions to be performed. Such a solution is possibly suitable for small systems, but is most inefficient in large sophisticated systems. Indeed, a much better solution is to introduce a central switching installation, as seen in Fig. 8.1(b), which has the prime task of controlling the traffic in the network. Here all the computers have direct

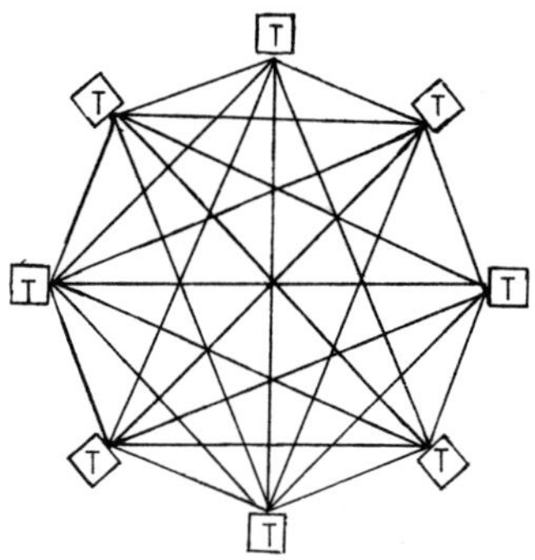

(a) NON-SWITCHED, FULLY CONNECTING CONFIGURATION

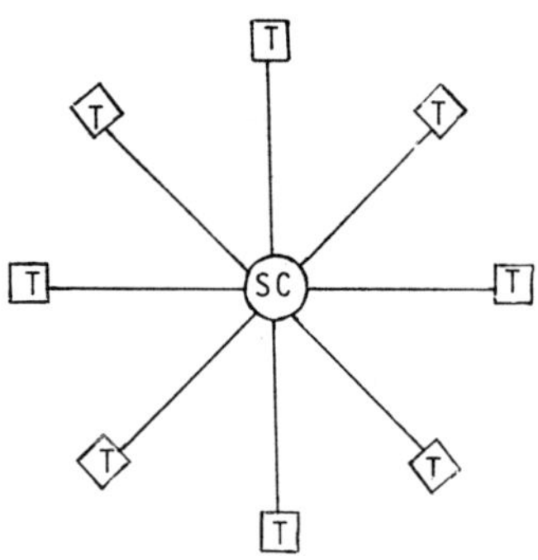

(b) SWITCHING CENTRE IN THE HUB OF STAR CONFIGURATION

Fig. 8.1 Various network configuration for direct terminal-to-terminal communication

communication channels only with the centre, and it is thus the function of the centre to switch the messages between the various computer installations. This switching centre (SC) is then located at the hub of the communication network, in a configuration known as a star network. The reduction in the communication channels of the star configuration, as compared with the fully connected configuration is very significant, having great economical advantages. Nevertheless, such a configuration calls for the introduction of new communication equipment, namely that of data switching exchanges (DSE).

The outer nodes of the star network configuration may be terminals or computers, with the switching centre providing the means, whereas any message originating from any one node can reach any of the other nodes. The outer nodes may also be centres of local network, thereby extending the network. In this case, they may be multiplexors or concentrators collecting the data from a number of terminals, as seen in Fig. 8.2. In this extended network configuration too, each terminal can reach any other terminal in the network but only after they have been switched by the centre. The messages may have to pass through local centres before they are switched by the centre node to their destination. In the case where two adjacent local terminals communicate with each other, a message may have to pass twice through the same multiplexor or concentrator.

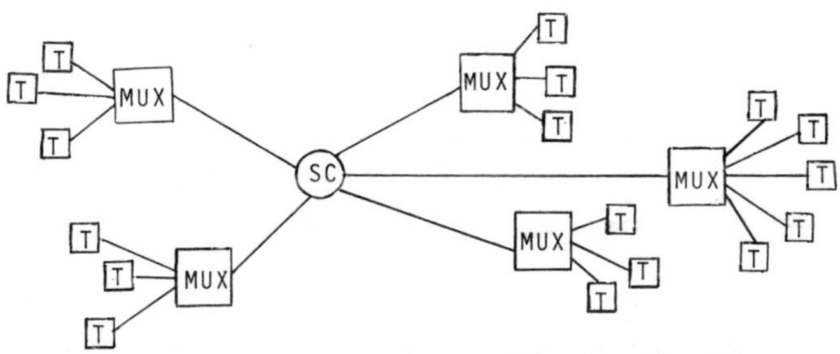

Fig. 8.2 Extended star network configuration

If there are M outer nodes (terminals or computers) connected to the star network, each node must be provided with communication facilities to transmit data to the remaining $M-1$ nodes. Since a node can be connected with only one other node at any given time, the non-switched network can provide $M/2$ simultaneous transmissions. The same facilities are provided by the switching centre which enables up to $M/2$ simultaneous connections to be performed. On the other hand, the local centres, when using multiplexors or concentrators, allow only one connection (i.e. transmission) to be performed at any given time period.

The star configuration is a form of a centralized communication network control, which provides the means for the interconnection of the independently operated outer nodes. In other words, while the network is centralized, the computer installation in the outer nodes can operate in a distributed system configuration. Such a facility is ideally suited for small system operations with low data rates of traffic flow. In large and sophisticated C^2 systems, however, which require traffic flow for both interaction and interoperability, the communication network, too, must be distributed. That is, in these large communication networks, there are no centre switching nodes in control of the traffic in the whole network but each switching centre is in control of all its local traffic and has the facility to communicate with all the other nodes in the system. Each switching centre is therefore only a node in the large network configuration performing common communication functions. This distributed network communication, as seen in Fig. 8.3, requires communication paths between the nodes, although not all and every node must be as directly connected, as the fully-connected configuration. In the distributed

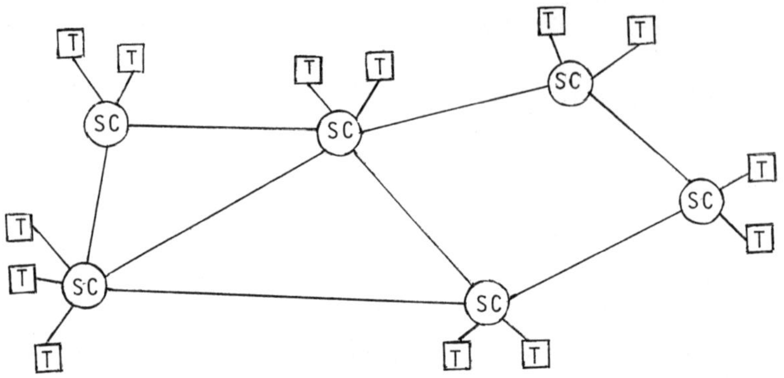

Fig. 8.3 Distributed network configuration

network configuration, each node can also act as an intermediate switching station for the other nodes in the network. Thus a message from one node may have to travel through a number of nodes on its route towards its destination.

The introduction of the distributed network configuration with the independently operating nodes, also introduces further communication facilities. One of these major facilities is that of routing, as now a message may have a number of path options it can choose from in order to reach its destination. It provides better efficiency utilization of both the channels and equipment. The switching centre nodes can now share the loads among the various paths or redirect the traffic on alternative routes when one of the channel fails. The distributed network is open-ended, and can grow indefinitely, with no single node acting as overall control. The channels between the nodes can be inserted or removed, according to the traffic needs and economical consideration. A direct channel is only added if the traffic over it can justify it financially. This type of distributed network configuration is most useful for C^2 system applications, for it can be indefinitely expandable, and can easily supply the demands of interoperability where independent C^2 systems exchange messages between them.

There are three basic concepts of switching centres, as seen in Fig. 8.4,

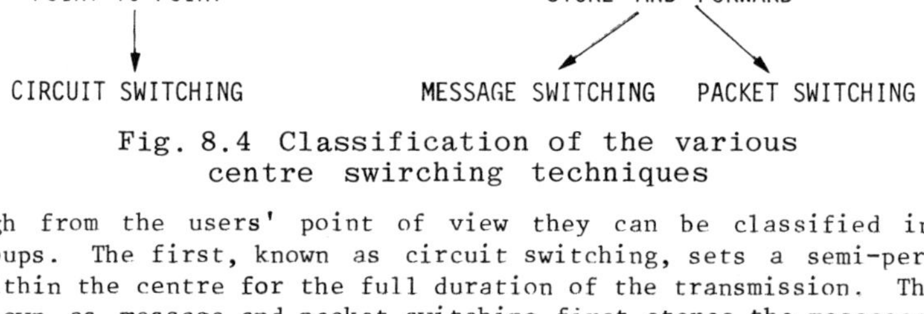

Fig. 8.4 Classification of the various centre swirching techniques

although from the users' point of view they can be classified in only two groups. The first, known as circuit switching, sets a semi-permanent path within the centre for the full duration of the transmission. The other two, known as message and packet switching, first stores the messages in the centre before they are routed to their destination according to the address given at the head of the message. In other words, circuit switching provides a physical path through the centre, as in point-to-point communication, while message and packet switching only provides a virtual path through the centre via its memory, using store-and-forward techniques. From a user's point of view, there is little difference between message and packet switching, although when regarded from the network side packet switching presents a much more efficient system.

Although the three switching centre techniques provide different types of communication facilities and are intended for different system applications, they have many similar characteristics. They all employ similar network configurations, and the nodes are all designed today with computers to perform their transmission routing tasks. Furthermore, they all use the same type of communication conveyances and the data transmitted over them may be fully digital. In this respect, one can refer to the centre switching centres as data switching equipment or exchange (DSE), as shown in Fig. 8.5. The DSE is therefore the means of connecting two DTE's, while the DCE is the interface unit between the terminal and the network.

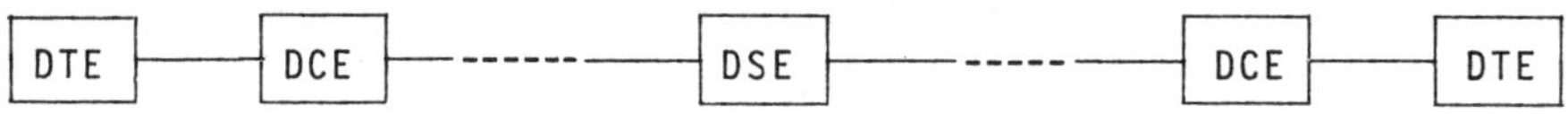

Fig. 8.5 The switching centre defined as a data switching exchange

Circuit switching is the common method used for voice telephone communication, although it can also provide data communication facilities over the dialled network. Message switching is used for slow speed terminal-to terminal communication, whereas packet switching is used for high speed computer-to-computer communication.

8.2 CIRCUIT SWITCHING

Circuit switching is a method of communication which completes a direct transmission circuit from sender to receiver for the full duration of the call. It therefore provides, on demand, an electrical path between the calling and called terminals for their exclusive use, until the connection is released. The most common circuit switching networks are automatic telephone exchanges, as provided by the common carrier companies. When using the dialled lines for the transmission of data, the signal must pass through these circuit switching exchanges en route to their destination. For this reason they will be described here. Admittedly, some of the circuit switching techniques discussed are out of date; they will, nevertheless, be presented, since many exchanges throughout the world still employ them and it will be many years to come before they are replaced by better and more advanced circuit switching techniques.

The most popular automatic telephone technique for many years was that suggested by Strowger, which uses a two motion mechanical selector. This double motion selector, as shown in Fig. 8.6, can be regarded as performing all the functions of N uni-selectors, arranged in layers, each with M outlets operating together. The first motion selects the required uni-selector layer and the second selects the required outlet of that particular layer. Such an equipment thereby provides the input to the double motion selector with one of the NM outlets. It is usual to design these double motion selectors with N and M being equal to 10, thus providing for each input a 100 outlets. The operation of the double motion selector is in two stages controlled by two separate mechanisms, one for horizontal operation and the other for vertical operation. The horizontal motion selects the required layer and the vertical motion takes over and then selects the required outlet.

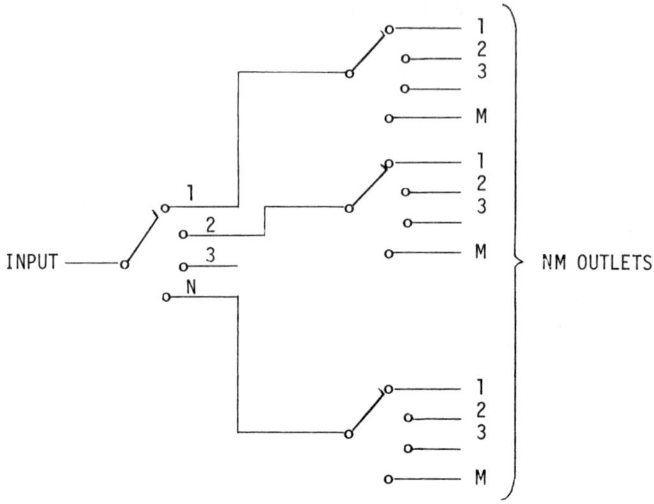

Fig. 8.6 Schematic illustration of the double motion selector's operation

It is usual to draw a representation of the double motion selector as shown in Fig. 8.7. Please note that only the layers are shown and not the layer outlets.

The automatic telephone exchange suggested by Strowger is better known by the term step-by-step, since it performs its operations in successive selection steps. In this scheme, as shown in Fig. 8.8, each digit dialled operates a different double motion selector, referred to as group selectors. When the calling subscriber lifts his telephone handset, the associated subscribers line equipment causes a uniselector to search for a free 1st group selector. That is, the uniselector steps a position at a time and checks if the outlet is free or busy by sensing the voltage level. If it is busy, it gets an impulse which steps it to the next stage, while if it is free, it holds it permanently in this position. When such a double motion selector is located, dialling

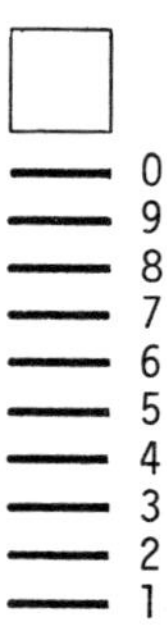

Fig. 8.7 Schematic representation of the double motion selector

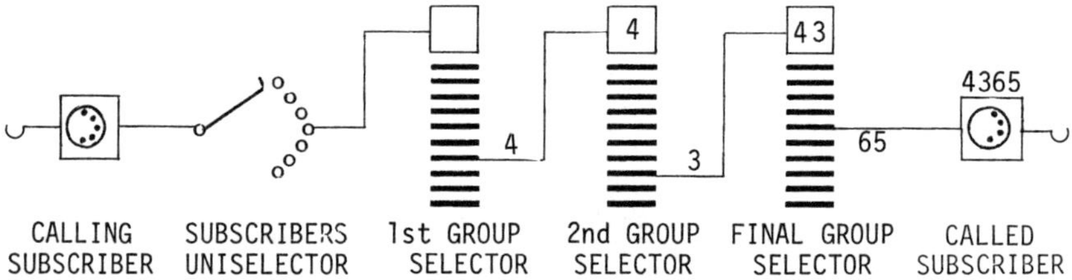

Fig. 8.8 The general arrangement of a 10,000 subscriber telephone exchange using Strowger's step-by-step technique

tone is returned to the subscriber, informing him that he can start his dialling process. The first digit dialled (say digit 4) steps the first double motion selector to the fourth layer and there it starts searching for a free outlet, which directs it to a 2nd group selector assigned to digit 4. When the second digit, say digit 3, is dialled, it selects the third layer of the 2nd group selector, and then searches again for a free outlet. The same operation is then repeated in each group selector, in accordance with the digits dialled, except for the last two digits. In a 10,000 line exchange only two group selectors are required since the last two digits directly select the called subscriber. The third digit, say digit 6, selects the sixth layer of the final selector, but then it does not need to search for a free outlet. Each of the 100 outlets of the final selector are each connected directly to a hundred different telephone sets. On dialling the last digit, say digit 5, the vertical motion selects the fifth outlet on the sixth layer, thus selecting subscriber 4365.

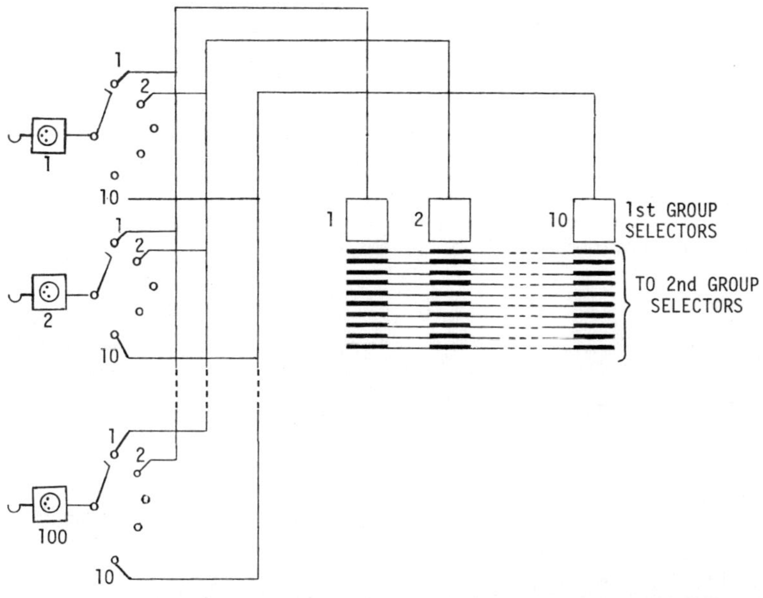

Fig. 8.9 Connection between selectors to provide common channelling of the calls

During the full duration of the call, all the four selectors (including the subscriber's uniselector) are held by the calling subscriber for his exclusive use, providing an electrical path over which he can transmit his message, i.e., over which he can talk. Although each call involves a lot of equipment, this does not mean that the exchange must provide enough equipment so that all the subscribers can communicate simultaneously. The exchange is designed on the assumption that only a certain percentage of the subscribers are active at any given time and that a different number of calls are directed to each location. Assume for example that there are a hundred subscribers, each with its one uniselector and they are all associated with the same first group selector, as seen in Fig. 8.9. Since it is not expected that all the hundred subscribers will be active, it can be presumed that only ten simultaneous calls are possible and therefore only ten first-group selectors are needed. In this case the outlets of all the subscriber uniselectors are compounded, with each common outlet leading to one of ten first-group selectors. The hundred outlets of the ten double motion selectors are also compounded, leading to 10 × 10 2nd group selectors. That is, the ten outlets of each layer are compounded and lead to ten second group selectors. In other words, there will be ten 2nd group selectors of digit 1, ten others for digit 2 and so on to digit 0. The ten 2nd group selectors of each digit may be associated at the same time with a hundred other 1st group selectors in the same way as described before. The same principles are applied to the connection between all group selectors.

The number of selector equipment in each exchange stage is defined by statistical recording of the number of calls. This is measured in traffic units (TU) specifying the intensity of the traffic in a particular period of time. This is then used to define the grade of service (GOS) specified by Erlang's formula which gives the relation between the proportion of lost calls, the number of outlets and the average volume of traffic in TU.

Another form of automatic telephone exchange employs the cross-bar space-division multiplexing system. This scheme uses common-control equipment which sets the call through ranks of cross-bar switches by means of markers. The advantage of the cross-bar switch is that a single unit can switch a number of calls simultaneously, thus reducing the number of individual required equipment. That is, in the step-by-step technique, each call requires a separate set of selector equipment, while with the cross-bar technique most of the equipment is common. Nevertheless, such exchanges are far more complex than the step-by-step scheme and therefore its application in the past has been limited. In the last few years, however, it has been found that the switching techniques used in space division multiplexing are ideal for implementation with digital switching.

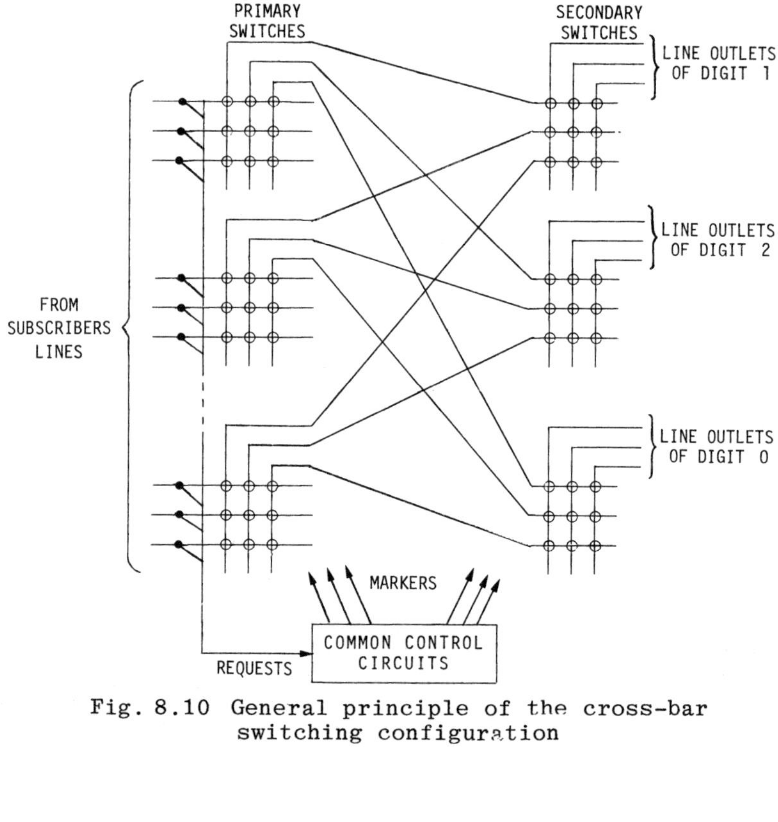

Fig. 8.10 General principle of the cross-bar switching configuration

Generally, one can replace each group selector by two ranks of cross-bar switches, as seen in Fig. 8.10. When there is an incoming call, the common control unit is alerted and accepts the dialled digits. The control unit marks an outlet in the primary switch in accordance with the digit dialled. That is, it selects a junction in the primary switch *P* between the subscriber and the secondary switch *S*. The control unit then selects a free outlet in the secondary switch to the required direction. The control unit therefore establishes a connection between the incoming lines to a free outgoing line in the required direction. There are many other variations of employing the cross-bar switches, using different configurations, but these are not discussed here.

The double cross-bar switch operates both as a router and as a distributor in a similar method to that of the double motion selector. Nevertheless, the cross-bar switch can allow many parallel transmissions with the markers preventing any conflict with calls to the same destination. As the call is progressed through the exchange, the control unit places a marker in each switch to hold the path throughout the transmission and prevent other calls from using the busy path. Fig. 8.11 shows the principle stages of a cross-bar exchange, with each double switch marking a different digit and the final double switch marking the last two digits. The description of the cross-bar exchange given here is based on a progressive step-by-step operation. This is not always the case, since the control unit can be used for many further functions, such as the translation of the digits dialled to the actual lines required. (The latter facility is not possible in the double motion selector, as will be later explained.)

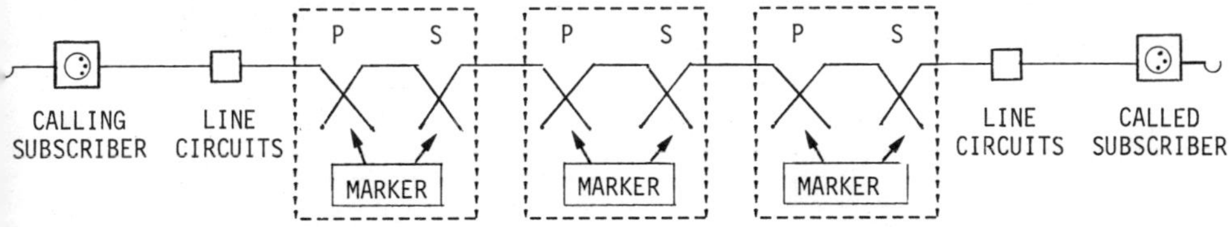

Fig. 8.11 General configuration of a cross-bar exchange

The subscribers' addressing scheme, i.e., its telephone numbering system scheme, plays an important part in the exchange design. There are two numbering schemes, open and closed. In the closed numbering scheme, all the subscribers receive the same number of digits, but with no subscriber's number containing similar digits. In the open numbering scheme any number of digits may be used, although again, no number with similar digits. Both schemes in general may be used according to the direction the calls are made. There are a number of hierarchical levels of exchanges; local, regional, national and international, each using a different numbering scheme, i.e., all the telephones in the area have a uniform number of digits. When dialling out of the area, extra digits are required which is defined by exchange levels in the hierarchical scales. There are accordingly three basic levels of exchanges: local exchange in the lower level, toll exchanges in the centre level and international exchanges in the top level. Today it is usual to dial 0 for toll and 00 for international, while direct uniform numbering schemes are used in the local exchanges.

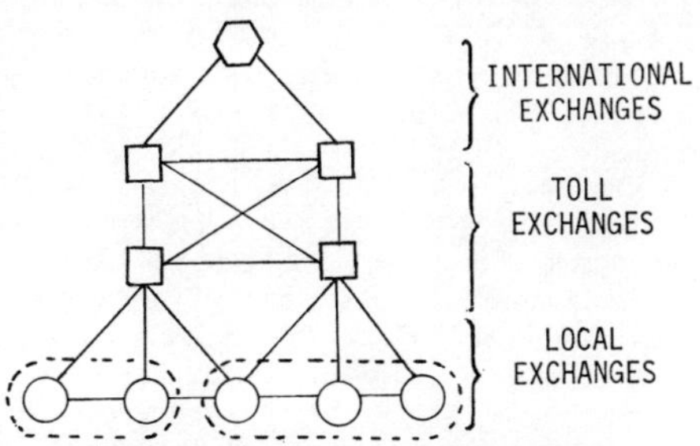

Fig. 8.12 Typical telephone hierarchical structure

In a regional area, there may be a large number of local exchanges which requires a uniform numbering scheme. If the area covers a large number of widely distributed exchanges, it is possible to use different digits to be added to each exchange. This may be practical if the traffic between the exchanges is small, but not in large cities, which require a closed numbering scheme. Take for example the London area where there are many local exchanges, and nevertheless, where everyone dials from to any subscriber in the area, the same digits are required, irrespective of its route to its destination. A transmission path may be made through the same exchange or may require to go through a number of exchanges. In all cases the same numbering addressing scheme must be used, although, it is up to the calling exchange to translate the number dialled to the actual number defined by the route to the called exchange. In the cross-bar exchange the translation is performed by the common control unit, while in the step-by-step exchange, additional equipment must be introduced. This additional equipment, shown in Fig. 8.13, is referred to as 'register-translator', or 'director' system. The first three digits dialled, which specify the required exchange, are inserted in a register and then translated to a different number of digits ranging from nought to seven. These new digits can thus direct the call to its proper destination; that is, once the translation is completed and the connection made, the register-translator can be by-passed.

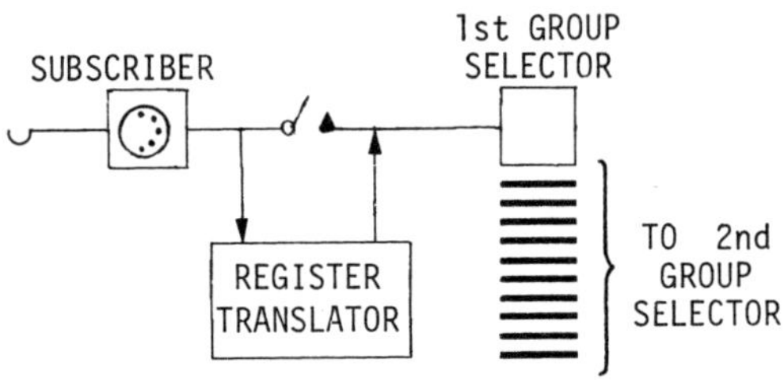

Fig. 8.13 The basic arrangement of a director exchange

Data transmission can be provided by the circuit switching techniques described here. The two terminals are connected by means of dialling which establishes a physical path between them. After the connection is made, the two terminals can communicate in either simplex, half duplex or duplex modes, but having the maximum data rate limited to about 1200 BPS.

8.3 DIGITAL CIRCUIT SWITCHING

Digital circuit switching provides a direct communication path between the calling and called terminal for the duration of the transmission. As distinct from the circuit switching techniques discussed in the previous section, the transmission duration may be staggered, having the same path shared with other terminal transmissions, although not at the same time duration. That is, each terminal is provided with a direct communication path for the duration of a given time slot.

For speech communication, the voice is transmitted in the form of PCM signals which are compressed into the corresponding time slots. The fact that the transmission is staggered does not affect the characteristics of the voice, provided that in each time slot the correct voice samples are transmitted. This is made possible by transmitting the signals at a much higher data rate than the voice sampling rate.

The most simplest form of digital circuit switching is that of the synchronous time division multiplexors, as shown in Fig. 8.14. The multiplexing and demultiplexing switches the common channel between a pair of communicating terminals. During the assigned time slots, the two terminals have a direct path over which they can transmit their messages. Such a network arrangement must be regarded as a primitive circuit switching scheme, since it only provides direct connection between two terminals but does not provide routing of messages between any other terminals.

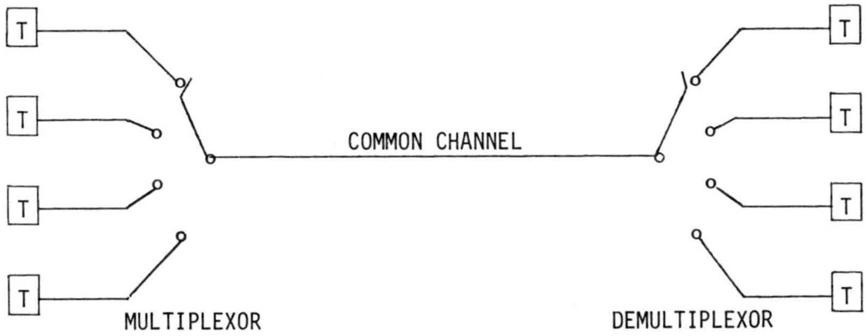

Fig. 8.14 STDM as a primitive circuit switching system

By digital circuit switching, a distinction must be made between the nature of the switching and the nature of the traffic, that is, between the control switching unit which sets up the call through the network to provide the physical transmission path and the actual data traffic which is transmitted over the path between the two terminals. In the earlier digital circuit switching systems, only the connection was made with digital techniques, whilst the transmission of the voice signals was still left to be performed over the conventional relay network. These techniques employed cross-bar switches with reed relay functions for the transmission path and electronic switching circuits in the control unit for marking the path. Today, the electronic exchanges are fully digital, with the voice signals transmitted as PCM signals integrated with digital switching, as seen in Fig. 8.15. The fully electronic exchange has the advantage over analog exchange of both performance and versatility. The PCM signals are not distorted in the network and therefore do not require special reshaping and amplification stages. It should be added that the PCM time slot pattern remains intact from the time it leaves the callers' A/D to the time it reaches its destinations A/D.

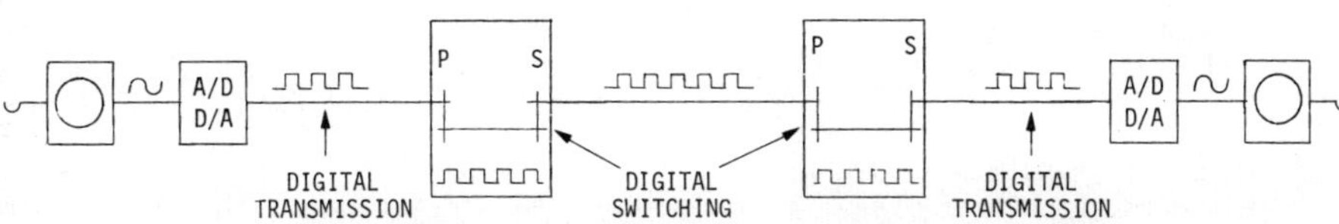

Fig. 8.15 Integrated digital transmission and switching in electronic circuit switching systems

As already pointed out, the simple STDM can only arrange the input data stream in a fixed prearranged order to time slots. This, however, does not provide any means of switching or routing the time slots between the various terminals. In fully digital telephone communication, means must be provided for the switching of the PCM streams in the direction towards its destination, taking advantage of the common switching and control equipment. In order that STDM techniques may be used, it is essential to provide means of 'interchanging' or 'shifting' of the time slot position, as otherwise no switching or routing is possible. This is provided by means of a 'time slot interchange' (TSI), as shown in Fig. 8.16, which has the function of interchanging the position of all the time slots, without any blocking, between its incoming stream and its outgoing stream. The TSI is simply a memory buffer of the size of two frames of N time slots. As one half of the memory

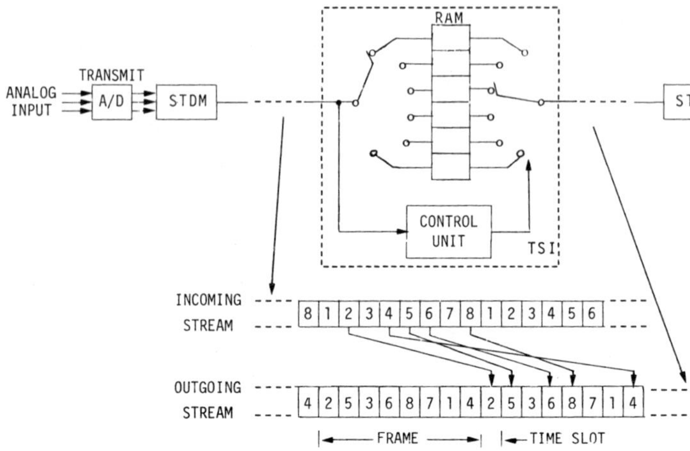

Fig. 8.16 Digital circuit switching using time-slot interchange

is being written into, the other half is read out. At the end of the frame the functions reverse and so the time delay of any time slot in the TSI is at least one frame period, but not longer that two time frames.

Several PCM signals are time multiplexed onto one common channel and then inserted into a high-speed random access memory (RAM). That is, each input time slot is written into the buffer memory in their sequential order with each time slot having its own fixed buffer cell. Reading out the time slots, however, is performed on a different sequence, thereby interchanging its relative position in the frame. This is formulated by the control unit which reads out the slots on the basis of an assignment procedure, dictated by the switching requirement for interconnecting the caller with the called terminals. Before the TSI starts its operation, it must therefore receive adequate instructions and it will then continue with the assigned interchanged pattern until other instructions are received. Since the number of time slots in the incoming and outgoing streams of the TSI are the same, the whole operation is without blocking. Nevertheless, there is always a time delay of at least one frame between the time the slot enters the TSI and the time it leaves it. This delay must not be so long as to be felt by the subscribers and its maximum period is defined by the number of PCM sources which are time multiplexed.

The time—slot—interchange together with the two time multiplexors as shown in Fig. 8.16, could be regarded as a small circuit-switching telephone exchange, since it provides the required switching facilities between any of the input subscribers with any of the output subscribers. Nevertheless, the size of such an exchange is limited since the allowed delay in the system must be restricted, and consequently the size of the buffer must also be restrained. For large telephone exchanges a number of TSI should be used combined with space division switching stages, as shown in Fig. 8.17. The space switches are implemented by the digital electronic circuits described in the previous chapter. Each input time slot has a choice of paths through the network, first through the TSI and then through the space switches, all of

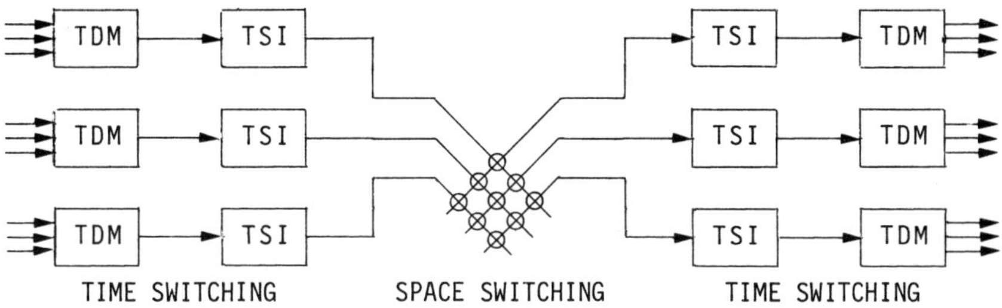

Fig. 8.17 Basic building blocks of digital circuit switching network architecture with time and space switching

which are implemented in the time domain. The TSI and space network is referred to as time-space-time (TST) switching network and they can be assembled in various combinations, one of which is shown in Fig. 8.18. It is also possible to form them as space-time-space (STS) combination, although the tendency today is to employ only two time stages with one or more space stages in between.

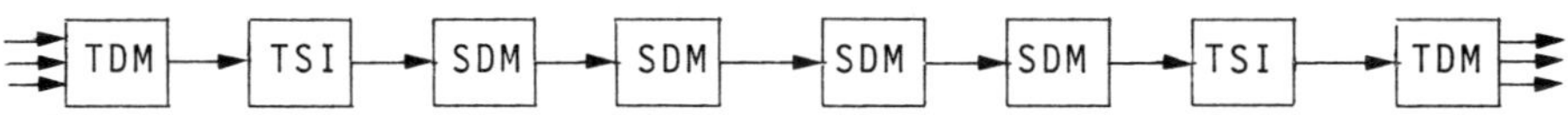

Fig. 8.18 Path of a message through a TSSSST circuit switch

The time-space combination is regarded as circuit switching although it produces a delay between the incoming and outgoing time slots. This is because the network still produces a semi-permanent connection between the calling and called terminals for the full duration of the transmission. The time slots throughout the route are permanently allocated to a given terminal transmission. That is, this operation should by no means be compared with store-and-forward switching, as there the transmission is indirect with the messages first stored and then retransmitted only according to dynamic conditions dictated by the traffic queues.

The control of the whole digital circuit switching operation is generally performed by a computer using time sharing techniques. The computer sets up the call through the exchange and then 'marks' it for the duration of the transmission, but it does not participate in the actual data communication. The employment of the computers, as the intelligence required for the telephone exchange control units, enables many further communication facilities to be introduced, which were not possible in the electromechanical exchange, such as conference calls and temporary change of the subscriber's number when he is on holiday.

8.4 MESSAGE SWITCHING

Circuit switching schemes sre primarily intended to handle voice communication, but they can adequately handle message transmission. Indeed, messages may be transmitted over a dialled network, it having the advantage that network access lines are easily available. Furthermore, the same line may be used for both data and voice, although the data rate is limited. Nevertheless, where comprehensive terminal-to-terminal communication is required, circuit switching proves to be very inefficient. Message transmission durations, and these could sometimes be even shorter than the relatively long duration that is required to make the connection in circuit switching schemes. For this reason, in systems where there is a high message traffic, specially designed communication network is required.

One means of solving message communication has been provided by the Telex system. This communication system enables each subscriber to dial and transmit messages to any other subscriber around the world who is connected to this network. The techniques involved in the telex communication are based on circuit switching and consequently suffer from long delays in obtaining the connection. For every message dispatched, a proper electrical path must be re-established. Whoever has used the telex communication system knows that the chances are high of obtaining a wrong subscriber, or for the called subscriber to be busy. It is also very likely that the capacity of the route towards the destination is in full use and is unable to accept any further calls. Furthermore, the telex system provides very little network services with all the communication tasks left to the subscriber. In private message communication systems, which is an indispensable administrative tool in many C^2 systems, such as banks or police forces, the telex solution is most inaffective.

Many large organizations, which may have a number of outgoing telex lines for external communication, may also require a number of private teleprinter lines for internal communication. For this purpose, they set up a teleprinter room, with all the lines leading to it, which performs as an intermediate dispatch centre. The operation of this room is essentially manual, with the operator performing the switching tasks. In this system, referred to as 'torn tape', special receiving and transmitting equipment, usually teleprinters, is in use exclusively for each communication channel. The operator, on receiving a message, lifts it off one teleprinter and transfers it manually to another for onward transmission. Message switching via the 'torn tape' centres is flexible but far from effective. The process is extremely time consuming with likelyhood of messages going astray undetected.

Message switching is the name given to the function of transferring the messages over the communication network without need for human intervention. Message switching schemes are intended for private systems, whereby the human operator is replaced by an automated system which performs all its routine tasks most efficiently. Applications where such a technique is required are wide spread, e.g., credit cards, airports, banks, etc.

Message switching centres are intended to control the transmission of digital information from one remote location to any one of the other locations associated with this private network. The centre may, if necessary, reroute a message, retain it until its destination is ready to accept it, or transmit it simultaneously to a number of destinations. In other words, message switching provides network efficiency which was not possible with circuit switching, coupled with the intelligence contained in the message heading. In circuit switching, only a physical path is provided, while the data contents is ignored. In message switching, it is the information within the message which causes the 'switching' of its routing operation towards its destination, a feature which provides high speed, accuracy and time saving. That is, the switching centre checks the message heading and having identified the address, routes the message to its appropriate destination.

Message switching employs store and forward techniques, with the entire message transmitted to the centre, where it is stored for a period of time, usually very short, and is then retransmitted towards its destination. The most important feature of message switching is therefore that the terminals are not directly connected with each other over a fixed electrical path, but indirectly through the centre to which all the termainals are linked. The centre is alone responsible for forwarding the messages, and must therefore be always available to accept all the messages dispatched from the terminals. In other words, the centre has the characteristics of the concentrator, but in addition it also provides switching tasks. The terminals can transmit their messages at their maximum speed, irrespective of whether there is an available free output route from the centre or not. The destination terminal may be out of order or busy accepting messages from other sources, but this is of no concern of the transmitting terminal. It is for the centre to place the messages in queues and retransmit them when the path to the terminal becomes available. The terminals associated with the message switching network may be of various types, operating at different speeds and using different communication formats and procedures. It is up to the centre to provide suitable transmission by translating the incoming messages to the format at the destination terminals.

The message switching centre is usually situated in the centre of the star network, as seen in Fig. 8.19, with lines leading directly to each terminal. Multiplexors and/or concentrators may be used in this network, but only as transparent intermediate stations connecting the terminals with the centre.

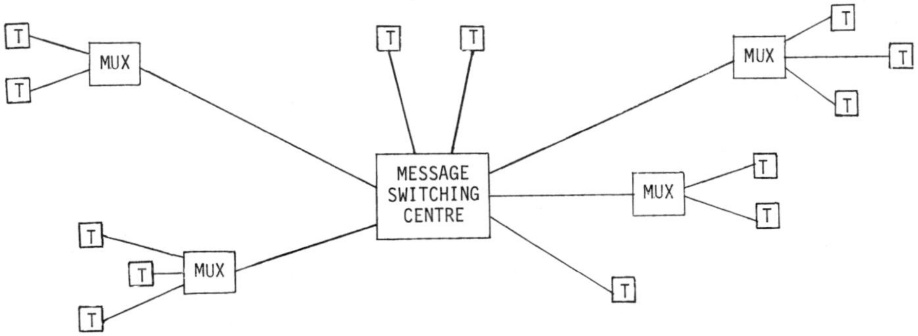

Fig. 8.19 Typical example of a message switching network

Message switching centres are intended not only as passive switches but as active administrative controllers of the traffic between the terminals. As the message enters the centre, it receives a unique number which facilitates the checking of the message to assure that it has not been lost en route. The message numbering schemes also serves for retrieving it at a later stage. That is, an important feature of the message switching scheme is that it is usual to store all the messages for long periods, even after the message contents have been successfully dispatched to their destination. If a query subsequently arises concerning the contents, or the destination of any message, the messages could easily be reproduced. The real-time transactions of the messages are performed in the main and the backing memories, while any further storage of the message is performed in the auxiliary memory where it can be reached off-line. Normally the messages are required to be retained for one 24 hours, but there are systems which may require longer periods. This feature is most important for some C^2 systems applications enabling them to keep an accurate log of all the traffic contents for further analysis. The provision of this additional storage capacity could, however, be prone to be extremely costly.

The structure of the switching centre generally consists of a duplex processing arrangement, as seen in Fig. 8.20. In this configuration, both halves of the centre are in full operation, performing identical tasks, although with only one processor in control of the output lines, and the other acting as standby. That is, both halves of the dual system accept the incoming messages, process them and place them in output queues. Nevertheless, only the active processor can output the messages on to the lines.

Each processor has its own main and backing memories together with all

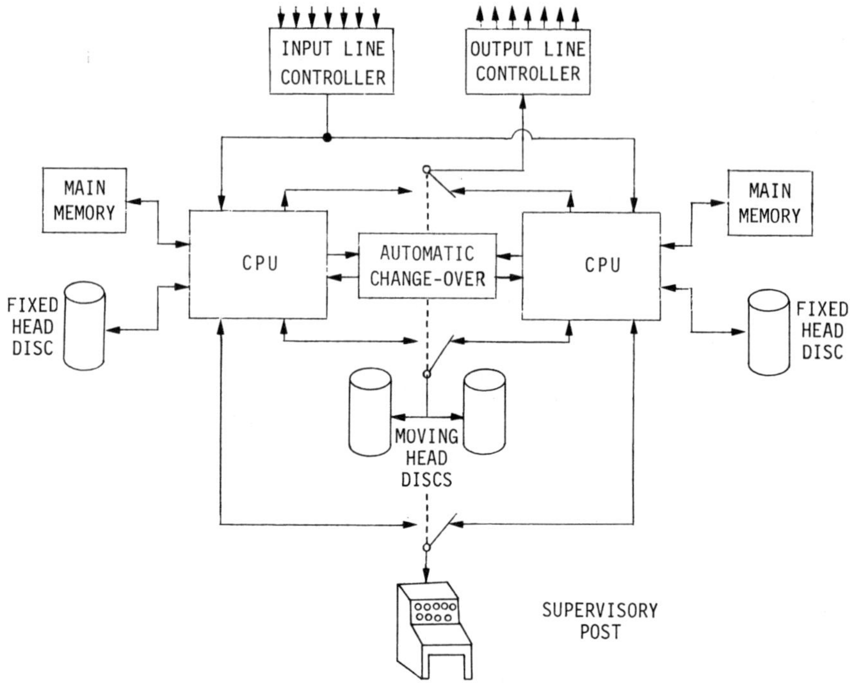

Fig. 8.20 Block diagram of the message switching centre

the required peripheral equipment; accordingly both system halves may operate separately in real-time environment, with both processors operating simultaneously, building up identical information patterns in their respective memories. The auxiliary memory, which holds the messages after their dispatch, plays no active part in the successive transfer of the messages, and there is therefore, no need to duplicate it; a single memory unit can serve both processors.

The automatic change-over unit is a separate equipment which checks the current active processor operation and switches over to the standby processor when a failure is recorded. In effect this unit continuously monitors all equipment both in the active and in the standby processor. When a failure occurs in either system, the supervisor is informed, but it is only when a failure is recorded in the active processor that a switch-over is performed. The automatic change over unit is an equipment which continuously receives OK pulses from various parts of the processors indicating the successful operation of the processors. Should a failure develop, one of the OK pulses will be inhibited and a failure will be recorded. If this failure occurs on the active processor, the automatic change-over unit will direct the standby processor to become active by switching over all the auxiliary equipment.

The automatic change-over unit must also keep both processors in synchronization, on a block basis, to assure that no information is lost on switch-over. As the standby processor always obtains a duplicate of the input data on all lines and builds up an identical data base, the traffic is not disrupted when a change-over takes place. When the standby processor becomes active, it uses the information in its own data base to reconstruct the recent history of the processing, and can then resume normal operation.

As the messages arrive into the system, usually a character at a time, they are assembled into blocks and transferred to the backing memory (fixed-head disks). As soon as the complete message is received, the heading of the message is verified in the main memories and then analysed. As a result of this analysis, the procured routing information and the messages are placed in queues on the relevant outgoing lines. The messages are not moved from the backing memory when they are placed in the queue; instead, a message pointer is placed on a suitable log table. Once the lines become free, the messages are lifted off the backing memory by the processor, a block at a time, and transferred back to the main memory. From there they are transmitted to the line, via the line control module, a character at a time. After the messages have been successfully transmitted to their appropriate destination, there is no need to hold them in the backing memory. That being the case, the copy of the dispatched messages can be transferred to the auxiliary memory for long term storage.

The supervisory console must be common for the full duplex processing systems, as the same operator is responsible for the whole system operation. That is, the system must be regarded as a single structure, it being of no importance for the message transfer which processor is in operation.

In any automatic communication system the attention of the supervisors and that of the terminal operators must be drawn to unusual operational effects which are not anticipated. The supervisor provides interrupt handling, error recovery procedures, and supervises the control of all the centre operations. The computer must be able to recognize those circumstances which require attention, and request the intervention of the supervisor. The supervisory console unit must be provided with facilities for instructing the computer to take whatever actions required to deal with the problems involved. The supervisor must also be able to request information from the computer on the operational status of the system.

8.5 MESSAGE SWITCHING VERSUS CONCERNTRATION, CIRCUIT SWITCHING AND LOOP TRANSMISSION

Message switching is a method of handling data messages over the communication network, where the entire message is transmitted to an intermediate point, and there stored for a period of time and then retransmitted towards its destination. This particular process of message handling could easily be compared with other forms of communication equipment and it is important therefore to define the switching and transmission facilities provided by message switching in comparison with the other communication network techniques.

The description given above for message switching could well be applied also to the concentration techniques. Both message switching and concentration provide a communication service to the terminals associated with them, whereby the terminals can dispatch their messages as soon as they are ready That is, the terminals transmit their messages without waiting for a direct free path or waiting to check whether the destination terminals are ready to accept the messages. The messages are stored in the equipment and then placed in queues on the outgoing channels, in their order of arrival or in accordance to priority. In other words, both message switching and concentration schemes use store and forward techniques in handling the messages. Nevertheless, both these communication techniques are different, with the message switching equipment providing many more services, above all providing switching facilities impossible with concentrators. The following table presents the main points of similarity and difference between these two communication techniques.

TABLE 8.1

FUNCTION	CONCENTRATION	MESSAGE SWITCHING
Equipment technique	Store and forward	Store and forward
Means of communication	Indirect via a computer	Indirect via a computer
Connection between terminals	No connection possible	Connection provided by equipment
Start of terminals transmission	Immediate as soon as message is ready	Immediate as soon as message is ready
Action to be taken when destination is busy	No concern of the sending terminal	No concern of the sending terminal
Off line storage of message	Generally not available	Available for a given period, for back-checking of messages
Communication procedures	Equipment fully transparent	Equipment used as an intermediate processor
Type of data and speed	All the terminals usually transmit at a fixed rate using the same procedures	Each terminal may transmit at a different speed or procedure
Message acknowledgement	Not required	Generally provided
Communication services	Intermediate handling	Intermediate switching
Additional communication services	None available	In accordance with demands
Size of computer	Small	Relatively large
Function	Increase of common channel efficiency	Increase of network efficiency
Application	Concentration of traffic on a shared channel	Switching of messages between terminals

Concentrators may be regarded as traffic cops controlling the data flow in one of two directions only. The object of the message switching scheme is much wider than that of the concentrator; it is intended to act as a large communication organizer allowing many transmissions to be performed simultaneously in different directions. Whereas the concentrator serves as an intermediate stage between the terminals and the centre, the message switching equipment serves as the centre itself.

Message switching and circuit switching equipment both perform the task of communication centre operations and employ similar network configurations. They are both intended for terminal-to-terminal type of communication and they both employ computers in providing their switching services. Nevertheless, they are both intended for completely different types of application, and in particular, they both serve different types of terminals. Circuit switching is primarily designed for voice communication while message switching is designed for manually operated terminals. Circuit switching networks can transmit over its link data messages while message switching is not designed to cope with voice messages even when they are in digital form. The purpose of circuit switching is to provide a direct connection for long transactions while that of message switching is for only short transactions.

As mentioned above, message and circuit switching employ general purpose stored program computers to effect the switching of data between the terminals. Despite the superficial similarity of both switching schemes, they both employ the computers in a different method, as seen in Fig. 8.21. With circuit switching the computer only marks the path but does not participate in the data handling. That is, the direct physical circuit path is completely divorced from the control unit. In message switching, the computer plays an

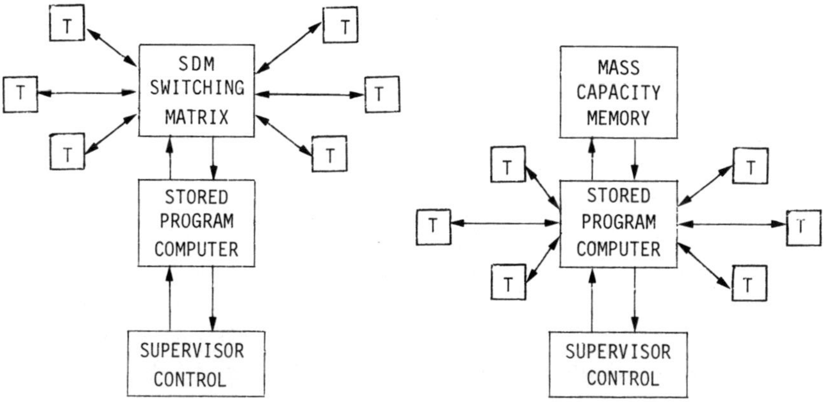

(a) CIRCUIT SWITCHING (b) MESSAGE SWITCHING

Fig. 8.21 The position of the computer in circuit and message switching centres

essential active part in the data handling operation. All the messages must be stored within the system's memory before they can be transmitted. That is, there is no possibility of a direct connection path between the two communication terminals.

TABLE 8.2

FUNCTION	CIRCUIT SWITCHING	MESSAGE SWITCHING
Means of communication	Direct between terminals	Indirect with aid of the centre
Form of data	Analog or digital	Digital only
Connection time	Relatively long	No time lags
Delay in connection	Delayed till a path can be located	No delays at terminal end
End to end delays	None once the connection is made	Depends on the queue of massages
Destination terminals busy	Requires closing down and a second try	Of no interest to the calling terminal
Utilization of the lines	Independent of the traffic	Smooth traffic flow
Efficient usage of equipment	Inadequate efficiency	Maximum efficiency
Sharing equipment resources	Dedicated to each connection	Common to all the connections
Data handling	None whatsoever	Storage, routing and error correction
Data storage	Not available	Available for back checking of messages
Information of the destination	Separate from message	In the message heading
Application	Long transactions	Short inquiry – response transactions
Connection information	Separate from message	Message headings

Both message and circuit switching may be rendered useful in C^2 systems. Circuit switching is essential for voice communication but could also be applicable in many real-time systems. If on-line real-time measurements are to be performed in a centre, only a direct physical path, as provided by circuit switching, can supply it. For short transactions, however, message switching is indispensable, as it provides the essential means of efficient communication.

Some of the differences between message and circuit switching schemes are presented in Table 8.2.

Of all the three communication schemes compared with message switching, the loop transmission scheme is the most interesting. Although message switching and loop transmission employ completely different switching techniques, they are both intended for the same type of application, viz, terminal-to-terminal communication. Both schemes have much to offer in this respect and it is sometimes difficult to decide which is the better scheme. The advantage of loop transmission is that the initial installation cost is relatively low, while with message switching it could be extremely high. Because of the reliability problems encountered with loop transmission it is suggested to employ it for message communication within a building or compound, while message switching schemes are recommended for systems with widely spread terminals.

A summary of the comparison points of loop transmission and message switching is presented in Table 8.3.

In conclusion of this section it is important to add that many of the points of comparison brought for the message switching schemes could adequately be applied also for packet switching. From a user's point of view, the network handling of the message being invisible, message and packet switching schemes are very similar. Nevertheless, message and packet switching techniques are different as will be discussed further on.

TABLE 8.3

FUNCTION	LOOP TRANSMISSION	MESSAGE SWITCHING
Network operation	Distributed	Concentrated
Initial system size	Relatively small	Large
Expansion of system	Modular, in a linear order	In jumps and changes of centre equipment
Expansion possibilities	In accordance with the users demands	In accordance with the available capacity of centre equipment
Initial system cost	The price of the terminals in use only	The price of the centre
Intelligence of the terminals	Extremely intelligent	No intelligence required
Communication services	None available	In accordance with demand
Network configuration	Ring (loop)	Star
Network reliability	A single terminal malfunction disconnects the whole system	A single terminal malfunction does not affect the system
Centre reliability	A malfunction in any terminal disconnects the system unless preventive measures introduced	A malfunction in the centre disconnects the system
Application	For message communication between fast terminals centrally located	For message communication between slow terminals distributed over wide geographical areas

8.6 DISTRIBUTED MESSAGE SWITCHING NETWORKS

In certain respects the message switching operation could be compared with the postal mail service. In both, each message is switched through the network without the system providing a direct connection between sender and receiver. Indeed, the post box could be considered as corresponding to the store and forward concentrator while the postal van as corresponding to the the communication channel. The mail sorting office certainly performs all the functions expected of a message switching centre. A single automated sorting office must and does cope with the postal mail from an extremely large area; but, as more areas are added to it, its efficiency would be reduced. The same thing applies also to digital message switching.

The centralized message switching star network configuration may operate quite adequately for many C^2 systems where a single switching centre serves the whole system. Nevertheless, as the system expands, it demands the introduction of long parallel transmission channels which will result in the escallation of the communication network cost. Star switching networks have proved most efficient communication means for small and medium sized C^2 systems, but with the increase in the number of terminals and their traffic demands, the system may reach a level of saturation. Such a situation results in longer queues and consequently longer periods of delay for the messages in the system. Slow manually operated terminal users may possibly accept these delays, but as the terminals become more sophisticated, such a network operation becomes intolerable. Furthermore, any failure in the single switching centre will adversely affect the whole communication network operation. It will then be advisable to install a distributed switching network to effect a decentralization of resources in such a way that any failure does not disrupt the working of the whole network but only isolated local terminals.

In the suggested network configuration, a number of star networks operate together, as shown in Fig. 8.22. Each terminal is associated with one star message switching centre, as described before. That is, each star centre operates independently, bearing the responsibility for the operation of all the terminals associated with it. The various centres are usually operated on a regional basis so that they can serve all the local terminal traffic requirements. In order to provide complete terminal-to-terminal communication facilities, the various message switching centres are connected between them by special trunk channels which usually operate at higher data rates than the local channels. The independent star centres can now be regarded as switching nodes of a large network configuration. The advantage of such a configuration is that it is open-ended and can be continually expanded.

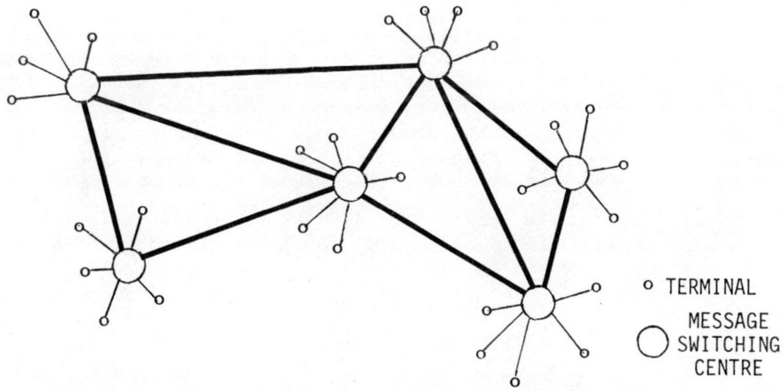

Fig. 8.22 Multi-star distributed message-switching network configuration

The message-switching centres handle the traffic in the trunk channels in the same way as they handle the local traffic. That is, all the messages are directed towards their destination in accordance with the address at the heading of the messages. The internal program of the centre would determine whether the message is local or one which must be transmitted to other centres.

The network configuration between the centres need not be fully connected but may use other centres as intermediate states between the sender and destination. Such an arrangement introduces a new network facility, namely, alternative routing. That is, each centre may have the choice of a number of routes for the message to flow between the sender and destination. The decision on the best route for a message to be transmitted over it depends on the channel operational status and its current traffic load.

In the distributed message switching network all channels are considered as of equal importance, for the centre handles all the messages, whether local or trunk, in the same way. Nevertheless, the centre can get rid of its trunk messages faster than its local messages because of the higher data rates on these trunk channels.

Although the new distributed network derives from the amalgamation of independent star networks, the whole combined configuration can be regarded as a single integrated network, and as such, the operational characteristics of the new network is different from the original single-star network. In the star network, all the transmissions transactions can only be performed with the centre's initiative, or in other words, all operations are based on centralized control. In the combined network, the control is distributed between the various centres, with each centre controlling only the channels associated directly with it. The centre therefore has a dual function of both controlling the local terminal traffic and providing relay station facilities between the other centres. A message originating from any terminal must first reach its local centre and only then may travel towards its destination, directly or via other centres.

Message switching network, whether centralized or distributed, has some basic features which are essential for the reliable operation of the system:

(a) Each terminal on the receipt of a message, must return an acknowledgement to the originating centre. Until a positive acknowledgement is returned, the centre must still hold full responsibility for the message. As the messages may have to pass through a number of centres and as there could be some delays before the acknowledgement is received, the originating centre must retain the full message in its backing memory for the whole waiting period before it may transfer the message to its auxiliary memory.

(b) The transmission of the messages between the terminals and the centre must be rapid with no overdue delays. A message transmitted to a centre in any channel must always be accepted by the centre. That is, it must always have enough memory capacity space for it. Since the centre cannot know in advance the size of the message, the demands on the capacity of the active memory is generally enormous.

The major disadvantage, then, of the distributed message switching network is the over-sized active mass memory capacity required in each centre to perform all the switching transactions. On top of all the real-time transactions, each centre must also be provided with huge auxiliary memory facilities to store all the 'old' messages which have already been transmitted. What is required is not only over-sized memory capacity but also extended processing abilities. The centres have the dual function of dealing with local and trunk traffic, both containing messages of variable length. For these reasons, this form of distributed message switching is not widely used. Such schemes are only found in some private networks, and even then only for slow speed manual terminal traffic, such as teleprinters. For high speed terminal traffic and for computer-to-computer communication, other communication techniques are necessary.

8.7 PACKET SWITCHING CONCEPT

Computer communication networks are intended to provide the facilities for resource sharing of the data files located in widely spread geographical areas. The network channels must be able to transfer effectively masses of data items at high speeds. The switching nodes that the messages must travel through, must be able to handle them skilfully without undue delays. The message switching networks, as described in the previous sections, are the most inefficient means of communication when applied to computer-type traffic. What makes it decidedly impractical are the characteristics of the centre switching in handling the computer message of unknown variable size.

Just over a decade ago the approach to message switching was radically changed, and a new technology was introduced. This was referred to as 'packet switching' and involved only fixed size messages. Although this change in message size may seem to be insignificant, it has completely revolutionized computer communication. While ten years ago most of the interactive data communication was circuit switched, today public packet switching networks are being generally set up.

Packet switching technology is aimed at solving the basic inefficiency of message switching, namely related to the variable message size and consequently the large memory volumes involved. In message switching it was required that each message should contain a character defining its size for use in the centre as means of checking whether the whole message has been received. A message switching centre cannot know in advance the length of the message and cannot therefore prepare its memory capacity in advance. Furthermore, it cannot define its queue length by the number of messages waiting to be transmitted but only by the sum of characters in all the messages. Today, the handling of the messages has become a simple operation. This has been achieved by simply adopting a fixed size for the messages flowing between the nodes and thus enabling the switching centres to deal with known quantities.

As it is obviously impossible to define a message size to fit all applications, a statistical message length is specified. The fixed size message, usually only a fraction of the original message, is called 'packet'. A packet, generally about one thousand characters in length; is that treated in the nodes as a single transmission unit. It requires a fixed memory space and its transmission period too is fixed, i.e., it requires a fixed size time slot.

Despite the fact that the only differences between message-switching and packet-switching is in the message size, the whole approach to the network operation has been changed. In message switching, the whole message is sent in one operation, while in packet switching it is the packet that is transmitted as one operation. If, in a packet switching network, a terminal dispatches a long message, the local node splits it up to fixed sized packets, adding to each its own identifying heading. Since all of the packets are part of the same message and are intended for the same final destination, they are theoretically all placed in the same queue.

The function of the switching node is to transmit all the packets as swiftly and smoothly as possible. In this respect it regards all packets as equal, irrespective of where they have come from or to which messages they originally belonged, This means in effect, that when there are alternative routes out of a node in the direction of the destination node, the packets from the same message may be split between the various outgoing channels. The packets (as with message switching) are transmitted on the various alternative channels, in accordance with the momentary traffic queues in the nodes. Therefore, packets may be transferred from one route to another so as to reduce the sudden queue load from one route to less loaded routes.

In message switching, the function of the centres is to provide a service to their associated local terminals with the trunk traffic regarded only as an extra task. In packet switching, the concept is changed, with the main functional task of the nodes being to provide a network communication service. Only the originating and destination nodes deal with messages, while en-route the nodes deal only with packets. Accordingly this means that the nodes in changing the routes do not check the packet connection with any particular message since their only concern is the network efficiency. This could result in the packets reaching their final node destination in a different sequence from that in which they had originally been transmitted. It is then necessary for the destination node to reorganize the messages back to their original sequence before they can be transmitted to their final destination (terminal or computer).

An example demonstrating the packet switching concept is illustrated in Fig. 8.23. In this example there are four alternative routes for a packet to travel between the originating node and the destination node. Each of these alternative routes is of equal importance (weight); that is, they all pass through two intermediate relay nodes. The traffic loads, however, in all these channels, are of different quantities because of the packets originating from other terminals in the network. Although each of these transmitted packets requires the same number of hops to reach its destination, they are each confronted with different delays in each node. Consequently they reach their destination in a different order from that of the original transmission. In the example, packet 1 is confronted with large queues in each of the nodes it passes through and therefore suffers the longest delays. Packets 3 and 4 are placed on the same route, but because of queues in the next node they are split up, resulting in packet 3 reaching its final destination before all the other packets.

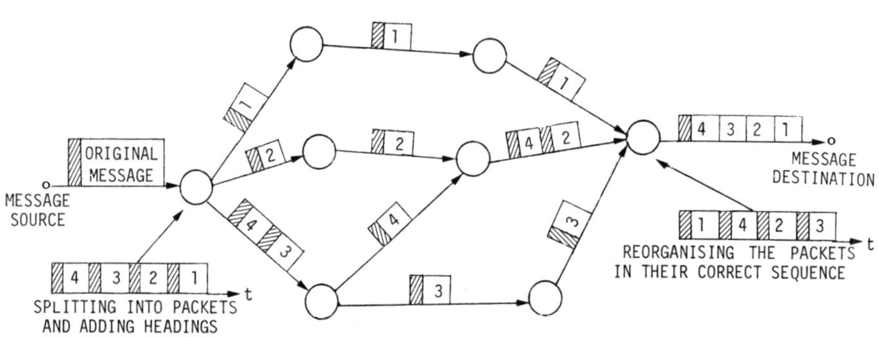

Fig. 8.23 Routing the packets in a packet switching network with the functional tasks of the various nodes

In both message and packet switching schemes the nodes operate as independent centres employing store and forward techniques; nevertheless, the handling of the message/packet in both schemes is different. Message switching network technology is based on the centralized concept with the originating node responsible for the message until it receives a positive acknowledgement that it has reached its final destination. The reason for this is that the message switching node acts as the terminal's agent and it is therefore fully responsible for the message passage throughout the network.

In packet switching technology a fully distributed concept is employed with each node seeing only its immediate neighbours. The nodes simply advance the packets to the next node in the direction of their destination according to special updated traffic tables, without concern for the packet beyond the next node. The responsibility of the nodes for a particular packet is only for the period it handled it. As soon as the node receives an acknowledgement from the next node, it can completely erase it from its memory, being now released from any further responsibility for it.

The packet switching concept is based on exploiting the nodes' memory capacity exclusively for the active packets. The memory can be cleared as soon as the packet is successfully accepted by another node. Fig. 8.24 demonstrates the packet transfer along a chain of nodes. To save time, the nodes may start retransmission of their packet at the time that they send an acknowledgement for its positive reception.

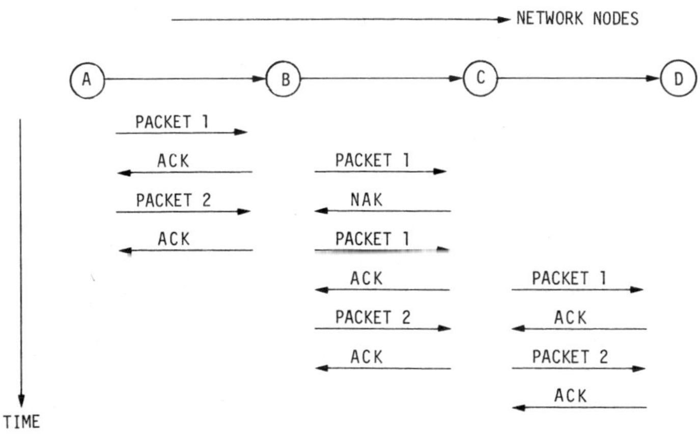

Fig. 8.24 The packet transfer concept between the various nodes

The functions of the packet switching node are to receive the packets, check their destination and retransmit them on the shortest available route, which is determined both by the number of hops needed to reach their destination and the queues in each of the nodes' outgoing alternative routes. The function of the nodes is therefore to provide relay services only, that is, without need to retain the packets or even to keep a record of them for later use. As a result, the packet switching centres operate independently with each requiring much smaller active and auxiliary memory capacity than is required for message switching and consequently the cost of the whole centre can be appreciately reduced.

In order to further increase the network efficiency, it is usual to operate the whole packet switching network with a common communication protocol. This means that the processing function of the nodes is further reduced as there is no need for any format translation and as all packets can be dealt with uniformly. The node computer, therefore, can also be made smaller and simpler.

8.8 PACKET SWITCHING VERSUS MESSAGE SWITCHING

Packet switching has been introduced as a special case of message switching, although, in fact, when analysing the technology employed in both schemes, it can be seen that their functional tasks are completely different. It is not just the size of the data stream which is transmitted as a single unit, but the whole performance of the communication switching node that is different. Message switching is intended for coordinating the traffic between the various terminals, each of which can operate at a different speed and format, while packet switching is adopted for coordinating the traffic between the various switching nodes, all of which operate with the same format and protocol. Packet switching is not concerned with the terminals but only with the packets transmitted.

Another basic difference between message and packet switching is in the functional operation of the switching node. Although both schemes use store and forward technology, the processing and memory requirements are different. Message switching requires a large computer installation while packet switching can operate adequately with a small or medium sized computer. Furthermore,

TABLE 8.4

FUNCTION	MESSAGE SWITCHING	PACKET SWITCHING
Original technique	Store and forward	Store and forward
Transmitted unit size	Unlimited variable length message	Fixed sized packets
Message routing	The whole message is transmitted over a single route	Each part of the message could be transmitted over a different route
Message overhead	Small, each message with one heading	High, each of the packets with a different heading
Number of channels per node	Large	Small
Provided node services	Maximum according to demand	Minimum but well defined
Node functional tasks	Service to the terminals	Service to the networks
Node memory facilities	Long storage duration even after message dispatch	Short storage duration only until an acknowledgement is received
Node memory size	Large active and auxiliary memories	Relatively small active memory
Acknowledgement	Between sender and receiver	Between the two communicating nodes only
Responsibility for the transmitted unit	Full responsibility is always retained by the sender node	Responsibility only towards the node handling the packet at any given time
Network operation	Centralized	Distributed
Channel data rates	Generally variable slow speed	Generally all the speeds are high and identical
Channel protocols	Generally variable depending on equipment	Fixed and identical
Channel traffic balance	Difficult because of variable message size	Maximum traffic balance achieved
Probability of message loss	Small; detected by numbering scheme	Small; detected when reassembling message
Retrieval of lost messages	Easy located in the originating node	Difficult since it is not retained in any of the network nodes
Network efficiency	Most inefficient with long delays	Most efficient with minimum delays
Application	Communication between slow speed terminals	Communication between fast speed computers

the processing functions and the system services provided by both schemes are also different. Message switching is intended for slow speed traffic, generally originating from manually operated terminals, while packet switching is intended for high speed traffic, generally originating from computers.

Although the network configurations of both schemes seem to be alike, this is not the case in practice. The main backbone of the packet switching network consists of high speed transmission channels which connect the various switching nodes, whereas the local channels connected to each node are onlyy one or two, each leading to a host computer. Terminals are generally not connected directly to the node but only via the host computer.

The differences between these two schemes is summed up in Table 8.4.

It was stated earlier in this chapter that from a user's point of view the differences between the two switching techniques are indistinguishable. This statement is not quite accurate, since packet switching provides a most efficient means of message communication; a statement which cannot be applied to message switching.

8.9 PACKET SWITCHING NODE OPERATION

Packet switching network is intended to provide communication facilities between various computer installations spread over wide geographical areas so that they may share their resources. The computers are not connected directly to the network but usually through a network node, as seen in Fig. 8.25. Although these network nodes generally provide a service to only one host computer, there are some nodes designed to serve two or three host computers. At all events, the number of host computers associated with any node is limited and no terminals are connected to the nodes but only via the host computer.

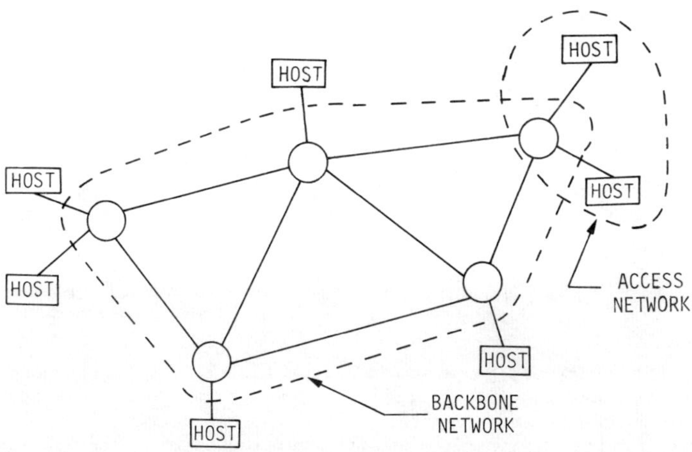

Fig. 8.25 General structure of a packet switching network

The network nodes are connected to the other nodes with long high speed communication channels and to their hosts with short channels. This allows the network configuration to be split into two sections. The nodes with their high speed channels are referred to as the backbone of the communication network and the nodes with the host computers as well as the terminals associated with them are referred to as the access network. The backbone network is usually controlled by a single governing authority that defines the data communication protocols within the network and controls the user's access to the network. The local network is completely isolated from the network backbone, and may operate as a separate independent communication system. Special standard protocols must define the interface between the access and backbone networks so as to prevent any conflicting operations. Since some packet switching networks provide world wide communication these standards are generally internationally accepted, with the one applying here being X.25 as specified by CCITT. This, however, does not apply to the standards used in either the local or the backbone network (only in their interface point) and each company selects whichever standard suits them. Furthermore, there are some packet switching networks which use different interface standards than the X.25.

As was shown in Fig. 8.25, the network nodes are at the interface point between the access and backbone networks and can be regarded as serving both networks. They are designed as independent centres employing communication processors and are referred to as information message processors (IMP). These IMP's have therefore a dual function: serving the local host computer and providing network packet communication. This dual alliance of the IMP complicates the design of the node, since it must combine in it two separate, possibly conflicting, tasks. These functions are defined below:

(a) The IMP's first function to serve its local host computer by acting as a front end processor (FEP), as shown in Fig. 8.26. In this rôle the FEP must release the host computer from all the long distance communication tasks so as to allow it to operate as a stand-alone system. The host computer may be small or large, operating with batch processing or time-sharing techniques and is usually in operation well before the FEP was added to provide it with communication network services. The host computer, in this respect, regards the FEP as one of its peripheral units which has the prime task of serving it alone, with no concern of any other tasks the FEP may perform in its 'spare' time.

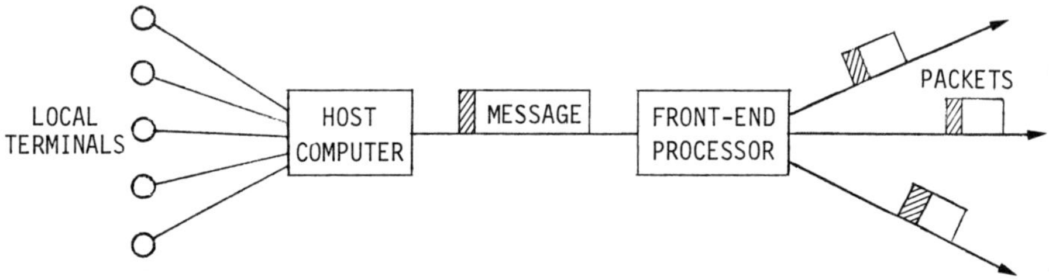

Fig. 8.26 The node's communication processor as the host front-end processor

The FEP receives the messages and instructions from its host computer and its first task then is to split the messages up to fixed sized packets. The FEP then adds to each packet a new corresponding heading, including information of its destination and its position in the original message. A task of the FEP is also to receive from the network packets which are addressed to its host computer. In this case, the FEP stores the packets until all the packets of the same message have arrived: it organizes the packets back in their original sequence and only then is the message transmitted to the host. It is possible however, that many packets may arrive for the host computer, with each belonging to a different message. When this occurs, it is the task of the FEP to store the packets and ensure that they are all correctly assembled.

(b) The IMP's second function is to serve as an independent switching node for routing the transit packets between the various communication channels, as seen in Fig. 8.27. The nodes receive packets from the neighbouring nodes, check them for any errors and then acknowledge their positive or negative acceptance. From the moment the node returns a positive acknowledgement it becomes fully responsible for the packet and has therefore to store it in its memory. The packets are correspondingly placed in one of the outgoing queues according to their destination and the present traffic conditions. When the outgoing channels become free, i.e., when the packets have reached the head of the queue, they are transmitted to the next node along the route. After a positive acknowledgement has been received from the next node, the packets may be erased from the memory in preparation for further packets.

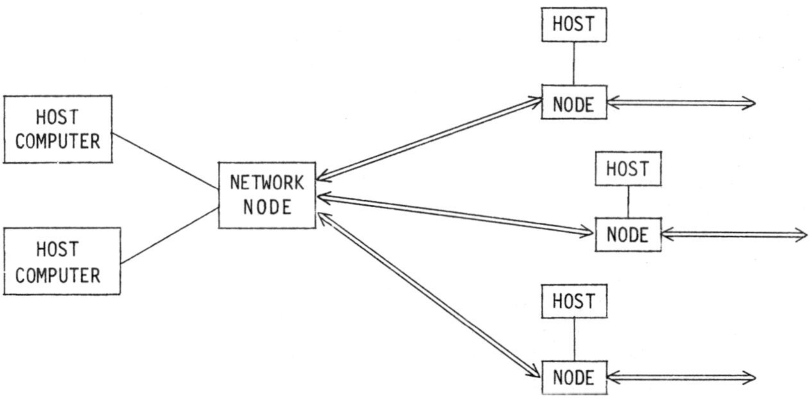

Fig. 8.27 The node's communication processor as a network relay switch

The network node, in its rôle as a network relay switch, must regard all the transit traffic as having priority over the host traffic. The efficiency of the whole traffic is based on fast transmission of the packets from any point in the network to any other point, and so any delay in one of the nodes would affect the whole network operation. Therefore, if the node is busy (with the task of assembling or disassembling the messages into packets as a service to its host) the transit traffic would be affected. Consequently, the tasks of the host computer must be given secondary priority services. No conflict of interests are allowed, since the backbone network operation is the essential link for successful packet switching. This is an interesting situation, as it is the host computer installation which pays for the setting up of the node.

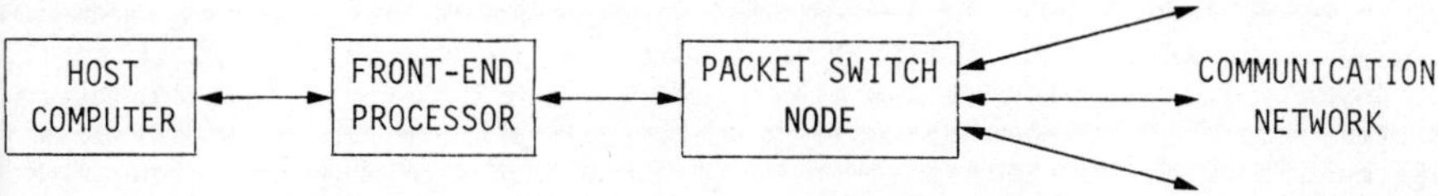

Fig. 8.28 A packet switching node with two communication processors

In large packet switching systems, the dual function of the node operations are segregated into two separate communication processors, as seen in Fig. 8.28. The packet switching node only handles the packets and is completely isolated from all the host operations. The packet switching node regards the hosts' front-end processor in the same way as it regards any of its neighbouring nodes. Any packets to its local host are transferred to its FEP using the same transmission techniques as if to other nodes. It is then up to the FEP, and not the switching node, to assemble and disassemble the messages into packets.

As mentioned above, the main task of the switching node is concerned with the network traffic. All the transit packets are stored in the memory while in queues waiting for their heading to be analysed and processed. After the packet destinations have been checked, the decision for the next step is based on the traffic tables and only then can the packets be placed in the outgoing queues. Since the destination checking can only be performed in serial and not in parallel, the packets must pass two stages of queuing, as seen in Fig. 8.29. The node must be designed so that it is never too busy to accept any further packets; that is, it must have adequate memory capacity. It is the nodes' processing ability of handling the queues which defines the packets' delay within the network.

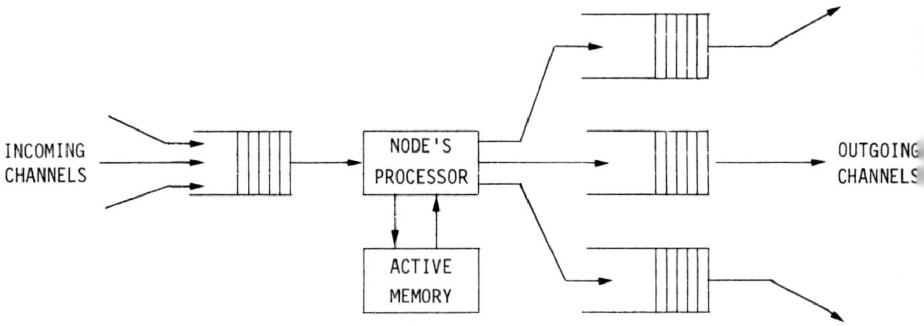

Fig. 8.29 Schematic diagram of the packet queues in the node

Determining the shortest route to the final destination is not made according to the shortest distance or to the least number of hops, but according to the traffic condition in each node and in the channels leading out of them. The distribution of the traffic in the network is random, since the number of packets originating from any source are undefined. Therefore, there is always the probability that some channels may be over loaded while others may be idle. It is the function of the network control to smooth out the traffic by distributing the load on all the channels. Channel capacity is fixed by the data rate that can be transmitted over it, while the over or under loading determines the queue length in the nodes. If no means are introduced to reduce unduly long queues and thereby smooth the traffic loads, some packets may be over delayed in the network.

Packet switching is based on a fully distributed network configuration, with no single node controlling the traffic in the network. That is, there is no single authority governing the traffic and smoothing the load, but the authority is distributed out between all the nodes. Each node, must therefore have some indication of the traffic conditions in the whole network so that it can make its decision on the best route for the packets to be transmitted. This is performed by special tables within each node, which are continuously updated according to the current traffic conditions in the whole network. Each node must have an indication of over-load conditions in any one of the channels or the malfunction of any node in the network. The traffic tables in the nodes are adjusted only by information supplied to each node by its closest adjacent neighbouring nodes. That is, there is no procedure for each node to supply information to all the other nodes in the network. Nevertheless, exchanging of information between adjacent nodes provides each node with the best possible updated network conditions. To demonstrate the process of updating these tables, an example of a network is presented in Fig. 8.30 where each of the channels and nodes receive a unique number.

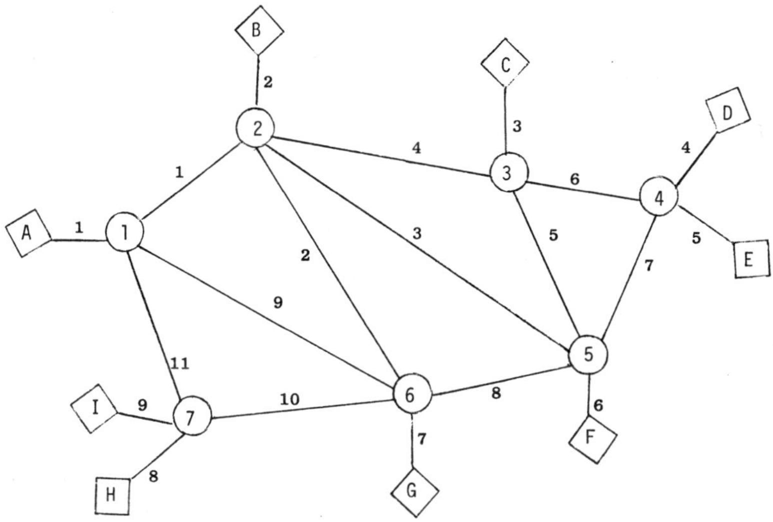

Fig. 8.30 An example of a packet switching network configuration

Each node contains three basic tables: (a) network delay table as seen from each node, (b) minimum delay table from each node to all the other nodes and (c) the node's queue table. Decisions on the shortest possible route are based on the minimum delay table, while the other two tables are employed only as means of updating the minimum table.

Each packet contains in its heading the address of its final destination. This address is formed in two parts, viz. the final node number and the host's channel number. For example, a packet to host *E* will always have the address 4, 5. The same address is used by any node which only defines the final destination but does not give any indication of a special route. It is for all the transit nodes en route to select the route they consider best, even if it is not the complete route that is suggested but only the next hop which appears to be the shortest route in accordance with the tables within the node.

Figure 8.31 brings an example of the tables within node 2, and this can serve for all the other nodes which have similar tables of their own. The main table, referred to as 'network delay table', indicates the delays of the packets for reaching all the other nodes in the whole network. A separate column is given to each of the nodes in the network and a separate row is provided for each outgoing channel from the node. It should be pointed out that the table does not give any indication of all the possible routes but only information of the channels leading out of the nodes. The numbers at the intersection of the columns and the rows of the table show the relative delays on a particular channel for reaching any node destination in the network. The route is then selected according to the minimum relative delay. For example, as seen in the table, the least delay from node 2 to node 7 is via channel 3, although this is not the shortest path.

The nodes send out periodically to all their immediate neighbouring nodes a summary of the main table in the form of the node's minimum delay table, giving details of delays but not of channels or routes. At the same time they receive from all their neighbours a similar table of summaries. With these summaries from all the nearby tables, a node adjusts its own table and keeps it continuously updated. When a node updates its table it must also consider its own queue of packets waiting to be transmitted. Take for example

CHANNEL \ NODE	1	2	3	4	5	6	7
1	7	0	6	4	11	21	20
2	6	0	10	8	9	22	8
3	2	0	14	20	7	20	6
4	5	0	20	22	8	15	30

NODE 2 NETWORK DELAY TABLE

2	0	6	4	7	15	6

NODE 2 MINIMUM DELAY TABLE

CHANNEL \ NODE	1	2	3	4	5	6	7
1	0	5	10	7	9	7	16
9	0	9	12	16	8	20	26
11	0	8	14	12	14	22	15

0	5	10	7	8	7	15

NODE 1 DELAY TABLE → CHANNEL 1

CHANNEL \ NODE		2	3	4	5	6	7
4	8	5	0	3	5	20	8
5	10	8	0	6	8	7	15
6	15	10	0	9	9	6	9

8	5	0	3	5	6	8

NODE 3 DELAY TABLE → CHANNEL 4

CHANNEL \ NODE	1	2	3	4	5	6	7
2	10	26	19	14	4	0	2
8	8	12	15	12	8	0	16
9	20	6	14	18	6	0	22
10	23	8	9	21	10	0	7

8	6	9	12	4	0	7

NODE 6 DELAY TABLE → CHANNEL 2

CHANNEL \ NODE	1	2	3	4	5	6	7
3	8	10	3	8	0	10	20
5	5	12	8	10	0	6	22
7	10	5	5	6	0	8	10
8	12	8	7	7	0	12	8

5	5	3	6	0	6	8

NODE 5 DELAY TABLE → CHANNEL 3

1	4
2	2
3	3
4	6

NODE 2 QUEUE TABLE →

CHANNEL \ NODE	1	2	3	4	5	6	7
1	4	0	14	11	12	11	19
2	10	0	11	14	6	2	9
3	8	0	6	9	3	9	11
4	14	0	6	9	11	12	14

NODE 2 UPDATED NETWORK DELAY TABLE

4	0	6	9	3	2	9

NODE 2 UPDATED MINIMUM DELAY TABLE

Fig. 8.31 The traffic tables in node 2 and the means of updating it

the delay from node 2 to node 7 where the minimum delay was over channel 3 which had a relative delay of 6. Node 2 receives from node 5 the information that the minimum queue length in that node is presently 8. Node 2 knows from its own queue tables that 3 packets are already waiting on the outgoing queue of channel 3 for dispatch. Node 2 therefore updates the delay to node 7 over channel 3 to 11 (= 3 + 8). The whole table is now updated in a similar manner, establishing the minimum delay to node 7 to be over channel 2. That is, any further packets reaching node 2 and addressed for node 7 will be placed in the queue of channel 2.

When a packet leaves a node, its destination is based on the route with the least weight at the time, in a particular channel. It could happen that by the time the packet reaches the next node, its table has been updated and the quickest route towards its destination is via the node it has just come from.

8.10 FLOW CONTROL IN PACKET-SWITCHING NETWORKS

The packet-switching concept provides a most efficient means of data communication, making it superior to both circuit-switching and message switching for computer type communication. Nevertheless, there are certain features in both circuit-switching and message-switching which have an advantage over packet-switching. Circuit-switching, after setting up the call, provides a direct physical link between sender and destination. Message-switching requires less transmission overhead than packet-switching and the whole message is transmitted in its original sequence over a single route. Packet-switching has the advantage of network and node efficiency by simply limiting the processing and transmission to the handling only of fixed sized packets. In the last few years considerable research has been carried out to find means of further increasing the efficiency of packet-switching by exploiting ideas deriving from both circuit and packet-switching.

The main disadvantage of packet-switching is the relatively high overhead of the packets which reduces the network utilization. This drawback comes from the fact that each packet must contain in its heading information of its destination as well as its sequential position in the message. The critical operation of packet-switching is caused, however, by the danger of network saturation. A packet-switching node must always be ready to accept all the packets sent to it and the nodes' memory is designed with enough memory space to retain them all. Nevertheless, as the network traffic increases, a situation may arise in which the nodes' memory capacity cannot be cleared in advance so as to prepare room for new packets. When such a situation arises, the network becomes congested with an overflow of packets beyond the capability of the nodes to handle or deliver. Increasing the memory capacity would only solve part of the trouble while increasing the node cost. Such a solution, however, will not reduce the throughput in the node nor reduce the required processing time. This means that as the network traffic increases, long delays of the packet transmission in the network will be caused.

To increase the network efficiency, two data flow interfaces between the host computer and the communication network have been proposed for packet-switching. The first, known as 'datagram interface', provides a service for the transmission of 'message type' packets in networks with relatively small traffic flow. The other, known as 'virtual-circuit' (VC) interface, provides a service for the transmission of many packets over the same path. Both types of interfaces are specified by international standards but used in varying details in different networks.

A datagram interface allows each user to enter each packet into the network independently of any other packet which has been or will be transmitted. In other words, the datagram is basically a self-contained packet which is accepted and treated by the network independently of all the other packets (similarly as to packets described in the previous section). Datagram has the advantage both of reliability and efficiency. It is reliable because it provides alternative routing when there is a channel failure, even after the start of transmission, i.e., even after the packet has entered the network. It is efficient since it transmits the packets freely through the network, allowing them to be routed independently over alternative routes. Datagram interface protocols distribute the traffic load over a number of channels so as to share out the data flow in the network.

Datagram uses a uniform addressing scheme throughout the network, permitting packets with the same address to be routed over alternative paths. Furthermore, no set-up time is required for the transmission of the datagram as is required with virtual circuits interface. This gives the datagram the advantage of providing the fastest message transport technique throughout the network. Nevertheless, datagram services do suffer delays in the destination node before they can be transferred to the host computer. Although the packets are transmitted in a given sequence according to their position in the message, sequential delivery cannot be guaranteed with datagram interface. As seen in Fig. 8.32, the packets must be held in the destination node until all the packets associated with one message have arrived and only then can they be transferred to the destination node. As each packet arrives at a different time, the waiting time may build up to long periods.

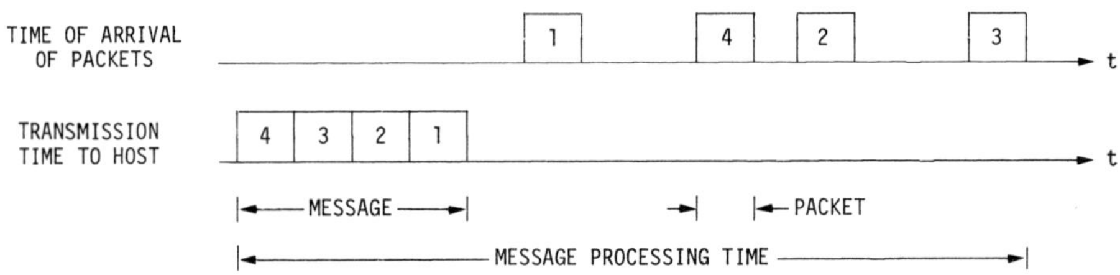

Fig. 8.32 Waiting period in the destination node when employing datagram interface

As stated before, the datagram also suffers from the requirement of a high overhead in each packet which reduces the factor of network utilization in comparison with message-switching. A further reduction of the network utilization is caused by the requirements of a separate acknowledgement for each packet when transferring the datagram between any two communicating nodes. Consequently, the datagram interface scheme, although efficient, could also be unduly expensive.

Virtual-circuit interface attempts to solve most of the datagram deficiencies by reducing both the packet and the network overheads and guaranteeing sequential delivery of the packets associated with the same message. Nevertheless, this is achieved by sacrificing the individual routing of the packets over alternative paths. Virtual-circuits interface protocols are ideal but only for very large packet-switching networks with high traffic flow.

Virtual-circuit interface takes advantage of both message-switching and circuit-switching techniques. It transmits the set of packets associated with one message as a single continuous operation: that is, one packet after the other with no gap between them. In other words, the transmission of packets in virtual-circuit is similar to that described for message-switching and accordingly *VC* requires less individual packet overhead. It also requires, as with circuit switching, the setting up of a transmission path throughout the network to allow fast transportation of the packets once the path has been set up. Despite any similarities of the virtual circuit to both message-switching and circuit-switching, *VC* must still be regarded as pure packet-switching. That is, all transmissions are based on fixed sized packets each requiring a standard sized time slot and held in the nodes in fixed sized memory buffers.

Virtual-circuit interface provides a logical path between the sender and destination set up for each message transmission. This requires the exchange of control information between the two communicating nodes for the purpose of having control of the data flow. As seen in Fig. 8.33, the connection between the two communicating DTE's appears as if they employ assigned channels; nevertheless, this is only over a virtual channel. That is, although virtual-circuit is like circuit switching, no physical path is set up between sender and destination. Furthermore, the same channels which transport the virtual-circuit packets may be shared with packets from other sources using either datagram or virtual-circuit interfaces.

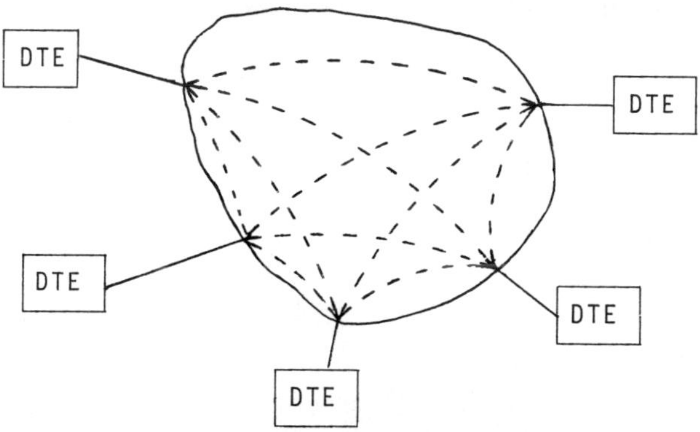

Fig. 8.33 Packet-switching network using virtual-circuit interface

A virtual-circuit must be set up before any data packet may be transmitted over it, as seen in Fig. 8.34. The local node, as a result of the host's demands, sends out a 'request' datagram to the destination node for permission for the transmission of *m* packets. This 'request' datagram is transferred through the network using datagram interface and having its path selected according to traffic load conditions. The path selected is the same path that the data packets will follow. The destination nodes, on receipt of this 'request' datagram, must prepare buffer memory space for *m* packets, and as soon as this space is made available, the destination node returns an 'accept' datagram. When the 'accept' datagram is received in the source node, a virtual circuit has been established between the two nodes and the packets can be transmitted over it with no further delays. A virtual-circuit, once set up, usually provides a duplex channel with data flowing in both directions. That is, it provides a bidirectional association between the two communicating nodes.

Once a virtual-circuit has been set up for one message, further sets of packets may be transferred between the two nodes using the same virtual circuit path, but this would still require a request packet stating the number of packets in the second message. For this purpose, the virtual-circuit must be cleared when the communication process is completed. This is performed by the destination node sending a clear-request packet in the reverse direction. This clear-request packet must be acknowledged by the source node and only then can the destination buffer be cleared to accept other packets from other

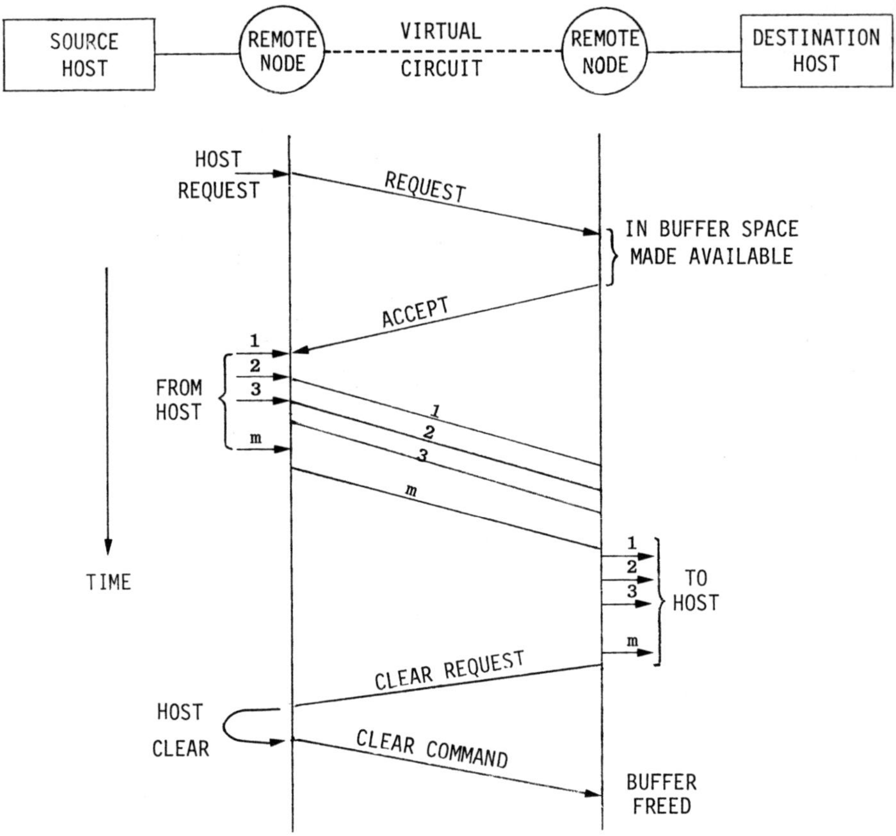

Fig. 8.34 Illustration of setting up and data transfer in packet vertical-circuit interface

services. It should be added that there are other forms of virtual-circuit interface communication protocols. These are defined in level 3 of X.25 CCITT standard.

Only the communicating DTE's can visualize the transmission path as a virtual-circuit. The network nodes still use store and forward techniques and so the same nodes in the network may have to deal simultaneously with a number of virtual-circuits. The nodes en route of the virtual-circuit provide a direct logical path connection without offering alternative paths for individual packets. Nevertheless, the nodes transmit each of the packets in accordance with traffic conditions. That is, the same channel may transmit packets from a number of sources, all interleaved together, as seen in Fig. 8.35. What remains is the sequencing of the packets in their right order although with possible delays between the packets associated with the same virtual-circuit. Virtual-circuits provide packet transmission with services similar to circuit-switching but with much higher network efficiency.

Packet switching has become universally accepted as the main means of computer communication, and many countries have set up public packet switching networks. Although in principle they are all based on the same technical techniques, using the same accepted international standards, they are interpreted in varying details in the different networks. This creates a difficult

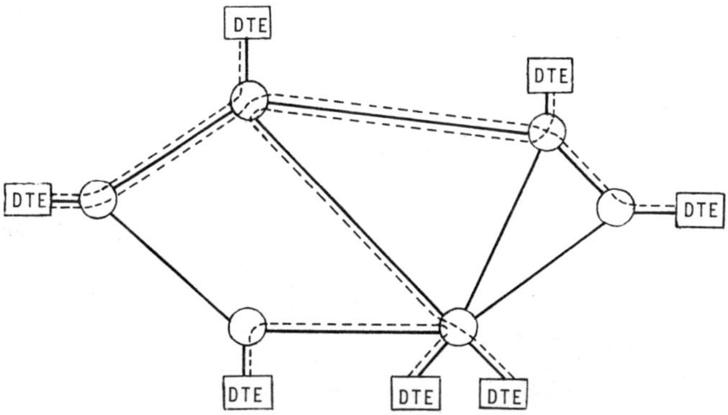

Fig. 8.35 Interleaving a number of virtual-circuits in the same network

when there is sometimes a need for interconnection between these different networks. To provide this requirement facility, a new network node has been introduced, referred to as a 'gateway'. This gateway (as seen in Fig. 8.36) has no host, its main task being to provide the means by which two or more separate networks are connected, to allow the transfer of packets from one network to another.

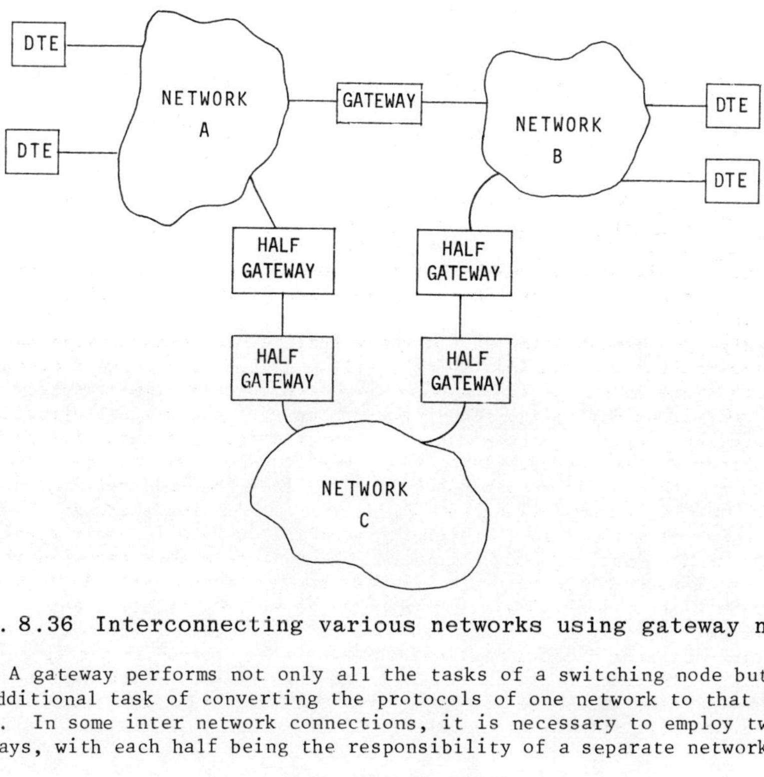

Fig. 8.36 Interconnecting various networks using gateway nodes

A gateway performs not only all the tasks of a switching node but also the additional task of converting the protocols of one network to that of the other. In some inter network connections, it is necessary to employ two half gateways, with each half being the responsibility of a separate network.

8.11 REFERENCES

(a) COMMUNICATION SWITCHING

1. Chandersekaran, C.S. and Shanker, K.S. *Towards Formally Specifying Communication Switches*, Trends and Applications, 1976: Computer Networks, IEEE 76 CH 1143-7C, pp.104-112.
2. Schwartz, M. *Computer-Communication Network Design and Analysis*, Prentice-Hall, 1977.

(b) CIRCUIT SWITCHING

3. Atkinson, J. *Telephony*, Sir Isaac Pitman & Son Ltd., 1952, vol.11, 860 pages.
4. Pollard, J.R. *Technological Progress in Telecommunication Switching*, Electronic and Power, August 1970, pp.305-308.
5. Leaky, D.M. *The Organisation of a Public Telephone System*, Electronic and Power, July 1971, pp.265-268.
6. Leaky, D.M. *Switching in Space and Time*, Proc. IEE, vol.124, no.1, January 1977, pp.17-24.
7. McDonald, J.C. *Techniques for Digital Switching*, IEEE Comm. Society Magazine, July 1978, pp.11-19.
8. Joel, A.E. *Circuit Switching – Unique Architecture and Applications*, Computer, vol.12, no.6, June 1979, pp.10-22.
9. Moralee, D. *British Telecommunication and the Information Age*, Electronic and Power, June 1979, pp.431-439.
10. Joel, A.E. *Digital Switching – How it has Developed*, IEEE Trans. vol. COM-27, no.7, July 1979, pp.948-959.
11. Skaperda, N.J. *Some Architectural Alternatives in the Design of a Digital Switch*, IEEE Trans., vol. COM-27, no.7, July 1979, pp.961-972.
12. Bruce, R.A., Giloth, P.K. and Siegel, E.H. *No. 4 ESS-Evolution of a Digital Switching System*, IEEE Trans., vol. COM-27, no.7, July 1979, pp.1001-1011.
13. Pitroda, S.G. *Telephones Go Digital*, IEEE Spectrum, vol.16, no.10, October 1979, pp.51-60.
14. Hall, C.D. *Digital Switching – The Enabler of Future Services*, IEEE Comm. Magazine, vol.18, no.3, May 1980, pp.24-29.
15. Joel, A.M. *What is Telecommunication Circuit Switching?* Proc. IEEE, vol.65, no.9, September 1977, pp.1237-1253.
16. Martin, J. *Teleprocessing Network Organisation*, Prentice Hall Inc., 1970, pp.233-245.
17. Martin, John, *Why System X?* Electronic and Power, vol.26, no.7, July 1980, pp.529-531.
18. Newman, J.R.W. *System X Architecture*, Electronic and Power, vol.26, no.7, July 1980, pp.536-540.

(c) MESSAGE SWITCHING

19. Hamsher, D.H., Editor in Chief, *Communication System Engineering Handbook* Chapter 7 by P. Schneider, *Message Switching Engineering* McGraw Hill Book Company, 1967, pp.7-33 to 7-42.
20. Richards, R.K. *Electronic Digital Systems*, John Wiley & Son Inc., 1966, pp.481-518.
21. Richards, R.K. *Digital Design*, John Wiley & Son Inc., 1971, pp.563-571.
22. Newport, C.B. and Ryziak, J. *Communication Processors*, Proc. IEEE, vol.60, no.11, November 1972, pp.1321-1332.
23. Doll, D.R. *Multiplexing and Concentration*, Proc. IEEE, vol.6, no.11, November 1972, pp.1313-1320.

24. Kahn, R.E. *Resources-Sharing Computer Communication Networks*, Proc. IEEE, vol.60, no.11, November 1972, pp.1397-1407.
25. Dell, F.R.E. *Features of a Proposed Synchronous Data Network*, IEEE Trans. on Communication, vol. COM-20, no.3, June 1972, pp.499-503.
26. Mattison, R.G. *Computer Processing for Data Communication*, Computer, vol.6, no.2, February 1973, pp.15-19.
27. Rudin, H., *Data Transmission: A Direction for Future Development*, IEEE Spectrum, February 1970, pp.79-85.
28. Decker, H. *Communication Processing for Large Data Networks*, Data Processing Magazine, November 1970,
29. Doll, D. *Planning Effective Data Communication System*, Data Processing Magazine, November 1970
30. Ball, C.J. *Communication and the Minicomputer*, Computer, vol.4, no.5, September/October 1971, pp.13-21.

(d) PACKET SWITCHING

31 Kahn, R.E. *Resource-Sharing Computer Communication Networks*, Proc. IEEE, vol.60, no.11, November 1972, pp.1397-1408.
32. Chou, W. *Computer Communication Network – The Parts That Make Up The Whole*, AFIPS, vol.44, 1975, pp.119-128.
33. Lam, S.S. *Store-and-Forward Buffer Requirements in a Packet Switching Network*, IEEE Trans. vol. COM-24, no.4, April 1976, pp.394-403.
34. Frisch, I.T. and Frank, H. *Computer Communication – How We Got Where We Got*, AFIPS, vol.44, 1975, pp.109-117.
35. Solomon, R.J. *Packet Networks*, Mini Micro Systems, March 1978, pp.48-53.
36. Roberts, L.G. *Data By The Packet*, IEEE Spectrum, February 1974, pp.46-51.
37. Kirstein, P.T. *Data Communication By Packet Switching*, Electronics and Power, 15 November, 1975, pp.503-508.
38. *Packet Data Communication*, vol.1 and 2, published by Future Systems Inc. with Digital Communication Corporation, 1977.
39. Ornstein, S.M. and Walden, D.C. *The Evolution of a High Performance Modular Packet Switch*, ICC, June 1975.
40. Frank, H. and Chou, W. *Topological Optimization of Computer Network*, Proc. IEEE, vol.60, no.11, November 1972, pp.1385-1397.
41. Roberts, L.G. *The Evolution of Packet Switching*, Proc. IEEE, vol.46, no.11, November 1978, pp.1307-1313.
42. Kleinrock, L. *Principles and Lessons in Packet Communication*, Proc. IEEE, vol.66, no.11, November 1978, pp.1320-1329.
43. Lickider, J.C.R. and Vezza, A. *Application of Information Networks*, Proc. IEEE, vol.66, no.11, November 1978, pp.1330-1346.
44. Sullivan, N. *TYMNET – Maintenance Considerations in a Very Large Network*, Proc. 5th Data Communication Symp., IEEE, Cat. No. 77 CH 1260-90, 1977, pp.3-1 to 3-3.
45. *ARPNET INFORMATION BROCHURE*, AD A052672, Defence Communication Agency, 1978, pp.1-27.
46. Schwartz, M. *TYMNET – A Tutorial Survey of a Computer Communication Network*, IEEE Communication Magazine, 1977, pp.20-24.
47. Harcharik, J.R. *TYMNET, Present and Future*, EASCON 1975, pp.124-A to 124-G.
48. Karp, D. and Seroussi, S., *A Communication Interface with Computer Network* IEEE Trans., vol. COM-20, no.3, June 1972, pp.550-556.
49. McQuillan, J.M., Falk, G. and Richer, I., *A Review of the Development and Performance of the ARPANET Routing Alogorithm*, IEEE Trans. vol. COM-26, no.12, December 1978, pp.1802-1810.
50. Cotton, I.W. *Computer Networks: Capabilities and Limitation*, Reprint in Packet Data Communication, vol.11, Future Systems Inc., May 1977.

51. Beere, M.P. and Sullivan, N.C. *TYMNET – A Serendipitous Evolution*, IEEE Trans. vol. COM-20, no.3, June 1972, pp.511-515.
52. Becher, W.D. and Aupperle, E.M. *The Communication Computer Hardware of the MERRIT Computer Network*, IEEE Trans., vol. COM-20, no.3, June 1972, pp.516-526.
53. Aupperle, E.M. *MERRIT Network Re-Examined*, Reprinted in Computer Network, IEEE Press, 1976, Edited Blacc and Cotton, pp.281-286.
54. Wood, D.C. *A survey of the Capabilities of 8 Packet Switching Networks*, Proc. 1975, Symp., Computer Network: Trends and Applications, IEEE 75 CH 0973-8C, pp.1-7.
55. Pouzin, L. *The Cyclade Network – Present State and Development Trends*, Proc. 1975 Symp. Computer Network: Trends and Applications, IEEE 75 CH 0973-8C, pp.8-13.
56. Schwartz, M., Boorstyne, R.R. and Pickholz, R.L. *Terminal-Oriented Computer-Communication Network*, Proc. IEEE, vol.60, November 1972, pp.1408-1423.
57. Cerf. V.G. and Kirstein, P.T. *Issues in Packet-Network Interconnection*, Proc. IEEE, vol.66, no.11, November 1978, pp.1386-1408.
58. Pandya, R.N. *Delay Analysis for DATAPAC – A Packet Switched Network with Two Priority Classes*, Proc. 5th Data Communication Symposium 1977, pp.3-14 to 3.21.
59. *Packet Switching and the P.O.*, Data Systems, November 1978, pp.22-23.
60. Pearson, D.F. and Wilkin, D. *Some Design Aspects of a Public Packet Switched Network*, Communication International, November and December 1974.
61. Barber, D.L.A. *EIN: A Focus for the Future*, EUROCON 77, Conf. Proc. on Communication, IEEE Cat. No. 77 CH 1257-5 Reg. 8, pt.2, pp.3.3.1.1 to 3.3.1.14.
62. Keon, H.P.L. *SAPONET: A Public Circuit and Packet Switching Network Facility Being Implemented by South Africa Post Office*, Information Technology, North Holland Publishing Comapny, 1978, pp.419-425.
63. Jestin, R.W. *DATAPAC 1000 – A Network Interface Machine Supporting Interactive Terminals*, ICC 1976, pt.2, pp.30-2 to 30-5.
64. Van den Toorn, L.P. and Jordaan, M.J. *COMSYS, A Privste Packet Switching Network*, Phillips Telecommunication Review, vol.36, no.3, August 1978, pp.149-164.
65. Stathopoulos, A. and Caley, H.F. *The AUTODIN 11 Network*, EASCON-77, pp.8-1A to 8.1I.
66. Shah, A.N. and Kotob, S. *Analysis of Flow Control in AUTODID 11*, NIC-77, pt.2, pp.37: 3-1 to 7.
67. Kleinrock, L. and Opderbeck, H., *Throughput in the ARPANET – Protocols and Measurements*, IEEE Trans., vol. COM-25, no.1, January 1977, pp.95-104.

(e) COMMUNICATION STANDARDS AND PROTOCOLS

68. McQuillan, J.M. and Cerf. V.G. *Tutorial: A Practical View of Computer Communication Protocols*, IEEE Computer Society, IEEE Catalog No. EHO 137-0, 1978, 258 pages.
69. McQuillan, J.M. *Enhanced Message Addressing Capabilities for Computer Network*, Proc. IEEE, vol.66, no.11, November 1978, pp.1517-1527.
70. Shah, A. *On Flow Control Mechanism in Packet-Switching Network – An Analytical Approach*, Computer Network, 1979, pp.37-41.
71. Rybcznski, A.M. and Weir, D.F. *Datapac X.25 Service Characteristics*, Proc. 5th Data Communication Symposium, 1977, pp.4-50 to 4.57.
72. Hovey, R.B. *Packet-Switching Networks Agree on Standard Interface*, Data Communication, May/June 1976, pp.75-85.
73. Folts, H.C. and Karp, H.R. *Data Communication Standards*, McGraw-Hill Publications Co., 1978,

74. Hummel, E. *State of CCITT Standardization on Public Data Networks*, Telecommunication Journal (ITU), vol.46, no.1, January 1979, pp.33-39.
75. Kelley, P.T.F. *Public Packet Switched Data Network, International Plans and Standards*, Proc. IEEE, vol.66, no.11, October 1978, pp.1539-1549.
76. Folts, H.C. *Status Report on New Standards for DTE/DCE Interface Protocols*, Computer Magazine, vol.12, no.9, September 1979, pp.12-19.
77. Walden, D.C. and McKenzie, A.A. *The Evolution of Host-to-Host Protocol Technology*, Computer Magazine, vol.12, no.9, September 1979, pp.29-38.
78. Day, J.D. *Resources Sharing Protocols*, Computer Magazine, vol.12, no.9, September 1979, pp.47-56.
79. Rybczynski, A.M., Wessler, R., Despres, R. and Wedlake, J. *A New Communication Protocol for Accessing Data Network – The International Packet Mode Interface*, NCC 1976, vol.45, pp.68-73.
80. Cotton, I.W, and Folts, H.C. *International Standards for Data Communication: A Status Report*, Proc. 5th Communication Symposium 1977, pp.4.26 to 4.36.
81. Sloman, M.S.C. *X.25 Explained*, Computer Communication, IPC Press, vol.1, no.6, December 1978, pp.310-327.
82. Cerf, V.G. and Kirstein, P.T. *Issues in Packet-Network Interconnection*, Proc. IEEE, vol.66, no.11, November 1978, pp.1386-1408.
83. Schwartz, M. and Stern, T.E. *Routing Techniques Used in Computer Communication Networks*, IEEE Trans., vol. COM-28, no.4, April 1980, pp.539-552.
84. Folts, H.C. *X.25 Transaction-Oriented Features – Datagram and Fast Select*, IEEE Trans., vol. COM-28, no.4, April 1980, pp.496-500.

CHAPTER 9

Multiple-Access Procedures

9.1 INTRODUCTION

The local area communication techniques presented in Chapter 7 used specific hardware equipment for the terminal traffic in a network controlled by a centre. In Chapter 8, the network was extended to cover wide areas using equipment to enable centre to centre communication, with the network control being either centralized or distributed. In this Chapter the local area network is anlysed again, to describe how the same feature of communication between centres could be provided for direct terminal to terminal communication in a local network by using software techniques. Such techniques are essential in C^2 systems which employ radio networks for control of mobile communication. It will also be shown that these software techniques could be employed for the control of the traffic in multidrop lines as introduced in Chapter 4 and discussed in Chapter 7.

With the growth of local area distributed processing systems, more efficient means of exploiting the common communication channels are being pursued. Initially, all the data flowing between the terminals and the computer were sent via point-to-point dedicated lines. For these applications, no computer power is required, as the data is transmitted as soon as it is ready at the terminals. The terminals used for such systems are rather simple and are thus described as 'dumb'. It is possible to cluster a number of these dumb terminals on a single communication channel by means of a multiplexor, although even then the point-to-point nature of the communication is retained. This is possible because the channel facility is divided between the terminals, with each terminal being allocated a separate dedicated part of the channel.

With the development of time sharing systems the function of the terminals changed from off-line input/output peripheral devices to man-machine real-time interactive devices. Here each computer centre is associated with a large number of terminals, which could communicate with it whenever the terminals have data to transmit. The traffic to and from each of the terminals is low and comes at random instances. In such systems it is wasteful to employ the point-to-point concept of allocating separate dedicated lines or channels to each terminal. A number of terminals can be attached to the same common communication channel by means of multi-drop and/or means of concentrators. In this case, communication procedures are introduced to enable the computer to control the data flow to and from each terminal. The computer assigns operation time to each terminal, either in turn or upon demand, so as to allow them to despatch their messages in an orderly manner. The terminals attached to these common communication channels can no longer be regarded as dumb, as some logic facilities must be inserted into the terminals whereby they can recognize their call number (address code) and act when called. In these systems the central computer is still the sole coordinator of the traffic along the common channel, with no initiating power given to the terminals.

With the advancement of distributed resources, more and more processing activity has been delegated to the terminals; this is made possible by incorporating within the terminals a minicomputer. These advanced terminals are regarded as intelligent since they offer the operator maximum local processing facilities, enabling him to perform on site many operations (such as editing and validity tests) without the need to disturb the central computer. The trend today is to extend the distributed intelligence to a more sophisticated form where the terminals are no longer minicomputers but could be small or large processing installations. In these advanced systems, each so-called terminal could operate alone, although it could operate in coordination with other so-called terminals so as to share their resources. In such systems, both the interactive terminals and/or processors at one installation may have access to and use the data that exist at other installations. This is usually achieved with the aid of switching computers, which become nodes in the communication network, while the processing computers and terminals are the host of these nodes. The nodes in this system could be circuit, message or packet switching centres, as explained in Chapter 8. Such a communication system can only be effective if the traffic flow between the nodes is smooth and at a high rate.

Multiple access provides another alternative to computer-to-computer (and terminal-to-terminal) communication to fill the operation gap for systems where the traffic between the computers (or so-called terminals) is low. In these systems each user may use his terminal either as a stand alone small installation computer with access to a larger system or as a local intelligent terminal of a large time-sharing system. In either case the 'terminal' must contain a buffer in which to queue the data prior to its transmission in the common channel until it is authorized to be transmitted.

In multiple-access channels any number of terminals, located over wide geographical areas, may have access to the common communication channel when they so require in order to transmit their messages. A multiple-access data channel may employ either radio or wire as its transmission medium. In fact, the thrust for the development of multiple access systems came from radio and satellite digital broadcasting communication systems. Most of the multiple-access systems presented here are basically intended for broadcasting systems but could easily be adapted to local distribution systems. The main feature of these systems, whether wire or radio, is that a signal transmitted by one terminal of the common channel may be received by all the other terminals operating on the same channel, regardless of their physical setting relative to the channel.

Both 'multiplexing' and 'multiple-access' are terms which seem to describe the same thing, i.e. the sharing of the communication channel spectrum between a number of terminals. This, in fact, is a misleading concept of both terms and the distinction must be clarified. Multiplexing is the division of the common transmission channel facility into multiple exclusive subchannels made available to each user by common equipment positioned at the two extreme ends of the channel. Multiple-access is the process by which the common transmission facility is made available to multiple users, with no restriction as to their physical position relative to the communications channel.

Multiplexing refers to the principle in which the hardware equipment placed at the two ends of the channel divides the transmission facility, while multiple access refers to the communication procedures, generally controlled by software programs stored in the terminal's memory, which provide the means of sharing the transmission facility.

Multiplexing is primarily used in wire communication, although it is also used in radio communication. In both cases, the multiplexing equipment is the star centre of a multipoint network. Multiple-access is predominantly intended for radio broadcasting communication, nevertheless, it could also be used for wire communication. When multiple-access is employed in wire communication, the terminals are generally connected in a multi-drop network configuration. These different network configurations are illustrated in Fig. 9.1.

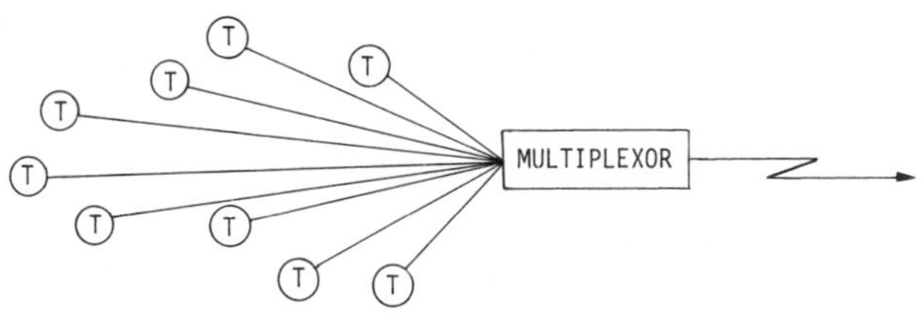

(a) MULTIPLEXING MULTIPOINT NETWORK

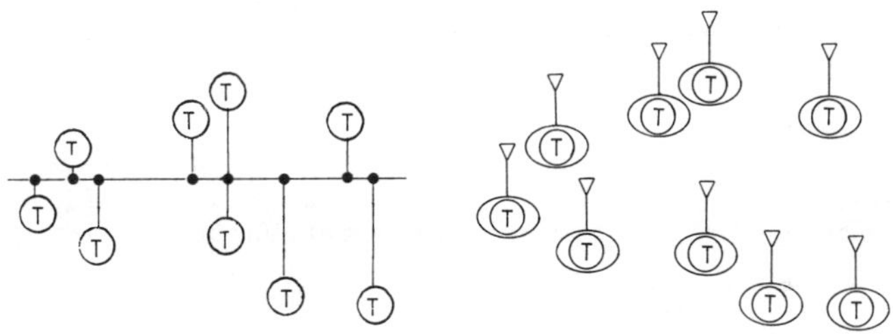

(b) MULTIPLE-ACCESS MULTIDROP NETWORK (c) MULTIPLE-ACCESS RADIO NETWORK

Fig. 9.1 Multiplexing and multiple-access network configuration

The pressing problems in employing a common communication channel, whether it is wire or radio, are to get each terminal to transmit its messages promptly whenever they are ready, without causing delays deriving from channel overloading, and to keep the traffic flowing smoothly without message collisions due to the absence of direct coordination between the terminals.

9.2 THE IMPORTANCE OF THE PROCEDURES

The task of the multiple-access procedure is to have the channel shared efficiently between the users (terminals) rather than to divide the channel into fixed portions. This can best be done only by orderly controlling the traffic in the channel.

In the case of a centralized computer communication system, the problem is relatively simple. The centre node controls the traffic flowing in the channel and permits the terminals to dispatch their messages only when the channel becomes free. In the case of a distributed local computer communication system, each terminal has its own processing power and may be responsible for initiating its own message transmission with no need to obtain permission from a centre computer. In these systems the control of data flow in the common communication channel calls for sophisticated procedures which provide wider initiative power to each terminal, taking advantage of their processing power.

The multiple-access commonucation channel is characterized by the random nature of the data flowing in the channel. The data originating from each terminal is generally very low; nevertheless, when it has a message to transmit, it has more-often-than-not a number of messages to dispatch. In other words the terminal acts in a random bursting manner. Furthermore, a message originating from one terminal is not directed to a particular terminal or centre but may be intended for any of the terminals associated with the common channel. The problem here is to ensure that there are no delays before the terminal can dispatch the messages and that the messages will always reach their destination without collision between messages originating from other terminals. Each terminal, when it has messages to transmit, will be able to dominate the channel for its own use. This, however, causes conflicting requirements between the terminals, and it is thus the function of the procedure to settle all disputes by providing some control of the channel. The procedure assigns the channel according to given rules which are intended to provide all the terminals with equal rights to the channel and with little to no centralized control.

It is possible to share the channel by assigning to each terminal a fixed portion of the channel. This provides ideal smooth transmission, ensuring that each messages reaches its destination with no danger of collision. Fixed assignment of the channel, however, presents low throughput due to an idling wasteful portion of the channel, when the traffic is low, and causes long delays in active terminals before they can despatch their messages. It is possible also to randomly assign the channel to each terminal in accordance with its needs. This solution is ideal since it ensures the terminal's complete independence with no delays before transmission. Random assignment of channel, however, has a very low saturation point with a high possibility of message collisions. These comments are summarized below:

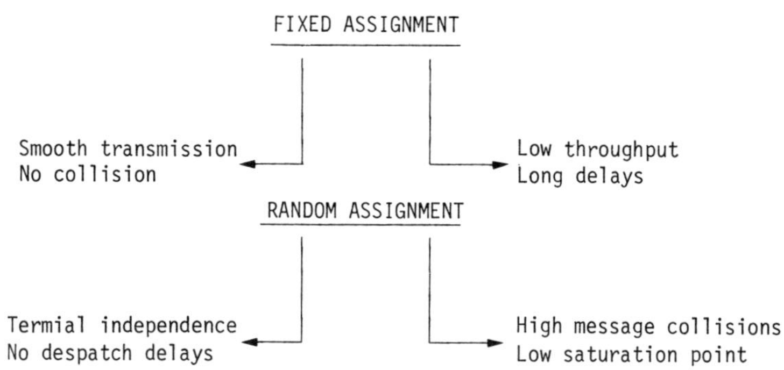

The solution to the above problem could be to dynamically assign the channel which both reduces the possibility of collision and decreases the idling wasted capacity. This too has an unavoidable price: it requires high overhead programming. Dynamic assignment could be achieved with centralized control whereby a centre computer allocates the channel according to traffic demand, as in contention procedures. Such a solution is ideal if most of the messages are directed to the centre, but for most multiple-access systems this exceeds the target of terminal independence. A procedure where each message may be intended for any of the other terminals is better achieved by distributed control with maximum power delegated to each terminal. This, however, will increase the processing power needed at each terminal.

The remarks above may give the impression that the solution is always through procedures which provide dynamic channel assignment either with centralized or distributed control. There is, however, another point to be considered and that is the nature of the traffic itself. In communication systems with high message traffic, the fixed assignment may be preferable, while in a low message traffic the random assignment may be preferable. Such solutions are only possible if the nature of the traffic is known, although in fact it could sometimes be high while at other times low. The traffic usually acts in a random bursting nature with the number of messages originating from the terminals in uneven distribution loads. One solution then is to employ a number of procedures which are selected in accordance with the traffic load or to design procedures which adapt themselves as a function of the traffic load. The procedures to be chosen for a particular application should always be decided upon in accordance with the nature of the application. There is no such thing as an ideal procedure, since each procedure has an unavoidable price of delay, collision, waste or overhead.

In this chapter a number of procedure choices are presented, ranging from tight static control to no control at all. In all the procedures presented, the messages are taken as being of fixed length, although they could be modified (at a price) to accept variable length. The common communication channel usually operates in a half duplex mode, i.e., each terminal may transmit or receive messages, but with both operations not simultaneously. This no doubt causes undue turn-around-time delays which must be considered when designing the system.

9.3 FIXED TIME DIVISION MULTIPLE ACCESS (F-TDMA)

Before discussing the various accepted procedures which enable the terminal to dispatch its messages into the common channel in the most efficient way, it is essential to introduce the basic means of sharing the channel, e.g. by fixed assignment.

The simplest and most robust procedure that can be implemented in a multiple access system is one where each user (terminal) is assigned a fixed permanent time slot in which it can despatch its messages. This means that the user has access to the whole common channel for the portion of the time allocated to it which is equal to the time slot duration. The time domain is divided into time frames, each of which is subdivided into M time slots of equal duration. The time slot duration of t_1 seconds is so selected that it is enough to transmit one full message of fixed duration. In broadcast systems, the time slot duration must also incorporate any propagation delays of the signal, to prevent overlapping of the messages. Each terminal is assigned one or more time slots in a frame and has access only to these particular slots.

The procedure as described may seem to be similar to what has been termed synchronous time division multiplexing (STDM), except that for multiple-access there is no centralized hardware equipment, although it does require added equipment in the terminal. Such a procedure has been called fixed time division multiple access (F-TDMA). Each terminal in the system has its own counter which counts the number of time periods t_1, from 1 to M and then repeats itself, as seen in Fig. 9.2. All the terminals' counters must be synchronized in such a manner that they all start simultaneously. Each terminal has its own decoder which selects its particular time slots within the frame. Hence the time slots allocated may be regarded as the terminals' personal property. In the duration of the assigned time slot, the terminal can transmit whatever it desires with no restriction as to format. This feature is important, as it enables the terminal, if so required, to code its message so that no other terminal, bar one, will be able to decode it.

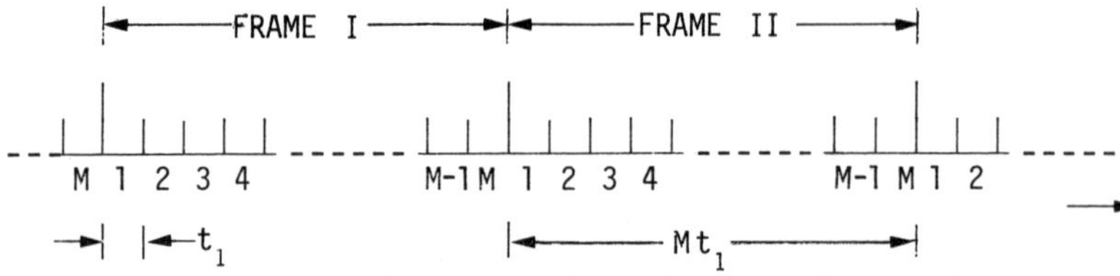

Fig. 9.2 Fixed time slot configuration

Generally each terminal is assigned one slot in a frame; thus if there are N terminals in a multiple-access system, the time frame will be of the size $M=N$. If priority is to be incorporated so as to enable terminals with heavier message load to have shorter waiting times, then these terminals may be assigned a number of slots within a frame. In this case $M>N$. Whatever slot allocation is given to the terminal, this allocation will be permanent without any direct means of changing it while the system is in operation.

A terminal with a message to despatch may have to wait as long as $(M-1)t_1$ seconds before it can start to transmit it, and then it has exactly t_1 seconds to do so. Even if only one terminal has a message to despatch, it still may have to wait the maximum Mt_1 second before it fully transmits it. If it has more than one message to despatch it still can transmit only one message per frame and must wait a complete time frame before it can transmit its second message.

This procedure is simple with no need for overhead to control the traffic flow in the common channel. In this respect the procedure may seem ideal, as it enables a large number of terminals (limited by N) to have access to the common channel without any danger that their transmission will be disturbed by any other terminal's transmission. If all the terminals had a message to transmit, the channel utilization would be 100%, since there are no wasted delays between the messages. Such a perfect scheduling is defined as $S=1$, where parameter S is the normalized channel message rate, i.e., the channel throuput.

Despite the maximum scheduling figure, F—TDMA is regarded as a wasteful procedure. Any time slot not being used by its terminal remains idle without any of the other terminals having access to these idle time slots. This means that the procedure presents low efficient utilization of the common channel when the traffic load is low. For this reason, F—TDMA is primarily suitable for communication systems where the traffic load despatched from all the terminals is relatively high and evenly distributed. For all other multiple-access application it is considered wasteful.

Although it has been stated that any number of terminals may be incorporated in the system, the number in fact is limited by the frame size of M time slots. Once the system is developed, no more terminals may be added without changing the hardware in each and every terminal, that is, increasing the counter size and altering or substituting the decoder.

Due to the fixed assignment nature of the procedure, F—TDMA requires accurate time synchronization of its transmission (i.e., coordinating the frame counter start) to achieve slot separation so as to avoid overlapping of messages. This means that the terminals require complex circuitry to implement the synchronization, although this problem exists in most of the procedures. Despite difficulty, the problem is solvable and in the satellite communication an accuracy of bands less than 200 nS has been achieved.

Another practical problem of F—TDMA worth mentioning is that of incorporating a terminal into the system once it is switched on. This terminal has a time slot assigned to it but its counter is not in synchronization with the other operative terminal. This problem is known as acquisition, as distinct from the previous synchronization problem which is known as tracking.

9.4 ASYNCHRONOUS TIME DIVISION MULTIPLE ACCESS (A-TDMA)

Similarly to asynchronous time division multiplexing techniques it is possible to employ a procedure which avoids the wasted idle time slots by providing transmission times only to those terminals which have messages to despatch. The operation in multiple access is identical with that performed with multiplexing, whereby each terminal is scanned in turn, except that with the multiple access technique the whole operation is performed by software. Such an asynchronous time division multiple access technique is generally referred to as polling. This technique is widely used as the conventional procedure in many of the existing wire data networks.

Various polling techniques have been proposed, two of which are commonly adopted in practice. In the roll-call polling each terminal is interrogated in turn by the central computer. In hub (or disturbed) polling the centralized computer only initiates the polling cycle by interrogating the first terminal. This terminal after transmitting its message, signals the next terminal in turn to transmit its messages. The interrogation is thus transferred from one terminal to another till it reaches the end of the line. This section is focused mainly on the roll-call polling.

In roll-call polling, the centre computer acts as a master station controlling the traffic in the common channel while all the remote terminals act as slave stations, since they cannot initiate any operations but must respond to all the centre's commands. Each remote terminal is allocated a unique code (address) which contains sufficient information for the terminal to recognize it and respond automatically to it. The particular addressed information flowing in the channel will reach all terminals simultaneously, although only a single terminal may respond to this unique address. That is, all the terminals will tend to decode the address, but only the addressed terminal will be able to react to the 'call'. (A detailed description of the format used for the polling procedure is given in Chapter 7.)

By having the centre station transmit the various terminal addresses, each terminal is asked in turn whether it has a message to despatch. The terminal on the receipt of the inquiry intended for it, must acknowledge whether it has a message to despatch; it can then (and only then) transmit the message without any further delay. If it has no message to transmit, it must still send an acknowledgement, in this case a negative acknowledgement (NAK) indicating that it has no message to transmit. When the centre receives a message from the remote terminal, it must acknowledge (ACK) that the message has been received correctly with no errors involved. If any parts of the message show an error, a request is given by the centre station for the retransmission of the message. In many schemes it is then left to the terminal to end the transmission.

Polling, in comparison with F–TDMA, has considerably reduced the time delay before transmission; nevertheless, this delay may still be significant when the number of terminals is large. The main attraction of polling is the simplicity of the procedure and the flexible number of terminals that can be attached to the common channel. With polling, as opposed to F–TDMA, there is no initiation left to the terminal and it can only despatch its messages when it gets the centre's permission. Furthermore, the terminal has no means of informing the control that it has a message to despatch and in addition it must acknowledge the control's request, whether positive or negative, irrespective of the fact whether there are messages to be transmitted. Hence, the polling scheme suffers from delays between the time the message is ready for transmission and the time it can transmit the message. A message may have to wait a full cycle while the centre control station interrogates all the other terminals. This characteristic of question and answer causes another time consuming element which grossly affects the polling scheme. Each terminal, irrespective of whether it has a message to transmit, must normally be in the receive mode and only then after it recognizes its own address in the polling enquiry message, can it transmit its answer. The turn-around-time (t-a-t) of the terminal or the computer from its receive mode to its transmit mode (or vice versa) may be as much as 50 — 200 mS depending on the modem circuits and transmission media involved. The modem turn-around-time could be made to be as short as 0.5 mS although artificial delay may then have to be added to prevent line echoes. For each polling enquiry message two turn-around-times are introduced into the system. The first is the t-a-t between the instant the terminal recognizes its polling identification and the instant that it starts to transmit its answer, and the second t-a-t is between the instant it starts to transmit another polling enquiry, as can be seen in Fig. 9.3.

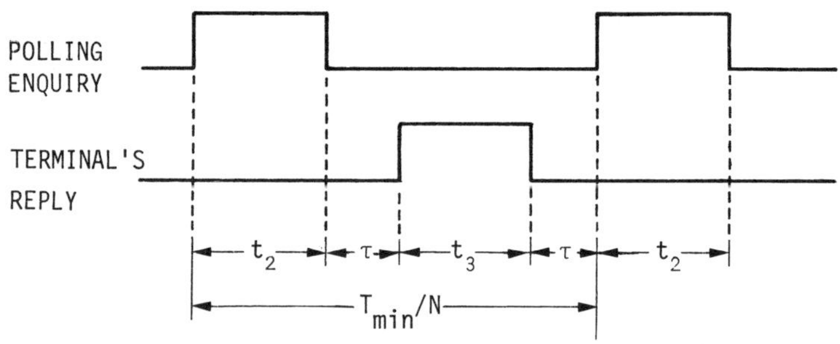

Fig. 9.3 Turn-around time in a roll-call polling scheme

If there are N terminals in the local distribution system, then the minimum time required for a complete roll-call polling cycle T with no information messages transmitted from any of the terminals, is as follows:

$$T_{\mathbf{max}} = t_{\mathbf{min}} = N(2\tau + t_2 + t_3)$$

where τ is the turn-around-time, t_2 is the time required for the polling enquiry message, and t_3 the time for the terminal's NAK reply message. This time T also represents the maximum time (i.e., $t_{\mathbf{max}}$) that a terminal will have to wait before it may unload its messages onto the common channel when only one terminal in the system has a message to despatch. If there are n messages to be transmitted in the system then the maximum waiting time will be:

$$t_{\mathbf{max}} = N(2\tau + t_2) + t_3\,(N - n) + t_1 n$$

where t_1 is the time required to transmit a single information message if the length of the message is constant, or

$$t'_{\mathbf{max}} = N(2\tau + t_2) + t_3\,(N - n) + \sum_{i=1}^{n} t_{(1)i}$$

if this is not the case.

In hub polling schemes, where the centre computer only initiates the sequence, the functions are similar although the cycle is shorter. At the end of each transmission, the active terminal must send an end of transmission signal with a poll enquiry signal to the next terminal in line. The second terminal must then reply either by the transmission of its messages or by a negative reply followed by a poll enquiry to the next terminal. This means that if a terminal has no message to despatch, it transmits an equiry poll including an address with the duration of t_2. Thus, in hub polling if no terminal has a message to transmit then the minimum time required for a complete cycle (which is also the maximum waiting time of a terminal) will be

$$T_{\mathbf{min}} = t_{\mathbf{max}} = N(t_2 + \tau) .$$

9.5 ADAPTIVE TIME DIVISION MULTIPLE ACCESS (AD-TDMA)

Fixed TDMA suffers from wasteful idle slots, a disadvantage which is overcome in asynchronous TDMA techniques. Nevertheless, A—TDMA necessitates centralized control, leaving no initiative to the terminal. Furthermore, A—TDMA calls for high overhead timing required for the polling enquiry and answer, which reduces the message throughput in the common channel. With F—TDMA there is no overhead and the initiative is left completely to the terminal. Both these disadvantages call for unavoidable charges to eliminate idle timing.

A procedure has been proposed which uses F-TDMA principles but adopts its slot allocation in accordance with the momentary traffic load. In the procedure known as adaptive TDMA each terminal is assigned time slots in the same manner as with F-TDMA, but here the procedure enables idle time slots to be exploited by active terminals. In this procedure, every terminal keeps a log of the traffic flowing in the common channel and recording the number of messages which are despatched from each terminal. It is assumed in this system that if a terminal has transmitted one message it is most probable that it has a number of further messages to be transmitted. By a statistical analysis of the past ten time frames it is possible to reallocate the time slots so that the active terminals could receive the use of extra slots. A simple program, available in each terminal, provides this service and ensures that all the active terminals are equally assigned extra slots.

This AD—TDMA procedure is intended to be applied in systems where there is a low probability of message despatch, although coming at random times with heavy load bursts. The redistribution of the time slots is performed with no extra overhead timing. For such applications, AD—TDMA is superior both to F—TDMA and to A—TDMA. If, however, the despatch probability is high, with small burst loads, A—TDMA is possibly preferable.

The adaptation of the time slot to the new traffic conditions takes a full time frame. That is, at the end of each present time frame, the proceeding time frame is re-assigned between the active terminals in accordance with their traffic in the past ten frames. This procedure can thus create a problem that a past inactive terminal when it becomes active finds its own time slot occupied by another terminal. When such an instant occurs, a collision must occur within the time slot between the messages transmitted by the two terminals. When this occurs, both messages are 'lost' (i.e., destroyed), nevertheless, they are retransmitted in the next time frame. The collision is recorded by all the terminals including the ones directly concerned, so that they could all include this in the next frame slot assignment. The slot which had contained the collision is returned (in the next frame) to its original owner and the other terminal is assigned another time slot. Such a program is not necessary if the inactive terminal when it becomes active finds its time slot free. It follows that the maximum time that a terminal may have to wait before it can transmit messages is two complete time frames, as compared with only one time frame in F—TDMA.

With AD—TDMA the contingency may still arise of idle time slots, although their numbers are grossly reduced in comparison with F—TDMA. This is because the assignment in AD—TDMA is performed in accordance with statistical methods rather than actual terminal requirements. Furthermore, the time slot duration in AD–TDMA must be increased in contrast to that of F–TDMA. In F–TDMA the messages originating from a particular terminal were identified by the time slot they were transmitted in. In AD–TDMA the time slot can be re-assigned and the messages in each slot may originate from any of the terminals. This means that each message must include the originator's address. It is important to add that this message addressing requirement is not restricted only to AD–TDMA, but is an essential requirement of all procedures, except that of F–TDMA.

There is another common procedure problem that could affect the operation of AD– TDMA, and this is the problem of the hidden terminal. Such a problem can occur in all procedures, including A–TDMA, where the channel control is distributed and each terminal assigns the slots as a function of traffic. This means that in order that for all the terminals to perform their common identical assignment they must all be in direct contact with each other, with no terminals being hidden from any of the others. Each terminal must receive all messages flowing in the common channel, without losing a single one. In wire communication, such a requirement is provided by the nature of the channel, although this is not so in radio communication. In the latter it is essential that all the terminals should be in hearing distance from each other, i.e., within 'line-of-sight'. With F–TDMA each terminal has its own assigned time slot and one can transmit in it whenever it requires to do so, with no need to coordinate with the other terminals. With A–TDMA roll-call polling, all the terminals must be in hearing distance from the centre computer only, with no need for all the other terminals to be in hearing distance from the active terminal. With A–TDMA hub polling there is no need for all the terminals to be in complete hearing distance from all the other terminals, although it is essential that every two adjacent terminals must be in hearing distance. With AD– TDMA, if all the terminals are not in good hearing distance the whole operation is disturbed; hence, if there are any hidden terminals in the system, this procedure cannot be employed.

9.6 FRAME ADDRESS TIME DIVISION MULTIPLE ACCESS (FA–TDMA)

Adaptive TDMA procedure performs the future slot assignment by a statistical analysis of the past slot exploitation. Such a procedure cannot achieve the best slot assignment, which is a function of the actual real-time terminal traffic requirements. With frame-address-TDMA procedure, each terminal informs all the other terminals of its actual requirement, i.e., its message queue. The term queue refers here to the number of messages in the terminal's buffer waiting to be despatched. With this information available at all the terminals, the slots can be assigned in accordance with the accurate real-time requirements. This procedure is intended for application in systems (similar to AD–TDMA), with medium despatch probability but having high spurting loads. It is obvious that FA–TDMA performs better than AD–TDMA, but this is at the expense of added overhead required in each frame for the request messages. It is worth adding that it has the same hidden terminal problems as AD–TDMA and it does not prevent idle periods although it reduces their number.

The time frame of FA–TDMA procedures comprises two subframes of unequal size, as seen in Fig. 9.4. In one subframe there are M time slots each of the duration t_1 which is long enough to transmit a single message (including the originating address and propagation delay). In the second subframe there are N time slots each of the duration t_4, which is long enough to transmit the queue length information (including propagation delay). Each one of the N terminals is assigned one short time slot in which it can transmit its request. There is no need to transmit the terminal's address since this is recognized by the small slot position. This means that the duration of t_4 may be long enough only to transmit a few bits, and hence $t_1 \gg t_4$.

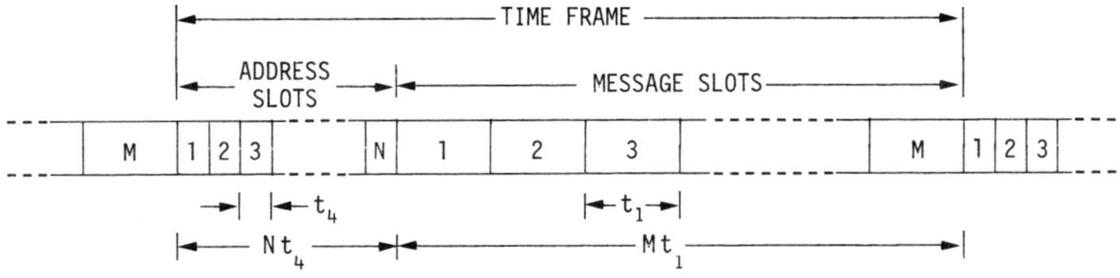

Fig. 9.4 Frame construction in FA–TDMA

The address subframe operates as fixed TDMA, while the message subframe operates as adaptive TDMA. The whole frame length is constant and equal to $Nt_4 + Mt_1$. The number of message time slots can be selected according to the application requirements, which need not necessarily be equal to the number of address slots, i.e., $N \neq M$.

Each terminal, even if it has no message to despatch, transmits the number of messages still in the queue ready for despatching, that is, after deducting those messages to be transmitted in the present frame. All the terminals receive the full traffic requirements and can dynamically assign the message slots of the following time frame. The assignment of the M common message slots is identically calculated in all the terminals in accordance with fixed programs based on simple tables. This program can take into account priority requirements or base their assignment so that all the terminals get an even chance of despatching at least one message

FA-TDMA is a procedure which adapts itself to traffic conditions and provides means to cope with bursts of high loads. If only one terminal has messages to despatch and it has a queue of k messages, it can transmit them in a single time frame, provided $k \leqslant M$. In F–TDMA the same queue load would have required k frames duration to transmit all the messages, i.e., kNt_1. Both fixed and frame-address TDMA procedures suffer from wasted idle times, although with FA–TDMA it is mainly in the address subframe.

9.7 RESERVATION TIME DIVISION MULTIPLE ACCESS (R–TDMA)

The main drawback of FA–TDMA is the additional overhead timing of the duration Nt_4 required for the requests. This overhead frame period acts as fixed TDMA with the same disadvantage of idle time slots corresponding to terminals with no messages to transmit. It is possible to reduce the overhead period by ordering only the active terminals to transmit their request and by transferring their queue information into the transmitted messages. This arrangement enables the active terminals to request at the end of their transmitted message the reservation of further time slots.

The time frame in R–TDMA is divided into N time slots, each assigned to the N terminals as in F–TDMA. The size of the time slot, however, must be expanded to include both the message and reservation request, i.e., $t_1 + t_4$. The frame size could be enlarged to M slots, with the M–N slots available to be distributed among the overloaded terminals.

Each active terminal when transmitting a message in its original assigned time slot sends a reservation request stating the number of messages to be despatched. Time slots not used by the original owners can be redistributed according to these reservation requests. In each time frame, the active terminals may transmit in a number of time slots, one of which must be the original slot assigned to it. Only in this slot can a reservation request be sent: otherwise the whole operation may be disturbed.

A terminal which was inactive and wants to become active will cause a collision in its originally assigned slot if this slot has been reassigned to another terminal. The queue information of the rejected terminal is not however lost in the collision since this information is sent only in the slots originally assigned to the terminal and not with messages sent in reassigned slots. In the following time frame both terminals are assigned slots with no possibility of further collision. Nevertheless, the previous active (rejected) terminal is assigned as many slots as it had reserved, while the previous inactive terminal is assigned only one single slot since his actual reservation request was lost in the collision.

If a terminal becomes 'hidden' to some of the terminals, it declares itself as unsynchronized. This declaration is received in at least one terminal, which then retransmits it in its own time slot so that it reaches all the terminals. When this occurs, all the terminals shift their operation from reservation mode to fixed mode. This means that as long as one or more terminals are hidden, the idle time slots cannot be reassigned but are retained by the terminals in accordance with the original assignment. Hidden terminals occur either when the terminals are distributed over a too wide area, or are operating in an unduly noisy radio spectrum. This extra feature of R—TDMA allows the operation to adapt itself not only to traffic conditions but also to radio communication conditions.

9.8 DISCRETE TIME DIVISION MULTIPLE ACCESS (D—TDMA)

Each TDMA procedure has its unavoidable cost, which must be assessed when considering the procedure for a particular application. Discrete TDMA has features very similar to F—TDMA, although it does not have the problem of wasted idle slots. F—TDMA is ideal for high average traffic, while D—TDMA adopts itself to traffic conditions and can cope both with high and low traffic load. It could also cope with a burst, although its best performance is when the number of messages in a burst is small.

The basic time frame is divided into N small discrete time slots which are assigned to each terminal, as with F—TDMA. The frame timing, however, is variable and changes in accordance with the message load. Each terminal contains a counter which is advanced once for every discrete time slot. The output of the counter is decoded separately for each terminal so that it can provide them each with a unique time reference in which they can start their transmission (in a similar scheme as with F—TDMA). The main feature of discrete TDMA is that the time slots are extremely short, indeed shorter than that of any of the previously described TDMA procedures.

All the terminals in the D—TDMA scheme must be in constant receiving position (i.e. with no hidden terminals) so that they can adapt themselves to the traffic conditions. When a terminal has a message to transmit,.it switches its operation mode from receive to transmit mode, which is performed in its assigned discrete time slot. The format of each message is headed by a start-of-transmission (SOT) character, which is recognized by all the other terminals as a stop-count instruction, although it still leaves them in the receive mode. The transmitting terminal has now got the full spectrum of the common channel. It can then transmit any number of messages, although it is better to limit the number of message transmission to one or at the most two. Once the terminal has completed its message transmission, it sends an end-of-transmission (EOT) character which is recognized by all the terminals as a restart count instruction, triggering their counters to continue counting from where they had stopped before. This EOT character timing coincides exactly with the last slot count pulse with no delay required, since all the terminals are in the receive mode and no turn-around-time is involved. (it is possible, although unessential, for the EOT character to be followed by the terminal reference code which identifies the required counter position). After the counter has been restarted the next terminal with a message to transmit can do so when its own unique referent time has been reached. The time pattern of messages transmitted in the common channel is shown in Fig. 9.5.

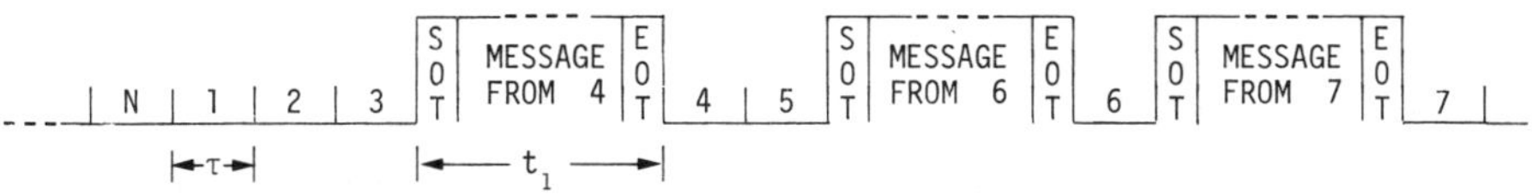

Fig. 9.5 Pulse pattern of D–TDMA in the common channel

The discrete time slot duration assigned to each terminal could be very short. It has to be of a duration long enough to transmit the SOT character and as a result stop the counters in all the terminals. In fact, it requires to be equal to the turn-around-time τ, as it has to be long enough to switch the transmitting terminal from receive to transmit mode for despatching the SOT stop-counter character. In a radio communication scheme this discrete time slot must include the propagation delay. If only one terminal has a message to despatch then its maximum possible waiting time will be $t_{max} = N\tau$.

The operation of D–TDMA is better understood with reference to the example given in Fig. 9.5. Each terminal steps its counter position every τ seconds unless it receives an SOT signal from any of the terminals. If a terminal sends an SOT signal, then all the counters stop and remain in the last count position. In our example where terminal 4 has a message to transmit, the count will remain in the 4 position as long as this terminal requires. As soon as the EOT signal is transmitted an extra delay of τ seconds is added and then all the terminals continue to advance simultaneously to position 5. Since terminal 5 has no message to despatch all the counters advance further to position 6. Terminal 6 has a message to transmit and the count is stopped till it transmits its EOT signal. Terminal 7 also has a message to despatch, but must wait τ seconds before it can transmit it. These τ seconds are necessary for it to receive the EOT signal from terminal 6 and only then change to transmit mode.

The main advantage of D–TDMA procedure is that there is no wasted time and that the channel is inactive unless there is a message transmitted in it. This means that this procedure adapts itself to operate both in low and high traffic conditions. If a number of terminals, or even all the terminals have messages to be despatched they can transmit them when their turn comes up with no added delay. This means that if all the terminals want to transmit messages, the D–TDMA system automatically converts to an F–TDMA system. This, however, can be achieved only if each terminal is allowed to transmit one single fixed size message. If more than one message is to be transmitted in one burst, longer queues may be developed in the other terminals in high traffic conditions. The number of messages to be transmitted in a burst depends on the application. If the traffic conditions are of low despatch probability then the burst load can be increased, but if there is high despatch probability the number of messages in each burst must be limited. This advantage of D–TDMA over F–TDMA is with the unavoidable price of the hidden terminal problem. D–TDMA cannot operate properly if all the terminals are not in direct hearing distance.

9.9 THE ALOHA SCHEMES

A completely different approach to time division multiple access is the random assignment which involves no control function at all. Such a procedure solves the problems of wasted idle time slots, hidden terminals, independence of the terminals, delay before transmission and overheads. Nevertheless, this procedure has the unavoidable price in the form of high probability of message collision which could also be regarded as causing wasted channel capacity.

This procedure is apparently one of the first radio broadcasting procedures developed to enable easy access to mobile terminals distributed over large geographical areas in hostile communication terrain. Although this procedure is primarily concerned with radio channels it could adequately be applied to wire channels. It was first described by Abramson in 1969 at the University of Hawaii, naming the procedure ALOHA, a name which has since been accepted for most random assignment procedures.

The ALOHA procedure does not attempt to coordinate between the terminals: in fact the messages may be transmitted at the instant they are generated. As a result, several terminals may transmit at the same instant, causing collisions in the channel of their messages. To ensure that the messages have reached their destination, an acknowledgement message (ACK) must be returned to the sender. If no ACK is received within an accepted period then the message must be transmitted again after a random delay. It is presumed that the absence of the ACK is due to a collision between two messages, and thus to prevent these two messages colliding again they are each delayed separately by an independent random period before being retransmitted. The sequence of events in the basic (pure) ALOHA scheme is shown in Fig. 9.6.

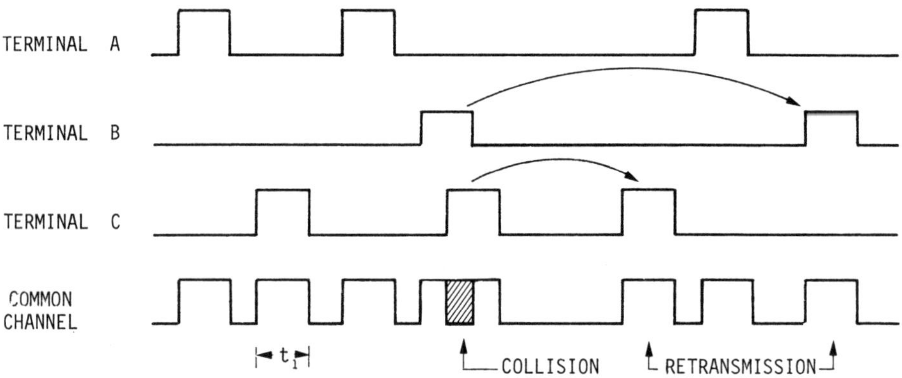

Fig. 9.6 Sequence of events in the pure ALOHA scheme

ALOHA is an ideal procedure for transmission of data in a common channel provided that the traffic could be guaranteed to be extremely small. Messages, however, are transmitted randomly and as a result several terminals may transmit at the same time duration. Collision of two messages may occur even if only a small part of the two messages overlap. In fact, even if only one bit of each message is affected then the two messages are received as damaged, as seen in Fig. 9.7. This means that the maximum duration D that can be affected in each collision is:

$$D_{\mathbf{max}} = 2(t_1 - 1_{\mathbf{bit}}) .$$

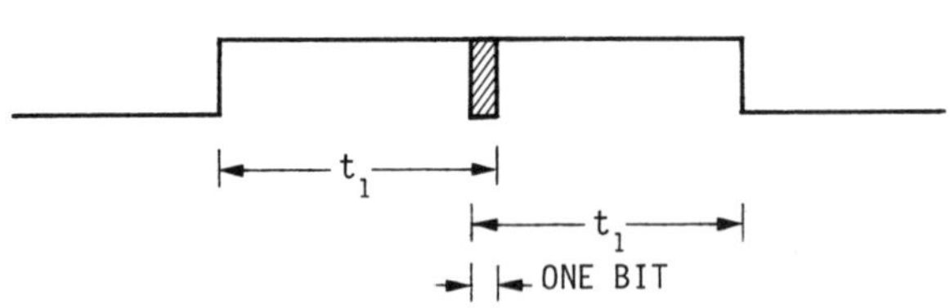

Fig. 9.7 Maximum collision duration in pure ALOHA procedure

Since the message length t_1 is much longer than a single bit, it can be said that $D_{max} = 2t_1$. On the other hand, the maximum delay before transmission being zero, $t_{max} = 0$, the messages are transmitted as soon as they are ready.

Owing to the long collision duration and its random aspect, the chances for collision is high even with very small traffic. Moreover, as the traffic increases, the collision causes more retransmission, which consequently further increases the traffic and thus magnifies the chances of collisions. The chain of events causes the channel to become saturated with a low throughput thus allowing no messages to reach their destination. Abramson calculated the throughput, using the exponential assumption of probability.

The channel throughput rate S is defined as the average number of successful messages R' transmitted in the channel with respect to time T, i.e., $S = R'T$. The throughput can be regarded also as the channel utilization. The maximum possible throughput that can be achieved is when a stream of messages, with no overlap or gaps between them, is transmitted into the common available channel, i.e., $S_{max} = 1$.

The channel offered traffic rate G is defined as the total number of messages R transmitted in the channel with respect to time T, i.e., $G = RT$.

The probability of successful transmission P_s is the ratio of the number of succeesful messages R' to the total number of messages transmitted R.

$$P_s = R/R' = S/G \qquad S = GP_s \ .$$

Since the collision duration is $2t_1$ and the arrivals are arbitrary in accordance with Poisson traffic exponential assumption, the probability that there is no collision in the available time period of $2t_1$ is given as

$$S = Ge^{-2G} .$$

Thus, the channel when using pure ALOHA achieves a maximum throughput of $S_{max} = 1/2e = 0.184$ at point $G = 1/2$. This is the saturation point with 18.4% of the full possible channel utilization. Any increase of the traffic beyond this point will only reduce the throughput.

By slightly modifying the ALOHA scgeme it is possible to raise the channel saturation point to 36.8%. This is achieved by reducing the collision duration, although this is at the expense of reducing the fully random aspect of the message transmission time, e.g., by limiting them to start transmission at the same corresponding times. The time domain is divided into equal length time slots, each of duration time enough to transmit a single message, including any propagation delays (i.e., t_1). The terminals can despatch their messages in any time slot (as seen in Fig. 9.8) provided they all start transmission at the leading edge of the slot and complete the message transmission before the end of the slot duration. This means that the basic random feature is still retained, as each message can be transmitted as soon as it becomes ready but it might have to be slightly delayed until the start of the next time slot, i.e., $t_{max} = t_1$.

Although the time domain is divided into slots of equal size this procedure is not a TDMA procedure. In all TDMA procedures the time slots are assigned to the terminals by fixed or dynamic rules. In the slotted ALOHA procedure the slots are not assigned at all. The terminals simply try to exploit the slots in a purely random order. Such a scheme is referred to as Slotted ALOHA, as distinct from the fully random scheme which is referred to as Pure ALOHA.

With slotted ALOHA, if two messages collide, they will overlap completely rather than partially. In other words, the time duration of each collision is exactly one message length, i.e., $D = t_1$. The throughput equation calculated, on the assumption of no collision in the vulnerable time period of t_1, becomes now:

$$S = Ge^{-G} .$$

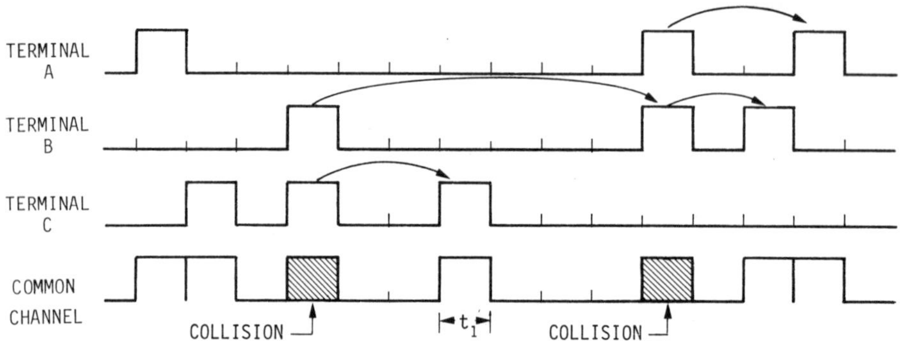

Fig. 9.8 Sequence of events in the slotted ALOHA scheme

With the simple change from pure to slotted ALOHA the throughput is increased by a factor of two since the maximum throughput is achieved to be $S_{\mathbf{max}} = 1/e = 0.364$, at point $G = 1$. This can be seen in Fig. 9.9, which is the diagram of the throughout S versus offered traffic G.

Despite the increase in throughput, slotted ALOHA still has low saturation probability and is thus applicable to systems with very low traffic loads.

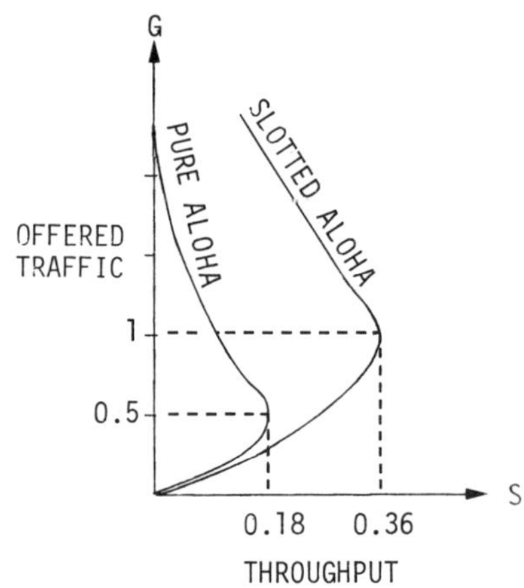

Fig. 9.9 Throughput function is pure and slotted ALOHA

It has already been stated that for the proper operation of the ALOHA scheme an acknowledgement must be returned to the sender to indicate the correct reception of the data message. This means in effect that for each message transmitted extra transmission period must be provided. This increases the traffic, which consequently augments the collision probability. The time duration t_3 of the acknowledged message is rather short as compared with the data message size, which has negligible effect in the pure ALOHA scheme. This, however, is not the case with slotted ALOHA, which requires that each message must start at the beginning of the time slot. This means that in S—ALOHA the successful transmission of a message may occupy two to three time slots, which thus drastically reduces the throughput. The acknowledgement message generally cannot be sent in the proceeding slot, but in the slot once removed, as seen in Fig. 9.10. This is because the destination terminal, which is in the receiving mode, must first decode the message to check that there are no errors before it can switch to transmission mode, i.e., an operation which is longer than the turn-around-time. Performing this operation causes the terminal to miss the next slot and it has to wait a full slot period before it can send the acknowledgement message.

To reduce the possibility of a collision between an acknowledgement message with another message, the acknowledgement can (if so required) get a priority over all new messages. To achieve this feature all the terminals must be in hearing distance from rach other and be able to detect whether there was a collision or that the message had reached its destination successfully. If the latter is the case, the slot following a successful message

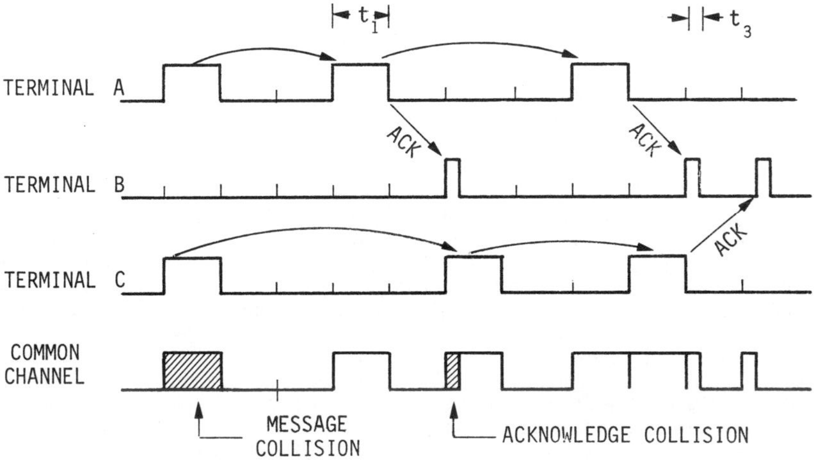

Fig. 9.10 Sequence of events in the slotted ALOHA with acknowledgment

transmission is frozen and may only be used for the acknowledgement. If, however, there was a collision, the following slot may still be available to random message transmission, as seen in Fig. 9.11. In order to reduce the period of the message transmission and its priority acknowledgement, the slot size may be increased to include the turn-around-time, i.e., $t_1 + \tau$. This will allow the acknowledgement to be transmitted in the first consecutive time slot. The slot size may be further increased to include the acknowledgement, i.e., $t_1 + \tau + t_3$, which will save message transmission period although it will increase the collision period.

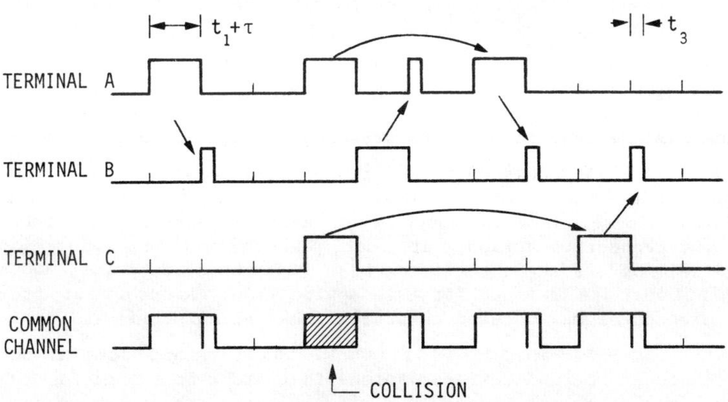

Fig. 9.11 Sequence of events in slotted ALOHA with priority acknowledgment

9.10 ADAPTIVE ALOHA

The ALOHA scheme is a most attractive procedure because of its simplicity: nevertheless, it has an unavoidable penalty of high probability of message collision. When such a collision occurs the messages are randomly delayed so as to prevent them colliding again when they are retransmitted. This means that to avoid further collisions the messages are distributionally spread out in the time domain. Despite this precaution, as the traffic increases the chances of collision increase too. Since each collision doubles the traffic, the number of collisions will increase as the traffic is increased further. The throughput can increase only to a given point where beyond it the number of collisions is so great that it will eventually cause the throughput to decline. It has already been shown that for S—ALOHA this point is 36% of the maximum possible channel utilization. Any further increase of the traffic will cause the throughput to diminish completely thus bringing the channel to a total standstill.

Once the channel is saturated, with no chance of any of the originating messages reaching their destination, the channel has very little chance of getting out of saturation. More messages are being generated, increasing the queues in each terminal to a point where all the buffers in each terminal also get saturated. When such a condition occurs the only possible solution is either to shut down the channel or to shift to another procedure. The first solution is a defeatist attitude while the second solution, although practical, misses the object of employing the ALOHA scheme. It should be remembered that the ALOHA scheme is intended for channels with low traffic density and no other procedure can offer this feature with so little programming. The solution to be adopted must be one which retains the ALOHA random features, which means in effect adopting the ALOHA scheme in such a way that the channel is prevented from getting into the saturation condition.

The terminals know what occurs in the channel even without providing them with special sensing and logic circuitry simply by calculating the numbe of retransmissions. The knowledge the terminal has of the traffic load condition in the channel depends on the processing power it has. Three degrees of knowledge may be available in the terminal when it is operating in an S—ALOHA procedure.

(a) The number of retransmissions required from a terminal to dispatch a single message till it reaches its destination.

(b) The terminal can make a statistical analysis of the traffic in the channel and distinguish between empty slots and busy slots.

(c) The terminal can distinguish between empty slots, busy slots, and collision slots. The last is practical since all the messages must start at the leading edge of a slot, and thus if a collision occurs the message heading with its destination address, becomes meaningless.

The S—ALOHA scheme operates with no fixed assigned time frame, which means that each terminal must record the traffic for the past M slots. Once the terminal senses that the channel is near its saturation point, it must take prompt preventive action. In fact, the prevention measures must be enforced well before the channel reaches its saturation point, and a threshold point must be defined for each application. A number of preventive control procedures may be used to prevent channel saturation.

(a) *Input Control Procedure* (ICP) where the terminal acts in accordance with the number of retransmissions that are required of it. When this number reaches the threshold limit, it stops its transmissions till it senses that the channel is less busy.

(b) *Retransmission Control Procedure* (RCP) which enforces extra delay before retransmission when the number is above a given threshold. It has already been explained that after each collision, the message is delayed for T seconds, where T is a random variable period. If the terminal senses that the number of retransmissions is large then the delay may be increased to KT, where K is any chosen number. When the traffic load returns to normal, the delay of KT may be reduced by gradually diminishing K.

(c) *Input Retransmission Control Procedure* (IRCP) which is a combination of both procedures. RCP reduces the possibility of channel saturation but cannot prevent it completely. ICP prevents channel saturation but does not adjust the traffic. Two threshold points are employed, the first for RCP: however, if the traffic still increases, till it reaches the second threshold then the ICP is employed.

(d) *Closed Loop Control Procedure* (CLCP) which bases the terminals' message transmission on the channel traffic condition (i.e., number of busy slots and/or collision slots) rather than on its own number of retransmissions. Such a procedure calls for higher programming overhead, although it then has the advantage of improved channel performance.

The four procedures described above solve the problem of channel saturation in the S—ALOHA scheme. This is achieved with the unavoidable cost of increased delay before transmission. Furthermore, AD—ALOHA does not solve the problem of channel efficiency, having the maximum throughput remaining at 36%.

9.11 RESERVATION ALOHA

The ALOHA schemes described so far are intended for random traffic loads with very low initiation probability. It is assumed that for these procedures the burst size is rather small so as not to change the random nature of the channel performance. In many applications this assumption is incorrect. In such cases the traffic may bring the channel to a saturation point even with only a few terminals being active, that is, if these terminals have bursts with a large number of messages. When this occurs, the traffic could increase rapidly with no time left for the input and/or retransmission control procedures to prevent the channel being saturated.

Random bursts of single messages, with Poisson distribution could give ample prior indication of the possible traffic increase. As large random burst are not dependent upon a specific pattern, high peaks may be reached in the traffic load, and this will bring the channel to a complete standstill before preventive measures may be used.

An ALOHA procedure to prevent the channel saturation in a bursting environment must ensure that each terminal can sense the number of collision in the channel, besides being able to calculate the number of its own retransmissions. In this procedure, a terminal which succeeds in transmitting a message in a slot with no collision is automatically assigned more slots, which are then exclusively reserved to it. Such a procedure has been called Reservation ALOHA. The procedure philosophy is based on the assumption that if a terminal has a single message to transmit, it is assumed that it has at least K more messages waiting to be despatched. Two such reservation procedures have been proposed.

If a terminal succeeds in transmitting a single message, then it will have the next K slots, immediately following the first slots, reserved to it. Once the $K+1$ slots have been used, the channel is available again to all the terminals for further random transmission. This procedure requires programming overhead in each terminal to keep a running account of the number of remaining reserved slots. Such a control procedure will automatically (following any initial successful transmission) freeze K slots even if the terminal does not have K messages to transmit. It should be added that with this form of traffic environment, K cannot be selected to be too small since the reservation size is the feature which prevents the channel saturating.

An alternative control procedure is to reserve only one slot at a time. Then, if this extra slot is used, to reserve the one following it, and so on till the terminal has despatched all of its messages. The reserved slot, in this procedure, is not the one immediately proceeding the successful slot but the M'th slot after it. Then if that slot is occupied the $2M$'th after it is reserved too. In this way all the slots in position iM are reserved for a particular terminal. Such an arrangement organizes the slots in fixed size frames of M slots. All the terminals know which slot is reserved. This is continued till the $(K+1)$ slot is found to be free, i.e., not used by the particular terminal. The $(K+2)M$ slot is again available to all the other terminals for random access. Those slots in the space of iM and $(i+1)M$ which are not reserved are continuously available to all the other terminals in this slotted ALOHA channel.

Theoretically, if all the terminals have messages to transmit and each message has reserved a slot, then the frame formed resembles a TDMA scheme. Such a situation cannot occur in this procedure because of the possible low throughput. Furthermore the frame is randomly organized with no fixed pattern, a condition which does not exist in TDMA.

This reservation ALOHA procedure is intended for low traffic load with random probability of bursts of variable sizes. Nevertheless, if there is a burst from one terminal with K messages to despatch, it will take KMt_1 seconds before they can all be transmitted. This timing period can only be inaugurated after a collision-free slot is retained. The asset of this procedure is that the terminals with smaller bursts can still have an even chance of transmitting their messages.

9.12 FIFO ALOHA

Both with adaptive and reservation ALOHA is a high probability of message collision, which means in effect, that the chances for the messages transmitted reaching their destination are small. An ALOHA procedure has been proposed which separates the problem of channel access from that of message dispatching. The channel, in this procedure, is still pursued by using all the S—ALOHA random features allowing each terminal to transmit as soon as it has a message to despatch. The difference, however, is that when the channel is approached, only a request for transmission is sent, but no message is despatched. The terminal can transmit its messages only when it can be assured that they will reach their destination with no disturbances. Furthermore, this procedure adopts the channel assignment in accordance with the traffic conditions.

The time domain is divided into slots of equal size in an identical manner to that performed with S—ALOHA. The only difference, however, is that the slot duration is now t_2, which is smaller than the previous slot size of t_1. Time period t_2 is long enough only to transmit the terminal's address and its queue length information, while t_1 is long enough to transmit a complete message.

If there are no messages to be despatched there are no changes in the channel performance as a S—ALOHA scheme. Any terminal with messages to despatch will randomly try to seize a slot and transmit a request in it. The first intercepted slot initiates the operation by which the messages can be transmitted, whereby M slots of t_2 duration are counted after the first seized slot. The time domain is then divided into wider duration slots each of a message size, i.e., t_1.

During the interval of Mt_2, before the start of the message slots, the other terminals can also request permission to transmit their messages. If there is a collision between two requests then they are both randomly delayed. This applies also to the first initiating slot. If the random delay is larger than the remainder of Mt_2 period, then the terminal must wait till the next request period and try again.

After the period of Mt_2, each terminal which succeeded in transmitting an uncollided request is assigned message slots in accordance with its queue length. If all the terminals together requested K messages then the number of message slots will be exactly K. The channel is divided between the terminals according to the simple rule of first-in-first-out (FIFO). This means that the first terminal that succeeds in transmitting an uncollided request gets K_1 message slots, then the second terminal gets K_2 slots and so on. This gives the maximum number of message slots to be

$$K = \sum_{i=1}^{n} K_i .$$

The number n is the maximum number of terminals that succeeded in requesting transmission permission and cannot be larger than the S–ALOHA throughput, i.e., $n_{max} = 0.36\,M$.

Once all the K messages, requested in period Mt_2, have been transmitted in period Kt_1, the channel time domain is again divided into slots of t_2 duration, ready for further requests.

The only advantages of this FIFO–ALOHA procedure is that it assures undisrupted message transmission in a random bursting environment. This means that if terminals have large bursts they have a higher probability of message despatch than any of the previous ALOHA procedures. This, however, is at the unavoidable cost of high programming overhead.

9.13 CARRIER SENSE MULTIPLE ACCESS (CSMA)

The low throughput of all the ALOHA procedures limits their applications to systems with very low traffic probability. An attractive modification on the ALOHA scheme is one where the terminal 'senses' the channel prior to the transmission to check whether it is occupied by any other terminal. That is, the terminal may transmit only if the channel is found to be idle. This procedure, referred to as Carrier Sense Multiple Access (CSMA), attempts to avoid message collision within the channel and can thus increase the message throughput to a saturation level of more than 80% channel efficiency. Nevertheless, it can still be regarded as random assignment of the channel resources, as in the ALOHA scheme. For this reason, some references call this procedure CSMA–ALOHA.

When the terminal has a message to despatch, it cannot off-load it as in the ALOHA scheme, but must first listen to the transmission in the common channel to sense whether another terminal is despatching its own messages. If the channel is sensed as free, then the terminal can despatch its messages with no further delay. If the channel is sensed as busy, i.e. if the channel is being used by other terminals, then the message transmission must be rescheduled to a later time and the same sensing procedure is invoked again.

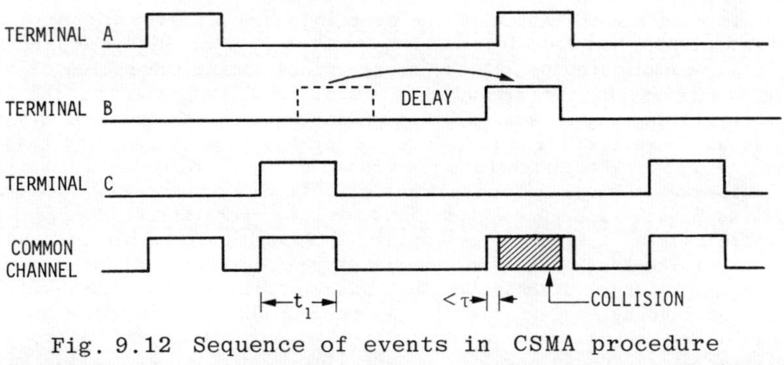

Fig. 9.12 Sequence of events in CSMA procedure

Theoretically, there cannot be any collision in CSMA procedure since no transmission may be initiated before the channel is sensed and proved to be free of other transmissions. In practice, however, although the CSMA procedure reduces the probability of collision it cannot prevent them completely. A terminal which has a message to despatch must be in the receive mode to sense whether the channel is clear. Only if it finds the channel free will it change its mode from receive to transmit and then off load its messages. If during the turn-around-time another terminal starts transmitting, it is unavoidable then for the two messages not to be in collision, as shown in Fig. 9.12.

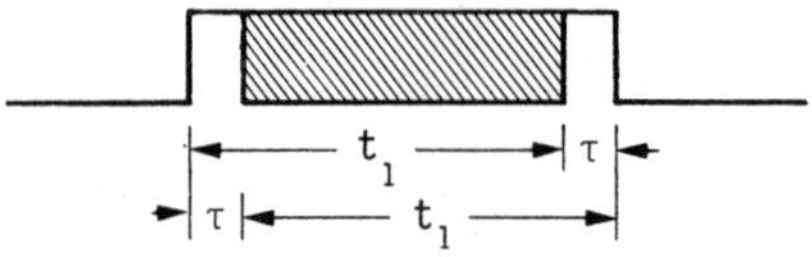

Fig. 9.13 Maximum collision duration in CSMA procedure

When a collision occurs both messages are destroyed, since the information they contained cannot be reconstructed except by retransmission. The maximum collision duration D in this case is equal to $D_{max} = t_1 + \tau$, as seen in Fig. 9.13. The time defined here as τ consists of both the turn-around-time and the propagation delay.

In the ALOHA procedure the throughput was calculated according to the probability of a successful transmission during the period D, where D is defined as the maximum collision period. It would be a mistake to make the same assumptions with CSMA, and the throughput there should be calculated according to the probability of a successful transmission during period τ only. This makes the CSMA throughput much higher than that of ALOHA.

It has already been stated that if the channel is sensed as busy, the transmission is delayed. There are various CSMA procedures depending on the terminal's performance after finding the channel occupied.

A non-persistent-CSMA delays the retransmissions by a random period. After this period the channel is sensed once more to see whether it is free or busy and the whole process is repeated. The advantage of this procedure is the reduction in the collision probability with the consequence increase in throughput. The delay before transmission, however, reduces the channel efficiency and has an indirect effect of limiting the throughput. In other words, such a procedure cannot present the full advantage of the channel.

A persistent-CSMA waits until the channel becomes free and then transmits its message. The advantage of such a scheme is the minimum delay between message transmissions, thus enabling the increase of the throughput. The disadvantage of such a system is that if two terminals had a message to despatch and both persisted in their efforts a collision is then inevitable. That is, there is a probability of 1 that a collision will occur; and for this reason the procedure is known as 1-Persistent CSMA, one which presents a worse throughput that the non-persistent CSMA.

It is possible to randomize the starting time of transmission in the persistent CSMA, thus reducing the danger of collision. With the 1-persistent there is an accumulation of all the sense-trials during the period of a full message transmission. By including an additional parameter p, the probability of collision can be reduced. The parameter p is the probability that a ready message persists: thus $(1-p)$ is the probability of delaying transmission by τ seconds. The parameter p is chosen so as to reduce the possibility of collision while keeping the ideal period between two message transmissions as short as possible. Such a scheme is known as p-Persistent CSMA.

There is also a slotted version of CSMA with both persistent and non-persistent procedures. The time span is divided into time slots of τ seconds; which include the turn-around-time and propagation delay. All transmissions start at the leading edge of the slot as in S—ALOHA. This means that the terminal senses the channel at the start of one slot, and if free starts transmission at the beginning of the next slot. There is very little advantage of S—CSMA in normal communication although it is extremely helpful in spread spectrum communication.

The maximum throughput figures for the various CSMA procedures have been calculated by Kleinrock and Tobagi, and their results are presented in Table 9.1. All the CSMA procedures produce better results than those encountered with the ALOHA procedures, by simply sensing the channel before transmission.

TABLE 9.1

PROCEDURE	MAXIMUM THROUHGPUT
Pure ALOHA	0.184
Slotted ALOHA	0.368
1-persistent CSMA	0.529
Slotted 1-persistent CSMA	0.531
0.1-persistent CSMA	0.791
Non-persistent CSMA	0.815
0.03-persistent CSMA	0.827
Slotted non-persistent CSMA	0.857

Collisions are unavoidable with CSMA although the probability for them has declined in comparison with ALOHA. Similarly to ALOHA there are CSMA systems which introduce priority acknowledgement to improve system reliability after collision. With ALOHA this additional feature was proposed only for the slotted version while with CSMA it could be applied to all its procedures.

Adding the acknowledgement feature calls for the modifying of the sensing process, as shown in Fig. 9.14. All acknowledgements are transmitted immediately after the message although there must be an idle period of τ between them to provide for propagation delay and turn-around-time in the destination terminal. This means that if a terminal has a message to transmit and finds the channel free it cannot be sure that the channel is actually available. A free channel in this form of the CSMA procedure may mean either no transmission or an interval between a message and its acknowledgement. To avoid collision between messages and acknowledgements, the terminal must sense the channel twice with an interval of τ between them (or sense the channel for a whole period of τ). Only if the terminal senses the channel free on both occasions can it transmit its message, although this requires a further τ seconds for the turn-around-time.

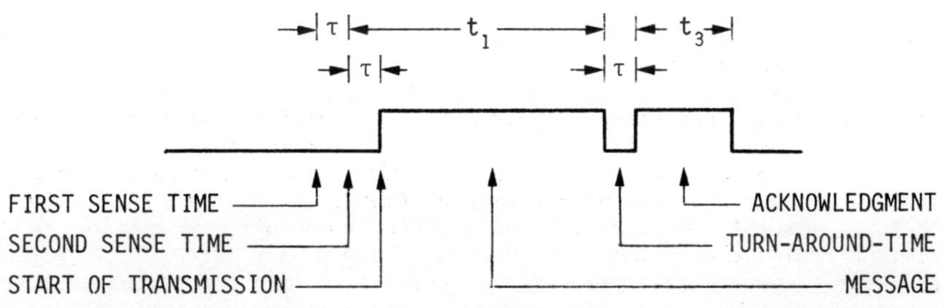

Fig. 9.14 CSMA procedure with priority acknowledgement

CSMA has been adopted for radio data transmission, but it could adequately be used in wire data transmission. When used in radio broadcast system it should be noted that it is very sensitive to hidden terminals. Also, in high interference environment, sensing the radio channel may result in false detection which could consequently cause unnecessary transmission delays.

9.14 LISTEN BEFORE TRANSMISSION MULTIPLE ACCESS (LBTMA)

By sensing the carrier in CSMA, the throughput has increased considerably, thus enabling the efficient sharing of the common communication channel.

Furthermore, the channel can serve a large population of terminals with bursting loads. Nevertheless, there is still a small probability of collision, which makes CSMA procedure unsuitable for systems where collisions are not permissible. Such systems occur where retransmissions are not possible.

A procedure which has similar features to that of CSMA but ensures that no collisions are possible is known as listen-before-transmission multiple access (LBTMA). This procedure exploits both CSMA and TDMA techniques and could be compared in operation with discrete TDMA. That is, LBTMA has the same delay before transmission as that of D-TDMA but it is much easier to implement, requiring minimum logic circuitry similar to that required in S-CSMA.

The collision danger period in CSMA is during the turn-around-time between the instant the terminal senses the channel free and the instant it starts transmitting. The time period τ is a random period and any number of terminals may start transmitting in this period and thereby causing a multiple collision. This LBTMA procedure simply ensures that no two terminals can sense the channel at the same time period of τ.

The time frame is divided into fixed time slots of the size of τ seconds which are shared between all of the N terminals. This means that the frame size is of $N\tau$ dration, provides each terminal with an exclusive time slot of τ duration in which it is allowed to sense if the channel is free or busy. Once a terminal starts transmission, it can transmit any number of messages, provided they are despatched one after the other.

A terminal with a message to despatch must first sense the channel in a similar manner to all slotted CSMA procedures. This, however, can be performed only in the slot allocated to it, as seen in Fig. 9.15, that is, only at the leading edge of the time slot. If the channel is sensed free, the terminal must start transmitting within the next τ seconds, to ensure it occupies the channel before the leading edge of the following time slot. If the channel is sensed busy the terminal must delay its transmission to a later stage. The CSMA procedure suggested here is that of 1-Persistent, which means in effect the terminal senses the channel again in the next slot allocated to it, i.e. in the following time frame.

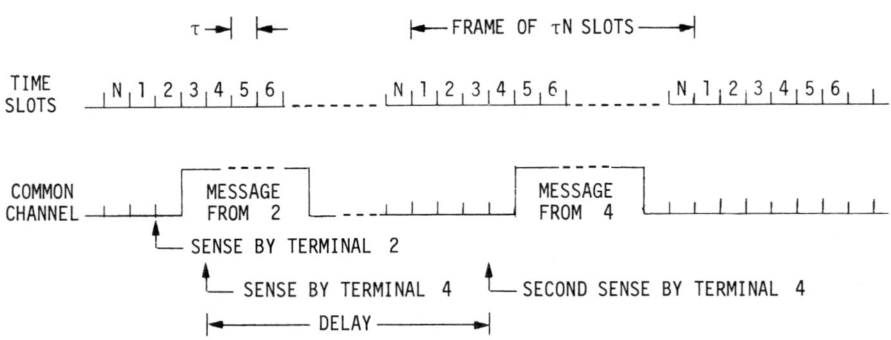

Fig. 9.15 Sequence of events in LBTMA procedure

As the frame size $N\tau$ is small, because $t_1 \gg \tau$, thus the delay before transmission is minimum and the wasted period negligible. The small size of the time frame also ensures that the spacing between the message is kept small. It should be added that the time frame is fixed (as in F-TDMA) and does not change with the message transmission or the traffic load. In the last feature, the LBTMA is inferior to D-TDMA as it requires continuous system synchronization, while with D-TDMA the system resynchronizes after each transmitted message.

9.15 CONFLICT FREE MULTIPLE ACCESS (CFMA)

The LBTMA procedure prevented any possibility of collision, but this was at the unavoidable price of full synchronization between the terminals. Furthermore, there was some overhead wasted period brought about by the requirement of the fixed time frame and because $t_1 \neq N\tau$. With $t_1 \neq N\tau$, there is a risk that a terminal with a message to dispatch may find the channel in use by another terminal in advance of its own allotted time slot.

A much better procedure is one where the frame timing is not fixed, but dynamically adjusted to the traffic load. In this procedure, termed conflict free (carrier sense) multiple access, the time slots are assigned to each terminal as in LBTMA, although here they depend on the terminal from which the last message was transmitted. The terminals sense the channel at the leading edge of each time slot and adjust their operation according to whether the channel is free or busy. The time domain is divided into time frames each of N time slots of fixed size, equal to τ seconds, i.e. of a duration equal to the turn-around-time and the propagation delay. Each terminal has a counter which keeps a record of the time slots so as to provide the terminal with a reference to which time slot it can start transmission. In this respect the procedure is similar to D–TDMA.

The terminals are in constant carrier sensing mode, and as long as the channel is free, the time slots keep changing. As soon as the channel is sensed busy the time slot counting is terminated. If a terminal has a message to despatch it must wait its turn and when the slot arrives it starts transmitting at the leading edge of its slot. All the terminals sense that the channel is busy and stop their counters well before the end of the time slot. Once the terminal has completed its transmission, the channel becomes free and the counters in all the terminals resume their count. A gap of τ seconds must be provided between the end of transmission and the beginning of the next transmissions. If terminal i is transmitting, then terminal $(i + 1)$ must have time for turn-around-time from sensing mode to transmitting mode. This means that all the counters do not resume counting as soon as the channel is sensed free, but after a short delay of τ. In effect this means that the counter must be reduced by one position (or inhibited) each time the terminal is sensed busy. The sequence of events for this procedure is given in Fig. 9.16. The reader should note that the slot time τ must be much larger than the raw propagation delay so as to synchronize with the end-of-message transmission.

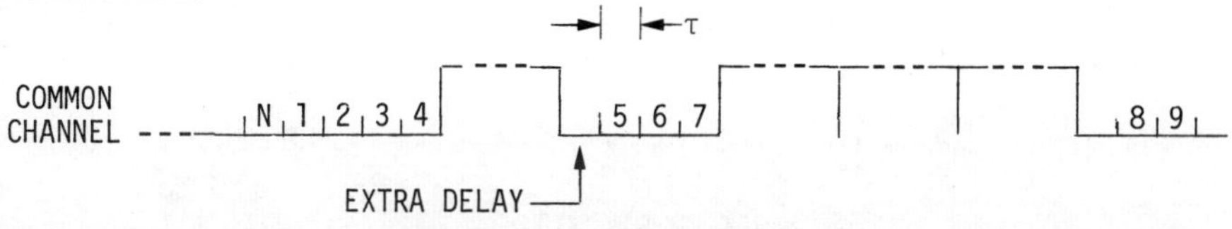

Fig. 9.16 Sequence of events in CFMA procedure

If a terminal has a burst containing a number of messages to despatch, it can do so, provided it leaves no gaps between the messages transmitted.

In the heading of each message the terminal must insert its own address so that all the other terminals know which one is transmitting. This requirement ensures that all the terminals can synchronize their clocks and also enables a terminal which was switched off to adjust its clock once it switches on.

This procedure solves most of the essential communication requirements for no collision, distributed control, channel efficiency with low or high traffic loads and message burst. Similarly to most CSMA procedures, all the terminals in CFMA must be in line of sight.

9.16 SPLIT CHANNEL MULTIPLE ACCESS (SCMA)

All the procedures proposed so far were aimed to provide efficient communication between terminals by fully exploiting the common channel. It was assumed that that are all in line of sight where, in fact, most of these procedures are not efficiently operative if any of the terminals are hidden from any of the others. The only procedures that may make allowances for the hidden terminal were F–TDMA and A–TDMA, and these at the unavoidable price of either wasted idle transmission time or wasted overhead timing. It should be added here that these problems occur only in radio communication, since with wire communication there are no hidden terminals.

Many procedures have been suggested to overcome the hidden terminal, although to date none are perfect. When all the terminals are in line of sight, the CFMA procedure has the least unavoidable price. With widely distributed terminals there is danger of the terminal losing synchronization with actual channel traffic, and so other solutions must be sought for. All the solutions presented have the unavoidable basic price of requiring two or more transmission channels. In most cases extra bandwidth is not available and the answer must be found within the channel itself. In this case the available channel bandwidth must split into subchannels. For this reason these procedures are referred to as split channel multiple access (SCMA).

The splitting of the common channel frequency bandwidth reduces the bandwidth available for the information message transmission. This can be seen in Fig. 9.17, where the voice transmission channel is split into two subchannels: the information and control channels. If the channel spectrum is fixed, as in most applications, the splitting of the channel also reduces the possible speed of transmission, which consequently increasees the time required to transmit a message. If the available channel is labelled B_c and the message subchannel as B_m, then the speed is reduced by the factor of B_m/B_c. Nevertheless, the bandwidth required for the control subchannel may sometimes be narrow, since the traffic in it is generally extremely small. In other words, the ratio of B_m/B_c equals a figure only slightly smaller than one, such that in this case the speed in the information channel is only slightly affected.

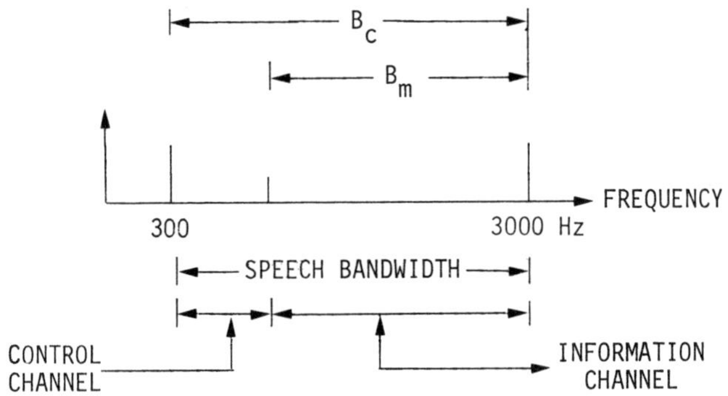

Fig. 9.17 Example of a split channel

Splitting the channel can also be regarded as frequency division multiple access (FDMA) although this term is generally referred to as independent channels operating in parallel, while with SCMA both subchannels operate in conjunction. Nevertheless, each subchannel in SCMA requires additional equipment for transmitting and receiving in the same way that they are required in FDMA.

The SCMA procedures, which are presented in the following sections, are based either on polling or contention schemes. In polling, the initiative for transmission remains in the centre with no independence given to the terminals. With contention, the terminal has the initiative although it must request permission for transmission from the centre. In both schemes the control is by a centre whose function is to coordinate with terminals hidden from each other but not from the centre. If the offered traffic is low, contention schemes are preferable, while if the offered traffic is high, polling schemes may be inevitable. Despite the general statements given above there are means of employing contention techniques (with one of the procedures already discussed) in channels with high offered traffic thus avoiding the need for polling.

Before discussing the various SCMA procedure it is essential to present other accepted solutions to the hidden channel problems. In these solutions the channel operation may be extended beyond the line of sight. Each terminal may 'double up' as a repeater, relaying messages from distant terminals. In such a scheme each terminal reports which other terminal it can hear, along with the receiving quality. This process is known as labelling. In a fixed location system all the terminals know the labelling information while in a mobile communication system this labelling process is essential. Each terminal keeps a table of all the nearby terminals and the routes from them to the other terminals. In a mobile system this label must be updated at frequency intervals. A terminal transmitting in such a system inserts into its message heading which terminal is to act as a repeater. This is essential so as to avoid double transmitting of the same message by two or more repeaters. Such a communicating procedure sees the common channel as a network with the repeaters acting as nodes.

As has already been stated, the repeater solution to the hidden terminal may be employed with only TDMA type procedures. If better channel efficiency is required, with less wasted timing, SCMA procedures are preferable. Even in that solution repeaters are used to communicate between two hidden terminals.

9.17 STRING POLLING MULTIPLE ACCESS (SPMA)

The hidden terminal presents no problem when using A—TDMA roll-call polling procedure since all the traffic is controlled by the centre which is in hearing distance from all the terminals. Nevertheless, with the roll-call polling the channel utilization is inefficient with wasted overhead timing. By reverting to split channel techniques, the overhead timing may be reduced.

In this procedure each terminal is scanned in turn, as in poll-call polling. Yet it differs by not requiring a negative reply from terminals with no messages to despatch. This is achieved by transmitting the polling sequence in one subchannel and the messages in the other subchannel.

A string of terminal identification codes flows in the control channel, with no spacing gaps between them, as shown in Fig. 9.18. For this reason the procedure is referred to as string polling multiple access (SPMA). In this procedure an active response to a poll-call is required only from a terminal with a message to despatch. The duration of each roll-call transmitted in the control channel is t_2, which is long enough for each terminal to recognize ita own address code and then stop the centre from continuing its polling sequence.

Upon the recognition of the particular polling address code, the terminal with the message to despatch will raise the carrier in the information channel by transmitting a sequence of all-ones, followed by the message itself. The centre, after the detection of the terminal's request, will transmit all-zeroes in the control channel. The all-zeroes in the control

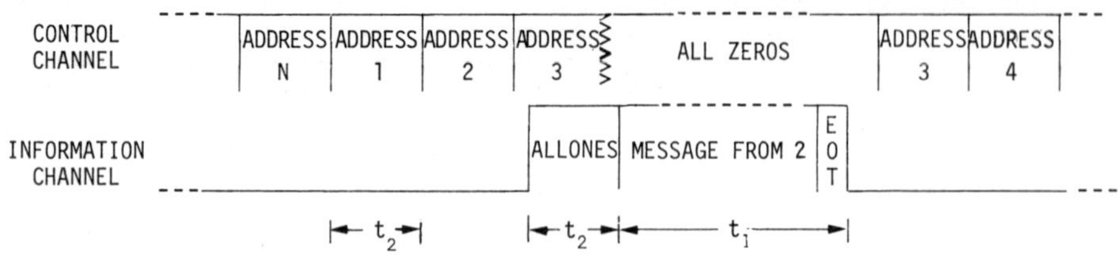

Fig. 9.18 Data flow in the two subchannels of SPMA procedure

channel cancel the last poll address code and informs all the other terminals that a message is being transmitted in the information channel.

The terminal cannot start its transmission in the information channel before it has received and decoded its full address code. In other words, the terminal can only start sending its all-ones request while the next address is being transmitted in the control channel. To ensure that only the last but one address is effective, the all-zeroes must destroy the last address. This is achieved by having each poll address code ending with 2-3 bits in state 1. Then when the all-zeroes is transmitted the 1's are complemented, preventing the full address from being transmitted. The terminal cannot start its message transmission before it can detect the all-zeroes in the control channel. This makes the all-ones duration equal or slightly larger than the address code of t_2.

A pronounced improvement of the split channel over single channel procedures is that there is no problem of the turn-around-time. The reason for this is that the sensing is performed in the control channel and the message is transmitted in the information channel. This means that with SPMA the duration of the all ones need not consider the turn–around–time, nevertheless, the propagation delay must be taken into account. This means that in effect the address code duration must be larger than twice the propagation delay.

If only one terminal has a message to despatch, then the maximum waiting time before it can transmit is the complete polling cycle and the all-ones period:

$$t_{\mathbf{min}} = (N+1)\ t_2.$$

The polling sequence starts again when the centre detects the end of transmission character at the end of the messages.

The SPMA was initially introduced for wire communication, yet it could adequately be employed for radio communication. It has the same delay before transmission as hub polling (which is better than that of roll-call polling) but it is better than hub polling, since it can operate with hidden terminals; however, the throughput is limited by the reduction of the information transmission channel. SPMA, as all polling schemes, is limited by the high overhead timing.

9.18 LISTEN WHILE TRANSMISSION MULTIPLE ACCESS (LWTMA)

The complex of SPMA and its high overhead control makes this procedure unattractive for many applications. A major drawback of this procedure is that the terminals have no initiative and must wait for the centre's permission for transmission. In an operating environment, where the traffic flow is low and the number of terminals is high, this overhead control and lack of initiative could be quite an obstruction. A solution in the form of CSMA is more attractive but this answer cannot be employed because of the hidden terminal problem.

A procedure based on CSMA features but using SCMA principles has been suggested, to achieve no overhead control and overcome the hidden terminal problem. In fact, this procedure can theoretically be regarded as superior to CSMA as it senses the channel not only before transmission but also while transmitting; for which reason it is called listen while transmission multiple access (LWTMA). In CSMA the probability of a collision can occur during the turn-around-time τ between the sensing mode and the transmission mode while the collision duration is $t_1 + \tau$. By having the terminal listening to the channel while transmitting, it is not possible to reduce the probability of collision but only to reduce the collision duration. A terminal which is transmitting can listen at the same time to the channel and if it senses that a collision has occurred it immediately stops its own transmission. By reducing the collision duration the throughput could be as high as 90%.

It is impossible for a terminal to sense the information channel while it is transmitting into it, as it is inevitable that the transmitter will saturate the receiver. Instead, two subchannels are used, one for transmitting and the other for receiving, with the signals in both subchannels well separated so as not to disturb each other. The terminals transmit their messages into one subchannel, as seen in Fig. 9.19, and then all the traffic in this subchannel is retransmited into the other subchannel by a repeater station. All the terminals need not be in line-of-sight from each other although the repeater station must have means of communicating with each of the terminals. The repeater station provides no control logic but simply retransmits into the second subchannel all that it can detect on the first subchannel.

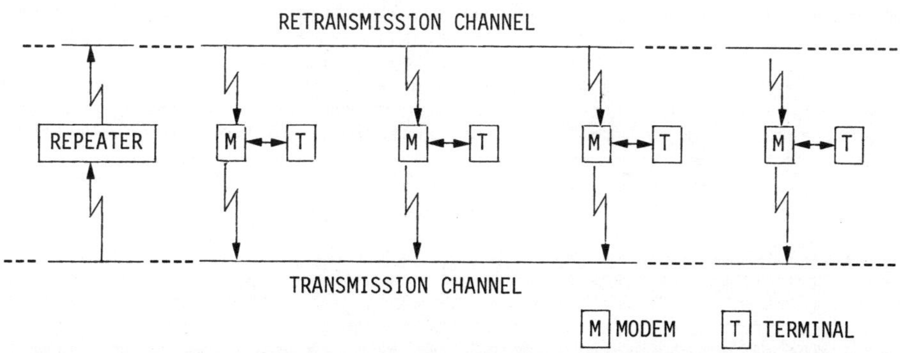

Fig. 9.19 The two subchannel structure of LWTMA procedure

The LWTMA procedure performs exactly as CSMA, that is, if a terminal senses the channel busy it can persist in transmitting or delay transmission and then sense the channel again. If a collision occurs it must delay its transmission by a random delay so as to avoid a second collision. LWTMA is superior to CSMA as it can operate with hidden terminals, although this is achieved by the unavoidable price of requiring two equal information subchannels. This means, in effect, that for the same bandwidth required for CSMA, it is possible with LWTMA to transmit at only half the rate of CSMA, i.e. $B_m = B_c/2$.

9.19 BUSY TONE MULTIPLE ACCESS (BTMA)

The LWTMA procedure seems to be the answer to all the communication requirements of the hidden terminal except that it has the major drawback of requiring two equal bandwidth subchannels both transmitting the same information, i.e., $B_c = 2B_m$. A procedure has been proposed which has the features of LWTMA whereby the terminal senses one subchannel and transmits into the other subchannel. The difference between the two procedures is that the information messages are transmitted only in one subchannel while in the other subchannel only an indication of the traffic is given. In other words, one subchannel is used for information and the other for control.

A centre station which is within the range of all the terminals operates as a refined repeater station. It senses all the transmissions in the information subchannel and if it detects any signals, transmits a sine wave in the control channel. This sine wave can be regarded as a busy tone, informing all the terminals that a message is being transmitted by one of the terminals. For this reason the procedure has been termed busy tone multiple access (BTMA).

The control subchannel bandwidth need not be very wide as no meaningful information is transmitted in it. This means that the offered traffic rate of BTMA is only slightly reduced to that of CSMA. In fact, BTMA can operate exactly as CSMA, except that now it can compensate for hidden terminals. Its operating procedure is identical with that of CSMA. The terminal with a message to despatch senses the control subchannel to detect if it is free. If the terminal detects no busy tone signal, it transmits its message in the information channel: otherwise it reschedules the message for transmission at a random time later. After this delay period it must sense the control channel once more and repeat the whole procedure again.

The centre station can only detect transmissions in the information subchannel, but it cannot detect collisions. This means that BTMA, in contrast to LWTMA, cannot sense its own transmissions since the terminal cannot determine from the busy tone whether it indicates its own transmission or that of another terminal. In consequence the chances of a collision in BTMA are higher than those of CSMA. An added reason is the fact that collisions may occur both in the centre station and in the control subchannel. The centre station takes a longer time before it can determine that it is not receiving noise but an actual message transission. There is always then the probability of incorrect detections resulting in false alarms, a danger which exists when the terminal signals received are very low. There is a further difficulty that confronts the terminals when employing the BTMA procedure. In CSMA the messages were transmitted using the entire channel bandwidth, which made the channel sensing a minor problem. In BTMA, the busy tone only uses a fraction of the channel bandwidth, and this makes its detection a longer process. This means that two terminals hidden from each other may start transmitting in a gap of 2τ in which both gather that the channel is free. The effect causes the throughput of BTMA to be 68% as compared with 81% in CSMA, when they are both operating in non-persistent modes.

9.20 SPLIT-CHANNEL RESERVATION MULTIPLE ACCESS (SRMA)

BTMA, as compared with CSMA, provides an efficient solution to the hidden terminal problem, but this at the unavoidable price of significant lowering of the throughput. Both CSMA and BTMA suffer from the probability of collision, although this is more pronounced in BTMA. The fact that there is a chance of collision limits the application of message burst, since the terminal cannot detect a collision while it is transmitting but can only discover this if it has not received an acknowledgement. In many systems (such as in real time applications) no message collisions are permitted, since this only degrades the performance of the system. To eliminate this extra problem of the hidden terminals and still retain the terminal's independence performance, a further extension of CSMA is considered, this time employing reservation techniques.

From the discussion so far it can be concluded that a system with hidden terminal would operate either with centre control and high overhead or with high collision probability and reduced throughput. Central control generally prevents collision by eliminating the terminal initiative. It is possible, however, to employ centre control and still retain some terminal initiative: this can be achieved by exploiting contention techniques. In such a system the centre is inactive unless it is approached by the terminal. This means that a terminal with a message to despatch must first request permission to transmit and only when it is granted can it start transmitting.

Contention technique is generally used as the input control sequence for the data transfer from a peripheral equipment to a computer. For this operation two control lines are used, one for the equipment so that it could raise a 'flag' to indicate to the computer its request for an interrupt and the other for the computer to order the equipment to transmit. The same contention operation could be applied in radio communication with hidden terminals although with minor modifications. This procedure which is based on split channel techniques is called split-channel reservation multiple access (SRMA).

Two SRMA procedures have been proposed; in one the channel is divided into two subchannels and in the second the channel is divided into three subchannels. The subchannels are of unequal bandwidth, with the wider subchannel for the information and the narrow ones for control applications. The terminal can transmit its messages into the information subchannel only if it had originally reserved the right of transmission and that the centre had granted the right. The advantage of both these SRMA procedures is that no collisions can occur in the information channel, although there is still a collision probability in the control channel.

The first procedure known as request message split channel reservation multiple access (RM—SRMA) is a simple request system in which the channel is divided into only two subchannels. A terminal with a message to despatch sends a request in the control subchannel. In this request the terminal sends its own code address and the number of messages it desires to transmit. The centre retains all the requests in a reservation queue and then grants permission in the order of first come first served. When the common information subchannel becomes available, the centre informs the terminal when its turn comes that it can transmit and then the terminal acts immediately without any danger of collision. Such a procedure raises two questions: how does the terminal send its request messages and how does the centre inform the terminal when it can start transmitting.

The traffic in the control subchannel results only from the terminal request mesagges and it can flow in only one direction, i.e., from the terminals to the centre. The traffic load in this subchannel is relatively low (as compared with the load in the information subchannel) since the request message is short and the number of request messages is small. This means that the control subchannel may be randomly assigned, employing any of the procedures already discussed such as ALOHA or CSMA, depending on the anticipated traffic load. It is suggested here to employ CFMA so as not to include any extra collision problems, although any procedure is adequate.

The traffic in the information subchannel results from data message transmission and instructions granting permission for the message transmission. Both the message and instructions are controlled by the centre and thus no special procedure is required for coordination. That is, after each instruction sent by the centre there is a flow of messages from one elected terminal.

The RM—SRMA procedure just described may possibly exhibit operation problems if the traffic load is too high. The terminal in this procedure, has no means of knowing whether its request message arrived safely in the centre without errors or collisions. An acknowledgement message in the control subchannel will increase the traffic there and may then reduce its usefulness. If after a given period the terminal is not awarded transmission rights it can presume the request has been lost en route. A request can get lost even when using CFMA in the control subchannel since sometimes there could be fading problems in the radio channel. When the terminal realizes it has not been grated transmission rights, it can send a second request message in the control subchannel. This may cause complications, if the lack of transmission rights is due to long queues caused by high offered traffic and not by the collision of the request messages. In this case the centre may get two identical requests for permission to transmit the same information messages. Another disadvantage of this procedure is the overhead instruction message of $(t_2 + 2\tau)$ duration added to information transmission.

The best means of informing the terminal that its request has been dealt with by the centre is to send it an acknowledgement. As already stated it is not practical to employ for this the same subchannel as the request message or the same subchannel as the information message. Instead, a second control subchannel is provided for the acknowledgement. Such a procedure has been termed Request Answer-to Request Message (RAM—SRMA). The terminals send their request for transmission rights in the first control subchannel and the centre answers these requests in the second control subchannel, as seen in Fig. 9.20. This time if the terminal does not get an acknowledgement in a given reasonable time then it is obvious the request is lost and not held up in a queue. In this case the terminal sends a second request, although it must further delay its second request by a random delay to prevent a repeated collision.

In the first control subchannel, used for the request messages, the procedure that can be employed may be either ALOHA or CSMA since the traffic load is small and an acknowledgement is guaranteed. In the second control subchannel, used for the answer to request, no procedures are necessary since all these messages are sent only from the centre. In the information subchannel the traffic consists of messages originated from any one of the N terminals, but again no procedure is necessary since the terminals may transmit only in the time slot period granted to them by the centre. It is possible, however, to regard the operation of the second and third subchannel as a form of split channel polling procedure.

With RM—SRMA procedure the system introduced overhead control messages in the information subchannel to inform the terminal when it could transmit. This consequently introduced wasted delays. With RAM—SRMA, however, there is no overhead control message in the information subchannel and there are no wasted delays due to control operation. All the control operations are performed in parallel by independent action. When the centre sends an acknowledgement to the terminal's request it includes in its answer message a notice as to when exactly it can transmit its message, which is calculated in accordance with the reservation queue in the centre. This notice can only be effective if it is related to a given time. In fact, the terminals are informed when they can transmit by defining the number of time slots they must wait before they can operate. This does not mean that a slotted procedure is used here or that the terminals need to be synchronized. Each terminal has a counter which steps every τ seconds provided by an independent clock source. The centre in the request-answer, sends the request K, which defines the exact number of time slots the terminal must wait before it can start transmitting. The value of K is calculated in the centre in accordance with the message queue from all the terminals, i.e., Mt_1, and adding the time duration of 2τ to each terminal source for space guards between transmissions. In the terminal, the value K is then translated to an actual delay period.

Both RM and RAM SRMA procedures present dynamic assignment of the channel, although they require centralized coordination of message flow. Nevertheless, the initiative of the message transmission process is left to the terminal.

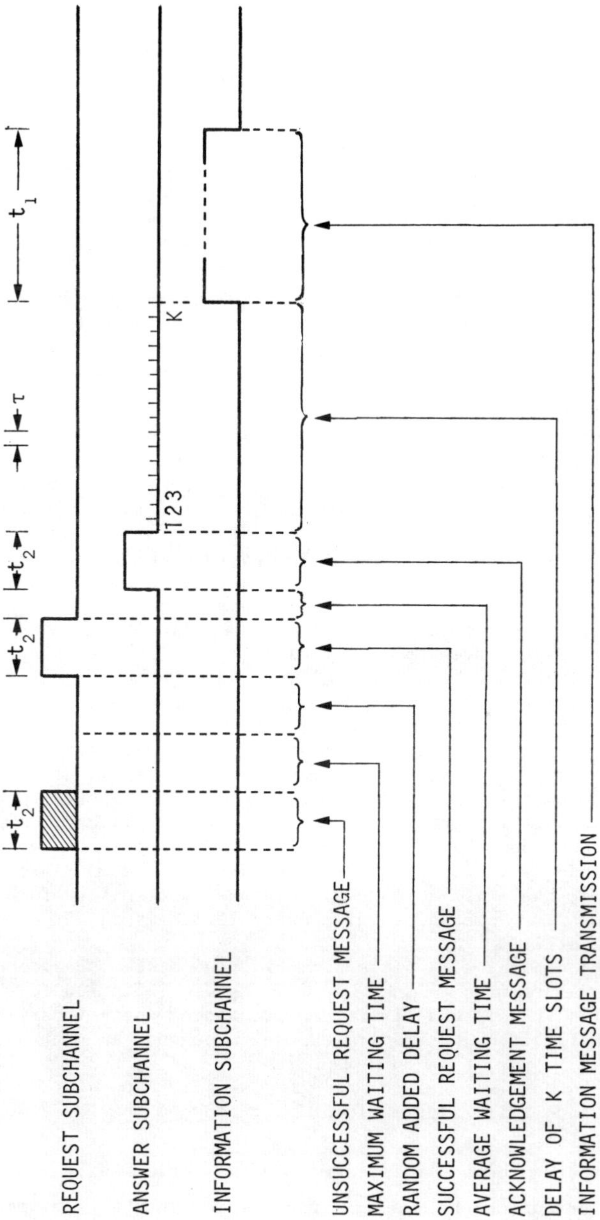

Fig. 9.20 Sequence of events in RAM—SRMA procedure

9.21 CONCLUDING COMMENTS

The full potential of the local communication network can only be realized with the common channels efficiently utilized and the terminals productively operated. These are two contradictory requirements with no single procedure answering both requirements. The best means of efficient exploiting of the common channel is by a centralized system which controls the traffic flow in the most orderly manner, although it entails wasted idle times and delays before transmission. The best means of efficiency operating the independent terminals is by a distributed system which provides maximum terminal freedom, although this presents many collisions and increases the probability of channel saturation. The best means of considering both channel and terminal requirements is possibly in a dynamic system, although this may introduce high overheads with some low collision probability and/or delay before transmission. The question still remains as to how much control to introduce since every procedure has its unavoidable price.

One cannot speak of the best ideal procedure, as each procedure has its limitations. The design goal is always the economical cost of transferring data from one node to another. It should be added that this is not only a function of the procedure, but it also depends largely on the terminals and channels employed. That is, what type of terminals are used, their numbers and their despatch traffic load probability; also, what type of channels are used, whether radio or wire and their operational bandwidth.

By providing the wrong procedure we may invest in a network that eventually turns out to be as much a liability as an asset. The number of points to consider is enormous and a single rule for selecting cannot be provided since no two applications are the same. Nevertheless, an effort has been made to collect the essential points of comparison and these are presented in Table 9.2. It should be added that the table gives only an indication, but does not present all the factors that need to be considered. It is impossible to describe all the perfect comparative solution points, and the designer should therefore regard the table only as an operational summary. Some of the essential points to be analyzed, mentioned before, are discussed here again.

The traffic load originated from each terminal is generally in a random form and can be defined for each application as low, average or high. The traffic load in many applications is evenly distributed in time although in other applications it may be clustered, depending on the time in the day. In consequence of the latter, the procedure may have to adjust itself to two extreme load conditions. The traffic load in some applications may not only be generated randomly and/or clustered but may also be in large and small bursts. That is, the traffic load may be non-symmetrical wherein some terminals may transmit many messages while other terminals may have no message transmission at all.

Most (although not all) large communication systems desire distributed control, with the terminal operation in complete isolation to the others. The best solution in this case is to provide random channel assignment (i.e., no control at all) although this limits the throughput and introduces collisions. In other words, this solution is only useful for very low offered traffic loads; otherwise the channel may be saturated and the traffic be reduced to a standstill. Increasing the throughput could be performed in three directions:

(a) Reducing the collision duration (as performed in S—ALOHA).

(b) Distributing the message transmission in the time domain (as performed in p-persistent CSMA).

(c) Limiting the spacing between the messages despatched (as performed in F—TDMA).

Any of the three directions may increase the throughput, although each one has its limitations. Even the combination of all three design suggestions cannot indicate the perfect procedure.

Note: To assist understanding the table, the following key is to be used:

- τ — turn-around-time
- t_1 — message despatch period
- t_2 — enquiry and/or address despatch period
- t_3 — reply (ACK or NAK) despatch period
- t_4 — queue length message despatch period
- T — cycle or frame duration
- N — number of terminals attached to the common channel
- M — number of time slots in a time frame
- t — waiting time before transmission
- D — collision duration

TABLE 9.2

PROCEDURE	CHANNEL LOADS FOR BEST PERFORMANCE	CHANNEL ASSIGNMENT	CHANNEL CONTROL	PROBABILITY OF IDLE TIMES	COLLISION PROBABILITY	TIME SYNCHRONIZATION
F-TDMA	High average traffic load	fixed	distributed	high	none	slot and frame
A-TDMA	High despatch probability with bursting loads	dynamic	centralized	none	none	none
Roll-call polling	High despatch probability with bursting loads	dynamic	centralized	none	none	none
Hub polling	High despatch probability with bursting loads	dynamic	centralized	none	none	none
AD-TDMA	Low despatch probability with burst of high loads	dynamic	distributed	small	low	slot frame and logic
FA-TDMA	Medium despatch probability with burst of high loads	dynamic	distributed	small	none	slot frame and logic
R-TDMA	Medium despatch probability with burst of high loads	dynamic	distributed	small	low	slot frame and logic
D-TDMA	Low or high traffic loads	dynamic	distributed	none	none	slot and frame
Pure ALOHA	Very low traffic loads	random	distributed	none	very high	none
S-ALOHA	Very low traffic loads	random	distributed	none	high	slots
AD-ALOHA	Low traffic loads	random	distributed	none	medium	slots and logic
R-ALOHA	Low traffic with bursting loads	random	distributed	small	high	slots and logic
FIFO-ALOHA	Low traffic with bursting loads	random	distributed	small	medium	slots and logic
CSMA	Medium traffic loads	random	distributed	none	small	none
S-CSMA	Medium traffic loads	dynamic	distributed	none	small	slots
LBTMA	Low or high traffic loads	dynamic	distributed	negligible	none	slots and frame
CFMA	Low or high traffic loads	dynamic	distributed	none	none	slots and frame
SPMA	Average bursting loads with high despatch probability	dynamic	centralized	none	none	none
LWTMA	Medium despatch probability with variable size bursts	dynamic	distributed	none	small	none
BIMA	Hidden terminals with medium traffic load probability	dynamic	centralized	none	none	none
RM-SRMA	Hidden terminals with low load probability	dynamic	centralized	none	none	none
RAM-SRMA	Hidden terminals witg medium load probability	dynamic	centralized	none	none	slot

TABLE 9.2

OVERHEAD TIMING	TRANSMISSION INITIATIVE	HIDDEN TERMINALS	NUMBER OF SUBCHANNELS	COLLISION DURATION	MAXIMUM DELAY IN A SINGLE ACTIVE TERMINAL	DELAY AFTER COLLISION	MAXIMUM THROUGHPUT	SLOT SIZE
none	terminal	good	1	0	Nt_1	–	very high	t_1
high	centre	good	1	0		–	high	–
high	centre	good	1	0	$N(2\tau + t_2 + t_3)$	–	average	–
high	centre	bad	1	0	$N(\tau + t_2)$	–	average	–
small	terminal	bad	1	T	$2Nt_1$	one frame	high	t_1
average	terminal	bad	1	0	$Nt_4 + Mt_1$	–	high	t_1 & t_4
small	terminal	good	1	T	$2n(t_1 + t_4)$	one frame	high	$t_1 + t_4$
none	terminal	bad	1	0	$N\tau$	–	very high	–
none	terminal	good	1	$2t_1$	0	random	very low	t_1
none	terminal	good	1	t_1	t_1	random	low	t_1
none	terminal	bad	1	t_1	t_1	random	low	t_1
none	terminal	bad	1	t_1	t_1	random	low	t_1
none	terminal	bad	1	t_1	Mt_1	random	low	t_1 & t_2
none	terminal	bad	1	$t_1 + \tau$	0	random	good	–
none	terminal	bad	1	$t_1 + \tau$	τ	random	good	τ
none	terminal	bad	1	0	$N\tau$	–	very high	τ
none	terminal	bad	1	0	$N\tau$	–	very high	τ
high	centre	good	2	0	$(N+1)\,t_2$	–	good	t_2
none	terminal	good	2		0	random	very good	–
none	terminal	good	2	$t_1 + 2\tau$	0	random	average	–
none	terminal	good	2	0	$2t_2$	–	average	–
none	terminal	good	3	0	$2t_2$	–	good	τ

9.22 REFERENCES

(a) COMMUNICATION PROCEDURES

1. Martin, J. *System Analysis for Data Transmission*, Prentice Hall, 1972.
2. Jackson, P.E. and Stubbs, C.D. *A Study of Multi-access Computer Communication*, 1969 Spring Joint Computer Conference AFIPS, vol.34, 1969, pp.491-504.
3. Fraser, A.G. *The Present Status and Future Trends in Computer/ Communication Technology*, Communication Society Magazine, vol.14, no.5, September 1976, pp.10-19.
4. IBM *General Information – Binary Synchronous Communication*, October 1970.
5. Frank, H., Gitman, I. and Van Slyke, R. *Packet Radio System – Network Considerations*, Nat. Computer Conf., AFIPS, vol. 44, 1975, pp.217-231.
6. Binder, R., McQuillan, J.M. and Rettberg, R.D., *The Impact of Multi-Access Satellites on Packet Switching Networks.* EASCON 75, pp.63A to 63F.
7. Swearingen, D.W. and Lipke, D.W. *Marisat Multiple Access Capabilities*, ICC 1976, vol.3, pp.51-10 to 51-13.
8. Chou, W., Frank, H. and Van Slyke, R., *Simulation of Centralized Computer Communication Systems*, IEEE Trans. vol. COM-23, no.9, September 1975, pp.994-1001.
9. Kahn, R.E. *The Organization of Computer Resources into a Packet Radio Network*, IEEE Trans. vol. COM-25, no.1, January 1977, pp.169-178.
10. Heitmeyer, C.L., Kullback, J.H. and Shore, J.E., *A Survey of Packet Switching Techniques for Broadcast Media*, Naval Research Laboratory, 1976, AD.AO 32154.
11. Gitman, I., Van Slyke, R.M. and Frank, H., *Routing in Packet Switching Broadcast Radio Networks*, IEEE Trans. vol. COM-24, no.8, August 1976, pp.926-930.

(b) TIME SHARING PROCEDURES

12. Pickholtz, R.L. and Vogelrang, W.F. *Dynamic Satellite Slot Allocations for Data Packets Vice Adaptive TDMA*, NIC 1977, pp.48:2-1 to 48:2-6.
13. Pierce, J.R. *Network for Block Switching of Data*, Bell Systems Technical Journal, vol.51, July-August 1972, pp.1135-1145.
14. Meisner, N.B., Segal, J.L. and Tanigawa, M.Y. *Dual-Mode Slotted TDMA Digital Bus*, Fifth Data Communication Symposium, September 1977, IEEE 77 CH 1260-9C, pp.5-14 to 5-18.
15. Meisner, N.B., Willard, D.G. and Hopkins, G.T. *Time Division Digital Bus Techniques Implemented on Coaxial Cable*, Proc. Computer Networking Symposium, December 1977, pp.112-117.
16. Lam, S.S. *Delay Analysis of a Time Division Multiple Access (TDMA) Channel*, IEEE Trans. vol. COM-25, no.12, December 1977, pp.1489-1494.

(c) ALOHA TYPE PROCEDURES

17. Abramson, N. *The ALOHA System – Another Alternative for Computer Communication*, AFIPS, vol.37, 1970 Fall Joint Computer Conference, pp.281-285.
18. Abramson, N. *Packet Switching with Satellites*, National Computer Conference, AFIPS, vol.42, 1973, pp.695-702.
19. Abramson, N. *The Throughput of Packet Broadcasting Channel*, IEEE Trans. vol. COM-25, no.1, January 1977, pp.117-128.
20. Binder, R., Abramson, N., Kuo, F., Okinaka, A. and Wax, D. *ALOHA Packet Broadcasting – A retrospect*, Nat. Computer Conf., AFIPS, vol.44, 1975, pp.203-215

21. Sastry, A.R.K. *A simple Reservation Scheme to Improve Excess Capacity of Slotted ALOHA Channels*, ICC 1977, vol.2, pp.23.3-117 to 22.3-120.
22. Ferguson, M.J. *On the Control, Stability and Waiting Time in a Slotted ALOHA Random Access System*, IEEE, Trans. vol. COM-23, no.11, November 1975, pp.1306-1311.
23. Carleial, A.B. and Hellman, M.E. *Bistable Behaviour of ALOHA-Type Systems*, IEEE Trans. vol. COM-23, no.4, April 1975, pp.401-409.
24. Metzner, J.J. *On Improving Utilization in ALOHA Networks*, IEEE Trans. vol. COM-24, no.4, April 1976, pp.447-448.
25. Gerla, M. and Kleinrock, L. *Closed Loop Controls for ALOHA Satellite Communications*, Fifth Data Communication Symposium, September 1977, IEEE 77 CH 1260-9C, pp.2-10 to 2-19.
26. Gitman, I., *On the Capacity of Slotted ALOHA Networks and Some Design Problems*, IEEE Trans. vol. COM-23, no.3, March 1975, pp.205-317.

(d) CARRIER SENSE PROCEDURES

27. Saito, T. *A Subscriber Carrier System Based Upon Frame Addressing System.* IEEE Trans. vol. COM-26, no.8, August 1978, pp.1287-1295.
28. Mark, J.W. *Global Scheduling Approach to Conflict-Free Multi-access via a Data Bus*, IEEE Trans. vol. COM-26, no.9, September 1978, pp.1342-1351.
29. Buchfield, J., Tomlinson, R. and Beeler, M. *Function and Structure of a Packet Radio Station*, Nat. Computer Conf., AFIPS, vol. 44, 1975, pp.245-251.
30. Fralick, S.C., Brandin, D.H., Kuo, F.F. amd Harrison, C. *Digital Terminals for Packet Broadcasting*, Nat. Computer Conf., AFIPS, vol.44, 1975, pp.253-261.
31. Lam, S.S. and Kleinrock, L., *Dynamic Control for a Packet Switched Multi-Access Broadcast Channel*, Nat. Computer Conf., AFIPS, vol.44, 1975, pp.143-153.
32. Lam, S.S. and Kleinrock, L. *Packet Switching in a Multi-access Broadcast Channel: Dynamic Control Procedures*, IEEE Trans. vol. COM-23, no.9, September 1975, pp.891-904.
33. Lam, S.S. *Satellite Multi-access Scheme for Data Traffic*, ICC 1977, vol.3, pp.37.1-19 to 37.1-24.
34. Kleinrock, L. and Tobagi, F.A. *Random Access Techniques for Data Transmission Over Packet-Switched Radio Channels*, Nat. Computer Conf., AFIPS, vol.44, 1975, pp.187-201.
35. Kleinrock, L. and Tobagi, F.A. *Packet Switching in Radio Channels: Part I — Carrier Sense Multiple-Access Modes and Their Throughput-Delay Characteristics*, IEEE Trans. vol. COM-23, no.12, December 1975, pp.1400-1416.
36. Tobagi, F.A. and Kleinrock, L. *Packet Switching in Radio Channels: Part II — The Hidden Terminal Problem in Carrier Sense Multiple-Access and the Busy-Tone Solution*, IEEE Trans. vol. COM-23, no.12, December 1975, pp.1417-1433.
37. Tobagi, F.A. and Kleinrock, L. *Packet Switching in Radio Channels: Part III — Polling and (Dynamic) Split Channel Reservation Multiple Access*, IEEE Trans. vol. COM-24, no.8, August 1976, pp.832-844.
38. Tobagi, F.A. and Kleinrock, L. *Packet Switching in Radio Channels: Part IV — Stability Considerations and Dynamic Control in Carrier Sense Multiple-access*, IEEE Trans. vol. COM-25, no.10, October 1977, pp.1103-1119.
39. Tobagi, F.A. and Kleinrock, L. *The Effect of Acknowledgement Traffic on the Capacity of Packet-Switched Radio Channels*, IEEE Trans. vol. COM-26, no.6, June 1978, pp.815-826.

40. Kleinrock, L. and Scholl, M. *Packet Switching in Radio Channels: New Conflict Free Multiple Access Schemes for a Small Number of Data Users*, ICC 1977, vol.2, pp.22.1-105 to 22.1-111.
41. Gerla, M., Nelsom, L. and Kleinrock, L. *Packet Satellite Multiple Access: Models and Measurements*, NIC 1977, pp.12:2-1 to 12:2-8.
42. Hansen, L.W. and Schwarts, M. *An Assigned-Slot Listen-Before-Transmission Protocol for a Multi-access Data Channel*, ICC 1977, vol.2, pp.22.2-112 to 22.2-116.

CHAPTER 10

Simultaneous Multiple-Access Transmission

10.1 INTRODUCTION

Means of sharing the common communication channel by a number of users (terminals) were discussed in the previous chapter. The common channel was defined as having a fixed operating bandwidth in which any number of terminals can transmit their messages. The facility through which these terminals, distributed over a wide area, may give access to the common channel was termed multiple access.

Multiple access communication, as pointed out in the previous chapter, allows only one transmission at any given time. It is the intention of this chapter to expand the multiple-access procedures by describing newer techniques which enable simultaneous communication from many sources over the same channel. Such techniques are not intended to increase network efficiency but to increase communication reliability. They are becoming an integral part of C^2 systems for special radio communication where direct undisturbed channels are required. An example showing the application of these techniques was given in Chapter 1 for ambulance communication.

Multiple-access procedures have been introduced to ensure smooth transmission of the messages to and from the various terminals through the medium of a common communication channel. The procedures are intended to introduce orderly control of the message flow by employing software means. Successful transmission is only possible when one and only one message is transmitted in the channel at any given time. Collision of two or more messages while being transmitted, causes the destruction of all the messages involved. Even if only a small part of two messages overlaps in time, both messages are rejected. This method of multiple-access, where the messages must flow in series could be regarded as a form of time multiplexing. In other words, all the multiple-access procedures are used as means of assigning time slots to the various terminals for transmitting in the common channel by either fixed, random or dynamic rules. All these procedures can be included under the extended concept of TDMA in which each terminal uses the same operating frequency band but not simultaneously.

Other means of sharing the common channel is assigning discrete frequency bands to the various terminals in which they can transmit their messages. In this form of multiple-access, a number of terminals transmit simultaneously but not in the same operating band. Such a scheme is referred to as frequency-division-multiple-access (FDMA) where each terminal, situated over a wide geographical area, is assigned a given frequency band. It is theoretically possible to design FDMA procedures where the channel bandwidth is assigned by either fixed, random or dynamic rules. Nevertheless, FDMA is generally employed in a fixed assignment mode. If, however, random or dynamic frequency assignment is used, the discrete frequency bands are achieved with the aid of a frequency synthesizer; although then, it is not a practical or an economical solution to assign more than one frequency band at any given time.

There is a completely different approach to time or frequency multiple-access operation, in a scheme which enables a large number of users (terminals) to have access to the full common channel bandwidth simultaneously. That is, many terminals spread over a wide geographical area may be operating in parallel by using the same frequency band, but without one terminal transmission disturbing or affecting the transmissions of other terminals. This is possible by coding each transmitted signal by means of a unique code which makes each signal distinguishable by its detailed coded structure and hence being referred to as code-division-multiple-access (CDMA). CDMA enables signals from many terminals to transmit their messages simultaneously and occupy the same frequency bandwidth.

When discussing the assignment of the transmission spectrum, it is usual to regard the spectrum in a two dimensional form of time and frequency, as seen in Fig. 10.1. It can be seen that FDMA provides simultaneous operations but not in the same operating band, whereas TDMA provides the same operating band but not simultaneously. Although it is shown that both multiple access techniques equally exploit the spectrum, in practice, TDMA permits more flexible and economical sharing of the spectrum than FDMA. This is because TDMA can adjust itself to traffic demands, while FDMA has wasted bandwidth due to inadequate frequency separation. TDMA, however, requires full coordination between the terminals while FDMA requires no real-time coordination among the terminals and can transmit either analog or digital information.

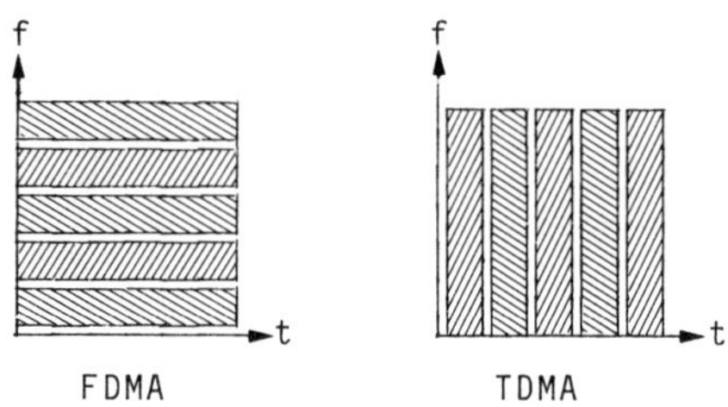

Fig. 10.1 Two dimension assignment of the spectrum according to frequency and time

CDMA gives the spectrum space pattern a third dimension, as seen in Fig. 10.2, thus enabling the assignment of the spectrum also according to code. That is, CDMA makes the full frequency bandwith available to all terminals simultaneously. It does not, however, exploit the spectrum resources as well as FDMA or TDMA, although it has other advantageous features, as will be explained later. All the three multiple-access schemes, shown by the three-dimension pattern, employ techniques which have disjoint frequency, time elements or distinct code.

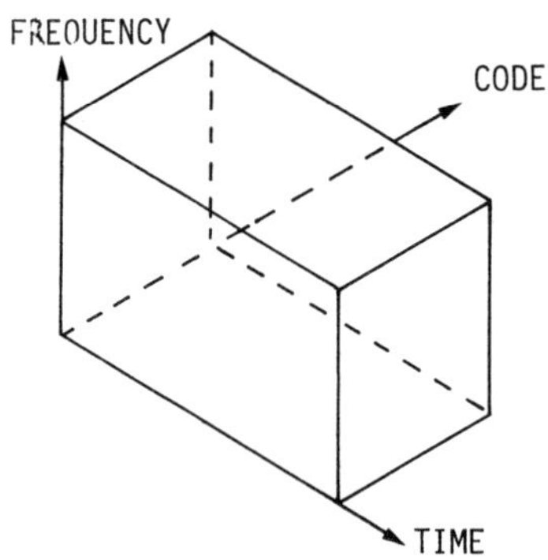

Fig. 10.2 Three dimension assignment of the transmitted spectrum

CDMA comes under various names, of which the most common form is spread-spectrum multiple-access (SSMA), although it must be noted here that not all the accepted SSMA schemes can be regarded as CDMA.

SSMA, operated as CDMA, is a multiple access scheme in which each terminal is assigned a particular code sequence which is modulated on a carrier along with digital data.

Before continuing, it is worth adding an important capability of CDMA which, unlike the conventional TDMA and FDMA, requires no precise time or frequency synchronization between the transmissions from the various terminals. In fact, each terminal can transmit whenever it desires to do so, with no limitations, i.e., in the same way that was described for pure ALOHA. CDMA, however, is superior to ALOHA since there is no danger of the message being destroyed due to collision.

10.2 SPREAD SPECTRUM PRINCIPLES

Spread spectrum communication techniques are the complete contrast to the conventional communication techniques currently in use. Instead of limiting the transmission channel bandwidth, as proposed in the chapter on digital modulation techniques, the aim here is to expand the channel bandwidth. Spread spectrum was originally introduced for purely military applications, as it has considerable advantages over the conventional communication techniques in providing both jamming resistance and cryptographic secrecy. Both these military advantages were not considered important enough for commercial application to justify the high expenditure on the complex circuitry required. Nevertheless, since the introduction of satellite communication, commercial transmissions have been open to espionage, and this has made the military spread spectrum features attractive also to commercial communication. Furthermore, in the last few years other features of spread spectrum have made this technique an important asset in commercial application.

As the radio communication spectrum is over-crowded with transmission channels and leaves no room for any additional channels, the requirement for extra channel space is constantly growing. One means of providing this need is by going increasingly to higher frequency bands. These frequencies, however, are also limited and are gradually being saturated, as well as causing increase in system cost. The only seemingly means left for possible increasing the number of channels is better exploitation of the existing channels. This is achieved by reducing the bandwidth available for each transmission which is made possible by employing better modulation schemes. The reduction of the available channel bandwidth and crowding a number of channels with little spacing between them is, however, also limited. There is a point in which any further channel reduction will consequently affect the transmission rate. It can easily be ascertained that the spectrum frequency band in the range of VF, VHF, UHF as well as SHF have reached a saturation point.

A completely different approach in increasing the number of communication channels in a given frequency band is provided by employing spread spectrum techniques. The principle of an approach of spread spectrum is the reverse of that proposed above. Instead of limiting the transmission channel, it is here suggested that the channel be expanded over a very wide bandwidth. Although this statement may seem to be in complete contradiction to the previous suggestion, this is not truly the case, since the spread spectrum technique reduces the transmitted power density and this allows many transmissions to be performed simultaneously.

The power density transmitted with a spread spectrum modulator is very much lower than that required in conventional transmissions. In conventional transmission the bandwidth required to transmit a given information is B_i with the power density of W_i. Since the power to transmit this information is constant, it is possible to reduce the power density and increase the bandwidth without affecting the transmitted information content, as seen in Fig. 10.3. In spread spectrum the bandwidth is increased to B_c and the power density reduced to W_c, with $B_c \gg B_i$ and $W_c \ll W_i$. In both cases the same

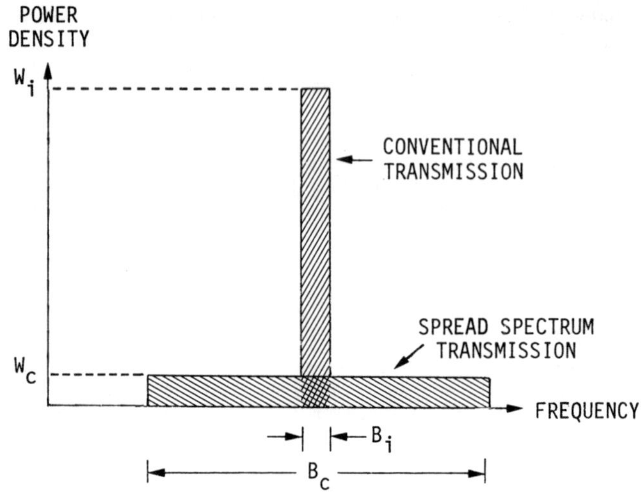

Fig. 10.3 Basic principle of spreading the spectrum

power is required to transmit the data, thus $B_c/B_i = W_i/W_c$

At the receiver of any communication system, the signal-to-noise ratio is affected by the noise in the available frequency band. In conventional communication the disturbing noise is only that present in bandwidth B_i while in spread spectrum the noise is that present in the full bandwidth of B_c. This means that although the same power is required in both schemes to transmit the same information contents, the signal level received in spread spectrum is much smaller and the noise is grossly increased. It may seem that the advantage of transmitting over a spread spectrum is diminished with the extremely poor signal-to-noise ratio received.

Despite the extremely poor signal-to-noise ratio of the spread spectrum signal, it is possible to compress the received information signal back to its original conventional bandwidth and reject most of the noise. This complex processing of the received signal is the most essential feature of spread spectrum, enabling the spread spectrum technique to perform in very noisy environments without this affecting the information transmission. It can be said that spread spectrum has the most efficient interference rejection capabilities not found in any other communication techniques. Any other conventional transmission which is simultaneously present in any part of the spread spectrum bandwidth will only affect a small part of the received signal (i.e., by the ratio of B_i/B_c). After the signal processing, this 'disturbing' signal will simply be detected as noise and will have a negligible effect on the 'true' signal. In fact, because of the interference rejection capabilities of spread spectrum, the signal detection is much more reliable than conventional signal detection.

The features of spread spectrum communication which make it attractive to commercial application could be summarized as follows:

(a) Spread spectrum transmission is ideal for operating in noisy environments;

(b) Spread spectrum transmission is not affected by other transmissions operating in the same bandwidth;

(c) Spread spectrum transmissions do not affect other transmissions operating in the same bandwidth because of the low power densities of the spread spectrum transmissions;

(d) Many simultaneous spread spectrum transmissions may be present in the same bandwidth without any one of them affecting the others;

(e) Spread spectrum provides good rejection of external interference and can operate simultaneously with other transmissions using the same frequency bandwidth;

(f) The data received in the spread spectrum system is much more reliable than that in a conventional system;

(g) The power density of the spread spectrum signal is much lower than that in conventional techniques. This makes the signal difficult to detect by a competitor using simple equipment;

(h) The wide bandwidth of the transmission makes it difficult for a competitor to jam or disturb the transmission.

The term spread spectrum is somewhat misleading. The reader may mistakenly consider that the transmitted information pulse, which requires the bandwidth of B_i, is spread over the wide bandwidth of B_c. This might possibly be true for analog signal transmission, but it is not the case for transmitting data. The digital pulse is not expanded but the same bit is encoded a number of times, each time in a different form and only then is the result transmitted. Since the same bit is transmitted a number of times, the transmitted rate is consequently increased and accordingly there will be a need for a wider bandwidth. In other words, the same information bit is not 'spread' but is first duplicated several times and then the result distributed over a wider spectrum.

The techniques employed to achieve spread spectrum are not new but in fact, are identical with those applied in digital modulation. That is, the transmission and reception of spread spectrum signals are performed in exactly the same manner as described for the modem operation used for data transmission. It can readily be said that the spread spectrum transmitter/receiver is a modem having the coded digits modulating the RF signal. This means, in effect, that the spread spectrum transmitter/receiver requires no IF stage, since the coded data directly modulates the RF carrier signal. If this is the case, the question may well be asked what the difference is between the principles of data communication and spread spectrum. The straightforward answer is that there is no operational difference between the modem used in data transmission and the receiver/transmitter in spread spectrum. They can both be regarded as identical although there might be differences in carrier frequency. The difference between the two techniques is not in the modulation-demodulation equipment, but in the way the information digits are processed before transmission. In data communication, the process is performed so that the transmission signal offers the maximum efficient utilization of the available bandwidth. This is made possible by having each modulation state represent a number of information bits, i.e., being equal to a bit, a dibit or even a fourbit. In spread spectrum the process is a reverse performance. There is no need for the maximum clustering of the data in a narrow band: on the contrary, the data are distributed over the widest possible band by transmitting the same information bit a number of times.

There are several spread spectrum techniques available to the communication designer for distributing the information data over a wide bandwidth. This chapter is confined only to the two most important techniques, that of direct sequence and frequency hopping.

10.3 PHASE CODED SPREAD SPECTRUM

The most common spread spectrum system used for communication is the direct sequence where each terminal is assigned a particular code sequence. The terminal's 'modem' modulates this code sequence on the carrier along with digital data. Before this system is elaborated, some further background is essential.

In both data communication and direct sequence spread spectrum, digital data is transmitted over the communication channel. The only difference between the two techniques is the transmitted bandwidth required, and consequently also the bit rate. With data communication the bit rate is R_i whereas with spread spectrum it is R_c, with $R_c \gg R_i$.

According to Shannon's theory, the maximum channel rate R_{max} is finite and this is referred to as the channel capacity C. That is, in a bandwidth of B Hz, it is possible to transmit $C = B \log_2 (1 + S/N)$ bits/second, where S/N is the received signal-to-noise ratio. This means that the capacity of the channel may be increased with the increase of the frequency band, even though the signal-to-noise ratio might be decreased. In effect, the rate of transmission R of a data source ($R = 1/T$ where T is equal to the clock timing) requires a given fixed bandwidth B. Any means of limiting the bandwidth below its minimum value will only introduce errors. That is, if we try to push too many bits per second over the channel, the errors begin to mount rapidly. Since we are dealing with radio transmission we can consider the available transmission spectrum to be unlimited. We can therefore assume that the required transmission bandwidth B is a direct function of the bit rate R.

When transmitting a carrier signal of $\cos \omega t$ then its transmitted spectrum is only that of a single frequency f_c, as seen in Fig. 10.4(a). If, however, the carrier signal is modulated with a data sequence $A(t)$, the transmitted signal $S(t) = A(t) \cos \omega t$ now occupies a bandwidth of B Hz. With a data rate of $R_1 = 1/T_1$ a bandwidth of B_1 is required, as seen in Fig. 10.4(b), and with a data rate of $R_2 = 1/T_2$ a bandwidth of B_2 is required, as seen in Fig. 10.4(c). Since $R_2 > R_1$ hence $B_2 > B_1$.

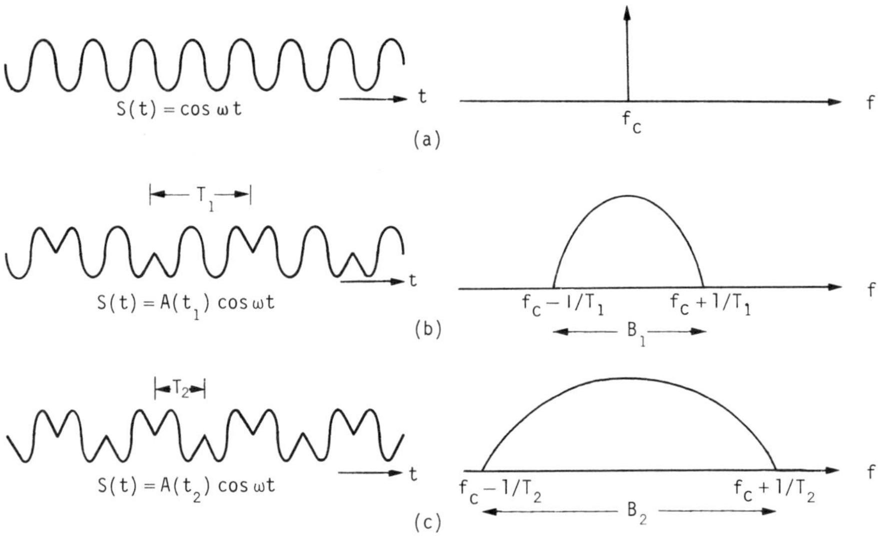

Fig. 10.4 The required transmission bandwidth

From the discussion above one can state that for the transmission of an information sequence with the bit rate of R_i, a bandwidth of B_i is required. By increasing the bandwidth to B_c, with $B_c \gg B_i$, it is possible then to transmit in the extended bandwidth a bit rate of R_c, which is much higher than the original bit rate, i.e., $R_c \gg R_i$. This increased data rate R_c is used to modulate the carrier signal and which thus consequently causes the spreading of the spectrum. What still remains is to explain how the data rate R_i can be transmitted at the faster rate of R_c and still retain both the data contents unaffected and with reference to the original clock timing T_i. This, however, will be explained in the following section.

There are many modulation techniques that may be used to modulate the carrier in a spread spectrum system. In fact, all the modulation techniques used for data transmission, such as ASK, FSK and PSK, may be employed. Phase shift keying, however, has been found to be the most efficient technique for code spread spectrum communication. In BPSK the phase of the carrier is shifted by 180° for each level change in the NRZ bit stream. This means that for any change in the data value from 1 to 0 or from 0 to 1 there is a shift of 180° in the carrier signal, as seen in Fig. 10.5. This modulation system was defined in Chapter 6 as a form of double-sided-amplitude-modulation-suppressed-carrier (DSBSC-AM). As such, the bandwidth required for transmission is equal to twice the bit rate of R_c, i.e., $B_c = 2R_c$.

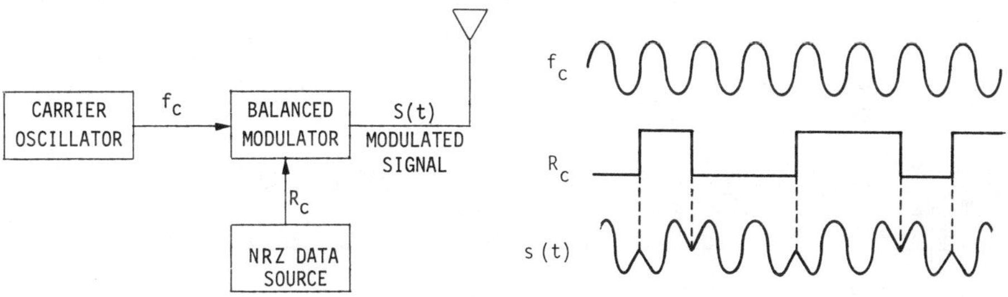

Fig. 10.5 Direct sequence spread spectrum modulation system

By using BPSK for spread spectrum applications, there are several benefits gained over other forms of digital modulation. Some of these benefits are summarized below:

(a) BPSK has a constant amplitude envelope modulated signal which is similar to that of FSK and efficiently exploits the spectrum in a similar manner to that of DSBSC-AM:

(b) Since a constant amplitude envelope is transmitted, the power density is spread evenly over the whole extended spectrum. This helps the efficient employment of the transmitter, an advantage which is not possible with OOK-ASK where the power density is a function of the data;

(c) The modulation circuits are simple, since the centre carrier frequency remains constant throughout the transmission period, an advantage which does not exist in FSK:

(d) Suppressing the carrier enhances the difficulties in detecting the presence of the signal by simple means ;

(e) Suppressing the carrier allows the transfer of the transmitted power to the side bands for transmission of the coded signal.

The benefits of BPSK, as listed above are identical with those mentioned for data cimmunication. The question may be asked again what the difference is between the two communication techniques. Here we can answer precisely. In data communication, the modulated signal $S(t)$ is directly related to the information rate $R_i = 1/T_i$, where T_i is the basic clock timing. In spread spectrum, the modulated signal $S(t)$ is related to the coded source with the rate of $R_c = 1/T_c$, with $R_c \gg R_i$.

10.4 DIRECT SEQUENCE SPREAD SPECTRUM

It is not accurate to state that the transmitted signal $S(t)$ is not related to the information rate, as there is a definite relation between the code and the information rate. That is, for each information data bit, there are N transmitted coded bits, viz., giving the ratio $N = R_c/R_i$. This ratio represents the factor by which the original bandwidth is spread, i.e., $N = B_i/B_c$. For practical spread spectrum systems it is usual to take N as being in the range of 500 – 2000.

As already mentioned, each data bit is transmitted N times, each time in a different coded form. In order to distinguish it from the data bit, each of the N coded bits is known as a 'chip'. The coded signal is achieved by modulo-two addition of a sequence of N chips to each of the data bits, as shown in Fig. 10.6. The result of this addition, i.e., $R_i \oplus R_c$, is used to modulate the carrier. In the example shown in Fig. 10.6, the ratio between the chip and data rates N is 7. While the data stream depends on the information contents being transmitted, the chip stream is repeated for each data bit. That is, a stream pattern of N chips is transmitted and then repeated; thus N can be regarded also as the chip sequence period. The modulo-two addition of the data and chip stream has the advantage of producing a signal representing the original data; this signal however is transmitted at a much higher rate, as requested for spread spectrum.

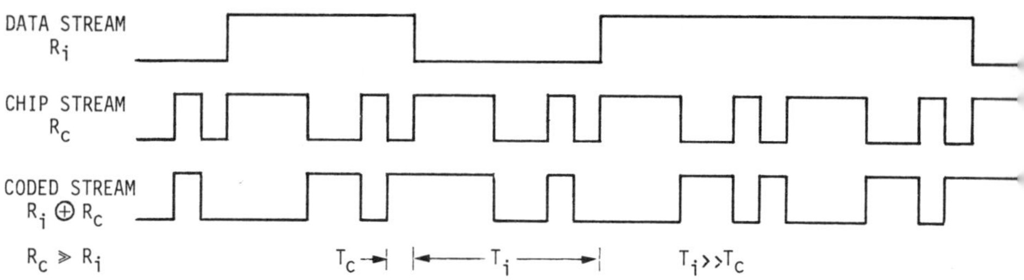

Fig. 10.6 Direct sequence spread spectrum waveforms

Modulo is a capacity or a unit of measurement; that is, it represents the maximum number that can be exhibited by the system. A modulo sum is a sum with respect to a modulus with the carry digits ignored. Modulo-2 is a form of binary notation except that for additions of two binary bits, there are no carry digits and for subtraction there are no borrow digits. This can be seen by the following example using the modulo-two addition equations of $0 \oplus 0 = 0$, $0 \oplus 1 = 1$, $1 \oplus 0 = 1$ and $1 \oplus 1 = 0$.

$$\begin{array}{rl} & 1\;0\;1\;1\;0\;1\;1 \quad - \; A \\ \oplus & 1\;1\;0\;1\;0\;1\;0 \quad - \; B \\ \hline & 0\;1\;1\;0\;0\;0\;1 \quad - \; C = A \oplus B \end{array}$$

Modulo-two addition has the advantage that a further addition of the result with one of the products is equal to the other product. This means that if A and B are modulo-two added, then the further addition of the result with A will produce B, or the result with B will produce A. This can be seen by continuing the above example.

	0 1 1 0 0 0 1 — C			0 1 1 0 0 0 1 — C
$\oplus$	1 0 1 1 0 1 1 — A		$\oplus$	1 1 0 1 0 1 0 — B
	1 1 0 1 0 1 0 — $B = C \oplus A$			1 0 1 1 0 1 1 — $A = C \oplus B$

The above (reverse operation feature of modulo-two addition) enables us to decipher the coded data signal back to its original form, by simply adding the received signal with an identical chip coded sequence (provided they are both synchronized). This can be represented by the following equation :

$$(R_i \oplus R_c) \oplus R_c = R_i.$$

The modulo-two addition is often called EXCLUSIVE-OR operation since it has the same properties, $A \oplus B = C$. If either A or B are of complementary values then $C = 1$, but when both A and B have equal values then $C = 0$.

In the technical literature the same operation may be represented by two disciplines. It is possible to replace the modulo-two addition (EXOR) by multiplication. In that case the conventional 1 and 0 binary states are replaced by the -1 and $+1$ signal levels, respectively. When using the signal level notation, the multiplication is referred to as a 'mixing' operation. The practical result of both operations is the same. This is made clear by the following table:

$1 \oplus 1 = 0$	$-1 \times -1 = +1$
$1 \oplus 0 = 1$	$-1 \times +1 = -1$
$0 \oplus 0 = 0$	$+1 \times +1 = +1$

It is usual to use the modulo-two addition operation when regarding the data as a sequence of square waves and the mixing multiplication operation when regarding the data as an analog wave.

The chip stream pattern of N bit period can take any sequence form of ones and zeros, provided that it can be reproduced identically for each data bit and that it is available both in the transmitter and receiver. As will be seen later, the best chip stream pattern is one which has pure random characteristics of ones and zeros, i.e., the number of ones equals the number of zeros with a non-repetitive pattern. It is possible to generate a sequence of the required chip length of N bits by a deterministic circuit with a pattern which looks random. This artificial sequence is not random since it has a finite periodicity, although within the period it has random characteristics. For this reason it is known as a pseudo-random sequence. Pseudo-random sequences are extensively used in error control, privacy encoding and other similar systems.

The chip pseudo-random sequence is produced by a linear feedback shift register of n stages. By suitably connecting the feedback gates, it is possible to achieve a sequence output stream with the maximum possibly length of $N = 2^n - 1$ bits. The length is measured by having a different pattern of ones and zeros for each consecutive n bits. Maximum length is achieved only if their are $2^n - 1$ different patterns (the pattern of n zeros is not considered). Only a sequence achieving the maximum length has the required random characteristics. A shift register which produces the required sequence is known both as maximum-length linear shift register and as a pseudo-random generator (PRG).

The data stream and the chip stream must be synchronized by being operated from the same clock reference source. This is achieved by counting N clock pulses, as seen in Fig. 10.7. The two streams are modulo-two added together by an EXOR gate producing the coded data stream with the same bit rate as the chip stream.

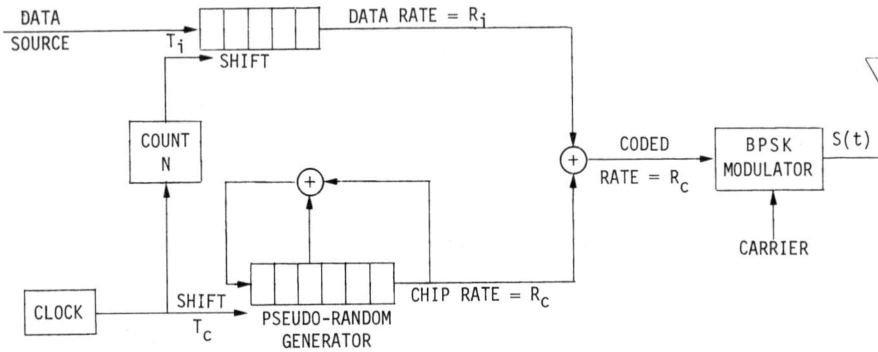

Fig. 10.7 Code sequence as means of expanding the spectrum

It can be seen that the original data stream, with the bit rate of R_i, has been coded thus producing a new stream with the bit rate of R_c. This means that by feeding the BPSK modulator with the coded stream the signal produced will have a bandwidth equal to B_c. This achieves the necessary requirements for spreading the original bandwidth of B_i over the wider bandwidth of B_c.

10.5 DETECTION OF THE SPREAD SPECTRUM SIGNAL

In the spread spectrum transmitter the data stream is both coded and modulated before it is transmitted into the spectrum. The transmission channel (i.e., spectrum) introduces unwanted noise into the system. Thus the channel may be regarded as a mixer in which the noise is added to the signal, i.e., $S(t) + X(t)$. Such noise is called 'white noise' because it is uniformally spread in all frequencies and has pure random features. In the case of spread spectrum, since the power density of the transmitted signal is low, the modulated data signal is embedded deep into the noise. If one does not know that there are transmitted signals in the spectrum, the signal $S(t) + X(t)$ seems like noise.

Recovering the data stream requires the received signal to be first separated from the noise, then demodulated and lastly decoded. Furthermore, one cannot decode the data stream before correlating the transmitted sequence with the corresponding chip sequence in the receiver. Such a serial sequence of events is extremely difficult to perform and other methods must preferably be used. In order to understand the receiver operation it is well first to assume that we have an ideal transmission channel with no losses or added noise and that both streams are fully correlated.

It has already been explained that the transmitted signal was achieved by two independent operations, the coding and the modulation processes. In the receiver too, these two operations can be regarded as separate processes with no direct relation between them, as seen in Fig. 10.8. That is, the signal is detected in the same way as with any BPSK modem, thus producing a stream of square pulses with the bit rate of R_c. This stream is identical to the transmitted data coded stream. By modulo-two addition of this received stream with the receiver internal pseudo-random sequence, the original data stream is reproduced, provided that the pseudo-random sequence is synchronized with the data. (At this stage, the two streams will be assumed to be fully synchronized.) It should be added that both the transmitter and receiver must employ the identical pattern of pseudo-random sequence.

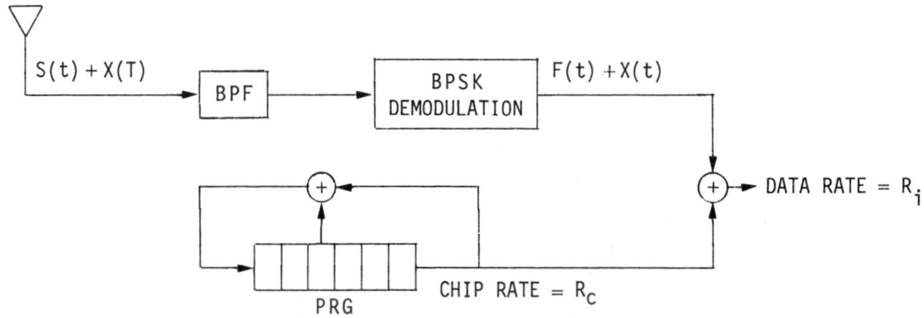

Fig. 10.8 Basic principle of spread spectrum receiver

The bandpass filter, shown in Fig. 10.8, is intended to limit the noise spectrum to the transmitted bandwidth. Thus the signal out of the demodulator is the coded data stream mixed with the noise limited to the spectrum bandwidth. The second half of the receiver is the decoder circuit which operates independently of the demodulator. The waveforms of the decoder circuit (neglecting the noise), are shown in Fig. 10.9. It can be seen from these waveforms that the decoding process is achieved simply by modulo-two adding the received semodulated signal to the internal pseudo-random stream.

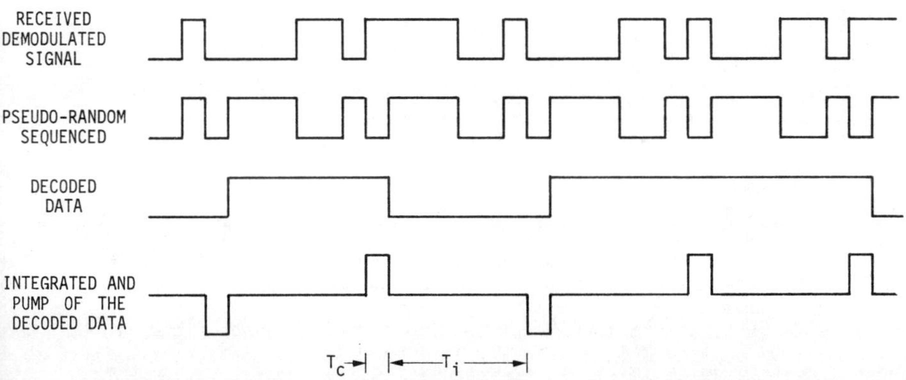

Fig. 10.9 Spread spectrum receiver-decoder waveforms

The process of decoding the data, using the modulo-two addition operation, is only possible if both streams are fully synchronized. Since the pseudo-random sequence generators in both the transmitter and receiver operate independently of each other, such a synchronization requirement cannot be guaranteed. Instead one must search for synchronization by shifting the receiver stream relative to the transmitter stream till synchronization is achieved, This forms an essential part of a process known as correlation. Correlation of digital data can be produced only if each and every bit in both sequences are compared for a complete chip period. By having the chip period corresponding to the required data bit period and both sequences correlated, then the result of each and every comparison bit process must give a positive result. At the end of the period if the number of positive results equals N, both sequences are correlated, but, if the result is less than N both sequences are uncorrelated. It will be shown later that when they are not correlated the result is much less than N. If after a complete chip period the two streams are found to be uncorrelated the internal chip stream is shifted by one bit and another correlation process is performed.

The correlation process as described could be regarded as a process where the received pulse after the modulo-two addition is scanned at the bit rate of R_c, i.e., every chip clock pulse T_c, and the results are summed up for the period of T_i which is the data bit period. If both sequences are synchronized a positive going pulse is received for a data '1' pulse and a negative going pulse for a data '0' pulse. No output will be obtained if both sequences were not synchronized. The waveform result of sampling the data pulse is also included in Fig. 10.9. The operation of sampling the data pulse and summing the result, a process which is essential for correlation, is shown in Fig. 10.10. The process could also be regarded as an integration and dump of the decoded data pulse, i.e.,

$$\sum_{j=1}^{N} F(jT_c) = \int_0^{T_i} F(T_c)\, dT_c \qquad \text{with } T_i = NT_c.$$

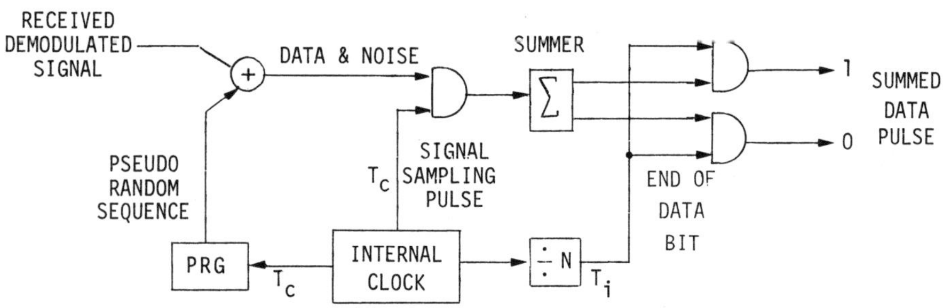

Fig. 10.10 Sample and summer circuit in the receiver

The process of sample and sum (integrate and dump) is required not only for correlation purposes; in fact it is the main operation for detecting the spread spectrum signal. Without this process it is impossible to recover the signal cleared from the noise.

The received data signal is difficult to detect because its amplitude is small and it is embedded deep in the noise. In other words, it presents a poor signal-to-noise ratio, as suggested in Fig. 10.11(a). With the aid of the sample and sum process, the data signal which has been spread over the whole time spectrum, is now compressed into a narrow time pulse while leaving the unwanted noise signal amplitude unchanged, as suggested in Fig. 10.11(b). The correlation process removes the coding from the data signal but has no effect on the noise. Since the internal clock T_c is N times faster than that of the data sequence timing T_i, each data bit period is sampled N times. This means that at the end of the data period, a significant output pulse is produced if the receiver sequence is correlated with that of the transmitter. If the sequences are not correlated only a small pulse is produced, representing the noise.

Sampling of the noise signal has no additive effect at all, since the noise has random characteristics with both positive and negative going pulses. Sampling a correlated data bit period signal produces pulses of a single direction. In the summer, the noise pulse samples cancel out, while the data pulse samples add up. Clearly, the wider the ratio of data period to the

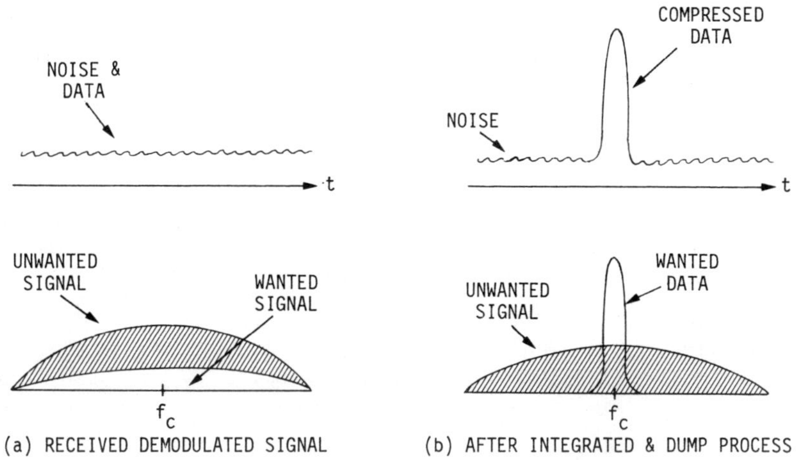

Fig. 10.11 Detection process of the wanted and unwanted signals

sampling period, i.e., T_i/T_c, the smaller the effect will be of the unwanted noise signal on the wanted signal at the output of the summer.

10.6 CORRELATION THEORY

The receiver decoding process as described regarded each operation as a single operation. In this operation, called correlation, the received signal is modulo-two added (multiplied) with the locally generated pseudo-random sequence and the resulting output integrated. By this method, synchronization is checked, the spread pulse is compressed and the noise rejected. The correlation process is a process where the signal is sampled every T_c and the results are summed up and checked once every T_i, i.e., after N samples. This can be expressed mathematically using modulo-two notation. The expression is known as the normalized correlation function ρ and is defined as follows:

$$\rho = 1/N \sum_{j=1}^{N} F_1(jT_c) \oplus \bar{F}_2(jT_c)$$

$$= 1/NT_c \int_0^{NT_c} [F_1(t) \oplus \bar{F}_2(t)]\, dt$$

$$= 1/T_i \int_0^{T_i} [F_1(t) \oplus \bar{F}_2(t)]\, dt$$

where $F_1(t)$ describes the transmitter P.R. sequence and $F_2(t)$ the receiver internal P.R. sequence. The internal sequence is complemented to provide a positive indication when the two sequences are summed.

By regarding the data bits 0 and 1 as voltage levels +1 and −1, the correlation equation can be rewritten as a multiplication operation.

$$\rho = 1/N \sum_{j=1}^{N} F_1(jT_c)\, F_2(jT_c) = 1/T_i \int_0^{T_i} [F_1(t)\, F_2(t)]\, dt.$$

As already stated, the received signal is not only a function of the data but it is also mixed with the noise, i.e., $F_1(t) + X(t)$ and this must be considered in the correlation process

$$\rho = 1/T_i \int_0^{T_i} [F_1(t) + X(t)]\, F_2(t)\, dt$$

$$= 1/T_i \int_0^{T_i} F_1(t)\, F_2(t)\, dt + 1/T_i \int_0^{T_i} F_2(t)\, X(t)\, dt$$

$$= \rho_{\mathbf{data}} + \rho_{\mathbf{noise}}$$

The left integration, i.e., $\rho_{\mathbf{data}}$, presents the sum of all the N data samples in a full period, while the right integration, i.e., $\rho_{\mathbf{noise}}$, presents the sum of all the noise samples. Since the noise $X(t)$ is random with symmetrical waves about the zero level, the average value of correlation to the noise with the pseudo-random sequence is zero, that is

$$\rho_{\mathbf{noise}} = \lim_{NT_c \to \infty} 1/NT_c \int_0^{NT_c} F_2(t)\, X(t)\, dt = 0\ .$$

If so, then it is obvious that the longer the integration time, i.e., the larger the ratio T_i/T_c, the smaller the effect of the noise.

Autocorrelation in this context refers to the degree of correspondence between a sequence and a phase shifted replica of itself. This can be expressed mathematically as a function of time

$$\rho(T_c) = \lim_{NT_c \to \infty} (1/NT_c) \int_0^{NT_c} F_1(t) \cdot F_2(t + kT_c)\, dt\ .$$

If $F_1(t)$ and $F_2(t)$ represent the same sequence, the function gives the autocorrelation, whereas when they represent separate functions, the cross-correlation is determined. In the case of identical sequences, the maximum correlation occurs when both sequences are fully synchronized and a peak output the occurs, i.e., for $F_1(t) = F_2(t)$ with $k = 0$ then $\rho = 1$. If the sequences are phase shifted by at least $+1$ or -1 bits then the correlation output is minimum, i.e., $F_1(t) = F_2(t + kT_c)$ with $|k| \geqslant 1$ then $\rho = -1/N$. This produces an output characteristic of a triangular shaped pulse whose base is two chips wide, as shown in Fig. 10.12.

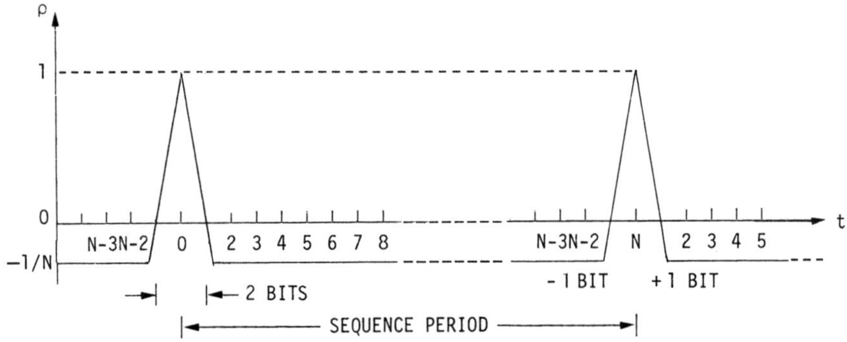

Fig. 10.12 Autocorrelation output for an N chip code

The received signal is sampled N times with the results added up, i.e.

$$\sum_{j=1}^{N} e_j .$$

This factor N, which was defined as the chip pseudo-random sequence length, i.e., $N = 2^n - 1$, also specifies the ratio by which the spectrum is spread in the transmitter, i.e., $N = B_c/B_i$. In the receiver, the same factor N is responsible for the noise rejection. It has been shown that the longer the number of samples per data period, the better the results achieved. In the receiver this same factor is referred to as 'processing gain', and could be given in decibels.

$$G_p = N = B_c/B_i = R_c/R_i = T_i/T_c$$

In most spread spectrum systems a processing gain in the range of 500 to 2000 is commonly used. It may seem that the processing gain could be increased to any desired level, but this is not practically advantageous. Any further increase of the ratio chip to data rates, above a given point, would not improve the noise rejection. In practical systems one must take into account both the system signal-to-noise ratio at the information output and the system implementation losses. The factor to consider is not the processing gain only, but the interference margin M_i :

$$M_i = G_p - [L_{sys} + (S/N)_{out}]$$

where L_{sys} is the system losses and $(S/N)_{out}$ is signal-to-noise ratio.

10.7 PRINCIPLES OF PSEUDO-RANDOM SEQUENCES

The spread spectrum operation as described here is based on characteristics of pseudo-random sequences. Before discussing code-division multiple access it is essential to present some further background of these sequences. It is convenient in digital systems to represent the series of binary digits as the coefficients of a polynomial in a dummy variable of X. If the data series consists of k bits, then the polynomial will be of the order of $(k-1)$:

$$H(X) = a_{k-1}X^{k-1} + a_{k-2}X^{k-2} + \ldots + a_i X^i + \ldots + a_2 X^2 + a_1 X^1 + a_0$$

where a_i may be either '0' or '1' only and $0 \leq i \leq (k-1)$.

The pseudo-random sequences are produced by what is called a maximum-length linear-feedback shift-register. If the maximum length of the sequence is N, the sequence can be represented by the following polynomial:

$$F(X) = a_{N-1}X^{N-1} + a_{N-2}X^{N-2} + \ldots + a_i X^i + \ldots + a_2 X^2 + a_1 X^1 + a_0$$

Such pseudo-random sequences have several important properties which make them attractive for spread spectrum applications. Some of these properties are presented below:

(1) The sequence length of a pseudo-random sequence is exactly $N = 2^n - 1$ chips, before it repeats itself. This is the maximum possible sequence length that can be achieved practically by the combination of n binary bits. An example of such a sequence with a 15 chips length is shown in Fig. 10.13. In this case the pseudo-random sequence is represented by the polynomial $F(X)$:

$$F(X) = 000100110101111$$

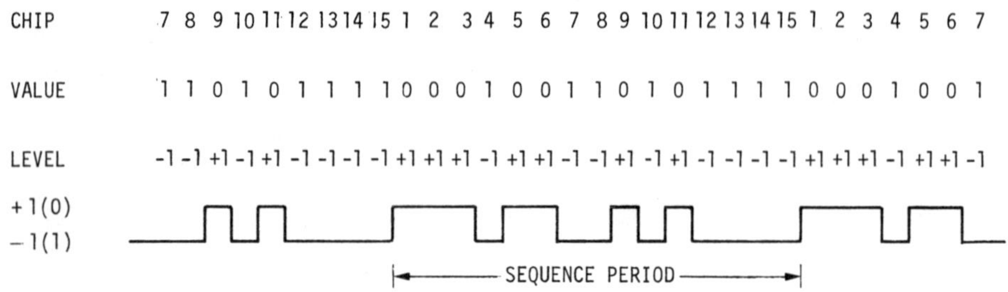

Fig. 10.13 Pseudo-random sequence of 15 chips

(2) A pseudo-random sequence of $2^n - 1$ in length is generated by a linear feedback shift register of n stages, as shown in Fig. 10.14. With any n stage shift register wired in a particular linear feedback mode, it is possible to generate at least one pseudo-random sequence; but not every linear feedback shift register produces a pseudo-random sequence. As the number of the shift register stages n is increased more pseudo-random sequences can be formed by adequately selecting the linear feedback circuits.

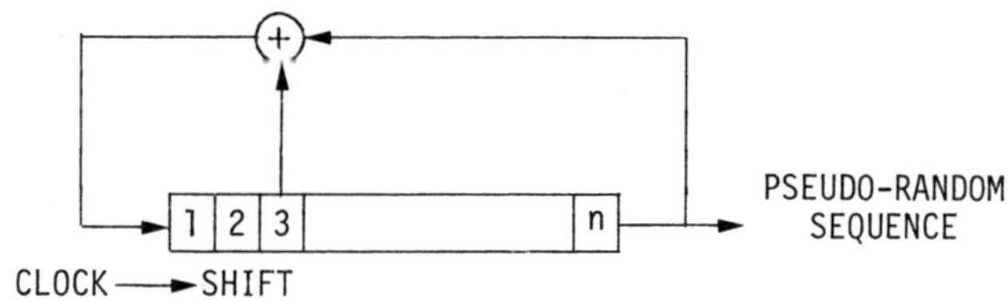

Fig. 10.14 Linear feedback shift register as a pseudo-random sequence generator

(3) An essential requirement of a pseudo-random sequence is that every group of n consecutive chips presents a different pattern of ones and zeroes. In every pseudo-random sequence there are exactly $2^n - 1$ such pattern groups.

The same 15-chip length sequence example (Fig. 10.13) can be presented in 15 pattern groups of n consecutive chips, as follows:

1 — 1110	9 — 0110
2 — 1100	10 — 1101
3 — 1000	11 — 1010
4 — 0001	12 — 0101
5 — 0010	13 — 1011
6 — 0100	14 — 0111
7 — 1001	15 — 1111
8 — 0011	1 — 1110

(4) Among the groups of consecutive runs of ones and zeroes in each sequence, one half of the runs of each kind are of the length one, one fourth of each kind are of the length two, one eighth of the length three, and so on. This means that the statistical distributions of the sequence are well defined and always the same, nevertheless, the

relative position of each run varies from code to code. This is known as the run property.

(5) In any pseudo-random sequence, there is no sequence run of n consecutive chips all having the value zero, although there is one group with a run of n consecutive chips all having the value of one.

(6) Any group of n consecutive chips may be defined as the start or the end of the sequence. For convenience only we shall define a run of n consecutive ones as the end of the chip sequence period, as shown in Fig. 10.15. This group can then be used both in the transmitter and the receiver for synchronization of the chip sequence with the data sequence.

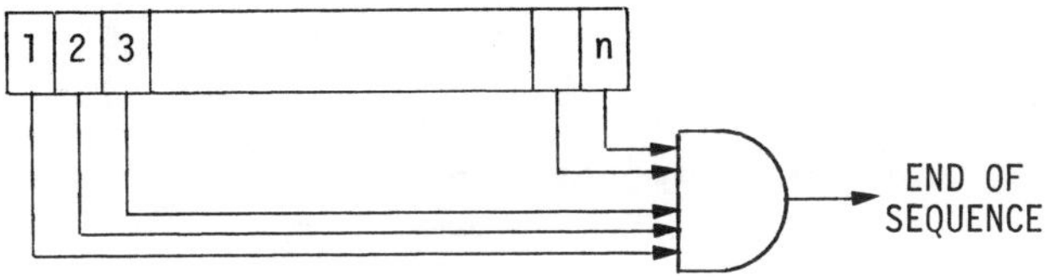

Fig. 10.15 Generating the end of sequence period

(7) In each sequence period the difference between the number of chips having the value 'one' with the chips having the value 'zero' is always one. (This property stems from the fact that there is no run with n zeros.) This can be expressed by the following equation:

$$\sum_{i=0}^{N-1} [F_i(X) = 1] - \sum_{i=0}^{N-1} [F_i(X) = 0] = 1.$$

This equation expresses the random property of the maximum length sequence. Any sequence which does not possess this property is not a pseudo-random sequence, and so this is known as the balance property.

(8) It follows from (7) that in any pseudo-random sequence the number of chips in a complete sequence is odd, with the number of ones even and the number of zeros odd.

(9) The modulo-two addition of two identical pseudo-random sequences which are fully synchronized is always equal to a sequence of zeros.

$$F_1(X) \oplus F_2(X) = 0 \quad \text{with} \quad F_1(X) = F_2(X)$$

For example:

```
  ..... 0 0 0 1 0 0 1 1 0 1 0 1 1 1 1 .....
⊕ ..... 0 0 0 1 0 0 1 1 0 1 0 1 1 1 1 .....
  -----------------------------------------
  ..... 0 0 0 0 0 0 0 0 0 0 0 0 0 0 0 .....
```

In the same way the results of the modulo-two addition of two identical pseudo-random sequences which are fully synchronized, although one is the complement of the other, is always equal to a sequence of ones.

$$F_1(X) \oplus \bar{F}_2(X) = 1 \quad \text{with} \quad F_1(X) = F_2(X)$$

For example:

```
  ..... 0 0 0 1 0 0 1 1 0 1 0 1 1 1 1 .....
⊕ ..... 1 1 1 0 1 1 0 0 1 0 1 0 0 0 0 .....
  -----------------------------------------
  ..... 1 1 1 1 1 1 1 1 1 1 1 1 1 1 1 .....
```

(10) The multiplication (mixing) of two identical pseudo-random sequence signals, which are fully synchronized, always gives a signal with a +1 level.

$$F_1(X) \times F_2(X) = +1 \quad \text{with} \quad F_1(X) = F_2(X)$$

For example:

```
  ..... +1 +1 +1 -1 +1 +1 -1 -1 +1 -1 +1 -1 -1 -1.....
× ..... +1 +1 +1 -1 +1 +1 -1 -1 +1 -1 +1 -1 -1 -1.....
  -----------------------------------------------------
        +1 +1 +1 +1 +1 +1 +1 +1 +1 +1 +1 +1 +1 +1
```

The above results, obtained in (9) and (10) were produced when the sequence was added or multiplied with itself.

(11) The modulo-two addition of two identical pseudo-random sequences, although equal but shifted in time, always give a result which is different from zero.

$$F(X) \oplus F(X-kt) \neq 0 \quad \text{with} \quad 1 \leqslant k \leqslant 2^n - 1$$

For example:

```
  ..... 000100110101111.....
⊕ ..... 001001101011110.....
  --------------------------
  ..... 001101011110001.....
```

It follows that the multiplication of two identical pseudo-random sequence signals shifted in time always gives a result which is different from one.

$$F(X) \times F(X-kt) \neq 1 \quad \text{with} \quad 1 \leqslant k \leqslant 2^n - 1$$

For example:

```
  ..... +1 +1 +1 -1 +1 +1 -1 -1 +1 -1 +1 -1 -1 -1 -1 .....
× ..... +1 +1 -1 +1 +1 -1 -1 +1 -1 +1 -1 -1 -1 -1 +1 .....
  ---------------------------------------------------------
  ..... +1 +1 -1 -1 +1 -1 +1 -1 -1 -1 -1 +1 +1 +1 -1 .....
```

(12) An analysis of the results obtained in (11) shows that modulo-two addition (multiplication) of two identical sequences, although shifted in time, produce another pseudo-random sequence. This property is known as the autocorrelation property which in this context refers to the degree of correspondence between a sequence and a time shifted replica of itself.

$$F(X) \oplus F(X-kt) = F(Y) \; , \quad F(X) \times F(X-kt) = F(Y).$$

The resultant sequence $F(Y)$ is always a pseudo-random sequence since the number of chips in both sequences, $F(X)$ and $F(X+kt)$, which are the same (at any given time) differ from those which complement each other by exactly one.

(13) A further analysis of the results obtained in (12) shows that the result of the modulo-two addition (multiplication) of two identical sequence shifted in time relative to each other does not only produce a pseudo-random sequence, but, in fact, produces a replica sequence of itself, shifted in time

$$F(X) \oplus F(X-kt) = F(X-pt) \; , \quad F(X) \times F(X-kt) = F(X-pt).$$

This property is an extension of the autocorrelation property.

(14) Two identical sequences which are shifted in time relative to each other are said to be orthogonal to each other. Orthogonal in this context is said to signify that two orthogonal signals when mixed together are non-interfering. Orthogonality indicates that the integral property of the products of two identical sequences shifted in time over a complete sequence period shall be zero

$$\int_0^T F(X) \times F(X-kt)\,dx = 0\,, \text{ i.e., } F(X) \perp F(X-kt)\,.$$

This property is an analogy to the orthogonal property of sinusoidals

$$\int_0^{2\pi} \sin X \cos X \, dX = 0\,.$$

It has already been shown, when discussing digital modulation techniques, that when modulating two data streams with the same carrier wave although shifted by 90°, the transmitted signals are non-interfering. The same orthogonal property applies also to pseudo-random sequences which allow the data to be mixed with two identical but shifted sequences, and then the transmitted sequence signals do not interfere with each other.

This property of pseudo-random sequences is the one which enables two spread spectrum transmissions, using the same sequence code although shifted in time, to occupy the same frequency band without interfering with each other. In fact, with any pseudo-random sequences of $2^n - 1$ in length, one can produce $2^n - 1$ orthogonal sequences, provided however, that they all have their clocks synchronized and each sequence is shifted by at least a chip relative to the others. This means that in theory, $2^n - 1$ transmission may operate in the same frequency band without interfering with each other.

(15) Modulo-two addition of two pseudo-random sequences which are generated by two non-identical generators produces a sequence which is not necessarily a pseudo-random sequence. (A pseudo-random sequence is defined as a sequence with a maximum length of $2^n - 1$ different runs of n chips). this can be seen in the following example taking two unidentical sequences of the same length

$$\begin{array}{lll} & \ldots\ 000100110101111\ \ldots & (\Sigma 1 = 8) - (\Sigma 0 = 7) = 1 \\ \oplus & \ldots\ 001000111101011\ \ldots & (\Sigma 1 = 8) - (\Sigma 0 = 7) = 1 \\ \hline & \ldots\ 001100001000100\ \ldots & (\Sigma 1 = 4) - (\Sigma 0 = 11) = -7 \end{array}$$

In the above example the sequences were proved to have non-random characteristics by summing the number of ones in each sequence period. This method is not suggested as conclusive, since the sum result may be correct, even though the resultant sequence is not pseudo-random. An example of this effect is shown below, using the same sequences although further shifted in time

$$\begin{array}{lll} & \ldots\ 000100110101111\ \ldots & (\Sigma 1 = 8) - (\Sigma 0 = 7) = 1 \\ \oplus & \ldots\ 001111010110010\ \ldots & (\Sigma 1 = 8) - (\Sigma 0 = 7) = 1 \\ \hline & \ldots\ 001011100011101\ \ldots & (\Sigma 1 = 8) - (\Sigma 0 = 7) = 1 \end{array}$$

The resultant sequence is not pseudo-random since it does not present a maximum length. Take for example the run 1110 which appears twice in the sequence.

(16) Modulo-two addition of two non-identical pseudo-random sequences produces a sequence which is always different from zero, but it is not necessarily pseudo-random.

$$F_1(X) \oplus F_2(X) = F_3(X) \neq 0 \quad \text{with} \quad F_1(X) \neq F_2(X)$$

This property, known as cross-correlation, is defined by the measure of agreement between two different sequences. It does not, however, behave in the same attractive way as auto-correlation. The two pseudo-random sequences, $F_1(X)$ and $F_2(X)$, are said to be 'quasi-orthogonal sequences'. Using quasi-orthogonal sequences for coding the data bits in spread spectrum systems does not guarantee the noise rejection that may be achieved by using fully orchogonal sequences.

(17) The correlation of two binary sequences when compared chip by chip is defined by the following equation

$$\rho = \frac{\text{number of agreements} - \text{number of disagreements}}{\text{number of chips in the period of a sequence}}$$

Using the above equation, we can define the correlation property of the three forms of modulo-two addition of two pseudo-random sequences:

(a) Two identical and matched sequences (i.e. identical and synchronized sequences):

$$F(X) \oplus F(X) = 0 \qquad \rho = 1$$

(Correlation of the two sequences results in no disagreements, thus giving $\rho = N/N = 1$).

(b) Two identical but shifted sequences (i.e., orthogonal sequences):

$$F(X) \oplus F(X+kt) = F(X+pt) \qquad \rho = -1/N$$

where

$$F(X) \perp F(X+kt) \perp F(X+pt)$$

i.e., k and p are different from 0 or 2^n and $k \neq p$.

(Auto-correlation of $F(X)$ and $F(X+kt)$ produces a sequence which is pseudo-random. This means that the number of chip agreements, i.e., $1 \oplus 1 = 0 \oplus 0 = 0$, less the number of chip disagreements, i.e. $1 \oplus 0 = 0 \oplus 1 = 1$ must be equal to 1.)

(c) Two non-identical sequences (i.e., quasi-orthogonal sequences):

$$F_1(X) \oplus F_2(X) = F_3(X) \qquad 1/N \leqslant |\rho| < 1$$

where

$$F_1(X) \neq F_2(X) \neq F_3(X).$$

10.3 SYNCHRONIZATION OF SPREAD SPECTRUM RECEIVER WITH THE TRANSMITTER

The major problem in designing spread spectrum systems is the synchronization of the receiver with the transmitter. Without synchronization no data may be detected and decoded. In spread spectrum systems there are three synchronization processes which must be matched before the receiver may decode the data.

a. Synchronization of the carrier wave;

b. Synchronization of the internal clocks;

c. Synchronization of the pseudo-random sequences, i.e., correlation.

Synchronization of the carrier frequency waves and the clock timings, essential for PSK demodulation, use conventional processes and need not be discussed here in detail. The frequency may be synchronized by phase-locked-loops, and the clock may be extracted from the carrier wave if a coherent transmission system is used. In spread spectrum systems it is usual, although not essential, to first demodulate the signal and then decode the data. This means that the problem of correlating the sequences may be separated from the signal synchronizaton.

Correlation of two identical sequences is a different problem from synchronization of the signals. Even if the internal clock of the receiver is synchronized with that of the transmitter, there is no guarantee that the sequences are synchronized. In any pseudo-random sequence of 2^n-1 chips in length, there are 2^n-1 possibilities of matching the internal sequence with that of the transmitter and each one of these possibilities may have to be tested before full correlation may be reached. Each test must take a full period of $(2^n-1)T_c$ seconds, which means in effect that the correlation process may require a period as long as $(2^n-1)^2\,T_c = N^2T_c = NT_i$ seconds.

The process of correlation of two identical sequences is performed in two steps:

a. A search process over the transmitted pseudo-random sequence till correlation acquisition is achieved. This is a coarse search till the correlation is narrowed to within two chip intervals.

b. Once acquisition is reached a finer correlation process is performed keeping synchronization within $\pm\frac{1}{2}$ a chip of the maximum auto-correlation point. In this process both sequences are kept synchronized with no errors allowed by means of continuous tracking of the transmitted sequence.

The synchronization process can perhaps be better understood by looking at the flow chart shown in Fig. 10.16. When starting the operation, the two sequences are compared chip by chip, for the period of a complete sequence.

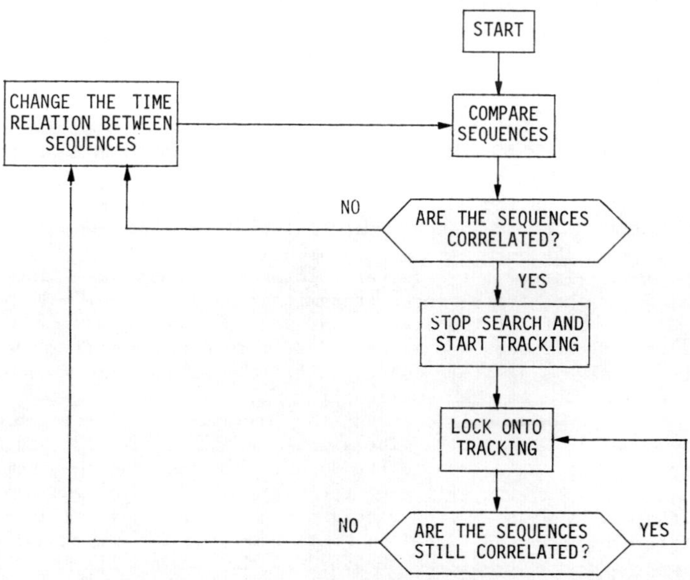

Fig. 10.16 Flow chart of a correlation process

If, at the end of the period the two sequences are found to be uncorrelated, then the internal sequence is delayed by one chip and the process is repeated. If the two sequences have acquired synchronization, the system must continue to track the transmitted signal, normally by means of early/late tracking loops.

The acquisition process is best followed with the aid of Fig. 10.17. The two sequences are compared chip by chip using a modulo-two addition operation and having the results accumulated. At the end of a period the accumulated signal is checked as to whether it is larger than a required threshold level. If the accumulated signal is below the required value, the internal receiver pseudo-random generator is delayed by a single chip and the whole process is repeated till a reasonable size output pulse is detected. This pulse also stops the search and the pseudo-random generator remains locked in this sequence correlation.

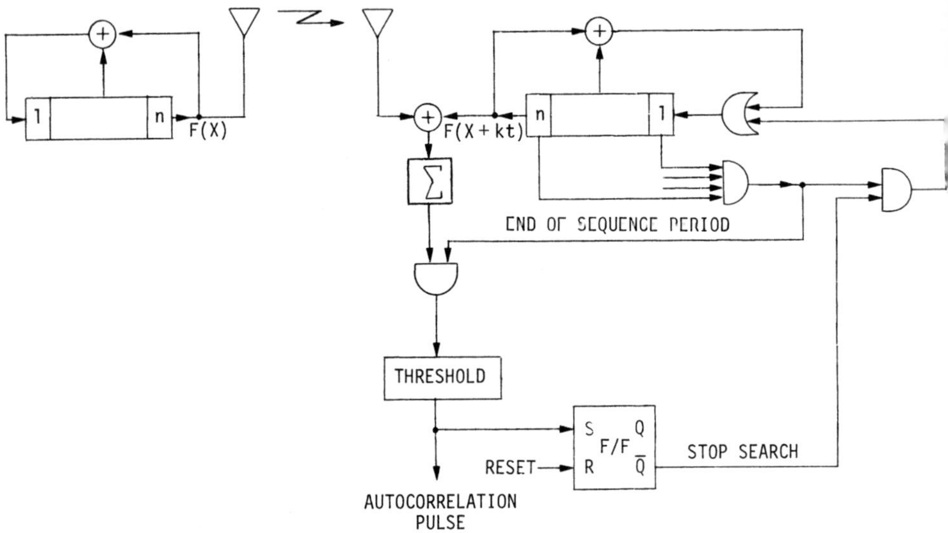

Fig. 10.17 Logical block diagram of pseudo-random synchronizer

The receiver, having no means of changing the transmitter sequence timing, changes its own sequence timing relative to that received, as shown in Fig. 10.18. At the end of every search sequence, if the sequences are found to be uncorrelated, the receiver's internal sequence is shifted by one chip. This means that for a single transmitter sequence of $2^n - 1$ chips there are up to $2^n - 1$ shifts of the internal sequence. The process is made possible as the transmitter sequence is continually repeated several times.

It is not always possible to demodulate the signal before decoding it, since the signal may be too small. If this is the case, its correlation process may be performed directly on the received pseudo-random modulated signal, as shown in Fig. 10.19. The signal is first passed through a band-pass filter to limit the search to the dynamic range of the spectrum. Before integrating the signal, the effect of the carrier modulation must be removed. This is performed by doubling the received signal frequency, using a square-law detector. The detected signal is then integrated for a complete sequence period. A positive output will be obtained by this method if the correlation of the two sequences is within a two-chip overlap of each other, as defined by the autocorrelation function. In other words, the two sequences may still

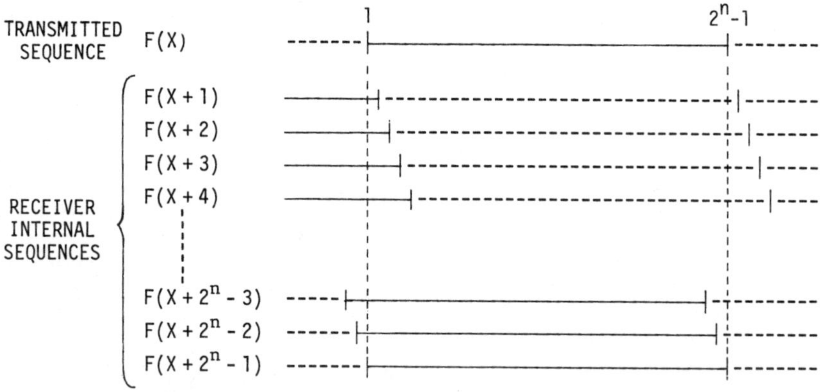

Fig. 10.18 Autocorrelation process of two identical sequences

be at the most two chips apart. For a finer adjustment of the two sequences, a tracking processor is used.

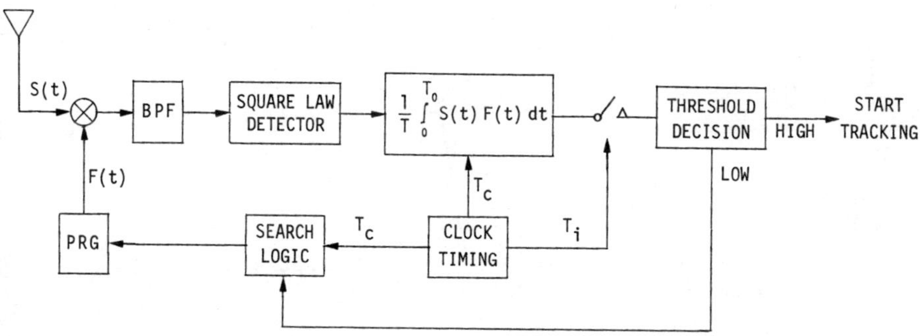

Fig. 10.19 Correlation acquisition detector circuit

Once the sequences are correlated, there is no need to continue the search process and the system can lock on the tracking process. This process uses two symmetrical tracking loops, as shown in Fig. 10.20, which correct the

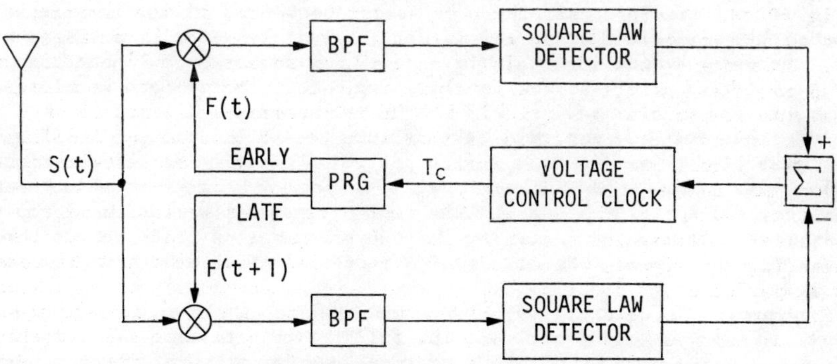

Fig. 10.20 Correlation tracking lock loop circuit

fluctuation of the internal sequence to a fine correlation with the transmitted signal. The internal sequence timing is controlled by a voltage controlled oscillator, which is adjusted by the results of both loops. Each tracking loop is multiplied separately by two identical pseudo-random sequences, separated by one chip interval. The two results obtained in both loops signify the early and late relationship of the two sequences with that of the transmitted signal. This tracking process enables us to keep the two sequences synchronized within plus or minus one half chip. This means in effect that the tracking process adjusts the clock timing by $\pm T_c/2$ whereas the acquisition process only brought the two sequences to a close synchronization of $\pm T_c$.

Figure 10.21 shows the combined autocorrelation function of both correlation lock loops. The maximum peak of each correlation process is obtained at the end of each sequence with an output characteristic of a triangular shape whose base is two chips wide. Since both sequence runs are one chip apart, the triangular peaks are also one chip apart. The relative position of the transmission sequence when both sequences are synchronized may lie anywhere within the two peaks. If each loop operated independently the relative position of the transmitted signal may be anywhere in the two triangles. By combining the results of the early and late correlation loops, the tracking circuit can always ensure that the synchronization will fall in the interval of the two peaks. This interval is characterized by having a rectangular shape whose base is one chip wide, although with a lower amplitude peak.

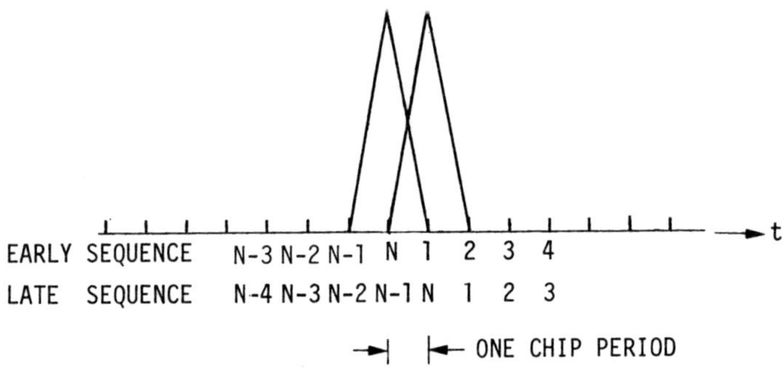

Fig. 10.21 Autocorrelation function of the double tracking loops

The time required for acquisition may be rather long, as the search may have to be performed over all the $2^n - 1$ sequences. This timing is a function of both the sequence length and timing. Take, for example, a system which has a sequence length of 1023 chips (i.e., generated by a linear feedback shift register of $n = 10$ stages) with timing of 50 nsec. Each complete integration process will require $1023 \times 50 \times 10^{-9} = 51.15$ µsec. A search over all the possible sequences, however, will take as long as $1023 \times 51.15 \times 10^{-6} = 52.3$ msec. It is possible to reduce this search period by reducing either the sequence length or its timing. This, however, will, consequently, require reduction in the transmitted spectrum bandwith. The search time may be reduced by employing other techniques, such as two search processes, a quick search through all the possible sequences and a slow process through selected promising sequences.

In order to ensure full autocorrelation of the two sequences, the integration must be performed over the full period. Nevertheless, when the two sequences are correlated, integration over a partial sequence period should also produce positive results. With partial results there is no guarantee of complete noise rejection and consequently it is difficult to distinguish between false and real correlated output. On the other hand, if the partial result is negative, obtained from the integration over part of the sequence period, this signifies that the two sequences are definitely not correlated.

This double search process uses two integration and dump circuits as shown in Fig. 10.22. The first integration is performed over a period D which is less than the sequence period $T_i\ (= NT_c)$, if, at the end of this period, the results are less than a given threshold σ_1 then the two sequences are obviously not correlated and the search over this sequence can be discarded. The internal receiver is thus stepped up by one chip and a new sequence is tested. If a positive result is obtained after integration over period D (i.e., the threshold is exceeded), there is a high probability that the sequences are correlated. Then, without changing the internal sequence phase, the integration time is increased and thus performed over the full sequence period T_i. The results obtained after the second integration are compared with a second threshold level, σ_2, which is much higher than the first one. If the results obtained exceed this second threshold, the two sequences are correlated within a two chip period.

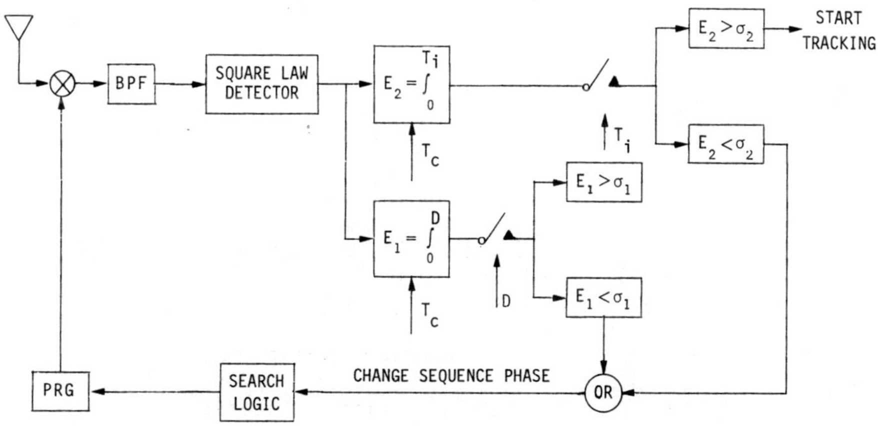

Fig. 10.22 Fast acquisition process circuit

It is theoretically possible to reduce the search period by employing a number of partial integration periods. In practice, however, two integration processes are sufficient. Fig. 10.23 presents a flow chart of this double integration process.

It is also possible to use the partial integration principles for the tracking process. This, however, provides only fringe benefits and at the price of complicated circuitry. The only advantage of this additional circuit is that the system reacts fast when synchronization is lost.

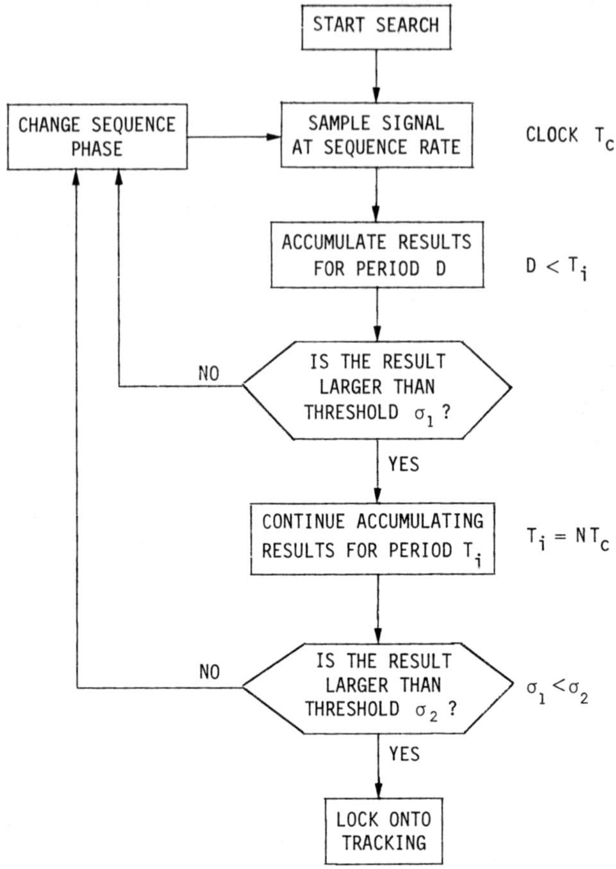

Fig. 10.23 Flow chart of the fast correlation process

10.9 CODE DIVISION MULTIPLE ACCESS (CDMA)

It has already been shown that one of the major features of spread spectrum is that its transmission does not disturb other transmissions using the same frequency bands and that other transmissions do not disturb spread spectrum transmissions. So far we have exploited this advantage for operating the system in a noisy environment. It is possible to use this feature of noise rejection to enable many spread spectrum transmissions to operate simultaneously over the same spectrum bandwidth. This can be achieved by providing each operating network with a different code sequence which is orthogonal to all the other code sequence operating in the same bandwidth. Such simultaneous transmission systems have been called spread spectrum multiple access (SSMA), phase coded SSMA, and code division multiple access (CDMA). The reader should note that the term SSMA refers in the literature also to satellite switched multiple access, and for this reason most sources refer to simultaneous coded spread spectrum transmissions as CDMA.

In this simultaneous multiple access system, each network codes its data using a different pseudo-random sequence, as shown in Fig. 10.24. Only terminals with the same sequence as that transmitted can detect the signals while rejecting all the other coded signals. Theoretically, any number of networks may operate in the same spectrum provided they all have different non-correlating sequences. Nevertheless, as the operative number of networks increases, the higher the spectrum noise. Since the data source comes from random operating terminals, the probability of them all operating together is small, and thus the number of parallel operating networks could be rather large.

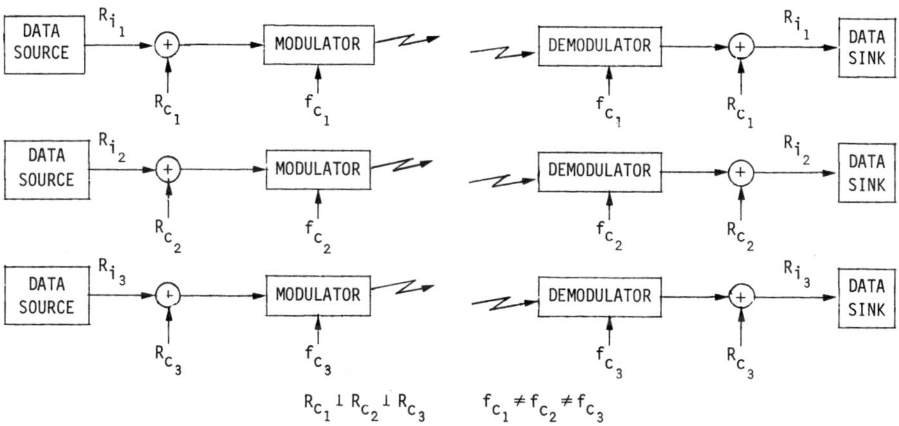

Fig. 10.24 CDMA parallel transmissions

It is practically difficult to synchronize all the terminals so that each network would be provided with pure orthogonal sequences; that is, each network will operate with the same basic sequence although shifted by at least one chip relative to the others. It is then probably possible to use the same sequence if each network is provided with a different sequence starting point. This would, however, allow two or at most three orthogonal sequences, but not the ideal number of $2^n - 1$ orthogonal sequences. Instead, use is made of quasi-orthogonal sequences. This allows many data sources to be coded by different sequences, and their transmissions to operate in the same spectrum band. However, since quasi-orthogonal sequences are used, the threshold margin of the cross-correlation is reduced following the lessening of mutual rejection.

It is possible to increase the number of simultaneous network transmissions by shifting the carrier frequency of each network by a small fraction. The correlation operation of any two signals modulated by two different carrier frequencies, i and j, will always be zero.

$$\rho = 1/NT_c \int_0^{NT_c} F_i(t)\,F_j(t) = 0 \quad \text{provided} \quad i \neq j\,.$$

This implies that shifting the carrier frequency provides means of obtaining orthogonal sequences. The minimum practical shift suggested is equal to the data rate (i.e., N times the sequence rate). For example, if the sequence period is 50 µsec then the minimum spacing between network carrier frequency is 20 KHz, a distance which still leaves wide room for manoevering.

The carrier shifting, coupled with full and quasi-orthogonal sequences, allows many networks to exploit the same frequency spectrum. Each network will regard the other network transmissions as noise. This also means that even if the same sequence is used in two networks, their transmitted signals are orthogonal to each other, provided they are both modulated on different carrier frequencies.

10.10 CODE DIVISION MULTIPLEXING (CDM)

The same principles of orthogonal code sequences for multiple-access system could also be employed for multiplexing. As already explained, multiple-access is used in a multidrop network and is defined as a process which makes the common transmission channel available to multiple users, with no restrictions as to their physical position relative to the channel. Multiplexing is used in a multipoint network and is defined as the division of the common transmission channel into fixed multiple exclusive sub-channels made available to each user by common equipment only at the extreme ends of the channel. In most cases one could regard multiplexing as a special case of multiple-access. When using FDM or TDM, the common equipment does not make their transmitted signals any different from those obtained from FDMA or TDMA. This, however, is not the case with the CDM.

With code division multiplexing (CDM) the common equipment places additional restrictions on the transmitted signals which make them different from those obtained from CDMA. These restrictions may be regarded as hard (amplitude) limiting of the sum of the transmitted coded signals. In CDM, as in CDMA, each active terminal codes its messages onto a unique pseudo-random sequence which is orthogonal to all the others. In addition, all the coded signals of CDM are summed up at one point (as shown in Fig. 10.25) and the resultant signal is transmitted. Before transmission, the signal is modulated by any of the techniques already suggested. It should be noted, however, that the transmitted signal is not a spread spectrum signal, but a coded multiplex signal.

(The summed signal obtained is similar to that which may be obtained in CDMA with a hard limiter, as for example, when using a repeater like a satellite.)

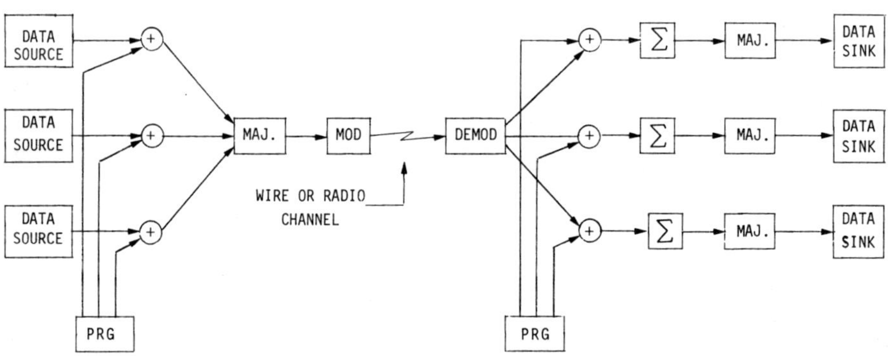

Fig. 10.25 Code division multiplexing transmission system

The advantage of CDM is that data streams derived from a number of separate sources are coded disjointly and then transmitted as a single signal stream. In other words, CDM acts as a concentrator in a similar method to that of any other multiplexing system. The feature strength of the combined transmitted signal is in its coding which has benefits both for error control and secrecy. The disadvantage of CDM is in the wasteful overhead data that is transmitted, i.e., it has a low ratio of useful data to that of transmitted data.

As seen in Fig. 10.25, the data stream derived from a number of separate sources, is modulo-two added to a pseudo-random sequence which is orthogonal to that used for the other sources. Since all the coded signals are summed up at one point, all the pseudo-random sequences may be generated by a single linear feedback shift register, although each sequence is delayed by at least one chip relative to the others. (It is possible to obtain n orthogonal sequences by taking the output from the n different stages; the other $2^n - 1 - n$ sequences can be derived by adding extra delay stages.) After each modulo-two addition operation, the coded data bit from all the parallel sources are compared and their majority value is transmitted. If there are m parallel data sources and their majority is larger than a given threshold (i.e., more ones than zeros) then a 1 valued bit is transmitted, while if the majority is smaller than the threshold, then a 0 is transmitted. The operation produces a new single data stream with $N = 2^n - 1$ bits, although this stream is not a pseudo-random sequence. Each bit in the transmitted stream can be defined as follows:

$$(\Sigma m = 1) \geqslant m/2 \quad \text{then} \quad k = 1$$

$$(\Sigma m = 1) < m/2 \quad \text{then} \quad k = 0$$

where k is the transmitted coded bit value and m is the number of parallel sources. The maximum number of data sources that can be coded simultaneously is less than half the pseudo-random sequence, i.e., $m < N/2$. This operation is possibly better understood with the following numerical example, where the number of parallel sources is taken as $m=5$ and with a pseudo-random sequence length of $N = 2^n - 1 = 15$.

TABLE 10.1

DATA SOURCES	DATA VALUE	ORTHOGONAL PSEUDO-RANDOM SEQUENCES	CODED DATA SEQUENCES	TRANSMITTED CODED SIGNAL
1	0	000111101011001	000111101011001	
2	1	001111010110010	110000101001101	
3	0	011110101100100	011110101100100	010101101001100
4	0	111101011001000	111101011001000	
5	1	111010110010001	000101001101110	

In the receiver the reverse operation is performed. That is, the received sequence of $2^n - 1$ bits is modulo-two added to m different orthogonal pseudo-random sequences. These sequences are the same as those used in the transmitter to code the data. As a result of this operation in the receiver, m different decoded sequences are obtained from the single received sequence, each with $N = 2^n - 1$ bits. Each sequence is fed into a different summer which accumulates the number of bits in the sequence which have the value one. If the result produces a majority of ones, the decoded data bit gets the value of 1, and if the majority is of zero value, the decoded data bit gets the value of 0. This can be defined as follows:

$(\Sigma p = 1) \geqslant N/2$ then data bit equals 1

$(\Sigma p = 1) < N/2$ then data bit equals 0

where p is the bit value of the decoded sequence and N is the sequence length. In order to clarify this receiver decoding operation the same numerical example is presented below :

TABLE 10.2

RECEIVED CODED SIGNAL	ORTHOGONAL PSEUDO-RANDOM SEQUENCES	DECODED DATA SEQUENCES	Σ1	DATA VALUE
	000111101011001	010010000010101	5	0
	001111010110010	011010111111110	11	1
010101101001100	011110101100100	001011000101000	5	0
	111101011001000	101000110000100	5	0
	111010110010001	101111011011101	11	1

CDM has been described here as a coding sequence of m separate data sources. Alternatively, it could be used to multiplex a single data source, but with coding m of its bits simultaneously, as shown in Fig. 10.26. Such a circuit can be used for coding its transmitted data for secrecy or for error correction. Errors in the decoded received sequence are soon detected because of the nature of the majority logic and by the fact that the pseudo-random sequences are known at the receiver. However, the maximum number of bits that may be coded is limited to $m < N/2$ and the number of bits that may be corrected is reduced with the increase of the ratio $2m/N$.

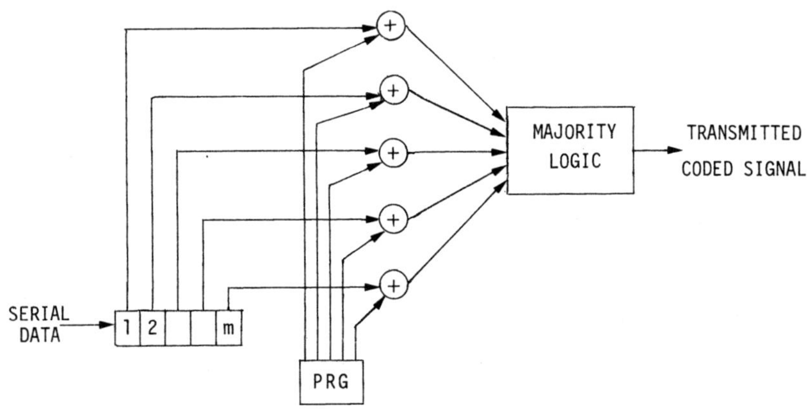

Fig. 10.26 CDM used to code a data sequence in an error detection system

Very little research has been directed to investigate code division multiplexing systems. As seen by the results obtained in the numerical example, the summed number of the ones in each sequence has a definite pattern, i.e., either 11 or 5. This indicates that it follows a specific function; no such function, has however, to the author's knowledge, been published. A research study carried out by the author shows that these figures are a function of both the number of simultaneous bits coded m, and their data value, i.e., whether the majority of the m parallel bits is one or zero. For example, in the case of $N = 15$ and $m = 5$ with the majority of the transmitted data bits equalling one then the accumulated figure received is 10 and 4 for data bit 1 and 0 respectively. It should also be noted that by majority decision of the value of the transmitted bit, the operation is not based any more on linear operation. Furthermore, the modulo-two addition of a pseudo-random sequence with a data bit equalling 1, produces a sequence which is no longer a pseudo-random sequence.

To conclude, it is important to add that CDM as such cannot be used for spread spectrum application, since the ratio of data bits to the chips is low, i.e., the number of parallel coded bits is usually taken as $m < (N/2) - 1$. If the CDM signal is to be used in a spread spectrum system, the CDM sequence must be regarded as a data source and as such it is then modulo-two added to a different pseudo-random sequence with a faster timing.

10.11 FREQUENCY HOPPING MULTIPLE ACCESS (FHMA)

Another form of simultaneous transmission using a common communication channel is known as frequency hopping multiple access (FHMA), although it has also been referred to by other names, such as multiple frequency code selected (frequency) shift keying (MFCSK). This technique, similar to CDMA, is a form of spread spectrum and is intended for operation in a noisy environment. FHMA, however, differs from CDMA in the way it expands the bandwidth and the means by which it rejects the noise interferences. In fact, FHMA coexistence properties with other transmissions are less advantageous than CDMA, since under certain conditions it can disturb other transmissions. Nevertheless, FHMA can operate over a wider dynamic range than is possible with CDMA and so has better rejection capabilities.

The centre feature of FHMA, is similar to CDMA, is the pseudo-random sequence generated both in the transmitter and receiver, which requires proper autocorrelation before the decoding process can commence. In CDMA, the pseudo-random sequence was operated directly on the bit sequence and the result modulated. In FHMA, however, the pseudo-random sequence and the bit sequence are each modulated separately and the results are then mixed together. It is therefore important to distinguish between the two modulation processes.

Frequency hopping can be regarded as a form of M-ary FSK, whereby at any given time period a different discrete frequency is transmitted. The frequency bandwidth B_H is divided into N equal sized sub-bandwidth ΔB_H. This can be seen in Fig. 10.27 where each sub-bandwidth is denoted by its centre frequency. As the transmitted signal is hopped between these frequencies thus it can be regarded as references for the instantaneous data modulation signal. If the power transmitted at each hop is ideally equal, the transmitted spectrum of these references produces a rectangular shaped envelope extending over the full spectrum bandwidth B_H. The frequency separation ΔB_H between the discrete reference frequencies is so selected that it must be at least as wide as the information bandwidth B_m.

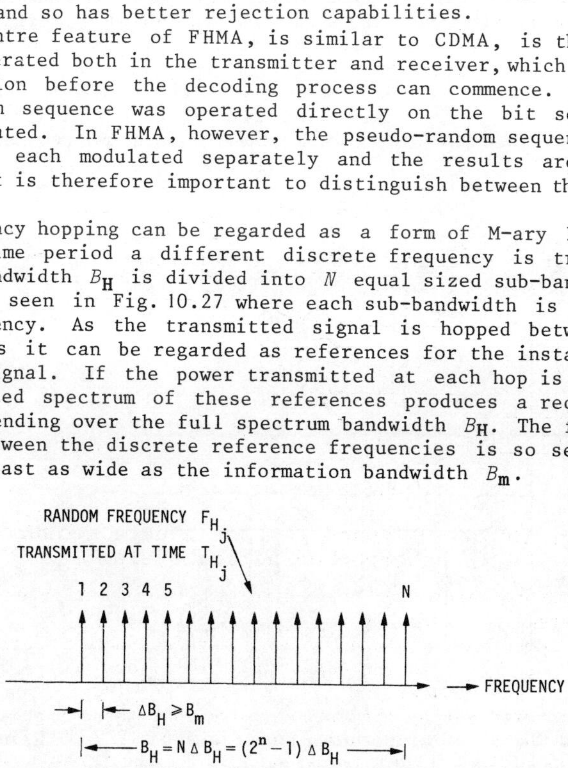

Fig. 10.27 Ideal frequency hopping spread spectrum

Each discrete frequency reference f_H is transmitted for a period T_H and at the end of this period the signal is hopped to another frequency. The period T_H is the time duration between the shift clock pulses of the pseudo-random sequence. The time function of FHMA, as seen in Fig. 10.28, is the same as any FSK pattern, except that with FHMA the signal is hopped between N different frequencies while with FSK it is keyed between two frequencies only. These N frequencies are selected by a pseudo-random sequence of $2^n - 1 = N$ chips generated by a linear feedback shift register with n stages. Each chip timing selects a different frequency.

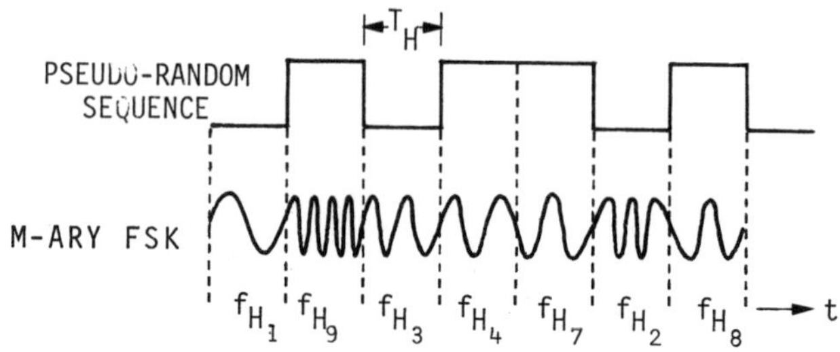

Fig. 10.28 Basic frequency hopping waveforms

With CDMA the spectrum is spread as a direct response to the individual chip value of the pseudo-random sequence. Since the chip can take the value of 1 or 0 only, nothing would be gained by using the serial generated chips for selecting the discrete frequencies of FHMA. Instead, a different property of pseudo-random sequences is employed which exploits the parallel generated output of the pseudo-random generator. It has already been stated that in a pseudo-random sequence, every group of n consecutive chips presents a different pattern of ones and zeros. In each pseudo-random sequence of the length of $2^n - 1$ chips there are exactly $2^n - 1$ different pattern groups before the same pattern repeats itself. This means that instead of taking the serial output from the linear-feedback shift register (as performed with CDMA), the parallel output is taken simultaneously from each of the n stages of the generator. Fig. 10.29 demonstrates a pseudo-random sequence with the length of $N = 2^n - 1 = 15$ chips where every combination of $n = 4$ presents a different pattern. Each of the N discrete patterns is labelled by a decimal equivalent number, ranging from 1 to N in a pseudo-random order.

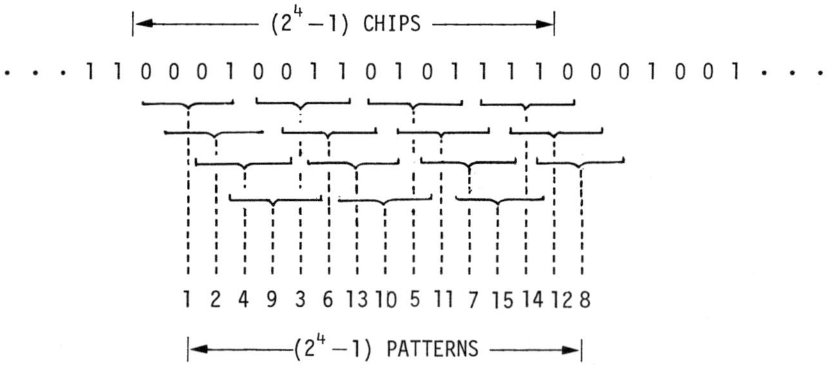

Fig. 10.29 Pseudo-random sequence of 2^n-1 chips organized in 2^n-1 different patterns

The various discrete frequencies are produced by a frequency synthesizer, as shown in Fig. 10.30. The n parallel generated chips are translated into $2^n - 1$ different codes which are used to select the discrete frequencies. Each time the generator steps up, a different pattern is generated causing another frequency to be produced.

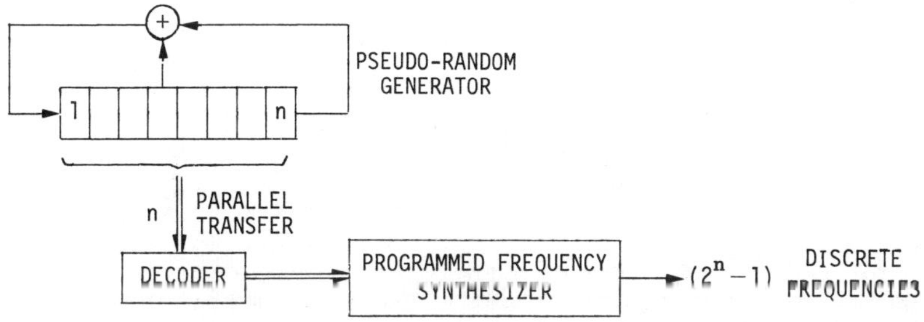

Fig. 10.30 Basic circuit for generating the discrete frequencies

Each one of the N discrete frequencies is used as a reference to modulate the digital data. Since the chip rate of the pseudo-random sequence is N times that of the data rate, it means in effect that the data bit is transmitted N times, although each time it is mixed with a different frequency reference. For this reason, each of the N discrete frequencies is transmitted with a power density N times less than if the data were transmitted by a single reference frequency. Nevertheless, the power spectral density at each frequency f_H during period T_H will be relatively high as compared with the same spectrum portion of CDMA. As a result FDMA could interfere with other transmissions.

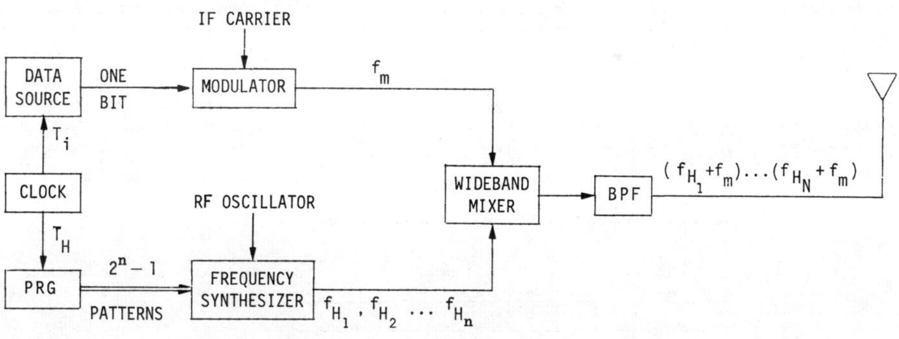

Fig. 10.31 Frequency hopping transmitter schematic circuit diagram

As already stated, the pseudo-random sequence and the data in FHMA are each modulated separately and only then mixed together, as seen in Fig. 10.31. That is, the data is used to modulate a fixed *IF* carrier frequency while the pseudo-random sequence is used to select an *RF* discrete reference frequency which hops the data coded signal. The carrier frequency need not necessarily be coherent with the discrete frequencies since due to the hopping the result may exceed the bandwidth coherence. Any digital modulation technique may be used to modulate the data sequence, although FSK and PSK are the most commonly used. The bandwidth required to modulate the data is defined as B_m, and so the frequency separation ΔB_H between the discrete frequencies must be at least as wide as the data bandwidth, i.e., $\Delta B_H \geqslant B_m$. The transmitted spectrum bandwidth is then equal to $B_H = (2^n - 1)\,\Delta B_H$.

Since the same modulation bit signal is used over all the N discrete frequencies f_H, they all deviate by the same fixed modulation frequency f_m. This can be represented by the following formula showing the transmitted signal, after mixing, for a given time period:

$$S(t) \text{ for period } T_{H_j} = \cos \omega_{H_j} t \cos \omega_m t = \tfrac{1}{2}\left[\cos(\omega_{H_j} + \omega_m)t + \cos(\omega_{H_j} - \omega_m)t\right].$$

In this way, each reference frequency will be transmitted for a given period T_H as

$$(f_{H_1} \pm f_m),\ (f_{H_2} \pm f_m),\ (f_{H_3} \pm f_m),\ \ldots,(f_{H_N} \pm f_m),$$

with f_m remaining constant for all the N frequencies. By suitably selecting the carrier and reference frequencies, it is possible to filter out the low frequencies obtained after mixing. That is, the lowest reference frequency must be larger than the modulated carrier frequency, i.e.,

$$f_{H_{min}} > f_{m_{max}}.$$

This means that the transmitted signal after filtering will be only

$$(f_{H_1} + f_m),\ (f_{H_2} + f_m),\ (f_{H_3} + f_m),\ \ldots,(f_{H_N} + f_m).$$

Figure 10.32 illustrates this frequency hopping operation over the total bandwidth, with a reference to the time domain. At each single hop period, T_H, only a bandwidth of ΔB_H is occupied, resulting in covering the total bandwidth B_H of the spectrum during a complete pseudo-random period.

In CDMA the data modulated signal, at any given period, is spread over the whole spectrum while with FHMA the data modulated signal at any given period is concentrated in a narrow bandwidth B_m. This means that with FHMA the pseudo-random sequence need not be as long as that used in CDMA and a length of 100 to 500 being usually sufficient.

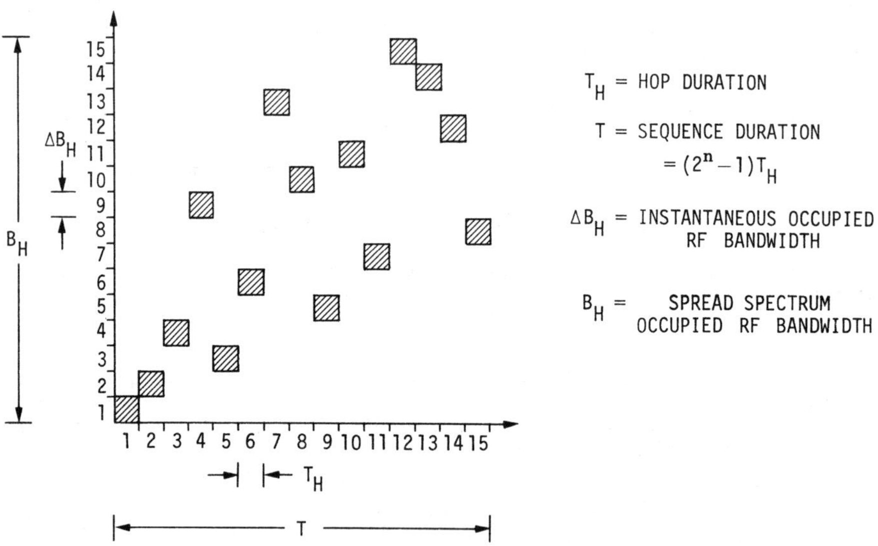

Fig. 10.32 Schematic diagram of frequency versus time in FHMA

In FHMA, as distinct from CDMA, it is possible to spread the spectrum with only part of the pseudo-random sequence. That is, it is possible to employ K discrete reference frequencies with $K \neq N$. This enables longer sequences to be employed, a property essential if FHMA is used for covert transmission, since each discrete frequency transmission can no longer be referred to a given bit. In all our presentations, however, we will presume that spectrum is spread by using a complete pseudo-random sequence length, i.e., $N = K$.

In the receiver, for decoding the transmitted signal, the reverse operation to that performed in the transmitter is exercised, as shown in Fig. 10.33. That is, the received signal is mixed again with locally generated reference frequencies which are a replica of those generated in the transmitter. Before this operation can be initiated, the receiver must be fully correlated with the transmitter.

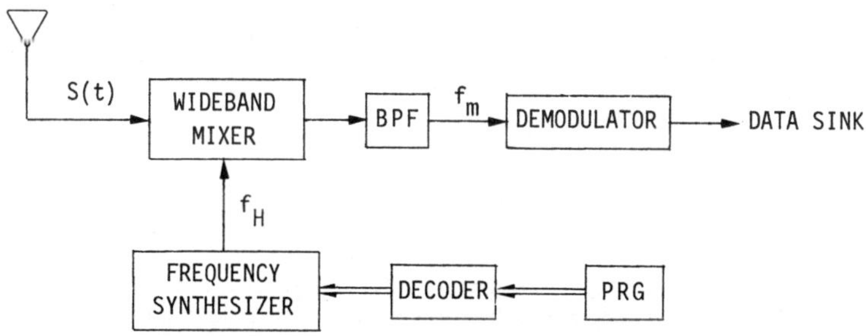

Fig. 10.33 Schematic circuit diagram of FHMA receiver

The mixing operation, at any given period T_{H_j}, is as follows:

$$\cos \omega_{H_j} t \cos (\omega_{H_j} + \omega_m)t = \tfrac{1}{2}[\cos (2\omega_{H_j} + \omega_m)t + \cos \omega_m t],$$

In this way it is possible to mix all the modulated frequencies with those locally generated, provided they are fully correlated; that is, f_H, is generated at the same time in both receiver and transmitter. Hence the mixing operation produces :

$$\left\{f_{H_1}, f_{H_2}, f_{H_3}, \ldots, f_{H_N}\right\} \times \left\{(f_{H_1} + f_m), (f_{H_2} + f_m), (f_{H_3} + f_m), \ldots, (f_{H_N} + f_m)\right\}$$
$$= \left\{(2f_{H_1} + f_m), (2f_{H_2} + f_m), (2f_{H_3} + f_m), \ldots, (2f_{H_N} + f_m)\right\}\left\{Nf_m\right\}.$$

Since the carrier frequency is much lower than the minimum discrete reference frequency, i.e.,

$$f_m < f_{H_{min}},$$

a simple bandpass filter will separate the modulated carrier frequency. The resultant signal can then be demodulated, thus reproducing the data signal.

As seen from the mixing operation, the process gain for frequency hopping is the same as that of CDMA, i.e. is equal to the number of discrete frequencies

$$G_p = B_H/\Delta B_H = N = (2^n - 1).$$

If the frequency separation is taken to be approximately equal to the data modulation bandwidth, $\Delta B_H \approx B_m$, then the processing gain is :

$$G_p = B_H/B_m = (2^n - 1)\,\Delta B_H/B_m \;.$$

Frequency hopping may also be used for multiple simultaneous transmission by simply providing each network with a unique code sequence which is orthogonal to the others.

10.12 CONCLUDING REMARKS

Simpultaneous multiple-access transmission systems have each, advantages and drawbacks which limit their application. In the last few years much work has been carried out on the lines of combing the various techniques in one system. Such schemes extend the application of each system and increase reliability. For example, the combination of FHMA with CDMA can support the overpopulated CDMA networks, thus yielding an increased throughput in the extended bandwidth. Another popular hybrid combination is CDMA or FHMA or both together with TDMA, which provides the efficiency of TDMA transmission associated with the security capabilities of CDMA transmission. Such hybrid systems have been implemented in satellite communication systems and have proved very successful. These systems could achieve reduced symbol distortions, but this only at the expense of complex circuitry.

10.13 REFERENCES

(a) SPREAD SPECTRUM

1. Dixon, R.C. *Spread Spectrum Systems*, John Wiley & Son, 1976, 314 pp.
2. Utlant, W.F. *Spread Spectrum-Principles and Possible Applications to Spectrum Utilization and Allocation*, IEEE Communication Society Magazine, September 1978, pp.21-30.
3. Mohanty, N.C. *Spread Spectrum and Time Division Multiple Access Satellite Communication*, IEEE Trans., vol. COM-25, no. 8, August 1977, pp.810-815.
4. Kahn, R.E., Gronemeyer, S.A., Burchfield, J., Kunzelman, R.C. *Advances in Packet Radio Technology*, Proc. IEEE, vol. 66, no. 11, November 1978, pp.1468-1496.
5. Ristenbatt, M.P., Daws, J.L. *Performance Criteria for Spread Spectrum Communication*, IEEE Trans., vol. COM-25, no. 8, August 1977, pp.756-762.
6. Viterbi, A.J. *Spread Spectrum Communication – Myths and Realities*, IEEE Comm. Mag., May 1979, pp.11-18.

(b) CODE DIVISION MULTIPLE ACCESS

7. Pursley, H.B. *Performance Evaluation of Phase Coded Spread Spectrum Multiple-Access Communication*, Trans. IEEE, vol. COM-25, no. 8, August 1977, pp.795-802.
8. Hopkins, P.M. *A Unified Analysis of Pseudo-noise Synchronization by Envelope Correlation*, IEEE Trans. vol. COM-25, no. 8, August 1977, pp.770-778.

(c) CODE DIVISION MULTIPLEXING

9. Wing, P.A. *Code Division Multiplexing*, Monitor-Proc. IREE, January/February 1970, pp.25-28.
10. Anderson, D.R. and Wintz, P.A. *Analysis of a Spread-Spectrum Multiple-Access System with a Hard Limiter*, IEEE Trans. vol. COM-17, no. 2, April 1969, pp.285-290.
11. Gordon, J.A. and Barrett, R., *Correlation-recovered Adaptive Majority Multiplexing*, Proc. IEEE, vol. 118, no. 3/4, March-April 1971, pp.417-422.
12. Tsao, S.H. *Generation of Delayed Replicas of Maximum Length Linear Binary Sequences*, Proc. IEE, vol. 111, no. 11, November 1964, pp.1803-1806.

Abbreviation Index

SCMA	Split Channel Multiple-Access	440-441
SDM	Space Division Multiplex	135,322,339-345,380-385
SDLC	Synchronous Data Link Control	207
SEL	Select Character	361
SFSK	Sinusoidal Frequency Shift Keying	312-313
SHF	Super High Frequency	230,236,238,242
SK	Shift Keying	260
SM	Statistical Multiplex	321,352,355
S/N	Signal to Noise Ratio	119
SOH	Start Of Heading Character	359
SOT	Start Of Transmission Character	426
SPM	Superposed Modulation	282-283
SPMA	String Polling Multiple-Access	441-442
SRMA	Split Channel Reservation Multiple-Access	444-447
SS	Satellite Switched	251-252
SS	Spread Spectrum	48,241,251-252
SSB	Single Side Band	263-264
SSBSC	Single Side Band Suppressed Carrier	243,246,263-264,325
SSMA	Satellite Switched Multiple-Access	480
SSMA	Spread Spectrum Multiple-Access	456-457,480
STDM	Synchronous Time Division Multiplexing	43,321,329-338,383-384,419
STDMA	Synchronous Time Division Multiple-Access	321-322,329-334
STS	Space-Time-Space	385
STX	Start of TEXT	94,359
SYN	Synchronous Character	358,366
T	Time	71
TAT	Turn Around Time	167,361,419.422
TDM	Time Division Multiplexing	45,54-58,168-171,226-228, 243,249,321-322 , 324-328
TDMA	Time Division Multiple Access	247,249,419,455-456,429
TFM	Tamed Frequency Modulation	312-313
TSI	Time Slot Interchange	343,383-385
TST	Time-Space-Time	385
TTY	Tele Typewriter	
TU	Traffic Unit	380
UHF	Ultra High Frequency	230,237,238,242
UNRZ	Unipolar Non-Return to Zero	90
UQPSK	Unbalanced Quadrature Phase Shift Keying	290
URZ	Unipolar Return to Zero	90
VC	Virtual Circuit	405-409
VCO	Voltage Control Oscillator	124
VCXO	Voltage Control Crystal Oscillator	365
VF	Voice Frequency	229
VHF	Very High Frequency	230,233,236,237,238
VLF	Very Low Frequency	230,234,235
VSB	Vestigial Side Band	263-265

Subject Index